FRESHWATER ECOSYSTEMS AND CLIMATE CHANGE IN NORTH AMERICA

ADVANCES IN HYDROLOGICAL PROCESSES

Series Editors M. G. Anderson
D. E. Walling
N. E. Peters

Terrain Analysis and Distributed Modelling in Hydrology

Edited by K. E. Beven and I. D. Moore

Scale Issues in Hydrological Modelling

Edited by J. D. Kalma and M. Sivapalan

Freshwater Ecosystems and Climate Change in North America
A Regional Assessment

Edited by C. E. Cushing

Water Quality Trends and Geochemical Mass Balance

Edited by N. E. Peters, O. P. Bricker and M. M. Kennedy

Distributed Hydrological Modelling: Applications of the TOPMODEL Concepts

Edited by K. Beven

FRESHWATER ECOSYSTEMS AND CLIMATE CHANGE IN NORTH AMERICA

A REGIONAL ASSESSMENT

Edited by

COLBERT E. CUSHING

Ecology Group, Battelle-Pacific Northwest National Laboratory, USA

JOHN WILEY & SONS

Chichester · New York · Weinheim · Brisbane · Singapore · Toronto

The papers in this volume were originally published in
Hydrological Processes—An International Journal, volume 11,
issue 8 (819–1067), 1997

Baffins Lane, Chichester,
West Sussex PO19 1UD, England
National 01243 779777
International (+44) 1243 779777
e-mail (for orders and customer service enquiries): cs-books @wiley.co.uk
Visit our Home Page on http://www.wiley.co.uk
or http://www.wiley.com

Other Wiley Editorial Offices

John Wiley & Sons, Inc., 605 Third Avenue,
New York, NY 10158-0012, USA

WILEY–VCH Verlag GmbH, Pappelallee 3,
D-69469 Weinheim, Germany

Jacaranda Wiley Ltd, 33 Park Road, Milton,
Queensland 4064, Australia

John Wiley & Sons (Asia) Pte Ltd, 2 Clementi Loop #02–01,
Jin Xing Distripark, Singapore 129809

John Wiley & Sons (Canada) Ltd, 22 Worcester Road,
Rexdale, Ontario M9W 1L1, Canada

British Library Cataloguing in Publication Data

A catalogue record for this book is available from the British Library

ISBN 0 471 978221

Typeset in 10/12 Times
Printed and bound in Great Britain by BPC Wheatons Ltd, Exeter

CONTENTS

PREFACE

This book presents the assessments of eight regional panels that were convened at the symposium entitled 'Regional Assessment of Freshwater Ecosystems and Climate Change in North America', held from 24–26 October, 1994 in Leesburg, Virginia and jointly sponsored by the American Society of Limnology and Oceanography (ASLO) and the North American Benthological Society (NABS).

The eight regional groups represented the following geographical areas for the purposes of this symposium: Laurentian Great Lakes and Precambrian Shield of the USA and Canada; Arctic and Sub-Arctic Regions of the USA and Canada; Rocky Mountains in the USA and Canada; Mid-Atlantic and New England Areas of the USA; Southeastern USA and Coastal Mexico; Pacific Coast Mountains and Western Great Basin; Great Plains of the USA and Canada; and Basin and Range Regions and Adjacent Arid and Semi-arid Regions of the USA and Mexico. The symposium was structured around a regional approach, and the panels were asked to address the central question, 'What would be the impact to and response of freshwater ecosystems in your region as a result of a hypothetical doubling of CO_2 in the atmosphere?' The panels, composed of experts in the fields of geochemistry, climatology, hydrology and aquatic ecology, were provided with the input from the Canadian General Circulation Models, (GCM) which used a doubling of atmospheric CO_2; they considered model predictions and various scenarios both prior to and during the symposium to develop and come to some conclusions as to the impact on and response of freshwater ecosystems in their respective regions. Following an oral presentation of their preliminary assessment, the panels then prepared full reports on their evaluations; these are published herein and as a special issue of *Hydrological Processes* (Vol. 11, No. 8, 1997). An introductory chapter explaining the approach for identifying regions by Dr George Leavesley and a concluding chapter by Dr David Schindler synthesizing the concepts and findings of the regional chapters is also included. Several oral papers were also presented during the symposium and these have been published as a special issue of *Limnology and Oceanography* (Vol. 41, No. 4, 1996).

The preface to the special issue of *Limnology and Oceanography* reviews the events leading to this symposium, including previous symposia jointly or individually sponsored by ASLO and NABS, development of governmental working groups addressing projected global climate change scenarios, etc. Suffice it to say here, that the present symposium was held at an auspicious time and has focused much needed attention on freshwater ecosystems.

The organizers of the symposium had three basic goals:

- evaluate current evidence for directional change in inland waters;
- examine potential important responses to climatic changes resulting from a hypothetical doubling of atmospheric carbon dioxide (or its radiative equivalent); and
- develop recommendations for experimental studies or augmentation of monitoring programmes.

The papers prepared by these panelists and published here represent their interpretation of what a doubling of atmospheric CO_2 would do to the freshwater ecosystems in their respective regions. Consensus was not always reached among all panel members for every aspect considered, and both agreements and disagreements are presented. As I wrote in the Conclusions to the Symposium Report, despite the expected challenges associated with the scenarios considered, some general patterns were identified. These included potential ecosystem responses linked to changes in transpiration and evapotranspiration rates which, in turn, were further linked to interactions with stream-flow and water temperature changes. For some regions, catchment-wide interactions had overriding influences on changes at the stream- and lake-ecosystem level. Direction of climatic change varied from straightforward predictions of warmer and drier conditions in the Mid-Atlantic and New England region to one of complete unpredictability in the Basin and Range, Arid Southwest and Mexico region. Historical and annual variations in climate, hydrology and ecosystem characteristics needed to be considered in the context of the models used. Further postulated responses varied widely, ranging from predictable detrimental impacts on species existing in warm waters already near lethal thermal conditions to dramatic increases in primary productivity and dependent food-webs in regions where increasing temperatures may extend ice-free conditions. Effects to endemic organisms appear to be mainly indirect through links to changes in other ecosystem components, such as changing nutrient regimes resulting from increased or altered runoff patterns. It is hoped that this book will provide a solid foundation upon which scientists can proceed and build in their efforts to provide other scientists, regulators, and decision-makers with meaningful information concerning the important phenomena associated with climate change.

C. E. Cushing
Editor

1

REGIONAL DELINEATION OF NORTH AMERICA FOR THE ASSESSMENT OF FRESHWATER ECOSYSTEMS AND CLIMATE CHANGE

G. H. LEAVESLEY, K. TURNER, F. A. D'AGNESE AND D. McKNIGHT

USGS, PO Box 25046, M5412, Denver Federal Center, Denver, CO 80225, USA

INTRODUCTION

A basic premise in organizing the Symposium on Climate Change and Freshwater Ecosystems was that a regional approach, rather than a continental approach, would enable stronger linkages with current climate models, better utilization of paleoclimate records, clearer identification of paths for future research and greater possibilities for reaching definitive conclusions. The efficacy of this regional approach was therefore dependent upon identifying useful regional boundaries. The boundary selection process involved the consideration of many relevant factors and the utilization of diverse data sources.

North America has an area of about 21 471 100 km^2 (Espenshade and Morrison, 1978) and is composed of the countries of Canada (*ca.* 9 976 680 km^2), the United States (*ca.* 9 520 840 km^2) and Mexico (*ca.* 1 973 580 km^2). The large latitudinal and elevational gradients found across North America strongly influence the variability of its climate in terms of the spatial and temporal distribution of ecologically important variables such as temperature, precipitation, and solar radiation. The magnitude and variance of these variables and their interactions with landscape variables, such as geology, soils and vegetation, affect many factors that influence the type and diversity of freshwater ecosystems. The addition of anthropogenic factors to these interactions further increases the number, complexity and diversity of these ecosystems.

While it is recognized that a large number of levels of detail are possible in a regional delineation of freshwater ecosystems in North America, it was determined that eight regions would provide a level of detail sufficient to meet the goals of the Symposium. These goals were to: (1) evaluate current evidence for directional change in inland waters, which broadly include lakes, wetlands, streams, rivers and estuaries; (2) examine potential important responses to climatic changes resulting from a hypothetical doubling of atmospheric carbon dioxide (or its radiative equivalent); and (3) develop recommendations for experimental studies or augmentation of monitoring programmes. The delineated regions are shown in Figure 1 and are listed in Table I.

REGIONAL DELINEATION PROCESS

The determination of the eight regions was made by analysing a number of databases and maps generated by previous investigations of the climatic, physiographical, hydrological and landscape characteristics of North America. A major consideration in the setting of regional size and boundaries was a goal of compatibility of scale with meso-scale general circulation models (GCMs).

Maps of several climatic and geographical characteristics were generated using data from the Global Ecosystems Database (GED) (US EPA and NOAA, 1991). These maps served as guides for the development of meaningful regions in the context of the Symposium goals. Once the regions were defined, these maps provided an overview of basic similarities and differences between regions, and of heterogeneities within regions. Within this context, a map of land surface elevation zones provided a critical foundation (Plate 1). The map of land surface elevation zones was developed using the gridded 5 minute elevation data from the

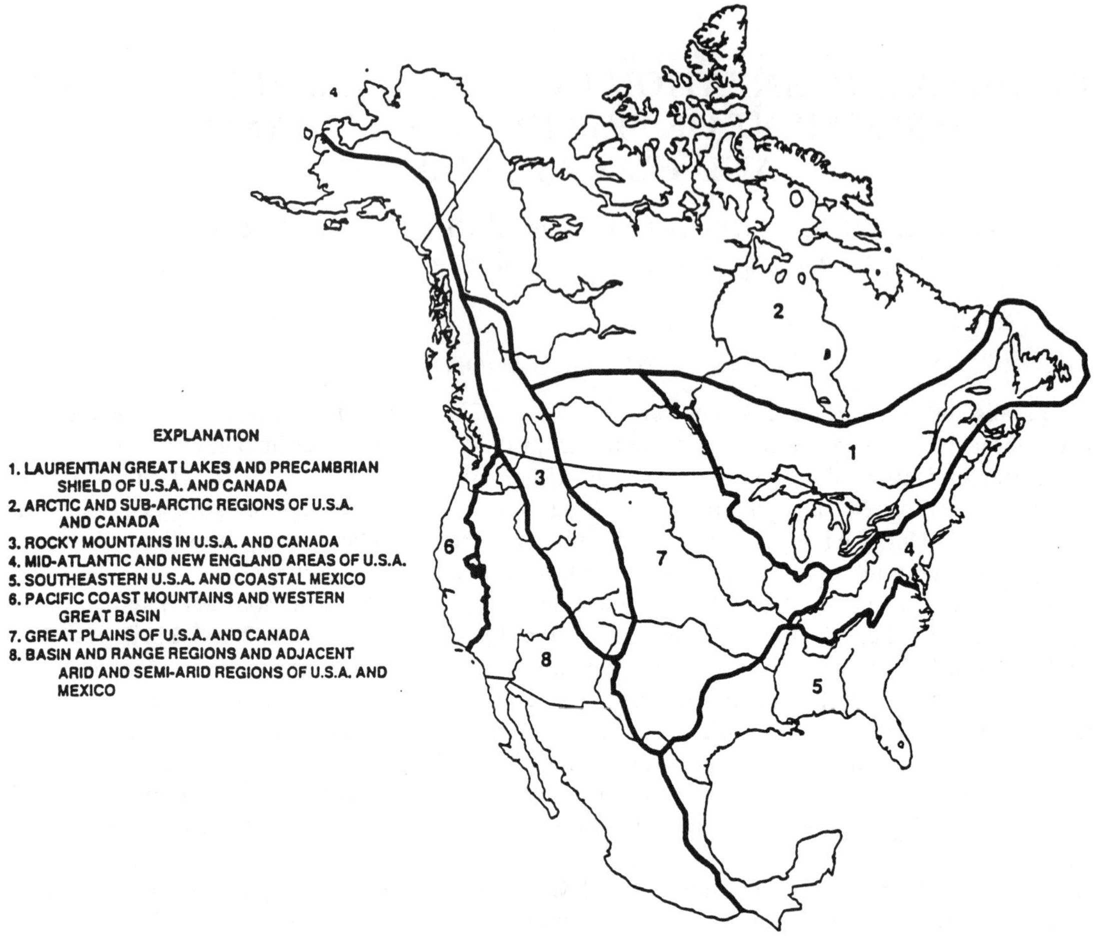

Figure 1. Location of the eight delineated regions of North America

ETOPO5 data file of the GED. A standard base map containing country boundaries, state boundaries within the US and major river systems was generated using the US Geological Survey 1 : 2 000 000 digital line graph (DLG) files and was added to this map and the others generated from the GED.

Precipitation and temperature regimes are basic climatic characteristics that directly and indirectly influence freshwater ecosystems. Precipitation depth, seasonality, intensity and form as rain or snow are

Table I. List of the eight regions delineated for the assessment of freshwater ecosystems and climate change

1. Laurentian Great Lakes and Precambrian Shield of USA and Canada
2. Arctic and subarctic regions of USA and Canada
3. Rocky Mountains in USA and Canada
4. Mid-Atlantic and New England areas of USA
5. South-eastern USA and coastal Mexico
6. Pacific Coast Mountains and Western Great Basin
7. Great Plains of USA and Canada
8. Basin and Range regions and adjacent arid and semi-arid regions of USA and Mexico

major factors that affect ecosystem type, composition and diversity. For the purposes of this Symposium, mean annual precipitation depth was assumed to provide sufficient detail for regional delineation and comparison. Mean annual temperature, however, was not considered an adequate measure because of the strong influence temperature extremes have on freshwater ecosystems. Instead, mean monthly temperatures for the months of January and July were selected to reflect better the inter- and intra-regional difference in winter and summer temperatures.

Maps of mean annual precipitation (Plate 2), mean January temperature (Plate 3) and mean July temperature (Plate 4) were generated using the average monthly temperature and precipitation data files of the GED. These files were created using data from Legates and Willmott (1989). The files contain 12 monthly averages for the period 1920–1980. The data are a 10-minute grid resampled from a 0.5-degree grid which was interpolated from 24 941 independent surface air temperature and 26 858 independent precipitation stations and ocean grid point averages around the world. Maps depicting the Strahler world climate zones and the Köppen–Geiger system of climate classification (Strahler and Strahler, 1987) were also used as input to the delineation process.

Maps of the physiographical divisions of North America were obtained from Strahler and Strahler (1987) and from the National Atlas of the United States (US Geological Survey, 1968). These included maps of physical divisions as delineated by Fenneman (1928) and Lobeck (1932). Major hydrological divisions were obtained for North America from maps in Espenshade and Morrison (1978) and for the United States from the US Geological Survey hydrological units map (US Geological Survey, 1987).

All maps were analysed together and a first estimate of regional boundaries was made. As noted by Omernik (1987) in his delineation of ecoregions of the United States, a key to the delineation process was 'distinguishing the regional homogeneity in a combination of characteristics from the heterogeneity in each characteristic.' Some regions, such as the Rocky Mountains of the United States and Canada were clearly and easily defined. Other boundaries, such as between the mid-Atlantic and south-eastern United States were less clear and a broader interpretation of differences in characteristics was employed. The initial boundary locations were reviewed by the Symposium Organizing Committee and the final boundary locations were established based on review comments from participants at the Symposium. The climatic, physiographical, hydrological and ecological characteristics of each region are discussed in the individual region reports.

REGIONAL CHARACTERISTICS

Within the goals of the Symposium we considered the full range of aquatic ecosystems, both inland and estuarine. Although each region contains a variety of freshwater ecosystems, some ecosystems are more important than others in a particular region. Within the eight delineated regions, aquatic ecosystems were classified at a general level and at a more detailed level using additional physical and chemical features. The three general classes used were lakes, wetlands and streams/rivers. There are overlaps between these general classes; for example, reservoirs are commonly found within large river systems and wetlands are commonly found with lakes.

The classification scheme and the relative importance of each class within each of the eight regions is summarized in Table II. Saline lakes and black water rivers were distinguished because such chemical characteristics have a pervasive influence on ecosystem properties. Saline in this scheme is meant to indicate lakes with high concentrations of dissolved solids. Prairie wetlands and intermittent streams were distinguished because their hydrological regimes also profoundly influence the ecosystem, and because of their potential sensitivity to climate change.

While this classification scheme is simple, it does illustrate some similarities and differences between the eight regions. Small lakes/ponds, large rivers and perennial streams are three types of inland waters that are common to all eight regions. However, large swamps, prairie wetlands and black water rivers are major

Table II. Inland waters, classification and occurrence by region

Classification	Regions							
	1	2	3	4	5	6	7	8
Lakes								
Large lakes	X	X	—	—	X	X	—	—
Saline lakes	—	X	—	—	X	X	X	X
Small lakes/ponds	X	X	X	X	X	X	X	X
Reservoirs	X	—	X	X	X	X	X	X
Wetlands								
Large riparian wetland	X	X	—	X	X	X	X	—
Large swamps	—	—	—	—	X	—	—	—
Prairie wetlands	—	—	—	—	—	X	X	—
Small bogs/fens	X	X	X	X	X	X	—	—
Streams/rivers								
Large rivers	X	X	X	X	X	X	X	X
Black water rivers	—	—	—	—	X	—	—	—
Alpine streams	—	X	X	—	—	X	—	—
Intermittent streams	—	—	X	—	—	X	X	X
Perennial streams	X	X	X	X	X	X	X	X
Coastal zone areas	X	X	—	X	X	X	—	X

types in only one or two regions. The Pacific Coast Mountains and Western Great Basin region (region 6) has the greatest variety of inland waters. The Mid-Atlantic and new England region (region 4) has the least variety with most inland waters classified as small lakes and ponds, bogs and fens, large rivers and perennial streams. This general classification was extended to include coastal zone areas and shows that all but two of the regions contain river systems that discharge into coastal zones and influence coastal ecosystems.

EXPERIMENTAL STUDIES

Recommendations for experimental studies or augmentation of monitoring programmes could potentially be implemented to some extent through existing research and monitoring programmes that include inland aquatic ecosystems. The National Science Foundation in the US operates the Long-Term Ecological Research (LTER) Network, which includes 15 sites on the North American continent (Figure 2). Many of these sites include some study of aquatic ecosystems. In Canada, research at the Experimental Lakes Area (ELA) in Ontario (Figure 2) has provided important understanding of the response of lake ecosystems to climatic change. Figure 2 indicates that coverage of the eight regions by the LTER network and ELA is sparse.

Another national programme in the US is the National Water Quality Assessment (NAWQA) program of the US Geological Survey. NAWQA has an emphasis on long-term monitoring of streams and rivers, and includes monitoring of all levels of these ecosystems (from algae to fish). The distribution of the 60 NAWQA basins (Figure 3) covers a large portion of the eight regions within the United States.

Clearly, within each region there are other long-term monitoring or research programmes funded by federal, regional or local agencies that will continue to provide important understanding of climatic influences on aquatic ecosystems. Greater coordination and scientific interaction at the regional level would facilitate the development of this understanding and provide the framework needed to support the development of future advances.

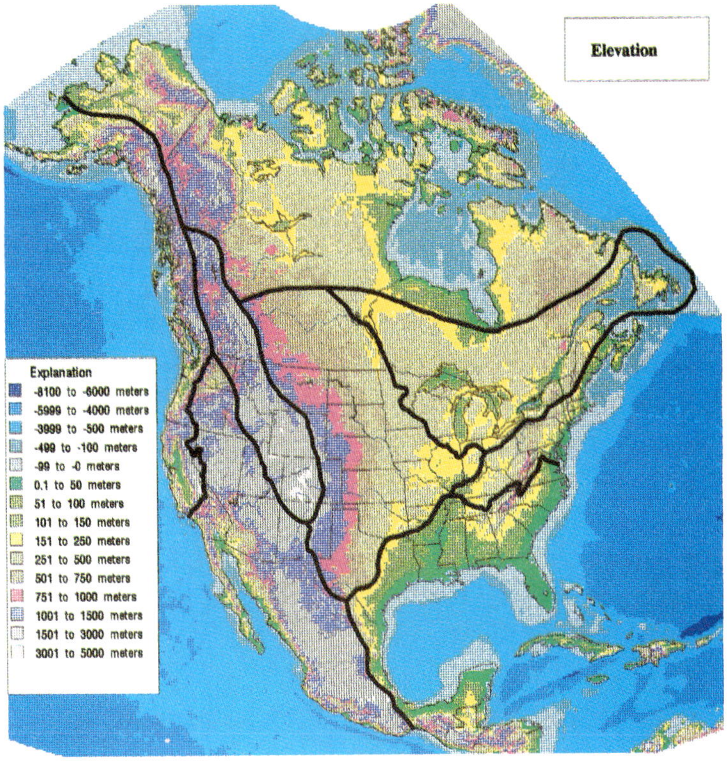

Plate 1. Land surface elevation zones

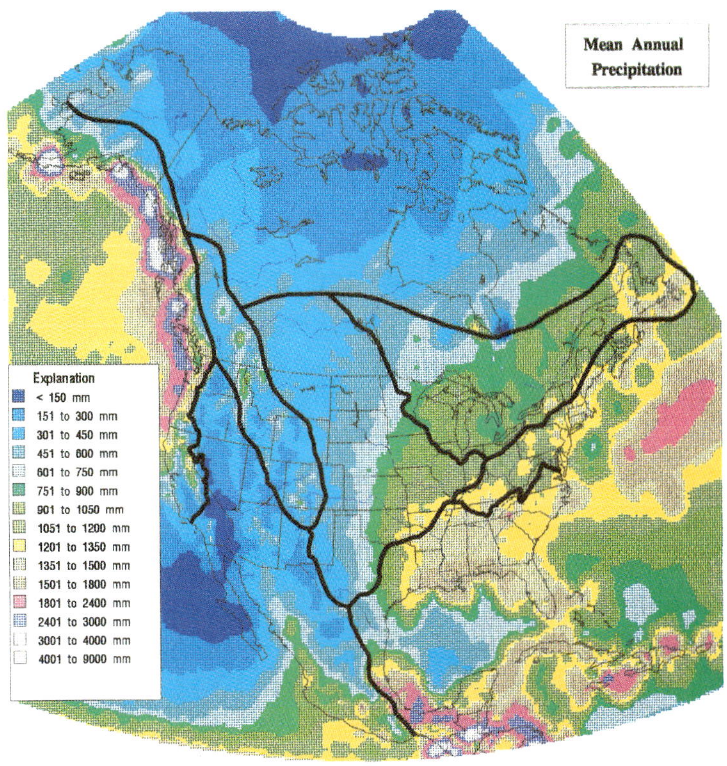

Plate 2. Mean annual precipitation zones

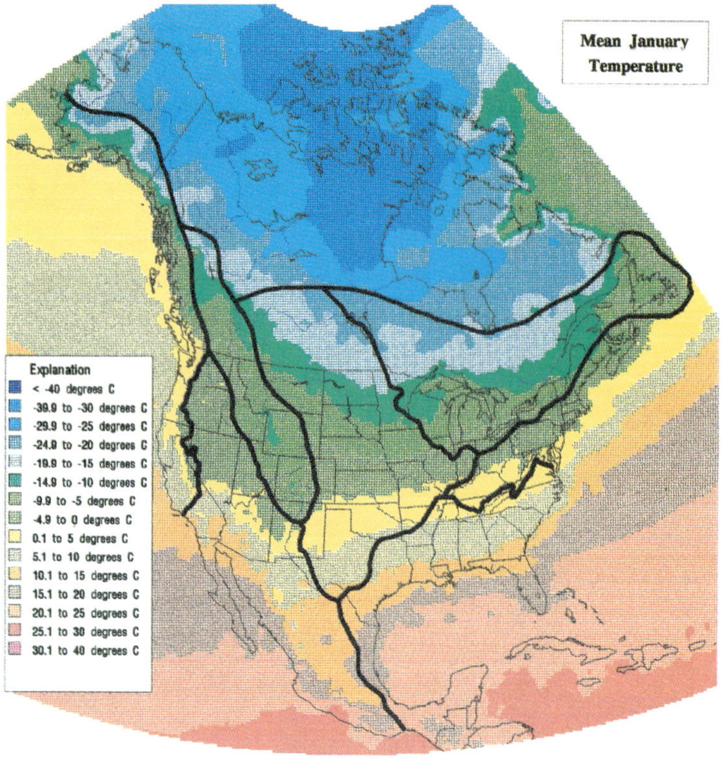

Plate 3. Mean January temperature zones

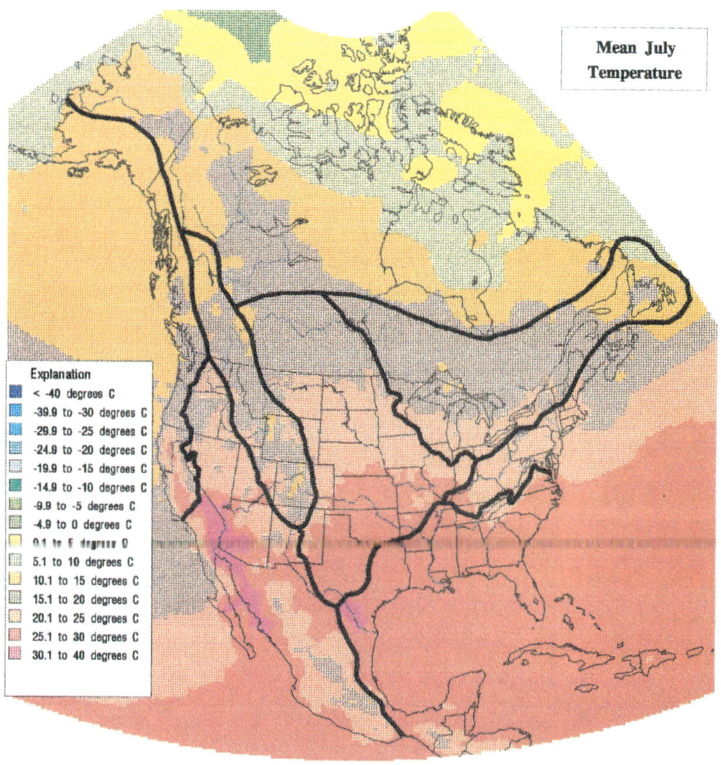

Plate 4. Mean July temperature zones

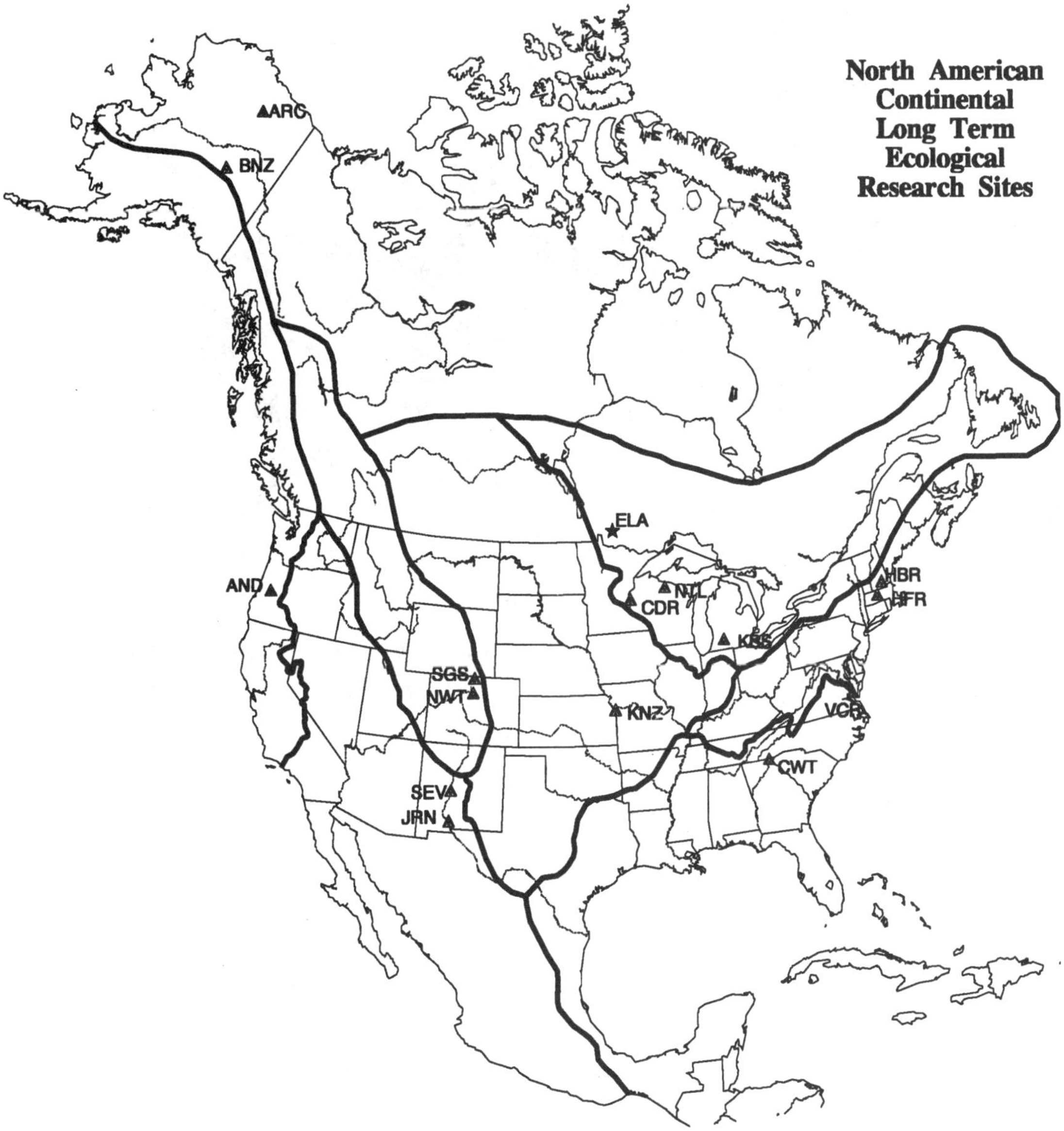

Figure 2. Locations of LTER network sites and the ELA site

SUMMARY

There is a recognition that the response of freshwater ecosystems to predicted climate changes will vary greatly on a regional basis. Spatial and temporal changes in climate variables such as precipitation, temperature and solar radiation will affect the physical and chemical environments of all ecosystems in North America. The eight regions identified for this Symposium provide a framework for a systematic and rigorous analysis of similarities and differences in freshwater ecosystems between regions and of the variations in their anticipated response to potential climate changes. Integration of the regional findings will provide a better understanding of ecosystem responses at the broader continental scale.

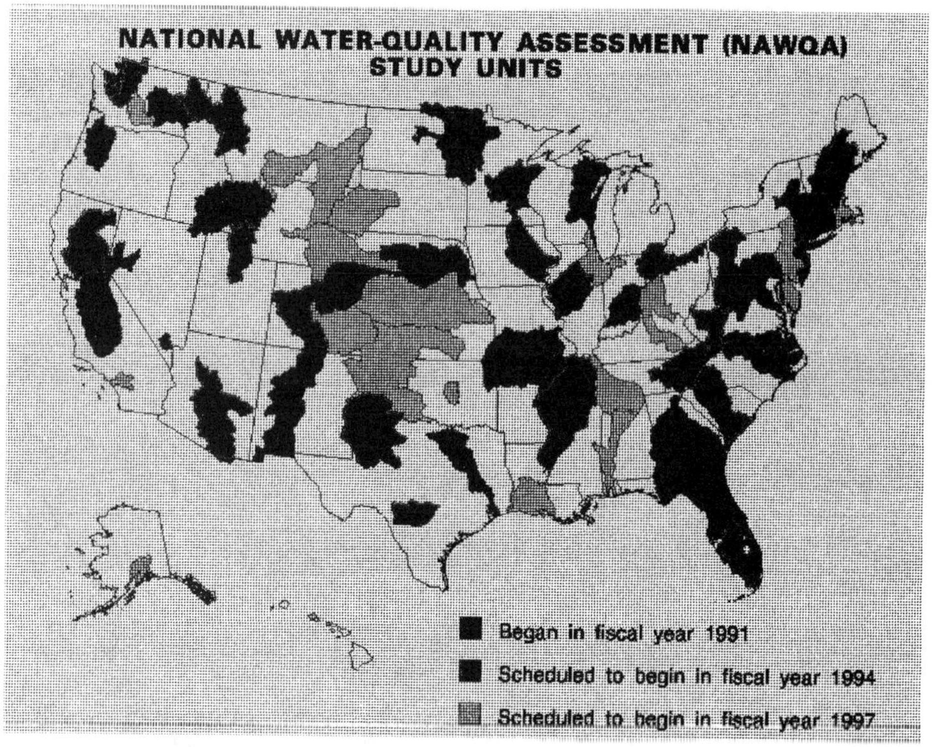

Figure 3. Locations of the NAWQA study basins

REFERENCES

Espenshade, E. B., Jr. and Morrison, J. L. 1978. *Goode's World Atlas*. Rand McNally and Co., Chicago.

Fenneman, N. M. 1928. 'Physiographic divisions of the United States,' *Ann. Assoc. Am. Geogr.* **18**, 261–353.

Legates, D. R. and Willmott, C. J. 1989. 'Global air temperature and precipitation,' National Centre for Atmospheric Research, Boulder, CO, Digital data, 13·5 MB.

Lobeck, A. K. 1932. *Physiographic Diagram of the United States*. Wisconsin Geography Press, Madison. 8 pp.

Omernik, J. M. 1987. 'Ecoregions of the conterminous United States,' *Ann. Assoc. Am. Geogr.* **77**(1), 118–125.

Strahler, A. N. and Strahler, A. H. 1987. *Modern Physical Geography*. John Wiley and Sons, New York.

US EPA (Environmental Protection Agency) and NOAA (National Oceanic and Atmospheric Administration), 1991. *Global Ecosystems Database*. National Geophysical Data Center, Boulder, CO, CD-ROM.

US Geological Survey, 1968. *National Atlas*. US Department of Interior, US Geological Survey, Reston, VA.

US Geological Survey, 1987. *Hydrologic Unit Maps*. US Geological Survey, Reston, VA. Water Supply Paper 2294. USGS, 63 p.

2

POTENTIAL EFFECTS OF CLIMATE CHANGES ON AQUATIC SYSTEMS: LAURENTIAN GREAT LAKES AND PRECAMBRIAN SHIELD REGION

J. J. MAGNUSON,[1] K. E. WEBSTER,[1,2] R. A. ASSEL,[3] C. J. BOWSER,[1] P. J. DILLON,[4] J. G. EATON,[5] H. E. EVANS,[4] E. J. FEE,[6] R. I. HALL,[7] L. R. MORTSCH,[8] D. W. SCHINDLER[9] AND F. H. QUINN[3]

[1] *Center for Limnology, University of Wisconsin-Madison, 680 N. Park St., Madison, WI, USA*
[2] *Wisconsin Department of Natural Resources, Madison, WI, USA*
[3] *NOAA Great Lakes Environmental Research Laboratory, Ann Arbor, MI, USA*
[4] *Dorset Research Centre, ON, Canada*
[5] *US Environmental Protection Agency, Duluth MN, USA*
[6] *Freshwater Institute, Winnipeg, MB, Canada*
[7] *University of Regina, Regina, SK, Canada*
[8] *Environmental Adaptation Research Group, Environment Canada, Burlington, ON, Canada*
[9] *University of Alberta, Edmonton, AB, Canada*

ABSTRACT

The region studied includes the Laurentian Great Lakes and a diversity of smaller glacial lakes, streams and wetlands south of permanent permafrost and towards the southern extent of Wisconsin glaciation. We emphasize lakes and quantitative implications. The region is warmer and wetter than it has been over most of the last 12 000 years. Since 1911 observed air temperatures have increased by about 0·11°C per decade in spring and 0·06°C in winter; annual precipitation has increased by about 2·1% per decade. Ice thaw phenologies since the 1850s indicate a late winter warming of about 2·5°C. In future scenarios for a doubled CO_2 climate, air temperature increases in summer and winter and precipitation decreases (summer) in western Ontario but increases (winter) in western Ontario, northern Minnesota, Wisconsin and Michigan. Such changes in climate have altered and would further alter hydrological and other physical features of lakes. Warmer climates, i.e. $2 \times CO_2$ climates, would lower net basin water supplies, stream flows and water levels owing to increased evaporation in excess of precipitation. Water levels have been responsive to drought and future scenarios for the Great Lakes simulate levels 0·2 to 2·5 m lower. Human adaptation to such changes is expensive. Warmer climates would decrease the spatial extent of ice cover on the Great Lakes; small lakes, especially to the south, would no longer freeze over every year. Temperature simulations for stratified lakes are 1–7°C warmer for surface waters, and 6°C cooler to 8°C warmer for deep waters. Thermocline depth would change (4 m shallower to 3·5 m deeper) with warmer climates alone; deepening owing to increases in light penetration would occur with reduced input of dissolved organic carbon (DOC) from dryer catchments. Dissolved oxygen would decrease below the thermocline. These physical changes would in turn affect the phytoplankton, zooplankton, benthos and fishes. Annual phytoplankton production may increase but many complex reactions of the phytoplankton community to altered temperatures, thermocline depths, light penetrations and nutrient inputs would be expected. Zooplankton biomass would increase, but, again, many complex interactions are expected.

Generally, the thermal habitat for warm-, cool- and even cold-water fishes would increase in size in deep stratified lakes, but would decrease in shallow unstratified lakes and in streams. Less dissolved oxygen below the thermocline of lakes would further degrade stratified lakes for cold water fishes. Growth and production would increase for fishes that are now in thermal environments cooler than their optimum but decrease for those that are at or above their optimum, provided they cannot move to a deeper or headwater thermal refuge. The zoogeographical boundary for fish species could move north by 500–600 km; invasions of warmer water fishes and extirpations of colder water fishes should increase. Aquatic ecosystems across the region do not necessarily exhibit coherent responses to climate changes and variability, even if they are in close proximity. Lakes, wetlands and streams respond differently, as do lakes of different depth or productivity. Differences in hydrology and the position in the hydrological flow system, in terrestrial vegetation and land use, in base climates and in the aquatic biota can all cause different responses. Climate change effects interact strongly with effects of other human-caused stresses such as eutrophication, acid precipitation, toxic chemicals and the spread of exotic organisms. Aquatic ecological systems in the region are sensitive to climate change and variation. Assessments of these potential effects are in an early stage and contain many uncertainties in the models and properties of aquatic ecological systems and of the climate system.

INTRODUCTION

The Laurentian Great Lakes and Precambrian Shield Region extends from the southern edge of the permafrost line to the southern extent of the Wisconsin glaciation (between 40–50°N) (Figure 1). Much of the current character of the region was shaped by geological and glacial processes, especially by the retreat of the Wisconsin glaciation beginning about 12 000 years ago. This water-rich region has low relief, cool to cold winters and warm to cool summers. A multitude and diversity of aquatic resources dominates the landscape, including the Laurentian Great Lakes, smaller glacial lakes, streams and wetlands.

The Great Lakes system has a surface area of 244 000 km^2 and a volume of 23×10^{15} litres, 20% of the world's fresh surface water. Lake Superior is the world's largest freshwater lake in surface area, while Huron is fifth and Michigan is sixth. Lake Ontario is the fourth deepest and fourteenth largest lake in the world. About 30 million people live in the Great Lakes Basin, including 9% of the total population of the US and 29% of that of Canada. The Great Lakes play an important role in the region's economy, supplying water for domestic uses, agriculture, power generation and industry; providing a vital inland shipping route for a variety of products; and supporting important recreational and commercial fisheries. The region includes high concentrations of smaller inland lakes. Over 376 000 lakes greater than 1 ha in area (about 18 000 000 ha total) occur in south-eastern Canada south of 52°N and east of 90°W. Similar high densities of lakes occur in the upper mid-western states of Wisconsin, Michigan and Minnesota. For example,

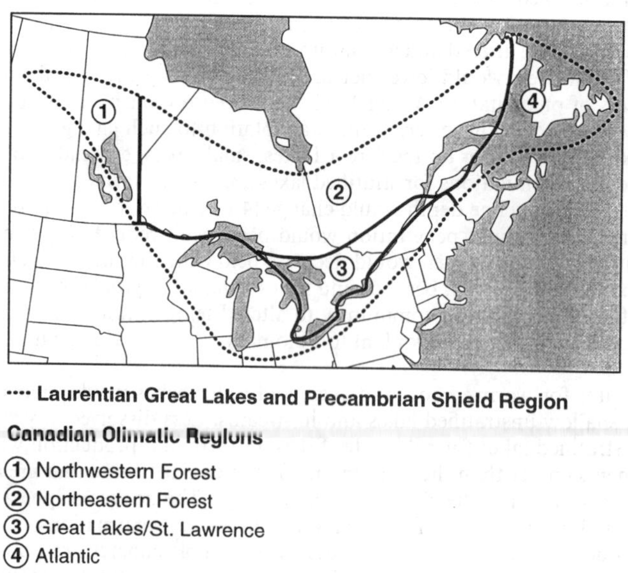

Figure 1. Major climatic areas of Canada used for analysis of temperature and precipitation trends

Wisconsin alone has 12 500 natural lakes covering 140 000 ha (Wisconsin Department of Natural Resources, 1995), or, on a smaller scale, nearly 2500 lakes occur in Vilas and Oneida Counties, Wisconsin, where they cover about 12% of the landscape. Wisconsin also contains about 2 000 000 ha of wetlands and perhaps 53 000 km of streams. These are important components of the landscape that are strong attractors for humans.

The physical, chemical and biological processes and attributes of these systems are sensitive to climate change as are the values placed on these aquatic systems by humans. Two overarching considerations shaped our thoughts; first, the wide array of limnological processes and phenomena expected to be influenced by climate changes; and, secondly, the heterogeneity or differences in responses expected among water bodies. Our interests were also drawn to quantitative estimates over qualitative ones. Our review builds on general overviews of the potential changes in aquatic ecosystems related to climatic warming (J. B. Smith, 1991; Carpenter *et al.*, 1992; Arnell *et al.*, 1996). We address potential changes in the climate itself; physical limnology, including ice phenologies, water temperatures and mixing regimes; hydrology, with respect to water levels, runoff, weathering and residence times; light availability, with respect to changes in water clarity; chemical limnology, such as dissolved organic carbon concentrations, nutrient availability and deep water oxygen concentrations; and distribution, growth and persistence of the fauna and flora. Watershed- and lake-specific factors were identified as being key in determining the heterogeneity of response between waters. Even adjacent lakes can respond differently owing to differences in morphology and position in the hydrological flow system.

PRESENT CLIMATE

The Laurentian Great Lakes and Precambrian Shield Region (Figure 1) contains a suite of climates; conditions range from the cold winters and cool summers of the boreal areas; to the eastern remnant of the prairies in Manitoba; to the warm, humid summers and cold winters of the south-western Great Lakes region; to the maritime influence in Newfoundland (Table I). This diversity reflects the latitudinal and longitudinal extent of the region, as well as the other climatic controlling factors of topography, proximity to water bodies, continentality and large-scale atmospheric circulation.

Most of the region lies in the zone of westerlies, where migrating cyclones and anticyclones predominate. In winter, storms travel from the west and south-west into the Great Lakes area and along the St Lawrence River valley, while in summer, they follow a more northerly route above Lake Superior and through Hudson Bay, northern Quebec and Labrador. Seasonal changes in the position of the polar jet stream affect the route of storm tracks and the predominant air masses. In winter, the more southerly extent of the jet stream allows cold, dry Arctic air to penetrate south, while in summer, the more northerly position of the jet stream allows warm, moist air from the Gulf of Mexico to intrude. Mean monthly and annual air temperatures generally decrease with increasing latitude, reflecting latitudinal differences in intensity and duration of solar radiation. Mean annual temperatures range from −1·1 to +2·2°C in the north, to +2·3 to +9·1°C near the Great Lakes, to 0·0 to +4·8°C on the east coast (Table I). The difference between mean January and July temperatures is greatest near Winnipeg, Manitoba (40°C), least near St. John's, Newfoundland (19°C), and intermediate near Windsor in southern Ontario (27°C) (Phillips, 1990). The southernmost limit of discontinuous permafrost coincides approximately with the −1·0°C mean annual air temperature (Mortsch, 1990). Scattered discontinuous permafrost occurs near James Bay; no permafrost is present in the Great Lakes basin and Newfoundland. The number of freeze-free days ranges from 140 in the north to 230 in the south. Most small lakes freeze over completely; the Great Lakes attain maximum ice cover in February or March but usually do not freeze over completely (Superior 75%, Michigan 45%, Huron 68%, Erie 90% and Ontario 24%; Assel *et al.*, 1983). Depending on flow, size and location, rivers may or may not freeze over. Hudson Bay and James Bay offer little warmth or moderating influence to the surrounding region because ice-out occurs in late June. Annual precipitation increases from west to east at a rate of approximately 40 mm every 100 km, from about 400 mm per year in Winnipeg, Manitoba, to 1400 mm on the eastern

Table I. Current climate statistics for sites in, or somewhat north of, the Great Lakes and Laurentian Shield Region. Data are from Atmospheric Environment Service (1982a–f) and McIver and Isaac (1989)

Climate station	Location	Mean annual daily temp. (°C)	Mean January temp. (°C)	Mean July temp. (°C)	Mean annual freeze-free days (days)	Mean annual growing season (days)	Mean annual growing degree–days (°C)	Total annual precip. (mm)	Total annual snowfall (cm)	Total January precip. (mm)	Total July precip. (mm)
Manitoba											
Winnipeg Int'l A	49°54′N 97°14′W	2·2	−19·3	19·6	170	159	1784·9	525·5	125·5	21·3	75·9
Ontario											
Moosonee	51°16′N 80°39′W	−1·1	−20·4	15·3	139	123	1107·4	727·7	239·3	40·7	95·9
Thunder Bay A	48°22′N 89°19′W	2·3	−15·4	17·6	161	147	1425·4	711·8	213·0	40·9	75·4
Wiarton	44°45′N 81°06′W	6·0	−7·1	18·5	141	N/A	1821·0	964·7	352·6	97·0	75·1
Windsor A	42°16′N 82°58′W	9·1	−4·9	22·2	231	213	2533·3	848·8	117·4	55·0	83·4
Quebec											
Sept-Isles A	50°13′N 66°16′W	1·1	−14·0	15·2	158	135	1015·6	1124·9	426·9	95·5	97·0
Newfoundland											
St John's A	47°37′N 52°45′W	4·8	−3·9	15·5	139	169	1196·1	1513·6	359·4	155·8	75·3
Goose (Bay) A	53°19′N 60°25′W	0·0	−16·4	15·8	151	135	1021·9	946·1	445·2	74·4	105·1

shore of Newfoundland (Phillips, 1990). Precipitation exceeds 1000 mm in the lee of the Great Lakes, along the St Lawrence River and in Newfoundland.

The mid-continental location of the Great Lakes Basin contributes to large seasonal air temperature contrasts between the warm summers and cold winters. The Atlantic Ocean cannot exert a significant moderating influence because air flow is primarily from the west. The polar jet stream is often located near or over the basin and is the focal point for the genesis and development of low pressure storm systems that are characterized by cloudy skies, windy conditions and precipitation. The frequent passage of these low pressure systems results in the day to day variability in weather conditions.

The Great Lakes are large heat reservoirs that moderate the surrounding climate. This moderation delays the onset of spring because large lakes are a heat sink, with water temperatures cooler than the surrounding land. The autumn is extended and first frost is delayed because the warmer waters slowly release stored heat. 'Lake effect' snowbelts occur along the leeward sides of the lakes. During late autumn and winter, when the water is still open, outbreaks of cold, dry Arctic air pass over the warm, moist waters of the Great Lakes and convective snowfall develops (Phillips and McCulloch, 1972; Kunkel *et al.*, 1993).

Newfoundland and Labrador have a maritime climate although the influence of the Labrador Sea and the Atlantic Ocean is minimal because the prevailing westerly winds are offshore. In spring and summer, winds can be onshore and the cold Labrador Current moderates temperatures and causes fog and low clouds. Newfoundland can be influenced occasionally by tropical cyclones.

PALAEOCLIMATES

A variety of palaeotechniques reveal a pattern of continuous climate change and variation over the last 12 000 years (Figure 2) (see also Grimm and Jacobson, 1992). A coarse scale interpretation is that climates were cooler than the present from 12 000 to 7000 BP and again from 3000 to 300 BP; a period of warmer climates existed from about 7000 to 3000 BP (Figure 2a). Temperatures were up to 7°C cooler and 3°C warmer. Precipitation at present is high relative to the last 12 000 years; only the years around 9000 BP were wetter than present (Figure 2b). Precipitation was up to 40 mm greater and 180 mm lower than at present. Below, we discuss the palaeoclimates and palaeolimnology for the three Canadian climate regions relevant to the Laurentian Great Lakes and Precambrian Shield Region (Figure 1).

Laurentian Great Lakes/St Lawrence

The earliest lacustrine and bog sediment records from the Laurentian Great Lakes and Precambrian Shield Region date to *ca.* 14 500 BP, with the development of glacial Lake Maumee in the Lake Erie basin, from where the Wisconsin ice sheet first waned northwards (Calkin and Feenstra, 1985). Most lake sediment records north of Lake Ontario and east of Lake Huron began somewhat later, *ca.* 12 000 to 11 500 BP, when the Wisconsin ice sheet retreated. Sediments of much longer duration exist only in a few cases; a sediment record from Ile de la Madeleine, Quebec, dates at least as far back as the Sagamonian climatic optimum (89–110 ka), and contains pollen suggesting a warmer climate than at any time during the Holocene (Dredge *et al.*, 1992). Climatic and vegetational trends were not synchronous throughout the Great Lakes region during the early postglacial period. Trends were initiated earlier in the south and later in the north as the ice retreated northwards and the Champlain Sea disappeared in approximately 11 000 BP (Anderson, 1987).

Pollen and lake level data, and the climatic estimates inferred from them, indicate that the greatest climatic changes took place between 12 000 and 9000 BP (Webb *et al.*, 1987). The initial vegetation following deglaciation was herb–shrub tundra, interspersed with open stands of spruce, poplar, juniper and shrub birch and alder (Anderson, 1987). This vegetation developed along the southern Great Lakes from shortly after deglaciation until about 11 200 BP, and persisted to the north of the Champlain Sea until about 10 000 BP (Anderson, 1987). Pollen transfer functions infer mean July temperatures ranging from 11 to 14°C during this period at Tonawa Lake (44°51′N, 77°11′W) near Bancroft, Ontario (McAndrews and Campbell, 1993), which are much lower than the present temperatures of about 19°C.

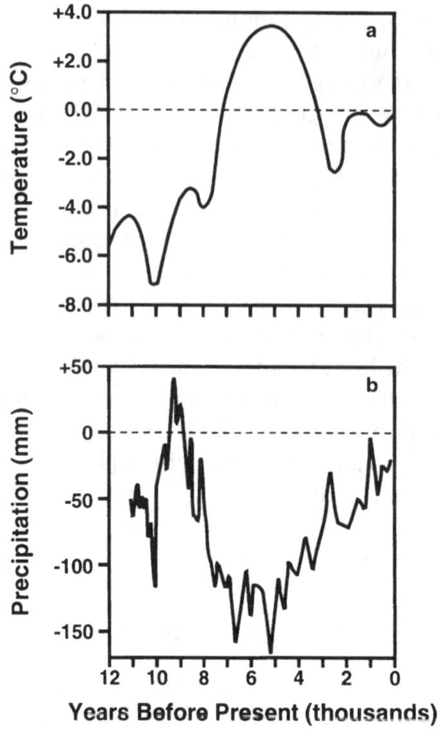

Figure 2. Mean annual palaeotemperature (a) for south-eastern Canada (modified from Terasmae, 1961) and precipitation estimates (b) from pollen data at Kirchner Marsh in Minnesota (modified from Webb *et al.*, 1987)

About 11 000 BP an open canopy spruce forest developed along the Great Lakes and migrated northwards with the margin of the Laurentian Ice Sheet, reaching the Precambrian Shield in the Ottawa Valley by about 10 200 BP (Anderson, 1987), and north of Sudbury, Ontario just after 10 000 BP (Liu, 1990). The spruce forest spread as climatic conditions warmed, but the climate remained cooler and wetter than today (Webb *et al.*, 1987). Pollen-inferred mean July temperatures ranged from 14 to 17°C in souther Ontario, and slightly cooler to the north (McAndrews and Campbell, 1993). The spruce period persisted for about 1000 years until birch and pine arrived from the south.

Up until the end of the spruce period, lake diatom assemblages usually consisted of benthic taxa dominated by *Fragilaria* sp. (Sreenivasa and Duthie, 1973; Smol and Boucherle, 1985; Hall, 1993). Precise inferences of limnological conditions are difficult because similar species occur today from oligotrophic to eutrophic conditions. The general absence of planktonic forms suggests that the lakes were ice covered for extended periods during summer and that thermal stratification did not develop. Apparently benthic diatoms were restricted largely to nearshore, ice-free moats in shallow margins of the lakes (Smol and Boucherle, 1985; Smol, 1982).

Pollen and lake level data indicate that 12 000 and 9000 BP were considerably cooler and wetter than the present (Terasmae, 1961; Webb *et al.*, 1987) (Figure 2). Seasonal differences in insolation and temperature were pronounced. The Laurentide Ice Sheet created unique climatic conditions that do not exist today. Mean annual temperatures warmed between 12 000 to 11 000 BP and again between 10 000 to 8000 BP, and generally ranged from 8–4°C below those at present.

Localized climatic shifts adjacent to the Laurentian Great Lakes were caused by the retreats and readvances of the ice sheet; this altered patterns of inflow and outflow to and from the lakes, as well as water levels, surface water temperatures and local climates (Lewis and Anderson, 1989; Anderson and Lewis,

1992). Lake Agassiz discharged cold glacial meltwater eastwards through the Great Lakes and the Gulf of St Lawrence during two distinct periods, 11 000–10 500 and 9600–8300 BP, when the Laurentide Ice Sheet migrated northwards (Teller and Thorliefson, 1983; Teller, 1985). Cold meltwater flowing into the Great Lakes most likely increased several-fold, cooled surface waters (Lewis and Anderson, 1989), cooled local climates and reduced the growing season around the upper Great Lakes and Gulf of St Lawrence (Lewis and Anderson, 1989, 1992).

Pollen assemblages and, thus, climatic and edaphic conditions, became largely synchronous everywhere in the Great Lakes/St. Lawrence area after 9500 BP, when a closed pine forest developed in response to a warmer and drier climate (Anderson, 1987). In southern Ontario, pollen-inferred July temperatures averaged 19–21°C, similar to those of today (McAndrews and Campbell, 1993). The warmer temperatures probably extended the growing season of phytoplankton in the deep water regions of lakes (Smol and Boucherle, 1985), and permitted planktonic diatom and chrysophyte communities to develop. Lake nutrient concentrations were low because watershed nutrient release was low from the immature pine forest.

The Hypsithermal, or mid-Holocene warm period, occurred between 8000 and 4000 BP. The pollen–climate transfer functions of Bartlein and Webb (1985) and Bartlein and Whitlock (1993, in Yu and McAndrews, 1994) indicated mean July temperatures 1–2°C warmer, and precipitation 10–20% less than at present. These conditions permitted hemlock, beech and thermophilous hardwood taxa (including maple, elm and oak) to migrate into southern Ontario. By 7500 BP, a forest similar to the present Great Lakes/St Lawrence forest became established in the southern parts of the region (Rowe, 1977). Independent evidence from stable isotopes (Edwards and Fritz, 1988) and lower lake levels (Webb *et al.*, 1987; Yu and McAndrews, 1994) also indicated a warm/dry climate between 7400 and 4000 BP.

About 3500 BP, during the Neoglacial cooling, the climate became cooler and moister (Anderson, 1989). Palaeoecological evidence for this cooling is not straightforward in the Great Lakes/St Lawrence region. Pollen-inferred temperatures did not decline at this time in southern Ontario (McAndrews and Campbell, 1993) but palaeolimnological evidence indicates that water levels rose in Rice Lake, which is attributable to cooler and wetter conditions as well as continued isostatic uplift (Yu and McAndrews, 1994). Perhaps the best evidence for Neoglacial cooling comes from sites farther north in the boreal forest, where the ecotonal boundary shifted southwards (see below).

Campbell and McAndrews (1991, 1993) presented pollen stratigraphical evidence of the Little Ice Age cooling between 1450 and 1850 AD; beech pollen declined at its northern limit just south of Lake Nippissing, and the climate optimum for white pine shifted 100–140 km southwards. Pollen transfer functions inferred that mean July temperatures declined by about 0·5°C during this period at Tonawa Lake, near Bancroft, Ontario.

North-eastern forest

A steep climatic gradient and ecotonal boundary between the boreal and Great Lakes/St Lawrence forest exists between Lake Huron and James Bay (Liu, 1990). Its position is controlled by continent-scale atmospheric circulation that determines the position of the arctic front. Thus, the boundary shifted markedly with postglacial climatic changes (Liu, 1990); climatic shifts did not occur synchronously across the entire area.

Lake sedimentation began after 10 000 BP, later than in the Great Lakes/St Lawrence area, because the glacier retreated later. Spruce forests began about 9500 BP near the southern edge of the area (Nina Lake, 46°36′N, 81°30′W; Liu, 1990) and was delayed until about 7500 BP 220 km farther north (Lake Six, 48°24′N, 81°19′W). The pine forest was also delayed relative to more southern sites. Pollen profiles near Sudbury and Sault Ste. Marie inferred relatively rapid increases in mean July temperatures, from 16 to 19°C, in the pine period (McAndrews and Campbell, 1993). Jack pine forest development occurred later at sites farther north, beginning about 9000 BP at Nina Lake, and not until about 7000 BP at Lake Six (Liu, 1990). The period of jack pine forest was shorter at more northern sites, as white pine became established and dominant everywhere by about 7000 BP (Liu, 1990).

Hypsithermal warming caused a marked northward migration of the boundary between the boreal and Great Lakes forests. By 7400 BP this ecotone had migrated north of Sudbury (Nina Lake). By 3000 BP the boundary advanced to its northernmost position of the entire postglacial period, i.e. 140 km north of its present position (Liu, 1990). Southward movement of the boundary in the area north of Sudbury provides strong evidence for Neoglacial cooling (Liu, 1990). The boundary migrated pass Lake Six (to the north) about 4000 BP and Nina Lake (to the south) about 2500 BP. Pollen-inferred mean July temperature dropped about 0·5–1°C both at Lake QC (46°50′N, 80°42′W) near Sudbury and R Lake (54°18′N, 84°34′W) near the north-western edge of James Bay (McAndrews and Campbell, 1993).

Atlantic

The timing and nature of climatic and vegetational shifts in eastern maritime Canada were somewhat different from Quebec and Ontario, possibly owing to oceanic influences and different soil development. Sediment records in New Brunswick and Nova Scotia began just before 12 500 BP (Wilson *et al.*, 1993). Pollen stratigraphies indicated that the region was initially colonized by herb–shrub tundra. Prior to 11 000 BP, spruce trees became established, but appeared unable to colonize sites with direct exposure to oceanic weather, i.e. Brier Island Bog Lake (Wilson *et al.*, 1993). Pollen profiles typically showed a decline in spruce abundance and an increase in tundra-like vegetation (Mott *et al.*, 1986) in the cool Younger Dryas (*ca.* 11 000–10 000 BP). Based on reconstructions using chironomid assemblages, Wilson *et al.* (1993) approximated midsummer water temperatures at 17–20°C prior to the Younger Dryas (before 11 000 BP) at Brier Island Bog Lake, Nova Scotia, and at Splan Lake, New Brunswick. During the Younger Dryas midsummer, water temperatures dropped to 13–16°C, and subsequently increased after 10 000 BP to 20–24°C (Wilson *et al.*, 1993). Diatom assemblages indicated that the lakes supported seasonally ice-free, open water, and possibly turbulent, wind-mixed conditions during the Younger Dryas (Wilson *et al.*, 1993).

In Labrador, initial vegetation developed between 10 500 and 9500 BP, and consisted of an open tundra dominated by sedges, grasses, herbs and shrubs (Engstrom and Hansen, 1985). The climate was cold, but generally warmer than present day tundra. From 9500 to 8000 BP, shrub taxa, including birch and alder, increased, creating a vegetation similar to the shrub tundra that exists today. Continuous vegetation changes occurred from 8000 to 6000 BP owing to the sequential rise and replacement of Labrador's major tree species. White spruce arrived at *ca.* 8000 BP, substantially later relative to sites farther south along the Gulf of St Lawrence where spruce appeared between 10 000 and 9000 BP (Engstrom and Hansen, 1985). This shift was attributed to progressive changes in soil conditions, towards more waterlogged peat, rather than to climatic factors.

A relatively stable pollen assemblage characteristic of the present boreal forest has persisted in Labrador from 6000 BP to present. However, pollen and geochemical evidence indicated a regional shift to colder conditions after 4000 BP, resulting in decreased tree density and thinner soil humus (Engstrom and Hansen, 1985). Interior sites were more strongly affected than coastal ones. This cooling, which began between 4000 and 3000 BP, corresponds with the period of Neoglacial cooling detected at a number of sites in the Laurentian Great Lakes and Precambrian Shield Region.

RECENT CLIMATE TRENDS

Four climatic regions delineated for analysis of the Canadian climate include portions of the Laurentian Great Lakes and Precambrian Shield Region, these are the North western Forest, the North eastern Forest, the Great Lakes/St Lawrence and the Atlantic (Figure 1). Daily temperature and precipitation (Mekis and Hogg, 1997) and monthly temperature and precipitation (Gullett *et al.*, 1992) data sets were developed to analyse climate trends across Canada. The daily data were a contribution to the 1995 Intergovernmental Panel on Climate Change (IPCC) report and the monthly data comprise the Historical Canadian Climate Database (HCCD). Daily precipitation data were corrected for under catch from wind and other factors, wetting loss, trace value correction and instrumentation changes (Mekis and Hogg, 1997; Metcalfe *et al.*, 1994). Monthly precipitation data were not corrected. Monthly temperature data were adjusted to account for discontinuities or 'inhomogeneities' from changes in observing programme, instrumentation, site conditions and other non-climatic effects (Gullet *et al.*, 1992; Skinner and Gullet, 1993). For both data sets, methodologies were developed for joining observation records of nearby stations to extend the time-series. Missing values were estimated (Gullet *et al.*, 1992; Mekis and Hogg, 1997). Monthly temperature and daily precipitation data were selected to provide the most reliable time-series. The 83 years from 1911 to 1993 generally comprised the daily precipitation time-series, while 1895–1993 were used for temperature time-series. In addition, for major Great Lake basins, we examined precipitation and temperature trends from 1900 to 1990. Two primary hydrometeorological variables affecting the Great Lakes/St Lawrence River Basin ecosystem are precipitation and air temperature. Monthly precipitation data (Croley and Hunter, 1994) were based on US Lake Survey District, Corps of Engineers areally weighted 'precipitation districts' for 1900–1929 (Quinn and Norton, 1982); a modified Thiessen approach using a 5-km grid for 1930–1947 (Quinn and Norton, 1982); and a Thiessen weighting approach using a 1-km grid and daily data for 1948–1990 (Croley and Hartmann, 1985). Air temperature data were an update of Bolsenga and Norton (1993) from gridded data of quality controlled land stations for the Great Lakes Basin. Primary sources were the Carbon Dioxide Information Analysis Center and the Canadian Climate Centre.

Annual and seasonal temperature and precipitation trends for each Canadian climate region (Figure 1) within the Laurentian Great Lakes and Precambrian Shield, are summarized in Table II. Both the Great Lakes/St Lawrence and North eastern Forest areas show temperature and precipitation trends (Figure 3). An analysis of the linear trend in the temperature time-series for the Great Lakes/St Lawrence area indicates an increase of 0·6°C from 1895 to 1993. Weak oscillations are suggested from visual inspection, with the values peaking near 1950. For the North eastern Forest, temperature increased 0·4°C from 1895 to 1993; virtually all of the warming had occurred by 1940. Examination of seasonal data reveals that most of the increase in both areas was during the spring. The most significant temperature increases occurred in spring and winter. Across the Great Lakes basin as a whole the long-term trend in annual temperature is weak and inconsistent over time (Figure 4). However, spring air temperatures over the basin have warmed persistently, summer temperatures have a slow downward trend since the 1940s, autumn temperatures have declined since the 1960s and winter temperatures have been lower since the late 1950s (Bolsenga and Norton, 1993).

Precipitation trends are more variable in both the Great Lakes/St Lawrence (1911–1993) and the North-eastern Forest (1918–1993) (Figure 3, bottom). Annual precipitation increased 2·0 mm per decade, or 2·1% per decade, for the Great Lakes and 2·4% per decade for the North eastern Forest, corresponding to an increase of 21–24% per century. These two areas demonstrate the largest absolute rate of change of the 11 Canadian climatic regions. Scrutiny of both time-series (Figure 3, bottom) reveals that precipitation peaked in the late 1960s to early 1970s and has been declining since then.

Table II. Linear temperature and precipitation trends within climatic regions of the Great Lakes and Laurentian Shield Region. Data are from Climate Change Detection, Atmospheric Environment Service, Environment Canada

Climatic area	Temperature change (°C) 1895–1993					Precipitation (% over 10 years) 1911–1993				
	Annual	Spring	Summer	Autumn	Winter	Annual	Spring	Summer	Autumn	Winter
North-western Forest*	1·4	2·3	1·1	0·4	1·7	1·4	−0·9	2·5	2·4	−2·9
Northeastern Forest†	0·4	0·9	0·5	0·0	0·5	2·4	2·8	1·8	2·4	2·9
Great Lakes/St Lawrence	0·6	0·8	0·1	0·3	1·1	2·1	1·5	2·3	2·9	1·5
Atlantic	0·2	0·2	0·8	−0·1	0·2	1·0	0·8	0·5	0·8	1·5

* Precipitation period 1937–1993.
† Precipitation period 1918–1993.

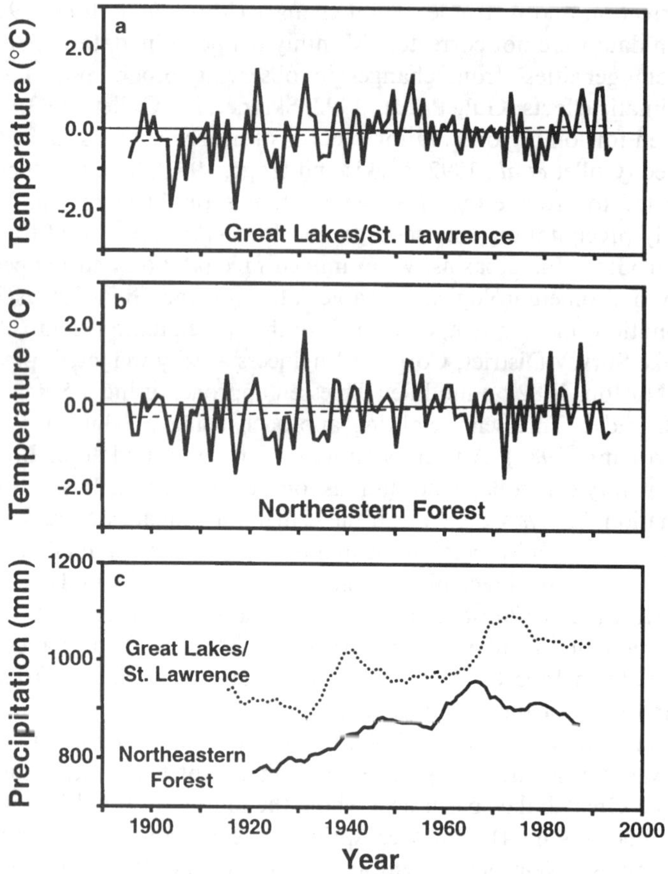

Figure 3. Departure from 1951–1980 average temperature (solid line) and linear trends (dashed lines) for: (a) Great Lakes Basin/ St Lawrence lowlands; and (b) North-eastern Forest. (c) Annual precipitation (9-year running mean) amount for these two regions

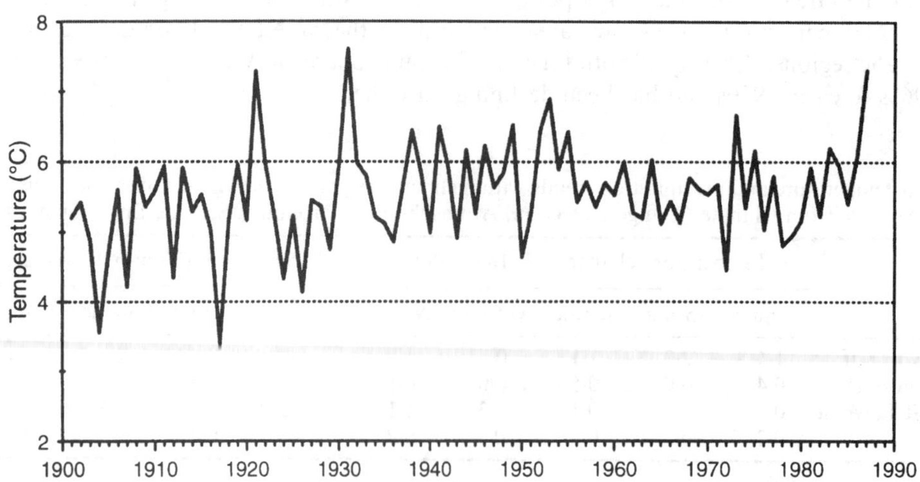

Figure 4. Annual temperatures for the Great Lakes Basin (1960–1990)

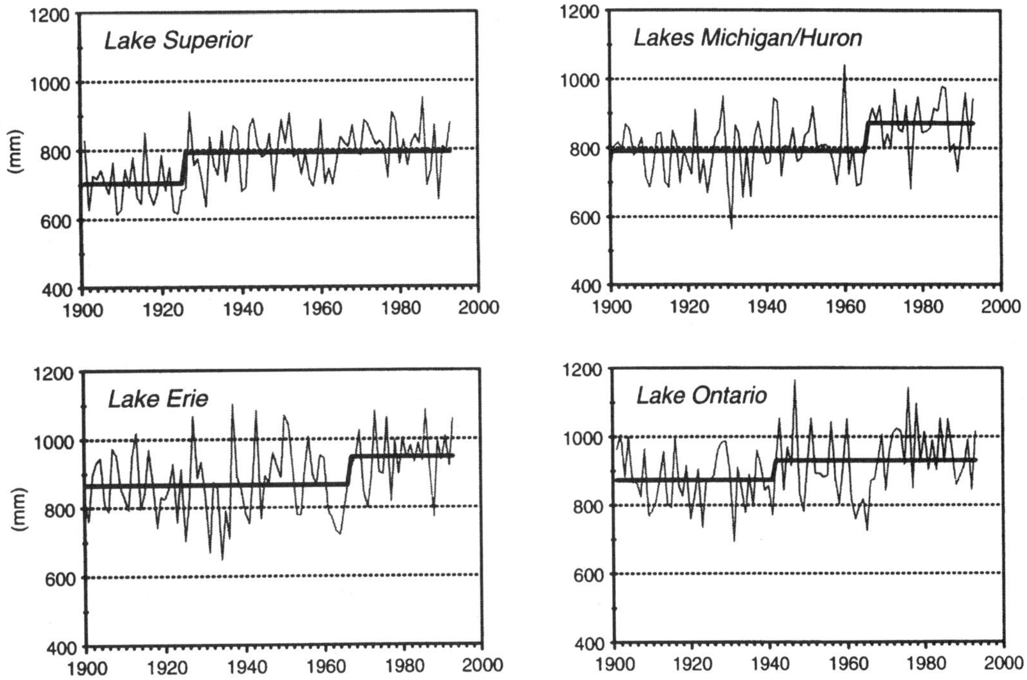

Figure 5. Annual precipitation and trends for Superior, Michigan/Huron, Erie and Ontario lake basins from early 1900s to the present. Heavy lines mark the time of step changes

Based on an update by Quinn (1981), several patterns emerge in seasonal and annual precipitation for each Great Lake basin. Annual precipitation time-series (Figure 5) indicate two distinct regimes for the Lake Michigan–Huron and Erie basins; a low precipitation regime from 1900 to about 1966 and a significantly higher precipitation regime from 1967 to the present. For the Lake Superior basin, a sudden increase in precipitation occurred in the late 1920s and was followed by a relatively stable period from 1930 to the present. The increase in precipitation around 1929 may be an artefact related to the number of stations in existence or to a change in the precipitation estimation methodology rather than a true change in regime. Lake Superior precipitation increased in all seasons from approximately 1930 (again, this may be an artefact), but during the 1980s, precipitation was lower for winter and spring in the Lake Superior Basin. For the Lake Ontario Basin, a low precipitation regime until the early 1940s was followed by a high precipitation regime. The increases in precipitation for Michigan and Huron, Erie, and Ontario (Figure 5), were primarily in summer and autumn.

Slivitsky and Mathier (1993) applied a variety of parametric and non-parametric time-series approaches to individual lake basins within the Great Lakes Basin using climate data from 1900 to 1988 for temperature, and 1990 to 1991 for precipitation. Most of the time-series contained statistically significant shifts in the mean during the century, suggesting that the changes were described well as steps or shifts rather than as linear trends.

CLIMATE CHANGE SCENARIOS

In this review, four general circulation models (GCMs) are used in various simulations of effects of increased greenhouse gases on climate. The Canadian Climate Centre general circulation model 2 (CCC GCMII) is discussed below in some detail. The others are the Goddard Institute for Space Studies model (GISS) (Hansen *et al.*, 1988), the Geophysical Fluid Dynamics Laboratory model (GFDL) (Manabe and Stouffer,

1980) and the Oregon State University model (OSU) (Schlesinger and Zhao, 1988). None of these models included the influence of aerosols and all were implemented with a step change from present to doubled CO_2 concentrations in the atmosphere (1 × CO_2 climates to 2 × CO_2 climates) rather than with a ramped change to doubled CO_2. Including such enhancements would most likely have reduced the magnitude of the simulated climate changes and their effects (Houghton *et al.*, 1996).

The second-generation atmospheric general circulation model from the Canadian Climate Centre of Environment Canada (CCC GCMII) was selected to illustrate a climate change scenario with an 'enhanced' greenhouse effect (Mortsch and Quinn, 1996). The CCC GCMII is a three-dimensional atmospheric GCM coupled to a mixed-layer ocean model that incorporates a thermodynamic sea ice component and allows a realistic simulation of the ocean surface temperature distribution and ice boundaries. Other improvements include greater horizontal resolution (3·75 degrees latitude by 3·75 degrees longitude), interactive cloudiness parameterization, more sophisticated radiative transfer treatment, including full diurnal and annual cycles, and incorporation of soil and vegetation information into land surface processes and hydrology (McFarlane *et al.*, 1992).

The CCC GCMII was used to test the sensitivity of the climate system to increased greenhouse gas concentrations in a 2 × CO_2, or doubled carbon dioxide, equilibrium response experiment (Boer *et al.*, 1992; Atmospheric Environment Service, 1994). First, a current climate simulation (1 × CO_2) was initiated where the average global carbon dioxide level was set at the present level; this model was run until a stable climate was produced. Secondly, the carbon dioxide gas component of the GCM was doubled and the model was run again until it reached a new equilibrium. The difference between the results of the 1 × CO_2 and the 2 × CO_2 runs provided an estimate of climate system sensitivity to a doubling of carbon dioxide. Although GCMs are the most advanced and most physically based tool for assessing the response of the climate system to significant increases in greenhouse gases, the results should not be interpreted as a prediction of climate change, but as a scenario. Climate scenarios provide a range of possible futures — 'what if' situations for exploring the implications of a changed climate.

The CCC GCMII 2 × CO_2 simulated a global mean surface temperature increase of 3·5°C; this is in the middle range of other GCM simulations (McFarlane, 1991; Boer *et al.*, 1992). The greatest warming was at high latitudes in winter. Precipitation increased by 3·8% over the globe and was statistically significant over 20% of the globe. The largest precipitation increase occurred over the oceans. Evaporation increased, soil moisture decreased by approximately 6·6% and summer dryness was enhanced in northern mid-latitudes. A major decrease in the area (53%) and mass (66%) of sea ice was simulated.

With scenarios for the Laurentian Great Lakes and Precambrian Shield Region, contours for increases in mean air temperature across the region ranged from +2 to +5°C in summer and +4 to +8°C in winter (Figure 6a and b). Contours for precipitation changes across the region ranged from −20 to +10% in summer and −10 to +20% in winter (Figure 6c and d). The predominant trend was for increased precipitation. Subregional differences in four scenarios for the Great Lakes Basin (CCC GCMII, GISS, GFDL and OSU) generally resulted in greater increases in temperature to the north in winter and to the west and south in summer, and greater increases in precipitation to the north-west in winter and to the north and east in spring (Mortsch and Quinn, 1996).

HYDROLOGY AND LAKE LEVELS

Climate changes influence the hydrological flows into and out of lakes, net basin water supplies and water levels. Thus, hydrological systems are potentially very sensitive to climate changes and can, in turn, influence many aspects of lake and stream ecosystems (Arnell *et al.*, 1996).

Laurentian Great Lakes

Historically, water levels of the Laurentian Great Lakes have fluctuated over a relatively small range, with approximately 1·8 m between record lows and highs. The Great Lakes system is naturally well regulated

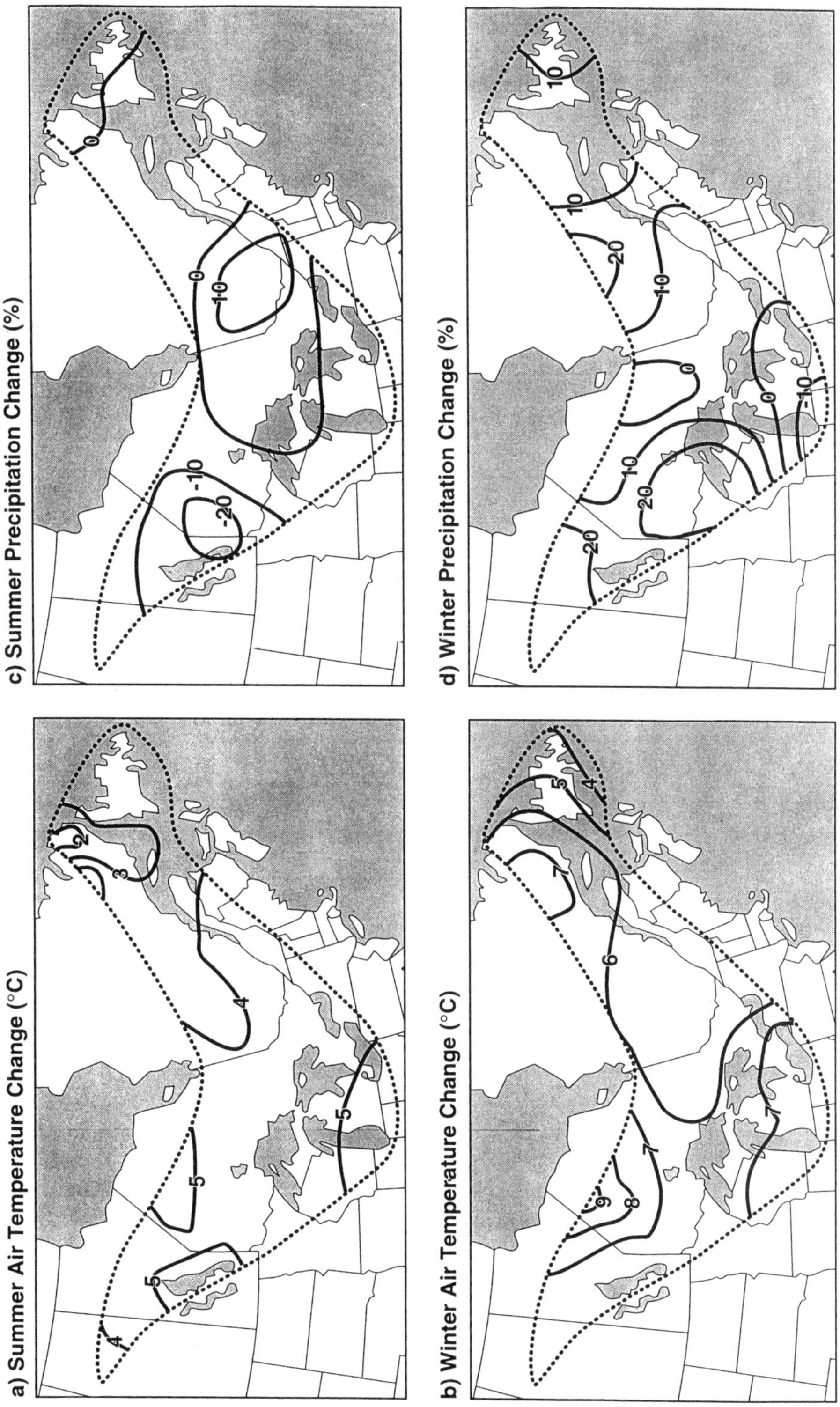

Figure 6. Changes in winter and summer mean air temperature (a,b) and winter and summer precipitation (c,d), comparing 2 × CO_2 with 1 × CO_2 scenarios

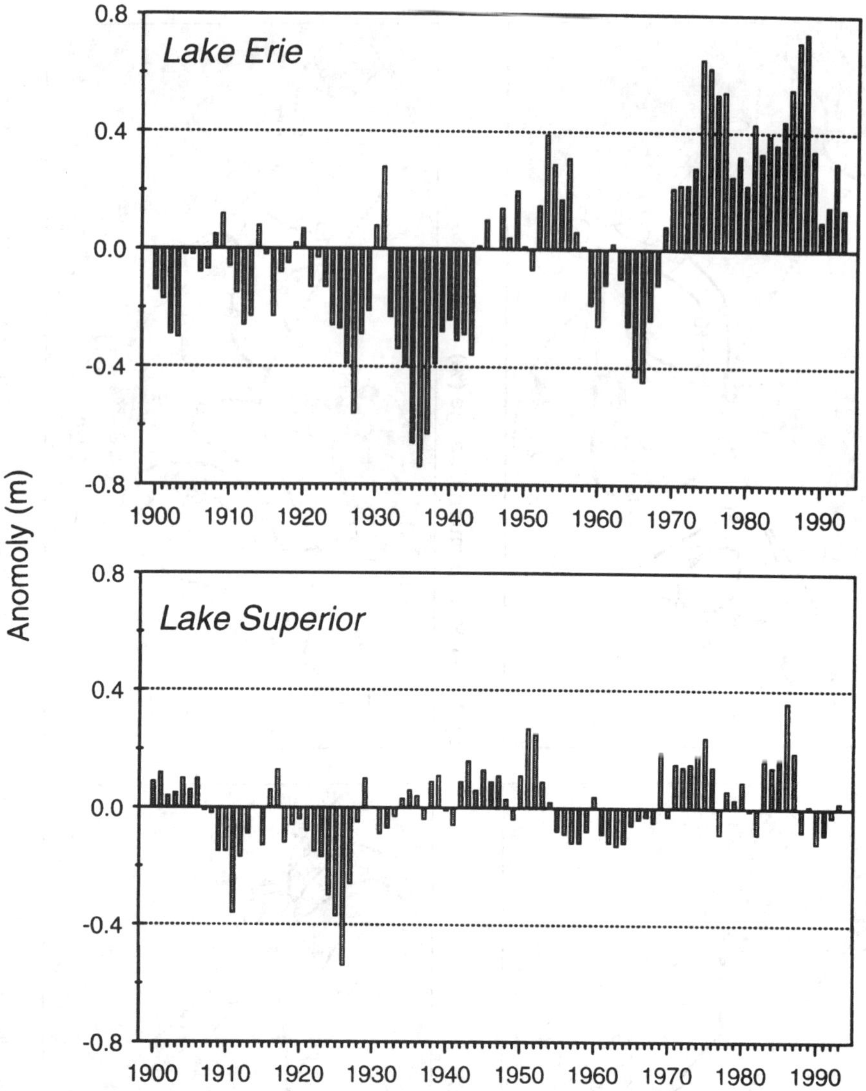

Figure 7. Water level anomalies (m) for lakes Erie and Superior

because it is large, has constructed outlet channels and two of the lakes (Superior and Ontario) have regulated outflows. The lakes have been in a high water-level regime since the late 1960s, culminating in record highs in 1986. A major decline in levels occurred in response to the 1987–1990 drought, but levels have continued to remain above the long-term mean for Michigan, Huron and Erie, and have been near or slightly below long-term mean levels for Superior. As examples, anomalies from the 1901–1989 long-term mean are shown for Erie and Superior (Figure 7). These above-average lake levels resulted from high water supplies beginning in the late 1960s.

In contrast to the general increases in water level over the last century, GCM scenarios for a doubling of CO_2 produced marked declines in water levels and outflows (Croley, 1990, 1993; Hartman, 1990; Mortsch and Quinn, 1996). Summaries of the hydrological model results by Mortsch and Quinn (1996) indicated declines from −0·23 to −2·48 m in water levels among all lakes and scenarios (CCC, GFDL, GISS and OSU). The greatest declines were for Michigan–Huron (−0·99 to −2·48 m among scenarios); the smallest

Table III. Average, annual, steady state Great Lakes Basin hydrology under base (1 × CO_2), transposition and 2 × CO_2 scenarios. Values in italics are the percentage change from the base case

Scenario	Overland precipitation (m^3/s)	Evapotranspiration (m^3/s)	Basin runoff (m^3/s)	Over lake precipitation (m^3/s)	Over lake evaporation (m^3/s)	Net basin supply (m^3/s)
Base case — 1 × CO_2	13855	7814	6206	6554	4958	7803
Transposition scenarios:						
#1 6°S × 10°W	14643 *+6%*	10201 *+31%*	4674 *−25%*	6767 *+3%*	7394 *+49%*	4048 *−48%*
#2 6°S × 0°W	17167 *+24%*	11198 *+43%*	6154 *−1%*	8169 *+25%*	6615 *+33%*	7708 *−1%*
#3 10°S × 11°W	16236 *+17%*	11563 *+48%*	4877 *−21%*	7379 *+13%*	8699 *+75%*	3556 *−54%*
#4 10°S × 5°W	20095 *+45%*	13907 *+78%*	6308 *+2%*	9482 *+45%*	8364 *+69%*	7426 *−5%*
2 × CO_2 scenarios						
CCC*	13637 *−2%*	7727 *+22%*	6090 *−32%*	6499 *0%*	*5352 +32%*	*7237 −46%*
GISS†	13871 *+2%*	9317 *+21%*	4658 *−24%*	6747 *+4%*	6821 *+27%*	4584 *−37%*
GFDL‡	13725 *+1%*	9176 *+19%*	4714 *−23%*	6501 *0%*	*7685 +44%*	*3530 −51%*
OSU§	14438 *+6%*	9204 *+19%*	5438 *−11%*	6903 *+6%*	6745 *+26%*	5596 *−23%*

* Canadian Climate Centre GCM (Croley, 1993).
† Goddard Institute for Space Studies GCM (Croley, 1990).
‡ Geophysical Fluid Dynamics Laboratory GCM (Croley, 1990).
§ Oregon State University GCM (Croley, 1990).

decline was for Superior (−0·23 to −0·47 m among scenarios); and Erie, Ontario and the St Lawrence River at Montreal were intermediate at about −1·3 m. Mean annual outflows also declined for all lakes; Lake Superior declined only −8% while the others ranged from −20 to −40% among scenarios, with a mean of about −30%. These declines in water level and outflows resulted from the simulated declines in inflows of −2 to −54% among lakes and scenarios. Greatest declines were for the Lake Erie Basin (−19 to −54%); smallest declines were for the Lake Superior Basin (−2 to −12%); and Michigan, Huron and Ontario were intermediate with declines ranging from −7 to −38%.

The seasonal pattern of hydrological parameters did not shift in the 2 × CO_2 simulations, nor were seasonal amplitudes accentuated a great deal (Croley, 1990; Hartman, 1990). However, compared with base conditions, runoff, temperature and evaporation were higher, and net basin supply was lower, in the 2 × CO_2 scenarios in all months. Net basin supply was highest in March and April when runoff was maximum and temperature and evaporation were low, and was lowest in August–October when runoff was low, and temperatures and evaporation were high.

To test further the sensitivity of the Laurentian Great Lakes hydrology, transposition climate scenarios were developed to introduce a change in both the mean and the variability of temperature and precipitation (Croley *et al.*, 1994, 1995; Mortsch and Quinn, 1996). Four climate regimes, selected from areas a few degrees south and west of the Great Lakes, were transposed to the Great Lakes Basin (Table III). Scenario 1 is warm and dry (6°S × 10°W); scenario 2 is warm and wet (6°S × 0°W); scenario 3 is very warm and dry (10°S × 11°W); and scenario 4 is very warm and wet (10°S × 5°W). Areal averages of the transposed climatic elements for 121 watersheds and over lake surfaces were derived by Thiessen weighting of daily data. The areally averaged transposed data were input to the hydrological models by the method used by Croley (1990, 1993). In these transposition scenarios, air temperature increased by +4 to +11°C and precipitation ranged from −20 to +70% of the current regime. Changes in Great Lakes water supplies resulting from the transposition were compared with the base case or current conditions and hydrology was expressed as total flows from all lake basins (Table III). Net basin supply declined by −1 to −54%; the decline was about 50% under the westernmost scenarios (1 and 3) because evaporation and evapotranspiration tended to be higher and precipitation lower than in the easternmost scenarios (2 and 4), where net basin supplies were close to the base case. For comparison, corresponding results are shown for four GCM scenarios (Table III).

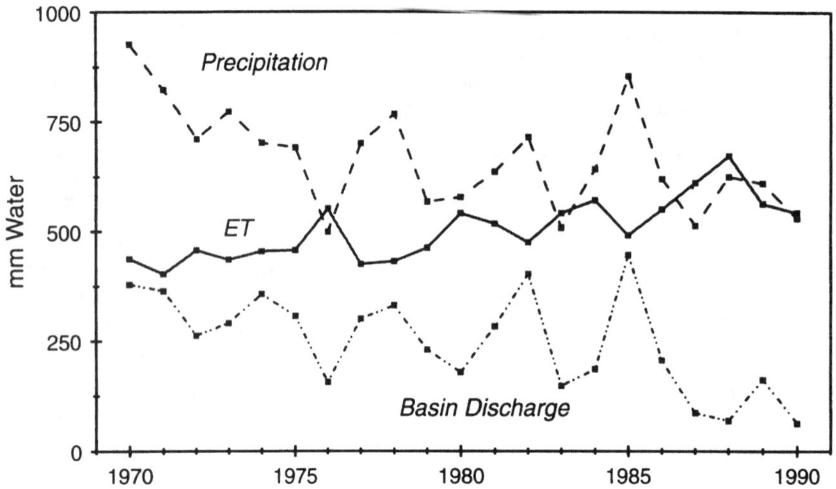

Figure 8. Hydrological changes observed for the Lake 240 basin during the 20-year period of warmer and drier conditions at ELA (adapted from Schindler *et al.*, 1996a)

Small inland lakes and streams

Comparable hydrological models for small lakes in the region have not been undertaken, but Mortsch and Quinn (1996) suggested that their water supplies would be expected to decline as well. They simulated changes in hydrological parameters (air temperature, precipitation, basin runoff and total basin moisture storage) for 121 sub-basins draining to the Laurentian Great Lakes. Moisture storage declined substantially in the doubled CO_2 scenarios, even though precipitation increased in the northern sub-basins; the declines occurred because temperature and evapotranspiration increased. Declines in basin runoff were greatest where simulated increases in temperature and declines in precipitation were greatest, namely in the southern half of the basin. Such reductions in tributary flow and soil moisture would affect the hydrology and related limnological properties and processes of groundwater, wetland, stream and lake systems.

The sensitivity of the water levels of small inland lakes to climatic changes is apparent from the observed responses to recent droughts. During the 1987–1990 drought in Wisconsin, water levels of lakes at the North Temperate Lakes Long-Term Ecological Research (LTER) site declined by up to 1 m. Lake level declines were largest in seepage lakes that were high in the landscape and received most of their water inputs from direct precipitation (Webster *et al.*, 1996). In contrast, declines of only a few cm were observed for drainage lakes low in the landscape, which received substantial inputs of both surface and groundwater.

Substantial hydrological shifts were also observed during the 20-year period of warmer and drier conditions in the 1970s and 1980s at the Experimental Lakes Area (ELA) in north-western Ontario. There, increased evaporation and decreased runoff resulted in the lengthening of water retention times in many lakes by a factor of about four over 20 years (Figure 8) (Schindler *et al.*, 1996a) at a rate of increase ranging from one to seven years per decade. Streams in the ELA also showed significant hydrological responses to drought; the number of days that first-order streams were dry during summer increased from less than 10 days per summer to about 140 days during the 20-year drought (Schindler *et al.*, 1996a).

The most dramatic effect of climate change will be the disappearance of lakes as a result of increased evaporation relative to precipitation. Short of this, salinities and nutrient concentrations would increase as lake volumes shrink while the sizes of drainage basins remain unchanged. In lakes with increased water renewal times, pollutant concentrations would rise even if inputs remained at present levels. Accompanying these abiotic changes would be a general shift towards species characteristic of eutrophic, saline and contaminated lakes.

Human adaptation and costs

Changes in lake levels, recently observed or simulated in 2 × CO_2 climate change scenarios, exceed those observed or simulated for sea level changes (Arnell *et al.*, 1996). Human responses to such large changes would be costly, particularly those related to shipping, dredging and replacement or refurbishing of shoreline structures in the Great Lakes (J. B. Smith, 1991). Changes in ice cover also influence shipping costs. In addition, higher demand and usage of water from the Laurentian Great Lakes would probably occur following a decrease in net basin water supplies.

For shipping at simulated water levels, 0·5–1·5 m lower than base levels, dredging costs would be incurred or ships would have to carry lighter loads (Keith *et al.*, 1989). If lighter loads were carried, costs per ton transported in 2 × CO_2 scenarios increased from 1·6 to 33% depending on the harbour (Duluth/Superior, Two Harbors and Whitefish Bay on Lake Superior, and Toledo, Cleveland and Buffalo on Lake Erie) and the scenario (GISS, GFDL and OSU). Cargoes would have to be reduced by 1·6–27% to get into the harbours without additional dredging. Dredging costs can be high, ranging from $0·0–31 million per harbour among harbours and scenarios without including the costs associated with shipping-related facilities. For the 101 km Illinois shoreline of Lake Michigan, including Chicago, $138–312 million would be needed over a 50-year period for dredging harbours to compensate for a 1·25–2·5 m decline in lake level (Changnon *et al.*, 1989). The cost of sheeting and bulkheads, slips and docks was estimated at an additional $113–203 million. Taken together these shipping costs for the Illinois shoreline total $251–515 million over a 50-year period. Increased dredging activities would also have implications for destruction of benthic habitats and resuspension of toxics in harbour sediments.

Keith *et al.* (1989) observed that, even with lighter loads, the same goods could be shipped over a season if the ice-free season were longer (see section on ice, below). For Buffalo, an increase in the shipping season of 99 days would be sufficient to compensate for the need for lighter loads with a 1·5 m decline in water level; simulated increases in the ice-free period more than compensated for the need for lighter loads in two (GISS and OSU) of the three scenarios. For Lake Superior ports, a slightly shorter increase in the ice-free season would be sufficient, a value achieved in all GCM scenarios. The bottom line projection for shipping costs for ports on Lakes Superior and Erie, as a consequence of reduced water levels plus the longer shipping season apparent in 2 × CO_2 scenarios, was 1–7·5% above present costs or about one-half of the increases in costs from water level reductions taken alone.

Additional costs, unrelated to shipping, have been estimated for the Illinois shoreline of Lake Michigan by Changnon *et al.* (1989) for 2 × CO_2 scenarios. These included costs to extend water intake structures for city water supplies ($16–17 million), to relocate beach facilities ($1–2 million) and to extend and modify storm water outfalls ($2–4 million). These costs are less than those associated with shipping. Historical responses to lower water in the Chicago area include relocation and encroachment to take advantage of the new beach areas; damage to these structures was extensive when water levels returned to higher levels.

Generation of electricity from hydroelectric facilities in the Great Lakes Basin would also be reduced in a drier and warmer climate. Presently, the capacity of the Great Lakes electric generation system is about 3·2 million kW for Ontario, 1·7 million kW for Quebec and 3·1 million kW for New York (Melo, 1989). The costs of replacing the hydroelectric power generated at Niagara and along the St Lawrence River following a 0·6 m decline in water level in Lakes Erie and Ontario is high. Long-term annual costs of replacing this capacity with nuclear or fossil fuel plants were estimated to be in the range of US $160 million in 1988 for New York (Crissman, 1989), and Can $1 billion for Ontario (Melo, 1989). The output from these hydropower facilities is of the same magnitude as that of the Tennessee Valley Authority.

ICE COVER

Ice cover on lakes and streams is not only a robust indicator of climate change and variation (Robertson *et al.*, 1992), but also represents a limnological response to climate change that influences the ecology of

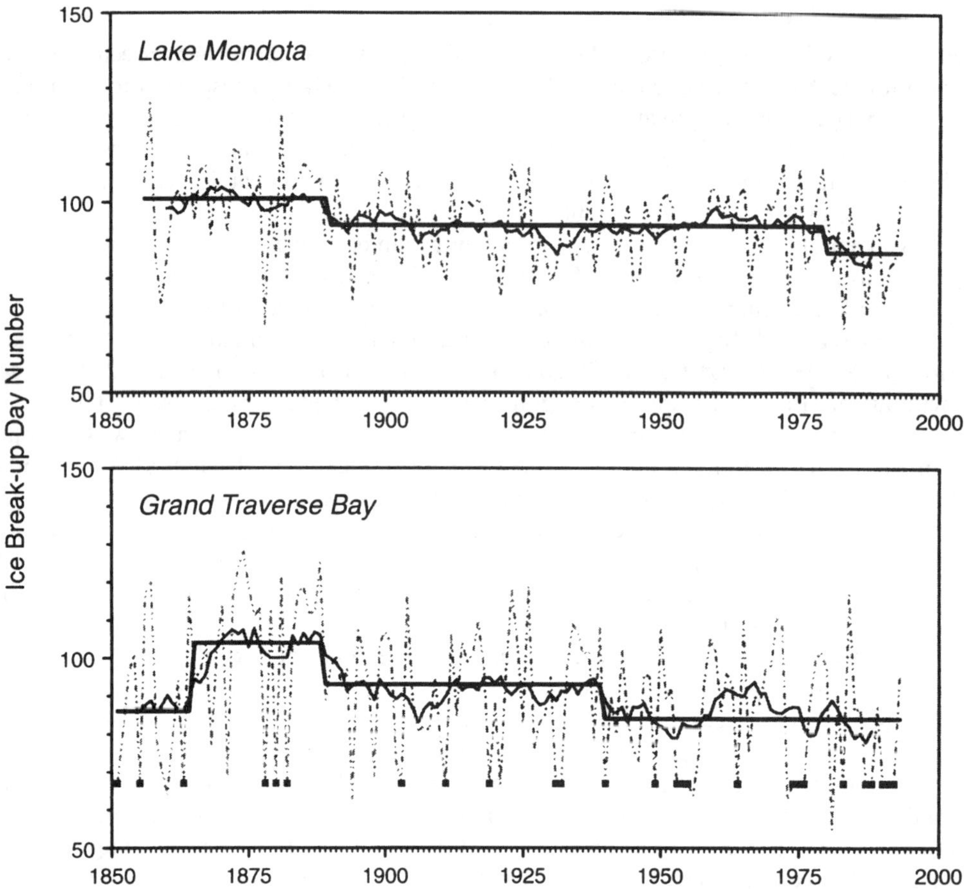

Figure 9. Ice-off dates for Lake Mendota and Grand Traverse Bay between 1851 and the present (dashed line) (adapted from Assel and Robertson, 1995). Heavy solid lines mark step changes in the long-term average, while the lighter solid line is a centred 10-year moving average. Years in which Grand Traverse Bay did not freeze over are marked by a solid square on 8 March

aquatic systems (Vanderploeg *et al.*, 1992). Ice phenologies, i.e. dates of freezing over (ice-on) and break-up (ice-off), are especially useful because shoreline residents and early navigators of lakes and streams have recorded these events longer and in more locations than other climatic data. Ice phenologies can also be measured by analysis of satellite images over large regions where there are no weather stations (Wynne and Lillesand, 1993).

Past trends in ice phenology

In general, the average duration of ice cover has decreased in a series of step changes in the Laurentian Great Lakes and Precambrian Shield Region over the last century or so, indicating early and late winter warming of 1·5–2·7°C. Continuous lake ice phenologies date back to 1823 in the Laurentian Great Lakes and 1855 for Lake Mendota, Wisconsin. Analyses of Lake Mendota and Grand Traverse Bay, Lake Michigan (*ca.* 320 km apart), indicated later ice-on dates and earlier ice-off dates (Figure 9) (Assel and Robertson, 1995). A step change of eight days (Mendota) and 12 days (Grand Traverse) in average ice-on date occurred in about the 1888–1889 winter in both systems, representing an increase of about 1·5°C in early winter air temperatures. Two step changes in average ice-off dates, totalling 14 days (Mendota) and 20 days (Grand Traverse), have occurred since the 1870s representing an increase in late winter air temperatures of

2·2°C (Mendota) and 2·7°C (Grand Traverse). The first step changes were coincident for both sites, *ca.* 1888–1889, while the second occurred in 1939–1940 (Grand Traverse) and 1979–1980 (Lake Mendota). The transition to later average ice-on and earlier average ice-off dates in the late 1800s corresponded with the end of the Little Ice Age.

For the Great Lakes, we simulated annual maximum ice cover averaged by decade from 1900 to 1990 using a regional winter air temperature model for annual maximum ice cover of the combined surface areas (Assel *et al.*, 1985). Average November–February monthly air temperatures were used from Duluth, Minnesota; Sault Ste. Marie, Michigan; Detroit, Michigan; and Buffalo, New York. Average ice cover decreased from approximately 61% maximum coverage for the decades ending in 1910 and 1920, to near 50% for the decades ending in 1940, 1950 and 1960. Ice cover then increased over the next two decades [1970 (61%) and 1980 (64%)] before declining again in the 1980s [1990 (55%)]. These trends are consistent with simulations from degree–day ice cover models developed for Lakes Erie and Superior (Assel, 1990). The later models suggested that decadal midwinter ice cover averaged about 10–25% greater for both 1900–1925 and 1960–1983 than for the intervening years (1926–1959). Contemporary midwinter ice cover averages 90% for Lake Erie and 56% for Lake Superior (Assel *et al.*, 1983).

For the shallow shore zones of the Great Lakes, the timing of ice-on, ice-off and percentage of lake area covered by ice are comparable to those of smaller inland lakes. However, because the Great Lakes have greater thermal inertia and longer wind fetch than the smaller lakes (Assel, 1986), ice forms later and is more dynamic in midlake areas than inshore areas or in the smaller lakes. Ice usually forms in shallow bays and over most of Lake Erie in December and January. Ice forms in the deeper embayments and in midlake areas in February, reaches its maximum extent in February or March and is usually lost by the end of April (Assel *et al.*, 1983). The Great Lakes do not freeze over completely because of their large heat storage capacity, the action of winds and their exposure to relatively moderate winter temperatures, particularly towards the south. During mild winters, midlake areas remain relatively ice free (Assel *et al.*, 1985) and annual maximum ice cover is typically less than 35%. During severe winters midlake ice can form as early as January, can last into May and can cover 80% of the combined surface area of the Great Lakes (Assel *et al.*, 1996).

Changes and variations in ice-off dates were coherent between lakes across the Laurentian Great Lakes/ Precambrian Shield Region. The recent change to earlier ice-off dates for Lake Mendota was consistent with data from 19 other Wisconsin lakes (Anderson *et al.*, 1996) and with Lake 239 in north-western Ontario (Schindler *et al.*, 1990, 1996a). Rates of change were 0·7–8·4 (mean = 4·3) days per decade for northern Wisconsin, 0·6–12·2 (mean = 8·1) days per decade in southern Wisconsin and 7·6 days per decade for Lake 239. Locations around the Laurentian Great Lakes from Chequamegon Bay (Lake Superior), Grand Traverse Bay (Lake Michigan) and Buffalo (Lake Erie) also showed an earlier ice-off during the same 21 years (Assel *et al.*, 1995), and earlier ice-off dates were apparent in inland Canadian lakes from western Manitoba to central Quebec (Reycraft and Skinner, 1993). Ice-on dates, however, had no obvious trend over the same 21 years (Robertson *et al.*, 1992; Reycraft and Skinner, 1993; Assel *et al.*, 1995; Schindler *et al.*, 1996a).

In contrast to the apparent coherence in general trends of lake ice phenologies in Wisconsin and across the region, Wynne *et al.* (1996) found low average coherence in inter-year patterns among 62 inland lakes spanning a wide range of latitudes in the western part of the region. Using satellite-derived ice-off dates from 1987 to 1994, they measured coherence as the correlation coefficient between the time-series for each lake pair. In this analysis, two lakes with quite different mean ice-off dates could still be coherent, i.e. have the same inter-year pattern in ice-off dates. While some lake pairs were highly coherent ($r = +0{\cdot}9996$) others were negatively coherent ($r = -0{\cdot}89$), and average coherence was low ($r = +0{\cdot}45$) for the area as a whole. There was a geographical pattern to the coherence. First, adjacent lakes tended to be coherent. More interestingly, lakes tended to be more coherent if their mean ice-off dates were more similar. Furthermore, lakes that were not close but had high coherence tended to be at similar latitudes; for example, lakes in southern Wisconsin were highly coherent with those in eastern South Dakota. These results were interpreted to mean that lakes at the same location or latitude with similar ice-off dates, integrated the same seasonal weather, while those that were neither close nor at the same latitude, integrated different seasonal weather.

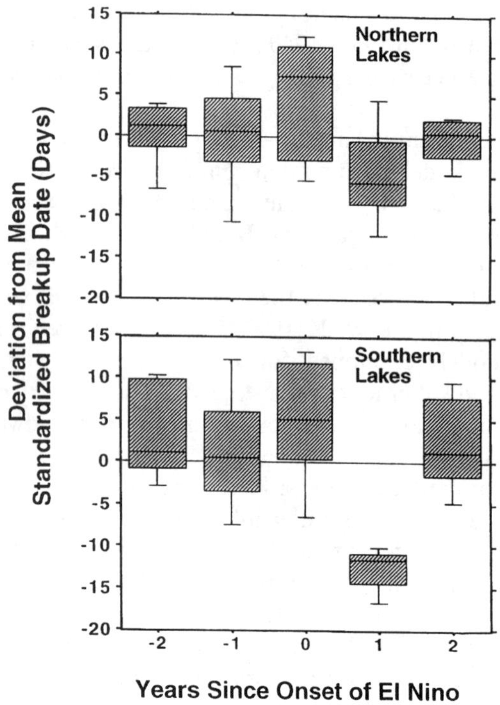

Figure 10. Effect of El Niño events on ice break-up dates for lakes in northern and southern Wisconsin (modified from Anderson *et al.*, 1996)

This point was also made by Assel and Robertson (1995) when comparing Grand Traverse Bay and Lake Mendota and by Anderson *et al.* (1996) when comparing El Niño effects on ice-out for Wisconsin lakes (Figure 10); lakes freezing or thawing on different dates integrate different climate windows and thus can have different inter-year patterns of ice phenologies. This suggests that lakes at different latitudes can be used to represent finer seasonal time-scales than, simply, early winter for ice-on or late winter for ice-off.

Ice formation and break-up are dependent on many climatic forcing variables, such as air temperature, solar radiation, wind and snow depth. Yet, air temperature alone often provides a reasonable prediction of ice phenologies. Using a moving average of air temperatures, McFadden (1965) observed that large lakes in central Canada and Lake Mendota in Wisconsin, froze when the 40-day average reached 0°C and broke up when it exceeded +5°C. Using a sensible heat transfer model developed by Robertson *et al.* (1992), Assel and Robertson (1995) found that ice-on changed by 5·3 and 8·4 days and ice-off changed by 6·4 and 7·1 days per 1°C change in air temperature for Lake Mendota and Grand Traverse Bay, respectively; conversely, air temperature increased by 0·18 and 0·12°C for each day change in ice-on date and by 0·15 and 0·14°C for each day change in ice-off date.

A more complete process model for lake ice phenologies was developed recently by Vavrus *et al.* (1996), patterned after a thermodynamic sea ice model (Maykut and Untersteiner, 1971; Parkinson and Washington, 1979). Inputs to the model are mean lake depth, air temperature and moisture, wind speed, solar radiation, snowfall and cloudiness, energy transfers between the air–snow, or air–ice, snow–ice and ice–water are included. When driven with hourly meteorological data, the model simulated the annual ice-on and ice-off dates for Lake Mendota between 1961 to 1990 with a median absolute error of two days (ice-on) and four days (ice-off). In the model, ice-on and ice-off dates responded strongly to changes in mean annual air temperature, ice-off was more sensitive to temperature changes than was ice-on and both events were more sensitive to climate warming than to climate cooling.

The model developed by Vavrus *et al.* (1996) was also sensitive to changes in snowfall. A 50% increase in snowfall increased ice-off by six days and a 50% decrease in snowfall decreased ice-off by four days; a no-snow scenario, however, again increased ice-off by one day. Snowfall was also an important factor in the spatial analyses of the satellite-derived time-series discussed above (Wynne *et al.*, 1996). The model of Wynne *et al.*'s (1996) ($R^2 = 0{\cdot}93$) was:

$$\text{Ice-off date} = (4{\cdot}47 \times \text{latitude}) + (0{\cdot}5 \times \text{snowfall in cm}) - 112$$

Simulations of changes in ice phenologies with climate warming predicted the occurrence of ice-free years for Wisconsin's Long-Term Ecological Research (LTER) lakes, which, historically, have had ice cover every year (Robertson *et al.*, 1992; DeStasio *et al.*, 1996). For Lake Mendota, in southern Wisconsin, a 1°C warming would shorten ice duration by about 11 days (Robertson *et al.*, 1992); a 4°C warming would result in frequent thawing and refreezing events; a 5°C warming would reduce mean ice cover by 64 days and result in no ice cover in 1 out of every 15–30 years. Robertson's model was parameterized for three lakes in northern Wisconsin (Crystal, Sparkling and Trout) and phenologies calculated with temperature increases for $2 \times CO_2$ scenarios; ice did not form in 33% of 27 runs (DeStasio *et al.*, 1996). The 27 runs were for three model scenarios (GFDL, GISS and OSU), three lakes and three base climates (a cool, intermediate and warm year). Using the same analysis for Lake Mendota in southern Wisconsin, ice did not form in 89% of the nine simulations. For the northern Wisconsin lakes, even when the lakes were simulated to freeze, the ice-free season was increased by 16–64 days.

For Lakes Erie and Superior, ice phenologies were simulated with a single $1 \times CO_2$ scenario over a 30-year base (1951–1980), and with $2 \times CO_2$ scenarios for the same years from three GCMs (GFDL, GISS and OSU) (Assel, 1991). With the $1 \times CO_2$ scenario ice cover formed every year for Lake Superior and for 29 of the 30 years for Lake Erie. With the $2 \times CO_2$ scenarios, ice-free winters appeared with the GFDL scenarios for Lake Erie in 17% of the 30 years and for Lake Superior in 43% of the years; with the other two scenarios the simulations were, for GISS, 10% (Superior) and 3% (Erie) and, for OSU, 7% (Superior) and 0% (Erie). Average ice cover duration for the 1951–1980 base period ($1 \times CO_2$) ranged from 13 to 16 weeks. Ice duration with $2 \times CO_2$ scenarios was reduced to 5 to 13 weeks; midlake areas were ice free in most $2 \times CO_2$ winters.

PHYSICAL LIMNOLOGY, WATER CLARITY AND DISSOLVED OXYGEN

Lake and stream temperatures are responsive to climate warming in the Laurentian Great Lakes and Precambrian Shield Region. Lake mixing and stratification would be altered directly by the warming, and indirectly by changes in water clarity, responding to decreased inputs of dissolved organic carbon (DOC) or changes in trophic state. Physical changes would in turn influence the dissolved oxygen availability in the deep stratified waters, with many biological and biogeochemical consequences. Refuges from the warmer temperatures may be available in lakes that have an oxygenated hypolimnion and in some smaller streams that receive significant groundwater and shading from riparian vegetation.

Lakes

Lake thermal regimes respond to climate change because they are controlled by solar radiation, wind velocity, air temperature and humidity. McCormick (1990) (see Figure 11 for Lake Michigan); Hondzo and Stefan (1991, 1993); Stefan *et al.* (1993a,b, 1996); and DeStasio *et al.* (1996) used several GCMs (CCC, GISS, GFDL, OSU) to simulate the effects of a doubling of CO_2 on physical limnology. Their simulations predicted temperature changes of 1–7°C warmer in surface mixed waters (epilimnion), of 8°C warmer to a surprising 6°C cooler in deep hypolimnetic waters; sharper thermoclines, from 4 m shallower to 3·5 m deeper; longer durations of summer stratification; and decreased durations of ice cover or frequencies of winters with ice cover (see section on ice phenology above).

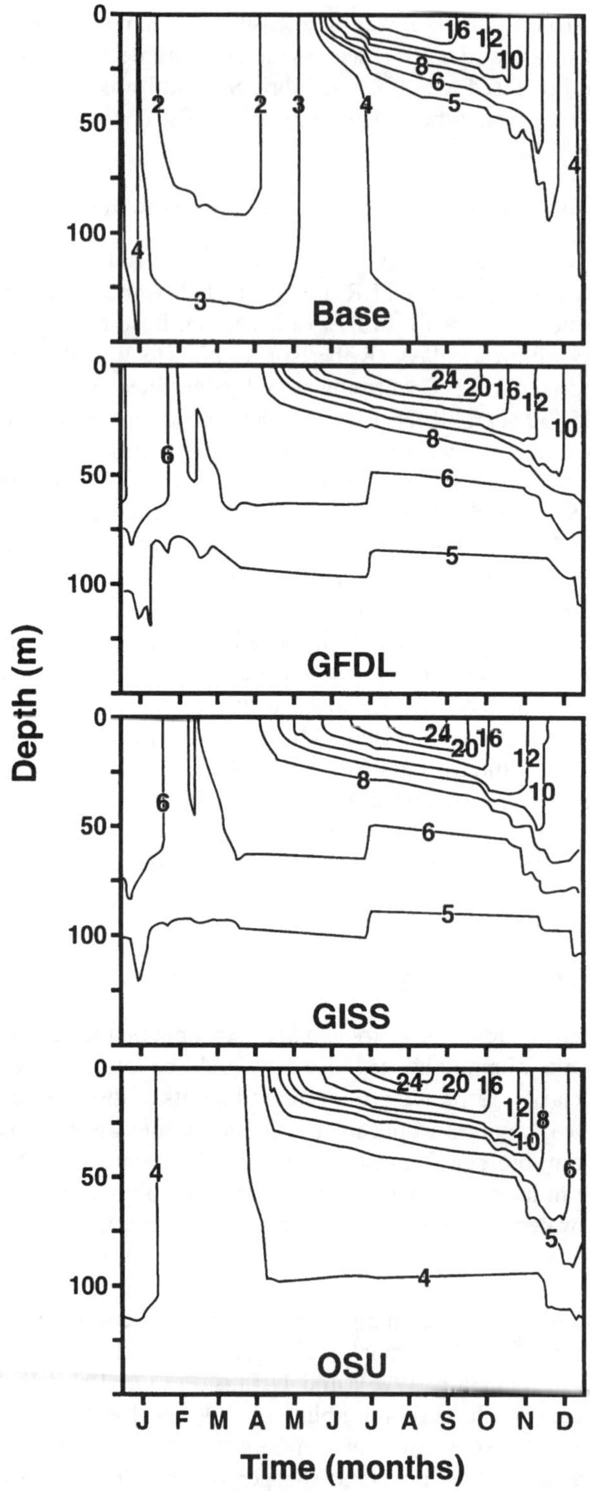

Figure 11. Simulated temperature isotherms (°C) for Lake Michigan under base and 2 × CO_2 scenarios from three global climate models (modified from McCormick, 1990)

The simulated changes varied among scenarios and lakes; differences in simulated responses were generally greater among GCMs than between lakes or base year, at least for the four Wisconsin lakes (Crystal, Sparkling, Trout and Mendota) used by DeStasio *et al.* (1996). Simulated increases in surface temperatures of four Wisconsin lakes averaged +4·2 to +4·8°C across four GCMs with a cool, intermediate and warm base climate. However, temperature simulations for bottom waters were more variable, +0·8 to +4·6°C, among the four lakes (DeStasio *et al.*, 1996). For 27 Minnesota lake classes (three maximum depths, three surface areas and three water clarities) 25-year averages of surface temperatures simulated with the GISS GCM were about 3°C warmer, compared with a 4·4°C air temperature rise (Hondzo and Stefan, 1993); increases in hypolimnetic temperatures ranged up to +3°C (Stefan *et al.*, 1996). The largest changes in water temperature occurred in spring and autumn.

Warming also would alter the mixing regimes of lakes. In the southern part of the region, lakes switching from ice covered to open in winter would become monomictic; i.e. they would mix in autumn, winter and spring, and stratify in summer. In the northern part, some lakes that are presently monomictic and mix only during summer, would stratify in summer and become dimictic, mixing in spring and autumn. Some large deep lakes, like Lake Michigan, that are dimictic would be less likely to mix completely (Figure 11) (McCormick, 1990).

Inter-lake differences in depth and transparency often have a greater influence on measured and simulated bottom temperatures or thermocline depths than does simulated climate warming (Stefan and Fang, 1995; DeStasio *et al.*, 1996; Fee *et al.*, 1996). Surface temperatures simulated with 1 × CO_2 climates changed little with lake depth or transparency (Hondzo and Stefan, 1993; DeStasio *et al.*, 1996), but bottom temperatures and thermocline depth did. For shallow lakes, bottom and surface temperatures were the same; for lakes 10 m deep bottom temperatures were less than half the surface temperatures; for those greater than 20 m deep, bottom temperatures were less than one-quarter of the surface temperatures. Simulations for less transparent lakes showed cooler hypolimnia and shallower thermoclines with 1 × CO_2 climates than they did for more transparent lakes.

Biogeochemical processes that alter lake transparency during warmer conditions may override the influence of physical factors on thermocline depths in many lakes. Because DOC is an important control on the transparency of many lakes in the region (Scully and Lean, 1994; Fee *et al.*, 1996), decreased DOC concentrations related to lower DOC export rates from catchments during warmer and drier conditions result in higher transparencies and thus deeper thermoclines (Figure 12) (Schindler *et al.*, 1996a), which is the opposite response to that predicted by some physically based models. Fee *et al.* (1996) suggested that for many lakes in the Experimental Lakes Area in north-western Ontario the expected magnitude of thermocline deepening related to changes in transparency was the same as the shallowing predicted by 2 × CO_2 scenarios. In fact, thermoclines deepened in a small ELA lake over a 20-year period when epilimnion temperatures rose by >2°C and DOC concentrations declined (Schindler *et al.*, 1990). Thermocline deepening in this small lake was probably affected by local forest fires (the basin was burned twice during the period of study) which increased the mean wind speed over the lake during the same period.

The effects of transparency and wind speed on thermocline depth are likely to be more important in small lakes than in large lakes. Epilimnion depth was, in general, positively related to lake area across a wide range of lake sizes (Fee *et al.*, 1996). Transparency was an important influence in those lakes smaller than 500 ha. Thus, the influence of increased transparency, driven by decreased DOC during warmer climates, may override other physical factors in smaller lakes. In contrast, decreased DOC concentrations in large lakes would probably be too small to modify significantly the mixing depth (Fee *et al.*, 1996). In addition, the over-water wind speed in large lakes would not be affected significantly by changes of the surface roughness of the watershed related to fires or land use modification.

Changes in temperature, stratification and light would be expected to alter the availability of dissolved oxygen in lakes. Longer summer stratification would increase the likelihood that deep hypolimnetic waters would be depleted of oxygen. Simulations with 2 × CO_2 scenarios for Lake Erie (Blumberg and Di Toro, 1990) predict summer oxygen declines of 1 mg/l in upper layers and 1–2 mg/l in deeper layers, as well as the

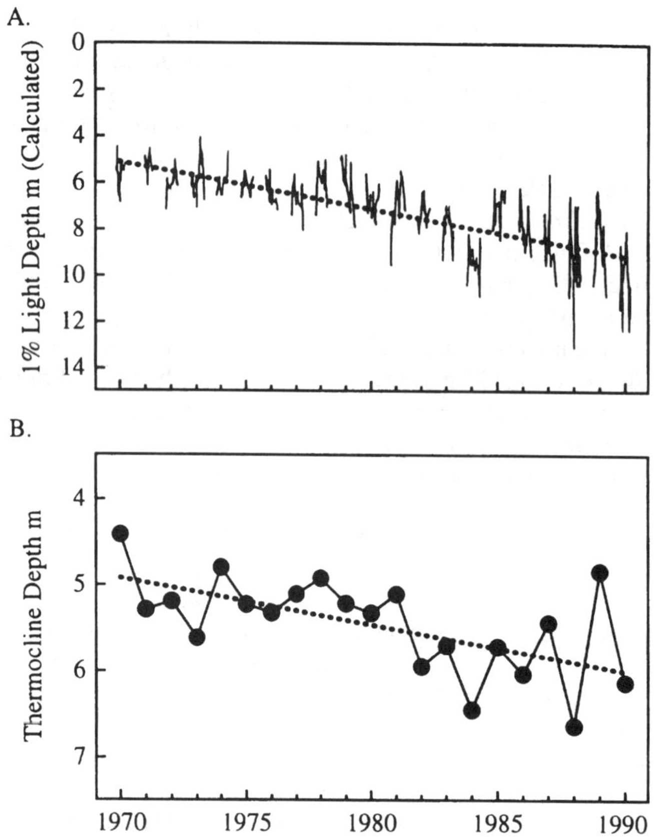

Figure 12. (a) Changes in penetration of photosynthetically active radiation (PAR) in Lake 239 as the depth of the isopleth representing 1% of surface light (equivalent to photic zone depth) (modified from Schindler *et al.*, 1996a); (b) changes in thermocline depth in Lake 239

development of a more extensive area of anoxia. In these simulations, warmer temperatures increased bacterial activity in deep waters and sediments; a change in thermocline depth was not the cause. Similar responses were obtained for smaller Minnesota lakes in model simulations by Stefan *et al.* (1993a), Stefan and Fang (1995) and Stefan and Fang (1994). Oxygen concentrations declined by 2 mg/l in surface waters and by as much as 8 mg/l in deep hypolimnetic waters, while summer oxygen depletion lasted up to two months longer compared with base climate scenarios. These oxygen declines occurred more rapidly and were longer lasting in eutrophic, compared with oligotrophic, lakes. In winter, shorter ice cover durations would decrease the likelihood of winter kill resulting from oxygen depletion in shallow lakes, although this may be countered by predicted declines in lake level and volume, which would lower the total content of dissolved oxygen available at ice-on.

Streams

Mean weekly temperatures tracked air temperatures closely in six Minnesota streams (Stefan and Preud'homme, 1993) according to the relationship:

$$T_w(\text{week}) = 2{\cdot}85 + 0{\cdot}866 T_a(\text{week})$$

where T_w(week) is the weekly mean water temperature in °C and T_a(week) is the weekly mean air temperature in °C. The regression slopes for individual rivers ranged from 0·669 for the Vemillion River to

1·026 for the Roseau River; intercepts ranged from 1·40°C for the Straight River to 4·49°C for the Mississippi River at St. Paul. The relationship for daily data was similar but had a larger standard deviation and a time lag of 1–4 days:

$$T_w(\text{day}) = 4{\cdot}90 + 0{\cdot}733T_a(\text{day})$$

The regression slopes for individual rivers in Minnesota ranged from 0·604 for the Vermillion River to 0·874 for the Roseau River; intercepts ranged from 3·96°C for the Straight River to 7·37°C for the Mississippi River at St. Paul.

Other factors besides air temperature influence stream temperatures, such as solar radiation, relative humidity, wind speed, water depth, groundwater inflow, artificial heat inputs and the thermal conductivity of the sediments. Of these, radiation is the most important (Sinokrot and Stefan, 1993); thus, diel variation and shading from riparian vegetation are important. Deterministic models including shading and wind sheltering can simulate water temperatures with a standard deviation of about 1°C, which is considerably smaller than the above regressions, but such models have much greater data demands; approximations can be achieved rapidly with the regressions (Stefan and Preud'homme, 1993).

A modified version of the MNSTREM model was used to simulate the potential affects of doubling atmospheric CO_2 for five Minnesota streams (Stefan and Sinokrot, 1993). Stream temperatures were projected to increase 2·4–4·7°C with scenario inputs from four GCMs (GISS, GFDL, UKMO and OSU) and 2 × CO_2 climates. If there was no riparian shading, additional increases up to 6°C were projected.

Increases in groundwater temperatures with climate warming are expected to be about the same magnitude as that predicted for mean annual air temperatures (Meisner *et al.*, 1988). As a result, the cooling effect of groundwater on streams would be reduced, indirectly contributing to stream warming. The combined effect of an increase in groundwater temperature of 4·8°C, with the direct effects of warmer air temperatures, on stream waters in south-central Ontario were simulated to decrease the length of headwater trout streams that remained below 24°C during July and August by 30 and 40%, respectively (Meisner, 1990). For one stream, the increase in water temperature was almost equally attributable to groundwater inputs and to air temperatures; in the other stream, nearly all of the increase in temperature was accountable by groundwater inputs. The effect of climate change on stream temperatures below dams will be more pronounced where the water release is from the reservoir epilimnion rather than the hypolimnion. Cold hypolimnetic water release is felt as far as 48 km downstream in small, shaded Minnesota streams. This distance is projected to be shortened by 25–50% under a 2 × CO_2 climate scenario (Sinokrot *et al.*, 1995).

BIOGEOCHEMISTRY

Effects of climate change on chemical properties of lakes are a function of the alteration of hydrological flows and subsequent changes in lake volume and water level, inputs of solutes and increases in hypolimnetic anoxia and summer stratification. Inferences are made primarily from a 20-year period of increasingly warm and dry weather at the Experimental Lakes Area (ELA) in north western Ontario and a late 1980s drought at the North Temperate Lakes LTER site in Wisconsin.

During the 1970s and 1980s at ELA, mean annual air temperatures increased by about 2°C, precipitation generally declined and evaporation increased (Schindler *et al.*, 1990, 1996a). As a result, stream flow and the flushing rates of lakes underwent a more or less continuous decline over the 20 years. While the cause of these climatic trends is unknown, the magnitude and rate of warming are similar to what might be expected from a doubling of CO_2. Thus the record provides an interesting glimpse of how climate change may affect small lakes in boreal regions of the Precambrian Shield. The effects of increased forest fires on lake chemistry during the period, an indirect effect of climate warming, were also important. Forest fires increased dramatically in Canada as the weather warmed, precipitation declined and evapotranspiration increased. Not surprisingly, a 1974 forest fire burned the eastern half of the Lake 239 catchment, including

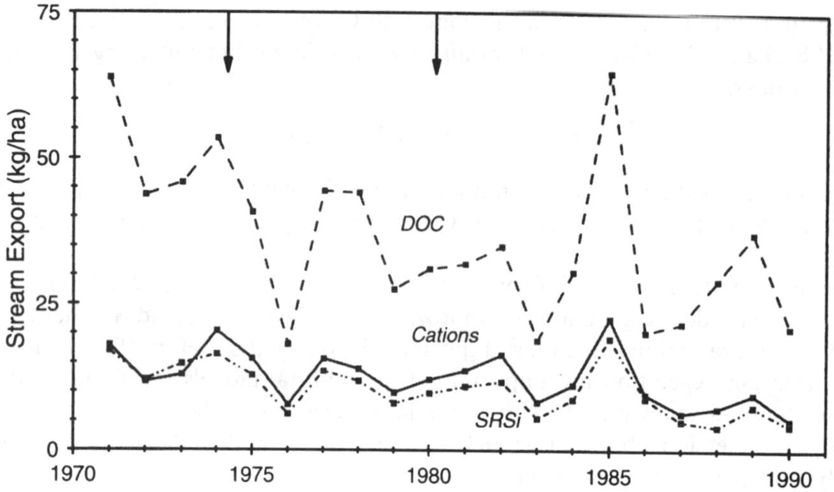

Figure 13. Annual exports of base cations (Ca + Mg + Na + K), silica and DOC from the catchment of Lake 239 (from Schindler *et al.*, 1996a). Rates of decrease are all significantly different from zero. Arrows indicate when major fires occured

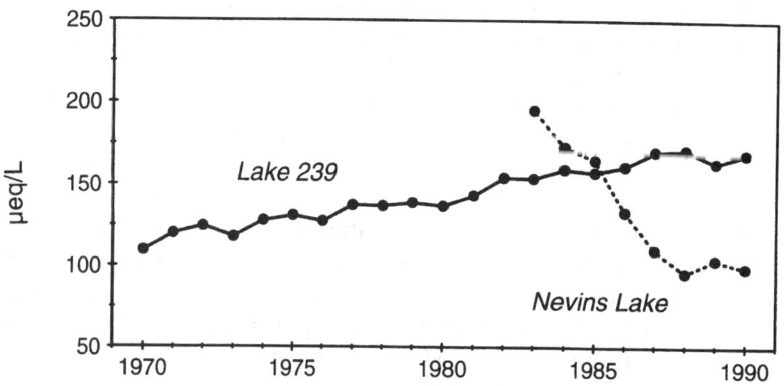

Figure 14. Temporal patterns in calcium in Lake 239 and Nevins Lake during drought. Data for Lake 239 are from Schindler *et al.* (1990) and for Nevins Lake are adapted from Webster *et al.* (1990)

the subbasins of two first-order streams, and a 1980 fire burned a third stream basin as well as the same subbasins again.

During these warmer and drier conditions, inputs of silica, DOC, base cations and phosphorus from streams to ELA lakes generally declined (Figures 13 and 14) (Bayley *et al.*, 1992a; Schindler *et al.*, 1996a). The decreased yields of soluble reactive silica from the east subbasin of Lake 239, which contains fairly deep soils, suggested a reduction in weathering caused by drier conditions. Despite decreased inputs, concentrations of most inorganic solutes increased and lakes became more alkaline (Schindler *et al.*, 1990, 1996a), a trend that appears to be continuing. The increased concentration of more conservative solutes was driven by the large increases in water retention time, which effectively increased the loading rate of solutes (Schindler, 1997). For example, as the water renewal time of Lake 239 reached 20 years, it would take 60 years to attain 95% of a new steady state if no further changes occurred in water or chemical inputs. In contrast to this accumulation effect, longer water retention times increased the importance of consumptive in-lake processes such as biotic assimilation and sulfate reduction. Consequently, concentrations of more bioactive solutes, such as silica, total dissolved phosphorus and DOC, declined in Lakes 239 and 240 and alkalinity-generating processes, such as sulfate reduction, increased (Schindler *et al.*, 1996a). Internal sulfate reduction caused sulfate : cation ratios to decline rather quickly, despite long water residence times. This, plus the return of base cations from sediments, caused the alkalinity to increase by 30–50% and pH to increase a few tenths of a unit (Schindler *et al.*, 1990, 1996a). Total dissolved nitrogen increased in Lake 240 and initially in Lake 239, with subsequent declines presumably related to changes in in-lake processes (Schindler *et al.*, 1990, 1996a).

A less extensive drought in northern Wisconsin from 1987 to 1990 also altered the hydrological pathways of water and solute flow to lake ecosystems. In contrast to the ELA drainage lakes set in thin soils ($<0{\cdot}5$ m in thickness), the Wisconsin lakes are a mixture of seepage and drainage systems set in thick sandy outwash till deposited by receding glaciers. Many chemical, biological and physical attributes of the Wisconsin lakes are related to their position in the groundwater flow system, ranging from precipitation-dominated lakes high in the landscape to groundwater-dominated lakes low in the landscape (Kratz *et al.*, 1997). Changes in the cations, Ca and Mg, during drought were related to lake landscape position (Webster *et al.*, 1996). Cation concentrations increased during drought as observed for the ELA lakes, but differences in the direction of change in cation mass, which accounts for the effect of evapoconcentration, were related to lake landscape position. This suggests that lakes may diverge in their long-term response to warmer and drier conditions. Increases in cation mass for lakes low in the landscape during drought were attributed to proportionately high inputs of groundwater from deeper flow paths (which are more cation rich) and to a lessening of the diluting effect of precipitation. In contrast, cation mass declined for lakes high in the landscape as transient groundwater inflows were interrupted during drought. Such differences among lakes are apparent in long-term calcium data on Lake 239 at ELA and Nevins Lake in the upper peninsula of Michigan (Figure 14). In Nevins, cation and silica concentrations, alkalinity and pH rapidly declined during the same drought (Webster *et al.*, 1990). This lake has a short water retention time (<2 years) compared with other seepage lakes and its chemistry is extremely responsive to changes in groundwater inputs driven by climate (Krabbenhoft and Webster, 1995).

In addition to responses to altered inputs of solutes, the biogeochemical cycling of nutrients and other ions in lake ecosystems will be influenced greatly by the oxygen conditions in a lake's deep waters. Many compounds, such as iron phosphates, are more soluble under anoxic conditions; their availability to the water column from nutrient-rich sediments would be increased with increased anoxic conditions. Thus, recycling of nutrients from the sediments to the water column would be enhanced by climate warming in thermally stratified lakes during summer. This would be countered to an unknown extent by reductions in hydrological flow and associated nutrient export from the catchments.

Climate conditions also influence retention and cycling of solutes in peatlands. Pulsed exports of sulfate, acidity and cations occurred from a minerotrophic swamp located near Dorset (LaZerte, 1993) and from a peatland draining into Lake 239 at ELA (Bayley *et al.*, 1992a) after dry summers when accumulated sulfur was exposed and oxidized. The strength of these acid pulses was related to deposition history; under conditions of higher acid loading and, thus, increased storage of sulfur, the potential for acidification events at downstream sites is enhanced (Bayley *et al.*, 1986). The cycling of solutes, particularly carbon, within peatlands is a function of groundwater flow patterns, which, in turn, are influenced by climate (Siegel *et al.*, 1995). In a northern Minnesota peatland, groundwater flow advected dissolved organic compounds into deeper peat layers during wet periods, stimulating methanogenesis, while during dry periods a discharge regime dominated and organic compounds were advected upwards, limiting methanogenesis to the more refractory carbon of the peat itself (Siegel *et al.*, 1995).

The biogeochemical responses of small lakes, streams and peatlands to climate change reveal an intricate interaction between weather, catchments and lakes, and between physical, chemical and biological effects. Biological factors on the land and geomorphological legacies influence how aquatic ecosystems respond to climate change. For example, standing trees on the shoreline affect wind exposure and lake thermal properties; DOC export from catchments influences light penetration and mixing depth; and position in the hydrological flow system influences changes in solute concentrations. Biogeochemical responses are also a function of lake water retention time and the biological reactivity of a given solute.

AQUATIC ECOLOGY

Responses to the diverse changes in the physicochemical environments induced by changes in climate can occur in individual organisms, in populations of organisms, in communities of interacting species and in the structure of interactions between biotic and abiotic components of entire aquatic ecosystems, including the landscape in which the lakes, streams and wetlands are imbedded. Our review treats these responses for the phytoplankton, zooplankton, benthos and fishes. The phytoplankton are the microscopic algae that live in the water column and, through their photosynthetic production of new organic matter, form the major base for the food-web in lakes. The zooplankton are microscopic animals that live in the water column and, functionally, are either herbivores on the phytoplankton or predators on smaller zooplankton. The benthos are organisms that live on bottom substrates of lakes, wetlands and streams; they include a wide range of plants and animals, from large-rooted plants and algal mats to immature insects, clams and snails and bottom-dwelling fishes. The fishes are generally long lived and mobile; they can be herbivores, predators or omnivores; they play important structuring roles in aquatic communities; and they are valued for sport, food and biological diversity.

Phytoplankton

Increased epilimnion temperatures are unlikely to affect rates of phytoplankton primary production directly, but species composition would change, annual production would increase and phytoplankton diversity and biomass might increase. The phytoplankton community consists of hundreds of species with simple life cycles. Even though rates of photosynthesis of nutrient-sufficient algal cultures increase exponentially with increasing temperature (Eppley, 1972), the realized rate of photosynthesis in lakes is virtually independent of temperature for several reasons.

1. Algae rapidly adapt to the temperature at which they are grown; thus, realized rates of photosynthesis, even in isolated species, are either partially or fully independent of temperature (Davison, 1991).

2. In freshwater, *in situ* phytoplankton photosynthesis is typically either nutrient (phosphorus, nitrogen, silica) or light limited, especially during midsummer when the greatest temperature increases are expected (Schindler, 1977; Healey and Hendzel, 1980); these primary limitations prevent phytoplankton from achieving temperature-determined physiological limits (Fee *et al.*, 1987).

3. Some phytoplankton species are favoured by lower temperatures, while others are favoured by higher temperatures. Because life cycles are short, competition soon results in a community capable of photosynthesizing optimally at any temperature.

Results from a long-term study of phytoplankton photosynthesis in seven Canadian Shield lakes, ranging in size from 30 to 34 600 ha (Fee *et al.*, 1992), clearly support these ideas.

1. As in most unperturbed temperate zone lakes, variations of chlorophyll concentration and phytoplankton production were highest in the spring and lowest in midsummer; i.e. the inverse of variations of temperature but precisely the pattern of variation of nutrient availability.

2. Interannual variations of phytoplankton production were unrelated to epilimnion temperatures (which varied by $>3{\cdot}0$°C; i.e. similar to the range of the mean temperature rise predicted for a $2 \times CO_2$ climate), but were significantly and positively related with total rainfall, which is the main nutrient source to remote shield lakes (Fee *et al.*, 1994).

Clearly, nutrients, not temperature, are the factors of greatest importance in determining phytoplankton growth and abundance in shield lakes (Hecky and Kilham, 1988). This conclusion is supported by 24 years of phytoplankton community composition data at the species level from a variety of lakes in the Experimental Lakes Area, where year to year changes of community composition correlated strongly with chemical variables (pH, DOC, nutrients, total cations) but not at all with water temperature (Kasian *et al.*,

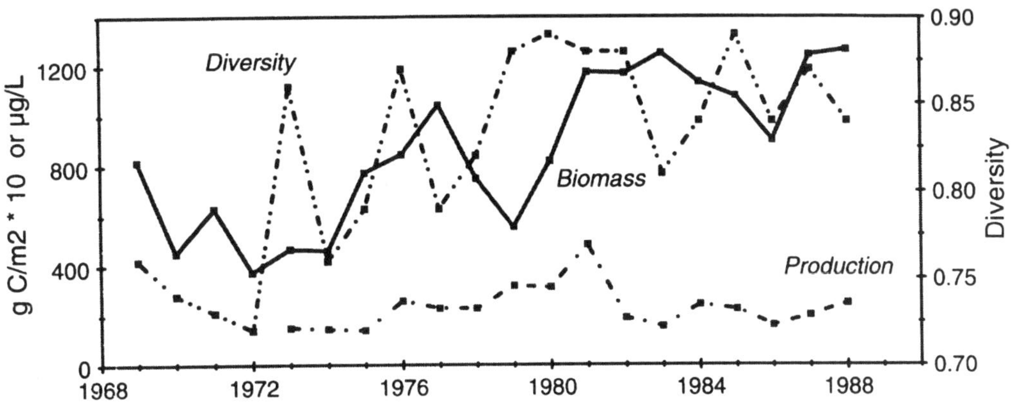

Figure 15. Change in phytoplankton biomass and diversity in Lake 239 (from Schindler *et al.*, 1990) and total and subepilimnetic production in lakes 239 and 240 (adapted from Schindler *et al.*, 1996a)

in press). The implication is that the new phytoplankton communities that are almost certain to appear in a $2 \times CO_2$ climate will be only secondarily related to elevated temperatures.

Although the direct effects of increased epilimnion temperatures on phytoplankton photosynthesis (primary production) will be minimal, some indirect effects are not so easily dismissed. Agreement is close between the temperature vs growth performance of major algal groups in cultures (Allen and Stanier, 1968; Canale and Vogel, 1974) and the phytoplankton community composition in lakes: blue-green algae dominate at the highest temperatures, and are succeeded by green algae, then flagellates and then diatoms as the temperature drops. Warmer epilimnia will thus favour blue-greens and greens at the expense of chrysophytes and diatoms. These climate-altered phytoplankton communities will have different trophic efficiencies than the present ones (Schindler, 1977), and ecosystemic effects could be profound. As Smayda (1980) points out, while there may be general temperature/growth tendencies in major taxonomic groups, detailed attempts to model seasonal succession of phytoplankton based on the temperature optima of individual species measured in isolation have not been successful. He concludes that the regulatory variable is not temperature *per se* but something strongly correlated with it.

Phytoplankton biomass and diversity did increase slightly during the 20-year period of increasing temperature at ELA (Schindler *et al.*, 1990, 1996a). However, consistent with the above, phytoplankton productivity did not appear to be related to climatic trends (Figure 15), and production below the thermocline did not increase, owing to the parallel deepening of the thermoclines and the euphotic zone.

Shallower epilimnia and sharper thermoclines from climate warming scenarios would cause epilimnetic phytoplankton to be more severely nutrient limited; shallowness by reducing light limitation and the residence time of particles in the mixed layer, sharpness by reducing the upward flux of hypolimnetic nutrients to the mixed layer (Fee *et al.*, 1994). While this may not affect phytoplankton productivity in shield lakes that are already nutrient limited (Guildford *et al.*, 1994), it would affect the elemental composition of phytoplankton and thus their food value to consumers (V. H. Smith, 1991; Sterner *et al.*, 1993), and consequently rates of bioaccumulation of organochlorines in the food web (Swackhamer and Skoglund, 1991) and the toxicity of contaminants to phytoplankton (Twiss and Nalewajko, 1992).

The expected decrease in DOC and increase in light transmission would have two important effects on phytoplankton. First, algal populations that are now light limited would grow faster. In contrast to epilimnetic phytoplankton that are rarely light limited (Fee *et al.*, 1992), phytoplankton that thrive deeper in the water column almost always are (DeNoyelles *et al.*, 1980; Shearer *et al.*, 1987), and would increase in productivity if water clarity increased. The influence of this potential increase might not affect the productivity of higher trophic levels because the characteristic species have large cells or are colonial (Fee, 1976); they may be too large for filter-feeding zooplankton to process.

Annual phytoplankton production would be expected to increase for lakes, provided that there were no large increases in cloudiness. Phytoplankton production in shield lakes was proportional to the length of the ice-free season (Fee *et al.*, 1992), i.e. a 25% increase from six to eight months resulted in 25% more phytoplankton production under an assumption that all other features of the lakes were unchanged. Obviously, this will only occur in lakes that are currently ice covered for part or all of the year. Other studies by Regier *et al.* (1990a), using analyses of empirical relations that removed the statistical influence of the morphoedapic index, also indicated that annual primary production increased with temperature with a Q_{10} of 3·7. However, annual primary production is markedly reduced by cloudy days at the North Temperate Lake LTER site in Wisconsin (Adams *et al.*, 1990).

Zooplankton

Zooplankton production or biomass may well be influenced by climate warming. Regier *et al.* (1990a), in an analysis of comparative lake data that statistically removed the influence of the morphoedapic index, observed an increase in lake zooplankton biomass with temperature, with a Q_{10} of 4·0. DeStasio *et al.* (1996) simulated differences in the vertical spatial overlap between zooplankton and zooplanktivorous fishes using a dynamic programming approach for two Wisconsin lakes and a 2 × CO_2 warming with four GCMs (CCC, GISS, GFDL and OSU). Results were mixed for Sparkling Lake, a small oligotrophic lake with a cold water zooplanktivore in northern Wisconsin; overlap in vertical distributions increased as much as 37% and decreased as much as 26% among GCM scenarios, but three of the four simulations were for decreased overlap. For Lake Mendota, a eutrophic lake with warm water zooplanktivores in southern Wisconsin, overlap consistently increased from 20 to 27% among the four GCM scenarios. The explanations for the difference between the two lakes were: (1) for Sparkling Lake, the fish were more restricted to the deeper cold waters as the surface waters warmed and the *Daphnia* could seek refuge in the warm epilimnion; and (2) for Lake Mendota, the warm water fish experienced an increase in warm waters and forced a portion of the *Daphnia* to seek refuge in the anoxic hypolimnion.

Temperature is the cue that stimulates the production of overwintering stages for zooplankton and their release from dormancy (Stirling and McQueen, 1986; Chen and Folt, 1996; Hairston, 1996). Warming events could alter the timing or occurrence of resting eggs in autumn, potentially leading to a loss of a cohort or of numbers in the 'seed bank'. For example, resting eggs hatch prematurely with a 12°C temperature increase in autumn (Chen and Folt, 1996). Local extinctions would be more likely when warm summer temperatures and anoxia reduce the hypolimnetic refuge required by particular species (Stemberger *et al.*, 1996).

Benthos

Although benthic algal photosynthesis has been widely reported to be no more than a few per cent of phytoplankton photosynthesis, even in small shield lakes (Schindler *et al.*, 1973), recent stable isotope evidence indicates that benthic algae are an important energy source for aquatic animals in lakes of all sizes and in many climates (Hecky and Hesslein, 1996). Either the methods used to measure photosynthesis in phytobenthos and phytoplankton do not yield comparable results, or benthic production is transferred to higher trophic levels much more efficiently than is planktonic production. Whatever the cause, it is important in the context of climate change because benthic photosynthesis in shield lakes is limited by dissolved inorganic carbon (Turner *et al.*, 1994), and should increase in response to a doubling of atmospheric CO_2. Phytoplankton photosynthesis, on the other hand, is not carbon limited (Schindler and Fee, 1973) and thus will not be directly affected by increased CO_2.

Benthic invertebrate species have fixed temperature thresholds for egg development and/or generation times that are a function of degree–days (Giberson *et al.*, 1991). Effects of temperature increases on benthic and other invertebrates were reviewed in Arnell *et al.* (1996); increased temperatures cause: faster life stage development and shorter generation times; greater rates of reproduction, ingestion, growth and respiration; lower rates of mortality, smaller body size and more generations per year. Given the high natural dispersal capability of benthic insects, rapid colonization of waters that are presently too cold for egg development can be expected. Furthermore, many insect species would mature more rapidly in waters that they already inhabit (e.g. in one year instead of two). These effects would not occur in species that inhabit the profundal zone of deep lakes, because temperatures there are expected to change little.

Fishes

Reviews and information on the potential effects of climate warming on fishes have been published recently in Arnell *et al.* (1996), Everett *et al.* (1996), Regier *et al.* (1996) and Wood and McDonald (1996). Useful earlier reviews include Coutant (1981), Meisner *et al.* (1987) and Magnuson *et al.* (1989) and the Regier *et al.* (1990b) symposium.

Thermal niche. Fishes are ectothermic heterotherms whose body temperatures track the habitat in which they live. Fry (1947, 1971) classified environmental factors, including temperature, by their effects on fishes as lethal, controlling and directive. Thus, while temperature can be lethal, it controls many physiological rates and behaviours within the lethal limits. Various rates are optimized at intermediate temperatures. Fishes are mobile and seek out their preferred temperatures that typically include these optima, provided those temperatures are available within their habitat. The thermal niche of a fish can be described in terms of lethal, controlling and directive effects on the particular species (Figure 16) (Magnuson *et al.*, 1979; Magnuson and DeStasio, 1996; Regier *et al.*, 1996). Warming of aquatic habitats would produce changes in fish species composition and productivity because fishes differ in their response to temperature depending on whether they are adapted for cold or warm waters.

The thermal niche was defined in terms of a directive factor as the final temperature preferendum ±2°C or ±5°C by Magnuson *et al.* (1979). Fish occupy temperatures within the 4°C range 67% of the time in laboratory temperature preference tanks, and all of the time within the 10°C range. These preferred temperatures grade continuously from low to high among species, but reasonable examples are near 10–15°C for cold water fishes, 21°C for a cool water fish and 30°C for a warm water fish (Figure 16) (Hokanson, 1977; Magnuson *et al.*, 1979). The adapted range of temperature is broader for survival, basal metabolism and digestion than for growth, egg development and active metabolism. Typically, species preferences and the thermal optima they include are near the warm edge of their lethal limits (Lin, 1995; Regier *et al.*, 1996); for example, the final temperature preferendum is about 5–10°C below the upper lethal

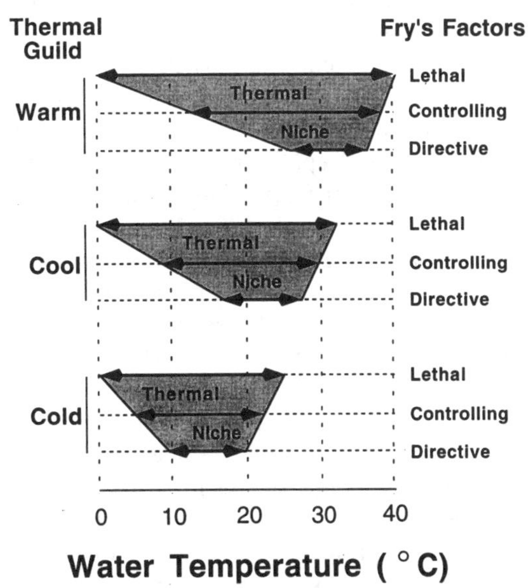

Figure 16. Niche dimension diagram for temperature showing Fry's factors for different thermal guilds

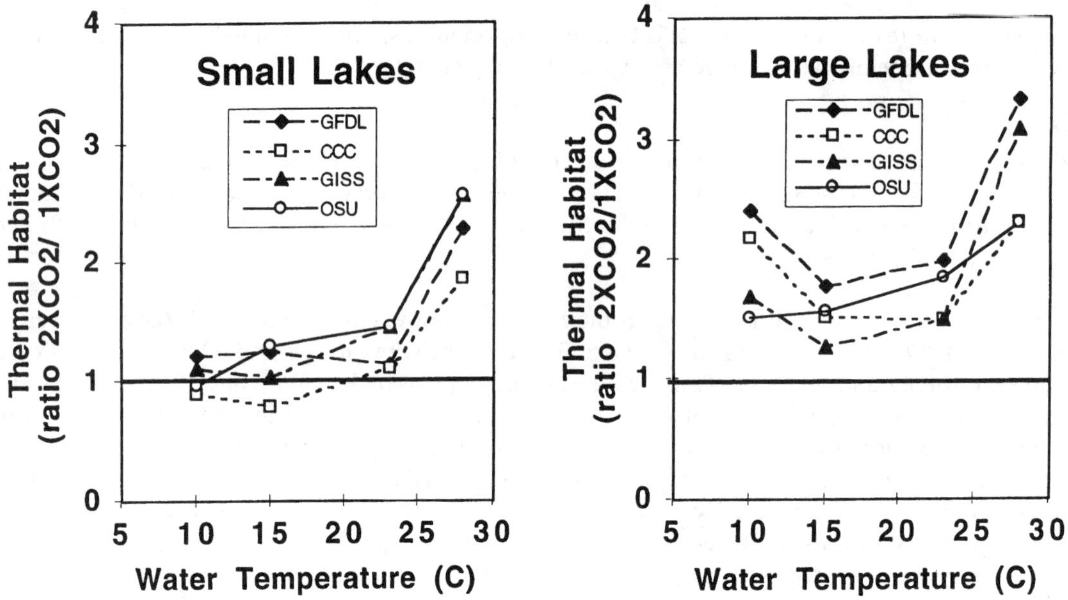

Figure 17. Simulated changes in thermal habitat in small (Crystal and Sparkling) and large (Michigan, Mendota and Trout) stratified temperate lakes for cold, cool and warm water fish under a 2 × CO_2 with four GCMs (adapted from Magnuson and DeStasio, 1996)

temperature. Regier *et al.* (1996) sum this up by stating that 'these ecotootherms seem to have evolved to live dangerously but not too dangerously'.

The thermal niche, measured as habitat prefences under laboratory conditions, is the fundamental or non-interactive niche (Magnuson and DeStasio, 1996), but in field situations the fish occupy their realized or interactive niche. *A priori*, fish would be expected to occupy their non-interactive niche if other environmental factors were benign and if preferred temperatures existed in the lake or stream. However, other factors are not necessarily benign; e.g. anoxia can develop in the hypolimnion or under the ice in shallow lakes and displace fishes to less than optimum temperatures. Also, temperatures within the non-interactive thermal niche often do not exist in an ecosystem at a particular season or latitude. Thus, the scenarios of climate change effects on fish thermal habitat must consider other factors influenced by global warming such as anoxia, stream flow and water clarity, as well as geographical location and date.

In global warming (2 × CO_2) scenarios for deep, thermally stratified lakes, thermal habitat (non-interactive thermal niche) typically increased, not only for warm-water fishes, but also for cool- and cold-water fishes (Figure 17) (Magnuson *et al.*, 1990; DeStasio *et al.*, 1996; Magnuson and DeStasio, 1996). For large and medium size lakes, habitat increased for all thermal guilds with four GCM scenarios (CCC, GISS, GFDL and OSU); for small lakes it increased for cool- and warm-water fishes in all scenarios, but for cold-water fishes it stayed about the same. The increases occurred because the length of the growing season was extended and because fish could move to deeper, cooler waters when surface waters exceeded preferred temperatures. Deep water refuges were maintained in model projections over large latitudinal ranges (McLain *et al.*, 1994). These scenarios of increased thermal habitat did not include the consequences of expected increase in anoxia in the hypolimnion. In Lake Erie scenarios, cold water refuges would not be available because the waters would be expected to be anoxic. Cold water habitat did decrease in small ELA lakes over 20 years of warming weather because the thermocline deepened and the anoxic zone at the bottom increased slightly (Schindler *et al.*, 1990, 1996a). Lake trout stopped reproducing in Lake 239 during the 1970s. Farther south, in Ohio ponds, cool bottom water habitat for northern pike was constricted in warmer years by lower dissolved oxygen and resulted in summer weight loss and reduced development (Headrick and Carline, 1993).

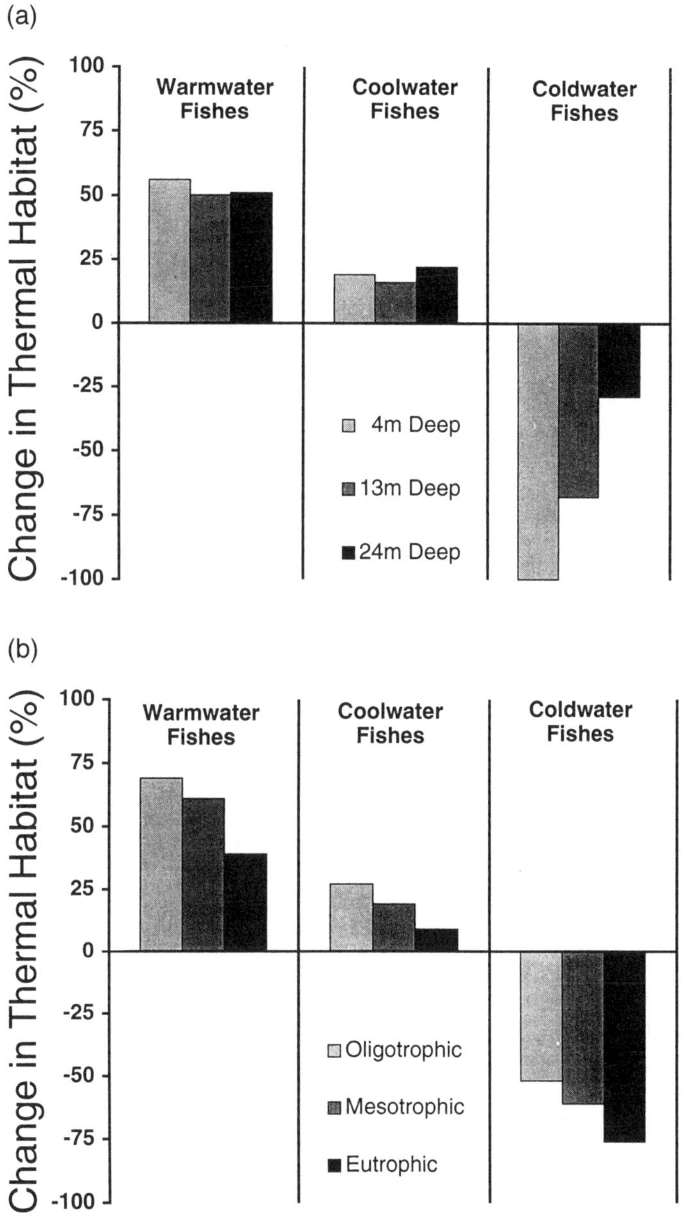

Figure 18. Changes in thermal habitat under 2 × CO_2 scenarios for warm, cool and cold water fishes in (a) lakes of different depths and (b) lakes of different trophic status (data from Stefan *et al.*, 1995)

Comprehensive warming scenarios for fish thermal niche in Minnesota lakes, including both temperature and dissolved oxygen, were developed for a wide range of lake maximum depths, from shallow (4 m) to deep (23 m); trophic status, from oligotrophic (Secchi depth 1·2 m) to eutrophic (Secchi depth 4·5 m); and area, from small (0·2 km^2) to large (10 km^2) (Stefan *et al.*, 1995). Temperatures that controlled growth of cold-, cool- and warm-water fishes were used to define changes in the favourable habitat. The simulations indicated that a loss of thermal habitat for cold-water fishes occurred across all lake types, but was progressively greater in shallow compared with deep lakes, and in eutrophic compared with oligotrophic lakes (Figure 18).

Thermal habitats increased for warm- and cool-water fishes across all lake types, but increases were progressively greater from eutrophic, to mesotrophic, to oligotrophic lakes (Figure 18).

Shallow ponds and streams would not have the possibility of the thermal refugia associated with the deep stratified waters of some lakes, thus, warming would be more likely to reduce fitness or even be lethal in shallow than in deep lakes. Even in thermally stratified lakes, simulated surface temperatures exceeded lethal levels even for warm-water fishes in some scenarios for selected Wisconsin lakes (DeStasio *et al.*, 1996). Simulations for stream fishes were conducted by Eaton and Scheller (1996) using simulated increases in the mean July, August and September air temperatures of 4·0–5·5°C, including 10 states at the southern part of our region. Based on their results, about 70% of the sites would become unsuitable for the cold-water rainbow trout and the cool-water white sucker, but the number of suitable sites for the warm-water largemouth bass appeared to about double. In two headwater streams with significant groundwater input in southern Ontario, a thermal refuge for cold-water fishes survived a 2 × CO_2 climate scenario, but they were reduced in size by 34 and 40% (Meisner, 1990).

Populations and production. The production of fishes in lakes would be expected to increase with warming. Maximum sustainable fishery yields increased with temperature ($Q_{10} = 2{\cdot}6$) in a statistical analysis of 43 lakes, many in our region (Regier *et al.*, 1990a). Yields of cold- and cool-water species were correlated with the size of the thermal habitat (non-interactive niche) in 21 large north temperate lakes (Christie and Regier, 1988).

Under 2 × CO_2 climates, process models that included temperature dependence of all life history stages produced increases in population size and productivity of smallmouth bass along cooler shorelines of the Laurentian Great Lakes (analyses by Shuter in Magnuson *et al.*, 1989). Warmer climate scenarios resulted in a longer adult growing season, younger age of maturity, longer young-of-year growing season, greater young-of-year winter survival, greater year-class strengths and larger fishable populations. The ratios between 2 × CO_2 : 1 × CO_2 for year-class strengths and fishable populations were greatest for northern Lake Superior, intermediate for Lake Huron and almost zero for Lake Erie. The smallmouth bass has a thermal biology at the boundary between cool- and warm-water fishes. Similar responses would be expected for other species. Temperature increases hatching success and growth of sea lamprey (Holmes, 1990), but expansion of the sea lamprey would be a problem rather than a benefit to human users of fishery resources. Simulated growth rates of cold-, cool- and warm-water fishes all increased in 2 × CO_2 scenarios along warm, cool and cold shorelines of the Great Lakes (Hill and Magnuson, 1990). Growth rates increased and age of smoltification decreased in 2 × CO_2 simulations for Atlantic salmon in eastern Canada (Minns *et al.*, 1995). While this would increase the production of salmon, associated scenarios of river runoff decreased the area of fluvial habitat for juvenile salmon by an average of 4% (range −40 to +16%).

Since climate scenarios indicate greater warming in winter than summer, it is important to consider factors that affect fish populations in winter. On the one hand, warmer winter temperatures would increase winter survival of warmer water species (Johnson and Evans, 1990), those subject to winterkill from anoxic conditions in shallow lakes (Tonn, 1990) and those exposed to low pH and elevated aluminium (Leino and McCormick, 1993). Alternatively, warmer winters would reduce the reproductive success of species that require a cold period for normal gonadal development (Jones *et al.*, 1972), or of autumn-spawning species that have higher overwinter egg survival with ice cover (Taylor *et al.*, 1987). The white perch invasion of the Great Lakes seemed to have been enhanced by higher winter survival of young-of-the-year fishes in warm winters (Johnson and Evans, 1990). Reductions in the duration or the complete absence of ice cover would reduce or prevent anoxic conditions from forming in shallow lakes and backwaters of rivers. In shallow northern Wisconsin lakes, fish communities dominated by central mudminnow, a species tolerant of low dissolved oxygen, would become dominated by northern pike or largemouth bass, species intolerant of low oxygen (Magnuson *et al.*, 1989; Tonn *et al.*, 1990). Only moderate increases in acidity and aluminium affect the electrolyte balance of juvenile largemouth bass at winter temperatures (Leino and McCormick, 1993). The spring-spawning yellow perch require a winter cold period for normal egg development (Jones *et al.*,

1972); the highest percentages of viable eggs produced were 93% after overwintering at 4°C; at 6°C only 65% were viable and at 8°C only 31%. In Grand Traverse Bay, Lake Michigan, year classes of lake whitefish were smaller by an order of magnitude in years without ice cover than in years with ice cover, apparently because eggs on the bottom were dislodged from favourable substrates by the increased turbulence and currents without ice cover (Taylor *et al.*, 1987; Freeberg *et al.*, 1990; Brown and Taylor, 1993).

Species near the northern limits of their range have good years of recruitment and growth that are positively related to summer temperatures (Pereira *et al.*, 1995a,b). Such relationships were observed for freshwater drum, walleye, sauger and possibly for smallmouth bass in long-term data from Minnesota. These growth and recruitment events related to summer temperatures provide another example that fish populations would be sensitive to warming scenarios.

Many fish species, but not all, would grow faster, become more productive and disperse northwards. The ability of young-of-year fishes to reach a sufficient size to overwinter successfully explains the present northern limit of smallmouth bass and yellow perch across our region and other areas of Canada (Shuter and Post, 1990). Simulations of a new northern limit under 2 × CO_2 climate scenarios using this criterion resulted in a potential northward dispersal of about 500 km. Simulations for a doubling of CO_2 from an empirical analysis of fish distributions in Ontario watersheds produced similar potentials for a northward dispersal for smallmouth bass and carp, *ca.* 500–600 km (Minns and Moore, 1995). Thus, warming would be expected to accelerate greatly the northward dispersal of fishes following the Wisconsin glaciation.

In open systems like the Great Lakes and in north/south rivers, as successively warmer species were favoured, changes in species composition, abundance and overall production would occur; other warm-water species would invade through connecting rivers and canals (Mandrak, 1989). Such changes would also be expected to extend north of the Laurentian Great Lakes, with the Great Lakes serving as new sources of colonists. With an air temperature increase of 4·5–5·5°C, simulated increases in species richness ranged from 12 to 60 species per tertiary watershed in Ontario using empirical statistical models (Minns and Moore, 1995). At an individual lake level, the mean scenario was for three new species per lake, which, when compared with the present richness of 3–6 species per lake across much of Ontario, is a large increase; the changes in community structure and dynamics would be dramatic. Dispersal of 33 species already occurring in Ontario but with distributions bounded by temperature, differ across the province (Figure 19); fewer invasions would occur to the north owing to temperature limitations even after the warming scenario, and fewer would occur to the south because many of the species already occur there. In more closed aquatic systems both in Ontario and across the region, such as seepage lakes without inlets or outlets or stream reaches above impassable waterfalls or dams, the arrival of warm-water fauna would also be inhibited. At the same time, to the south, especially in shallow habitats, cold- and even cool-water species would find themselves in progressively suboptimal environments and their populations would be expected to decline progressively from the south. Within the geographical region now inhabited by cold- and cool-water species, local geomorphological, hydrological and anthropogenic factors such as lake depth, groundwater inflow and waste heat loadings would result in populations disappearing in a salt and pepper fashion over broad regions, as simulated by Eaton and Scheller (1996).

IMPLICATIONS OF TERRESTRIAL–AQUATIC LINKAGES

Although the potential effects of climate change on terrestrial environments and aquatic systems have been discussed extensively in the literature, less is known about the interactions between terrestrial and aquatic environments with respect to a changing climate. Interactions among the atmospheric (chemistry and climate), terrestrial (vegetation and soils) and aquatic (flowing water, wetlands and lakes) biomes are complex (Figure 20). Climate (and thus climate change) affects atmospheric chemistry, vegetation, soils, hydrology and aquatic ecosystems; the converse is also true. Included among the terrestrial processes that would directly or indirectly affect aquatic systems in a warming climate are increased rates of fire and drought, changes in the physical and chemical properties of soils and altered hydrology in both space and

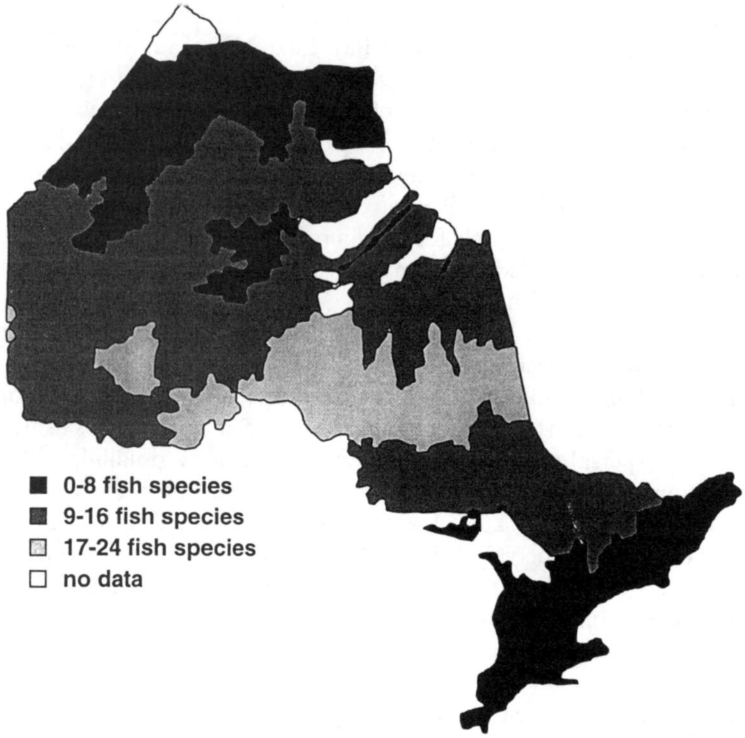

Figure 19. Tertiary watersheds in Ontario where 0–8, 9–16 and 17–24 of the 33 freshwater fish species with temperature-determined boundaries are predicted to be able to invade following climate warming of 4·5–5·5°C (modified from Minns and Moore, 1995)

time. Increased CO_2 concentrations will indirectly affect the terrestrial ecosystem through increased rates of photosynthesis, evapotranspiration, and respiration at higher temperatures. The effects of rising temperature on respiration may overwhelm the direct effects of rising CO_2 on photosynthesis and growth (Drake, 1992), although increases in CO_2 can reduce plant respiration (Bunce, 1994). Since plant metabolism and stomatal control of water loss depend on other climatic factors such as humidity and wind, an increase in evapotranspiration cannot necessarily be assumed (Rosenberg *et al.*, 1990 as cited by Poff, 1992).

Climate-related changes in precipitation, runoff and soil moisture will alter redox potentials and most likely cause declines in wetland coverage, both spatially and temporally. These changes will influence DOC production and export (Hemond, 1990; Sedell and Dahm, 1990) and the cycling of sulfur, phosphorus, nitrogen and metals (Devito and Dillon, 1993a,b), and will affect detrimentally the fauna that rely on wetlands for habitat. For example, the storage of sulfate and/or the oxidation of stored reduced sulfur usually occurs during dry seasons (Dillon and LaZerte, 1992), or in dry years when the water table recedes. At the beginning of the next season, or in a wet year, the stored sulfate is released back into the outflowing waters. The implication is that sulfur inputs to streams and lakes will be affected by climate change long after anthropogenic reductions in sulfate deposition occur. Decreased export of DOC from terrestrial to aquatic environments is expected to occur with climate warming and drying. DOC originates from the forest canopy, the forest floor, the mineral soil, wetlands and groundwater (Cronan, 1990). The quality of the DOC varies with source material and the chemical composition of the water leaching through it (Hynes, 1975; Wetzel, 1992). Higher carbon content in the terrestrial system might lead to higher DOC concentrations in mineral soils, which could counter, in part, the observed decreases in catchment export. Alternatively, soil warming could result in accelerated respiration of soil organic matter, which would otherwise be transported to streams as DOC, and thus decrease DOC inputs (Meyer and Pulliam, 1992).

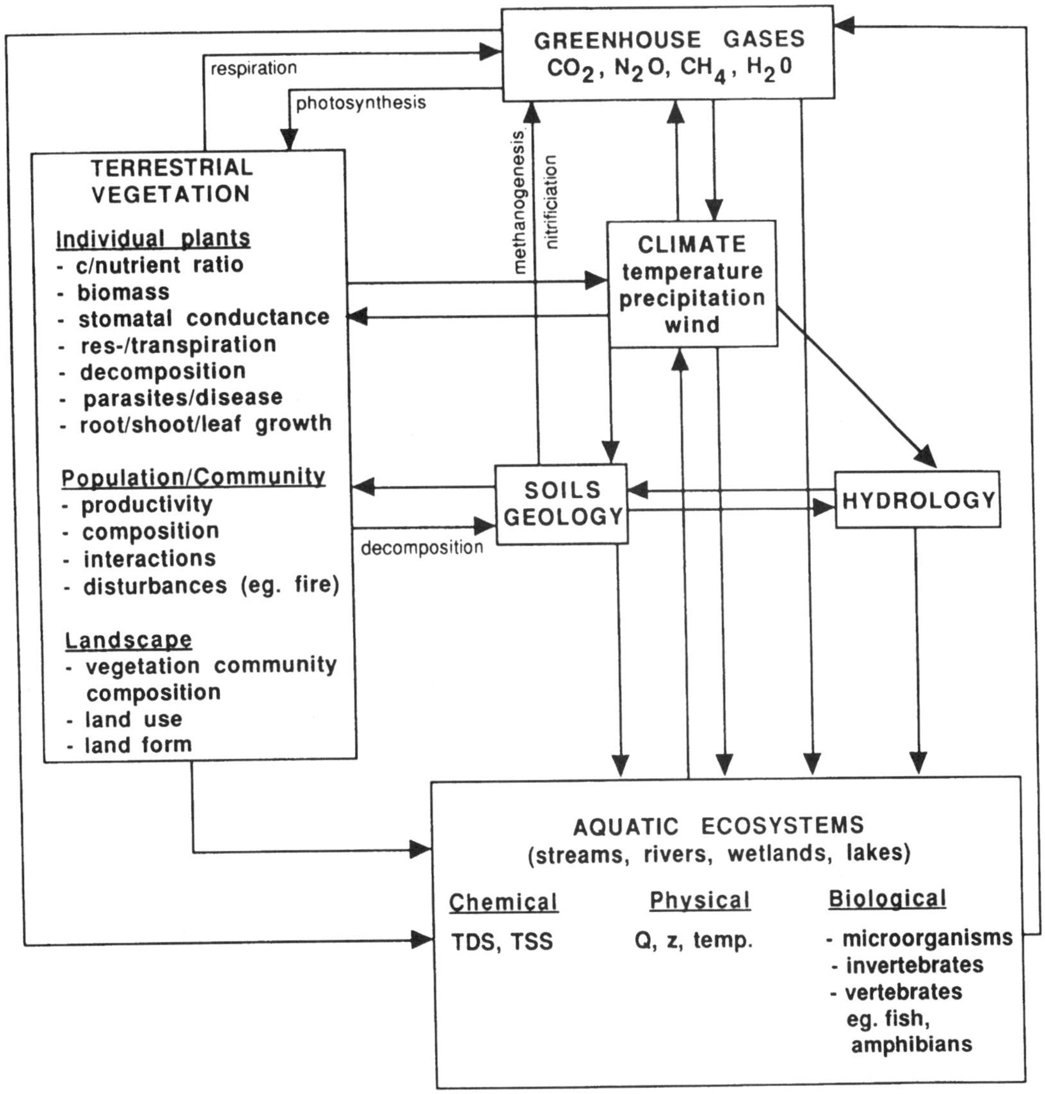

Figure 20. Terrestrial and aquatic linkages influenced by climate change

Other terrestrial effects relevant to freshwater ecosystems could result from changes in the ranges of plant species and species composition at a regional scale (Bazzaz, 1990). Rizzo and Wilken (1992) forecast that much of the moist boreal forest will be replaced by transitional grasslands and cool temperate forests. Northward range migrations of several tree species are projected (Peters, 1990). Unfortunately, while substantial changes in terrestrial vegetation patterns are predicted, the exact response differs between scenarios (Lenihan and Neilson, 1994). Changes to the terrestrial riparian zones of streams will affect exports of inorganic and organic matter to streams and lakes. Changes in plant species composition, decreases in foliar N content and increases in C : N ratios would affect the quality of the leaf litter both on the forest floor and in the streams. Increased production and transfer of plant biomass to the roots might affect nutrient uptake and the amount available to be leached from the soil. Finally, more fire and drought could increase the rate at which forests are 'opened up' for new trees, and thus short circuit the slower process of forest–gap overturn through old age mortality (Overpeck *et al.*, 1990). Inputs of woody debris to streams would increase, gaps left by fallen trees would undergo succession and the litter contributed by successional species would increase. In small streams, heavy litter accumulation from a longer dry season may cause deoxygenation (Hynes, 1975), which could affect transport of nutrients, DOC and metals.

The effect of climate change on carbon storage in the terrestrial ecosystem depends on plant species composition, soil water holding capacity and C : N dynamics. While most predict that, at least in the short term, carbon storage will increase because of higher productivity and C : N ratios (Post *et al.*, 1992; Rastetter *et al.*, 1992), rising temperatures might increase forest ecosystem respiration sufficiently for terrestrial ecosystems to become a source and not a sink for CO_2 (Drake, 1992). These predictions are complicated by expected increases of *ca.* 40% in the frequency and intensity of fires in boreal regions with global warming (Flannigan and Van Wagner, 1991, cited in Kasischke *et al.*, 1995). In the short term, the combined effects of changes in climate and fire regime would influence net carbon storage in the boreal forest through increases of 20–50% in the area burned, an increase in forest flammability and a decrease in carbon storage in the ground layer (Kasischke *et al.*, 1995). Over the long term (150–200 years), shifts in forest species composition to more deciduous trees would result in more carbon storage as living biomass and a decrease in forest flammability (Kasischke *et al.*, 1995).

INTERACTIONS WITH OTHER STRESSES

Lakes integrate the combined effects of human activities in their catchments and airsheds. Thus, they can be sensitive indicators of the added perturbations from climate warming. Many anthropogenic perturbations, already affecting aquatic systems in the Laurentian Great Lakes and Precambrian Shield Region, would interact with the potential effects of climate change. Climate change effects are not independent of other human stresses on the system, nor are the effects of these stresses independent of climate change effects. Interacting stresses include atmospheric deposition of acids and toxic organic chemicals; discharges of agricultural, industrial, and domestic wastes; forest fires and clear cutting; urbanization; pipeline and reservoir construction; mining; dredging; invasion of exotics; and so on. What most of these stresses have in common is that they change the fluxes of water and dissolved and suspended materials, leading to the generic problems of eutrophication and contamination. Thus, the stresses that we have been dealing with may be accentuated, or in some cases countered, by the added effects of climate change. Below we discuss some of the potential interactions between climate warming and these other human-caused stresses on aquatic systems, especially lakes.

Eutrophication

Regier *et al.* (1990a) made the case that the thermal effects of climate warming would superficially resemble those of eutrophication. Their empirical relations indicated that annual primary production of phytoplankton, zooplankton biomass and maximum sustained fishery yields all increased with temperature. Increased anoxia of the deep hypolimnetic waters, a feature of eutrophication, also increased in climate warming scenarios for both Lake Erie (Blumberg and DiToro, 1990) and small inland lakes scenarios (Stefan *et al.*, 1993a, 1995; Stefan and Fang, 1994). Faster growth rates for fishes in the Great Lakes were simulated in warming scenarios, provided adequate prey were available to fuel higher metabolic rates (Hill and Magnuson, 1990). So, superficially, the effects of eutrophication appear to parallel those of climate warming.

The influence of climate change on nutrient dynamics is less certain and more contradictory. Because climate change models predict lower runoff and, thus, a decline in nutrient export from catchments to lakes, a reduction in eutrophication might result. However, increases in the water residence time of lakes, as observed for many ELA lakes during warmer conditions (Schindler *et al.*, 1996a), may counteract the decreased inputs, potentially causing an overall increase in the effective loading rate. In addition, the increased frequency of severe weather events might result in higher pulsed inputs of nutrients during flooding and runoff events in agricultural and urban catchments, or at construction sites. The greater duration and extent of hypolimnetic anoxia would forecast an increase in recycling of nutrients from the sediments to the water column and increased blooms during spring and autumn mixis during the erosion of the thermocline in late summer and early autumn, and possibly in the metalimnion and hypolimnion of lakes that became even clearer owing to the decrease in DOC induced by climate change.

Acid precipitation and toxic substances

Climate warming interacts with acid deposition in complex ways. Alkalization of lakes, related to the increased importance of in-lake alkalinity-generation processes and decreased sulfate:cation ratios, was observed at ELA during the 20-year warming period (Schindler *et al.*, 1996a). In contrast, both episodic and long-term acidification of lakes and streams have been attributed to drought. Dry conditions expose the littoral zones of lakes and wetlands and enhance the oxidation of stored sulfur (Bayley *et al.*, 1992b; LaZerte, 1993; Yan *et al.*, 1996). Subsequent wet conditions then flush sulfate, cations and acidity into lakes and streams in pulsed acidification events. The acidity of these pulses is related to the deposition history; sites receiving higher acid loads export more acid waters (Bayley *et al.*, 1986; LaZerte, 1993). During drought, inputs of solute-rich groundwater to precipitation-dominated seepage lakes decrease (Anderson and Cheng, 1993; Wentz *et al.*, 1995), with the potential for rapid acidification (Webster *et al.*, 1990). A set of headwater, acid-sensitive lakes monitored in northern Minnesota, Wisconsin and Michigan failed to show a consistent increase in acid-neutralizing capacity (ANC) or pH as acid deposition rates declined in the region from 1983 to 1994 (Webster and Brezonik, 1995). General declines of SO_4^{2-} in the lakes were, instead, often associated with a decline in cations linked to decreased loading from deposition and decreased inputs of groundwater late in the 1980s.

Acidity, DOC and temperature changes related to climate warming could influence the transport and biological availability of a variety of toxic substances. The increased acidity of flow from certain landforms would have the potential for higher transport of metals, which become more soluble at lower pH levels. Decreases in DOC concentrations could reduce the chemical binding capacity of waters and cause the biotic effects of the toxins to increase (Connel and Miller, 1984; Moore *et al.*, 1995, 1997; Schindler *et al.*, 1996a). Increased transparency related to decreases in DOC could also increase the penetration of harmful UV-B radiation in lakes (Schindler *et al.*, 1996b; Yan *et al.*, 1996). Warming of lakes could increase the accumulation of methyl mercury in fishes; the ratio of methylation to demethylation increases with temperature and a high proportion of the variation in mercury content of fishes was associated with warmer temperatures (Bodaly *et al.*, 1993). Conversely, hydrological shifts that decrease the export of solutes from wetlands may likewise reduce the transport of mercury to lakes and streams; methyl mercury concentrations in lakes show a positive relationship with DOC and wetland area in the catchment (Mierle and Ingram, 1991; St. Louis *et al.*, 1994; Watras *et al.*, 1995). Heavy metal and pesticide accumulation increase at warmer water temperatures (Reinert *et al.*, 1974; Wood and McDonald, 1996). Zinc becomes more toxic to bluegill under conditions of reduced dissolved oxygen concentrations (Pickering, 1968), which are likely to accompany higher lake and stream temperatures. Acute exposure (24 h) of the fathead minnow to a synthetic pyrethyroid pesticide reduced its tolerance to both lethally high and low environmental temperatures (Heath *et al.*, 1994). Thus, toxic substances might be more problematic in some systems in a warmer and drier climate.

Exotic organisms

Climate warming and drying influence the movement of exotic organisms by: (1) providing warmer habitats that could favour invading rather than native organisms (Mandrak, 1989); (2) potential management actions to augment the communities of isolated lakes and stream sections by moving warmer water organisms northwards (Arnell *et al.*, 1996); and (3) reducing the connectivity of headwater lakes that can become more isolated as connecting streams or wetlands dry up (see Lodge, 1993). The poleward movement of freshwater communities in response to climate warming is expected to be just as dramatic as the poleward movement of terrestrial vegetation (Arnell *et al.*, 1996). Thus, many waters would become host to new species or 'exotics'. The northwards dispersal of organisms observed following the retreat of the Wisconsin glaciation would occur at a much faster rate. Twenty-seven fish species residing in the Great Lakes area have the potential to invade the Laurentian Great Lakes with climate warming, dramatically altering present Great Lakes fish communities (Mandrak, 1989). For isolated lakes and streams with barriers to dispersal from adjacent waters, pressures for the managed northwards dispersal would most likely be high, especially for game fishes and any threatened or endangered species. Such movements could have two positive features; increasing the productivity of a too-warm system occupied by species adapted for cooler waters, and saving a species about to be extirpated or become extinct. Alternatively, there are several disadvantages: reductions in the population viability of native species at a faster rate than that owing to climate change alone owing to species interactions with exotics, increasing rates of extirpation and increasing rates of species turnover, making species structure more chaotic and difficult to manage for human goals (see Magnuson, 1976; Regier *et al.*, 1996).

HETEROGENEITY IN RESPONSE AMONG WATER BODIES

Finally, we believe it is important to repeat what must be obvious from this article. Not only are lakes and other aquatic systems in the Laurentian Great Lakes and Precambrian Shield Region strong responders to climate change and variability, but also differences in response are paramount both across the region and among adjacent waters. Differences in morphometry or hydrology of lakes, streams and ponds can result in changes that are in opposite directions. Some responses differ within a lake or a stream, others are only apparent in comparisons among different lakes or streams; still others only are apparent in a network of lakes and streams or in analyses of large geographical regions. Changes in the edaphic, climatic, anthropogenic and morphological factors that typify the typologies of lakes all influence how the physical, chemical and biological limnology plays out in a warmer and drier climate. Hydrological considerations are especially significant in terms of: (1) solute and nutrient sources; (2) controls relative to the distribution of water flows among precipitation–groundwater–runoff; (3) influences of lake size on water residence times; and (4) the influence of catchment size relative to lake volume on biogeochemical responses. The dependence of water bodies on riparian and catchment attributes links lake and stream responses tightly to heterogeneity and changes in the surrounding landscape, whether these changes are induced vegetational transformations from climate change or altered land and water use by humans in the catchments. The altitudinal position of lakes and streams in the local landscape and hydrological flow system introduces heterogeneous responses among adjacent streams and lakes. Responses differ among seasons, not only because climate change does not occur uniformly over the year, but also because the array of sensitive ecological processes differs among seasons. Likewise, neither lake, stream, or wetland provide homogeneous entities of response to climate change. A vast array of interacting factors contributes to the responses of aquatic ecosystems to climate change, and suprises should be expected.

RESEARCH NEEDS

Five broad research needs have been identified.

1. *Long-term research* and monitoring should be maintained and expanded at key locations. Often, responses to climate are complex and unpredictable from first principles (Schindler, 1997).

2. *Models* of aquatic system behaviour should be improved and tested against long-term data and manipulative field experiments.

3. *Climate models should be improved* to include outputs of *wind and clouds* and at temporal and spatial scales more suitable for *subregional analyses*.

4. The heterogeneity of potential responses should be recognized and a *predictive understanding of this heterogeneity* should be developed.

5. A better understanding of the temperature responses of aquatic organisms and communities is needed, especially for winter.

ACKNOWLEDGEMENTS

Co-authors have contributed different sections and provided comments on the entire manuscript. Linda Mortsch and Frank Quinn are acknowledged for sections on climate, climate trends and future climate scenarios; Ray Assel for materials on Great Lakes Ice; Peter Dillon, Hayla Evans and Roland Hall for information on palaeoclimates and terrestrial ecosystems; Carl Bowser for his emphasis on the importance of hydrological processes and for helping to coordinate the effort; Everett Fee and David Schindler for information from ELA and north-western Ontario and biogeochemical and DOC responses; Everett Fee for the phytoplankton information; John Eaton for his contributions on the biological responses; Katherine Webster for contributions to biogeochemistry, general editing and figure development; John Magnuson for contributions on fish, interacting stresses, physical limnology, and general synthesis; Bill Feeney for drafting figures; and Alice Justice for assistance with production. We thank the American Society of Limnology and Oceanography and the North American Benthological Society for organizing the symposium from which our paper evolved and the US Environmental Protection Agency and the US Geological Survey for their sponsorship.

REFERENCES

Adams, M. S., Meinke, T. W., and Kratz, T. K. 1990. 'Primary productivity in three northern Wisconsin lakes', *Verh. Internat. Verein. Limnol.*, **24**, 432–437.

Allen, M. M. and Stanier, R. Y. 1968. 'Selective isolation of blue-green algae from water and soil', *J. Gen. Microbiol.*, **51**, 203–209.

Anderson, J. C. 1987. 'On the use of lake ice conditions to monitor climatic change', *Canadian Climate Centre Report No. 87-8*. AES, Downsview, p. 53.

Anderson, T. W. 1989. 'Vegetation changes over 12000 years: changes in eastern Ontario and adjacent areas gives evidence of global change', *GEOS (Energy, Mines and Resources Canada)*, **18**, 39–47.

Anderson, M. P. and Cheng, X. 1993. 'Long and short term transience in a groundwater/lake system in Wisconsin, USA', *J. Hydrol.*, **145**, 1–18.

Anderson, T. W. and Lewis, C. F. M. 1992. 'Climatic influences of deglacial drainage changes in southern Canada at 10 to 8 ka suggested by pollen evidence', *Geogr. Phys. Quat.*, **46**, 255–272.

Anderson, W., Robertson, D. M., and Magnuson, J. J. 1996. 'Evidence of recent warming and El Niño variations in ice breakup of Wisconsin lakes', *Limnol. Oceanogr.*, **41**, 815–821.

Arnell, N., Bates, B., Lang, H., Magnuson, J. J., and Mulholland, P. 1996. 'Hydrology and freshwater ecology', *in* Watson, R. T., Zinyowera, M. C., and Moss, R. H. (Eds), *Climate Change 1995: Impacts, Adaptations and Mitigation of Climate Change: Scientific-Technical Analyses*, Contribution of Working Group II to the Second Assessment Report of the Intergovernmental Panel on Climate Change. Cambridge University Press, New York. 880 pp.

Assel, R. A. 1986. 'Fall and winter thermal structure of Lake Superior', *J. Great Lakes Res.*, **12**, 251–262.

Assel, R. A. 1990. 'An ice-cover climatology for Lake Erie and Lake Superior for the winter seasons 1897–98 to 1982–83, *Int. J. Climatol.*, **10**, 731–748.

Assel, R. A. 1991. 'Implications of CO_2 global warming on Great Lakes ice cover', *Climate Change* **18**, 377–395.

Assel, R. A. and Robertson, D. M. 1995. 'Changes in winter air temperatures near Lake Michigan during 1851–1993, as determined from regional lake-ice records', *Limnol. Oceanogr.* **40**, 165–176.

Assel, R. A., Quinn, F. H., Leshkevich, G. A., and Bolslenga, S. J. 1983. 'Great Lakes ice atlas', *NOAA Atlas No. 4*. Great Lakes Environmental Research Lab., Ann Arbor, MI.

Assel, R. A., Snider, C. R., and Lawrence, R. 1985. 'Comparison of 1983 Great Lakes winter weather and ice conditions with previous years', *Monthly Weather Rev.*, **113**, 291–303.

Assel, R. A., Robertson, D. M., Hoff, M. H. and Selgeby, J. H. 1995. 'Climatic-change implications from long-term (1823–1994) ice records for the Laurentian Great Lakes', *Ann. Glaciol.*, **21**, 383–386.

Assel, R. A., Janowiak, J. E., Young, S. and Boyce, D. 1996. 'Winter 1994 weather and ice conditions for the Laurentian Great Lakes', *Bull. Am. Meteorol. Soc.*, **77**, 71–88.

Atmospheric Environment Service, 1982a. *Canadian Climate Normals 1951–80, Temperature and Precipitation — Quebec*. Supply and Services Canada, Ottawa.

Atmospheric Environment Service, 1982b. *Canadian Climate Normals 1951–80, Temperature and Precipitation — Atlantic Provinces*. Supply and Services Canada, Ottawa.

Atmospheric Environment Service, 1982c. *Canadian Climate Normals 1951–80, Temperature and Precipitation — Prairie Provinces*. Supply and Services Canada, Ottawa.

Atmospheric Environment Service, 1982d. *Canadian Climate Normals 1951–80, Temperature and Precipitation — Ontario*. Supply and Services Canada, Ottawa.

Atmospheric Environment Service, 1982e. *Canadian Climate Normals 1951–80, Degree Days*, Volume 4. Supply and Services Canada, Ottawa.

Atmospheric Environment Service, 1982f. *Canadian Climate Normals 1951–80, Frost*, Volume 6. Supply and Services Canada, Ottawa.

Atmospheric Environment Service, 1994. 'Modelling the global climate system', *Climate Change Digest, CCD 94-01*. Supply and Services Canada, Ottawa. p. 20.

Bartlein, P. J. and Webb, III, T. 1985. 'Mean July temperature at 6000 yr BP in eastern North America: regression equations for estimates from fossil-pollen data', *Syllogeus*, **55**, 301–342.

Bayley, S. E., Behr, R. S., and Kelly, C. A. 1986. 'Retention and release of sulfur from a freshwater wetland', *Water Air Soil Pollut.*, **31**, 101–114.

Bayley, S. E., Schindler, D. W., Parker, B. R., Stainton, M. P., and Beaty, K. G. 1992a. 'Effects of forest fire and drought on acidity of a base-poor boreal forest stream: similarities between climatic warming and acidic precipitation', *Biogeochemistry*, **17**, 191–204.

Bayley, S. E., Schindler, D. W., Beaty, K. G., Parker, B. R., and Stainton, M. P. 1992b. 'Effects of multiple fires on nutrient yields from streams draining boreal forest and fen watersheds: nitrogen and phosphorus', *Can. J. Fish. Aquat. Sci.*, **49**, 584–596.

Bazzaz, F. A. 1990. 'The response of natural ecosystems to the rising global CO_2 levels', *Annu. Rev. Ecol. Syst.*, **21**, 167–196.

Blumberg, A. F. and DiToro, D. M. 1990. 'Effects of climate warming on dissolved oxygen concentrations in Lake Erie', *Trans. Am. Fish. Soc.*, **119**, 210–223.

Bodaly, R. A., Rudd, J. W. M., Fudge, R. J. P., and Kelly, C. A. 1993. 'Mercury concentrations in fish related to size of remote Canadian Shield Lakes', *Can. J. Fish. Aquat. Sci.*, **50**, 980–987.

Boer, G. J., McFarlane, N. A., and Lazare, M. 1992. 'Greenhouse gas-induced climate change simulated with the CCC second-generation general circulation model', *J. Climate*, October 1992, 1045–1077.

Bolsenga, S. J. and Norton, D. C. 1993. 'Great Lakes air temperature trends for land stations, 1901–1987', *J. Great Lakes Res.*, **19**, 379–388.

Brown, R. W. and Taylor, W. W. 1993. 'Factors affecting the recruitment of Lake Whitefish in two areas of northern Lake Michigan', *J. Great Lakes Res.*, **19**, 418–428.

Bunce, J. A. 1994. 'Responses of respiration to increasing atmospheric carbon dioxide concentrations', *Physiol. Plantarum*, **90**, 427–430.

Calkin, P. E. and Feenstra, B. H. 1985. 'Evolution of the Erie Basin Great Lakes', *in* Karrow, P. F. and Calkin, P. E. Eds, *Quaternary Evolution of the Great Lakes. Geol. Assoc. Can., Spec. Pap.*, **30**, 149–170.

Campbell, I. D. and McAndrews, J. H. 1991. 'Cluster analysis of late Holocene pollen trends in Ontario', *Can. J. Bot.*, **69**, 1719–1730.

Campbell, I. D. and McAndrews, J. H., 1993. 'Forest disequilibrium caused by Little Ice Age cooling', *Nature*, **366**, 336–338.

Canale, R. P. and Vogel, A. H. 1974. 'Effects of temperature on phytoplankton growth', *J. Env. Eng. Div., Am. Soc. Civil Eng.*, **100**, 231–241.

Carpenter, S. R., Fisher, S. G., Grimm, N. B. and Kitchell, J. F. 1992. 'Global change and freshwater ecosystems', *Annu. Rev. Ecol. Syst.*, **23**, 119–139.

Changnon, S. A., Jr., Leffler, S. and Shealy, R. 1989. 'Impacts of extremes in Lake Michigan levels along Illinois shoreline: low levels, Report prepared for U.S. Environment Protection Agency, Washington, D.C.' *in* Smith, J. B. and Tirpak, D. A. (Eds). *The Potential Effects of Global Climate Change on the United States*. Hemisphere Publishing Corp., New York. pp. 3–1 to 3–48.

Chen, C. Y. and Folt, C. L. 1996. 'Consequences of fall warming for zooplankton overwintering success', *Limnol. Oceanogr.*, **41**, 1077–1086.

Christie, G. C. and Regier, H. A. 1988. 'Measures of optimal thermal habitat and their relationship to yields of four commercial fish species', *Can. J. Fish. Aquat. Sci.*, **45**, 301–314.

Connell, D. W. and Miller, G. J. 1984. *Chemistry and Ecotoxicology of Pollution*. John Wiley and Sons, Inc., New York. 444 pp.

Coutant, C. C. 1981. 'Foreseeable effects of CO_2-induced climate change: freshwater concerns', *Environ. Conserv.*, **8**, 285–297.

Crissman, R. D. 1989. 'Impacts on electricity generation in New York State', *in Report of the First U.S.–Canada Symposium on Impacts of Climate Change on the Great Lake Basin* US National Climate Program Office and Canadian Climate Centre. pp. 109–118.

Croley, T. E., II. 1990. 'Laurentian Great Lakes double-CO_2 climate change hydrological impacts', *Climat. Change*, **17**, 27–47.

Croley, T. E., II. 1993. 'CCC GCM $2 \times CO_2$ hydrological impacts on the Great Lakes, in *Climate, Climate Change, Water Level Forecasting and Frequency Analysis*, Supporting documents, Vol. 1, Water Supply Scenarios. Task Group 2, Working Committee 3, IJC, Levels Reference Study, Phase II.

Croley, T. E., II and Hartmann, H. C. 1985. 'Resolving Thiesson polygons', *J. Hydrol.*, **76**, 363–379.

Croley, T. E., II and Hunter, T. S. 1994. 'Great Lakes monthly hydrologic data', *NOAA Technical Memorandum ERL GLERL-83*, p. 83.

Croley, T. E., II, Quinn, F. H., Kunkel, K., and Changnon, S. J. 1994. 'Potential Great Lakes hydrology and lake level impacts resulting from global warming', *GLERL Report No. 931*. Great Lakes Environmental Research Laboratory, Ann Arbor, Michigan, p. 6.

Croley, T. E., II, Quinn, F. H., Kunkel, K. and Changnon, S. J. 1995. 'Potential Great Lakes hydrology and lake level impacts resulting from global warming', in *Preprint Vol: Sixth Conference on Global Change Studies*. American Meteorological Society, Boston, Massachusetts. pp. 67–72.

Cronan, C. S. 1990. 'Patterns of organic acid transport from forested watersheds to aquatic ecosystems', in Perdue, E. M. and Gjessing, E. T. (Eds), *Organic Acids in Aquatic Ecosystems*. John Wiley & Sons Ltd., New York. pp. 245–260.

Davison, I. R. 1991. 'Environmental effects on algal photosynthesis: temperature', *J. Phycol.*, **27**, 2–8.

DeNoyelles, F. Jr., Reinke, D., Knoechel, R., Treanor, D., and Altenhofen, C. 1980. 'Continuous culturing of natural phytoplankton communities in the Experimental Lakes Area: effects of enclosure, in situ incubation, light, phosphorus and cadmium', *Can. J. Fish. Aquat. Sci.*, **37**, 424–433.

DeStasio, B. T., Hill, D. K., Kleinhans, J. M., Nibbelink, N. P., and Magnuson, J. J. 1996. 'Potential effects of global climate change on small north temperate lakes: physics, fishes and plankton', *Limnol. Oceanogr.*, **41**, 1136–1149.

Devito, K. J. and Dillon, P. J. 1993a. 'The importance of runoff and winter anoxia to the P and N dynamics of a beaver pond', *Can. J. Fish. Aquat. Sci.*, **50**, 2222–2234.

Devito, K. J. and Dillon, P. J. 1993b. 'The influence of hydrologic conditions and peat oxia on the phosphorus and nitrogen dynamics of a conifer swamp', *Wat. Resour. Res.*, **29**, 2675–2685.

Dillon, P. J. and LaZerte, B. D. 1992. 'Response of the Plastic Lake catchment, Ontario to reduced sulfur deposition', *Environ. Pollut.*, **77**, 211–217.

Drake, B. G. 1992. 'The impact of rising CO_2 on ecosystem production', *Water Air Soil Pollut.*, **64**, 25–44.

Dredge, L. A., Mott, R. J., and Grant, D. R. 1992. 'Quaternary stratigraphy, paleoecology, and glacial geology, Iles de la Madeleine, Quebec', *Can. J. Earth Sci.*, **29**, 1981–1996.

Eaton, J. G. and Scheller, R. M. 1996. 'Effects of climate warming on fish thermal habitat in streams of the United States', *Limnol. Oceanogr.*, **41**, 1109–1115.

Edwards, T. W. D. and Fritz, P. 1988. 'Stable-isotope paleoclimate records for southern Ontario, Canada: comparison of results from marl and wood', *Can. J. Earth Sci.*, **25**, 1397–1406.

Engstrom, D. R. and Hansen, B. C. S. 1985. 'Postglacial vegetation change and soil development in southeastern Labrador as inferred from pollen and chemical stratigraphy', *Can. J. Bot.*, **63**, 543-561.

Eppley, R. W. 1972. 'Temperature and phytoplankton growth in the sea', *Fish. Bull.*, **70**, 1063–1085.

Everett, J. T., Troadec, J.-P., Regier, H. A., Krovnin, A., Okemwa, E., and Lluch Belda, D. 1996. 'Fisheries', in *The Intergovernmental Panel on Climate Change Working Group 2, Report on Impact Assessment and Adaptation Strategies*. Geneva, Switzerland. Chapter 8.

Fee, E. J. 1976. 'The vertical and seasonal distribution of chlorophyll in lakes of the Experimental Lakes Area, northwestern Ontario: implications for primary production estimates', *Limnol. Oceanogr.*, **21**, 767–783.

Fee, E. J., Hecky, R. E., and Welch, H. E. 1987. 'Phytoplankton photosynthesis parameters in central Canadian lakes', *J. Plankton Res.*, **9**, 305–316.

Fee, E. J., Shearer, J. A., DeBruyn, E. R., and Schindler, E. U. 1992. 'Effects of lake size on phytoplankton photosynthesis', *Can. J. Fish. Aquat. Sci.*, **49**, 2445–2459.

Fee, E. J., Hecky, R. E., Regehr, G. W., Hendzel, L. L., and Wilkinson, P. 1994. 'Effects of lake-size on nutrient availability in the mixed-layer during summer stratification', *Can. J. Fish. Aquat. Sci.*, **51**, 2756–2768.

Fee, E. J., Hecky, R. E., Kasian, S. E., and Cruikshank, 1996. 'Effects of lake size, water clarity, and climatic variability on mixing depths in Canadian Shield Lakes', *Limnol. Oceanogr.*, **41**, 912–920.

Freeberg, M. H., Taylor, W. W., and Brown, R. W. 1990. 'Effect of egg and larval survival on the year-class strength of lake whitefish I Grand Traverse Bay, Lake Michigan', *Trans. Am. Fish. Soc.*, **119**, 92–100.

Fry, F. E. J. 1947. 'Effects of the environment on animal activity', *Publ. Ontario Fish. Res. Lab.*, **68**, 1–62.

Fry, F. E. J. 1971. 'The effect of environmental factors on the physiology of fish', in Hoar, W. S. and Randall, D. J. (Eds), *Fish Physiology: Environmental Factors*. Academic Press, New York. pp. 1–98.

Giberson, D. J., Rosenberg, D. M., and Wiens, A. P. 1991. 'Changes in abundance of burrowing mayflies in Southern Indian Lake: lessons for environmental monitoring', *Ambio*, **20**, 139–142.

Grimm, E. C. and Jacobson, G. L., Jr. 1992. 'Fossil-pollen evidence for abrupt climate changes during the past 18 000 years in eastern North America', *Climate Dynam.*, **6**, 179–184.

Guildford, S. J., Hendzel, L. L., Kling, H. J., Fee, E. J., Robinson, G. G. C., Hecky, R. E., and Kasian, S. E. M. 1994. 'Effects of lake size on phytoplankton nutrient deficiency status', *Can. J. Fish. Aquat. Sci.*, **51**, 2769–2783.

Gullett, D., Skinner, W., and Vincent, L. 1992. 'Development of an historical Canadian climate database for temperature and other climate elements', *Climatol. Bull.*, **26**, 125–131.

Hairston, N. G. 1996. 'Zooplankton egg banks as biotic reservoirs in changing environments', *Limnol. Oceanogr.*, **41**, 1087–1092.

Hall, R. I. 1993. 'Paleolimnological analysis of lake-watershed interactions and long term lake trophic status;, *PhD Thesis*, Biology Dept., Queen's University, Kingston, Ontario, Canada. p. 323.

Hansen, J., Fung, I., Lacis, A., Rind, D., Ruddel, G., Lebedeff, S., Ruedy, R., and Stone, P. 1988. 'Global climate changes as forecast by the Goddard Institute for Space Studies three-dimensional model', *J. Geophys. Res.*, **93**, 9341–9364.

Hartmann, H. C. 1990. 'Climate change impacts on Laurentian Great Lakes levels', *Climate Change*, **17**, 49–67.

Headrick, M. R. and Carline, R. F. 1993. 'Restricted summer habitat and growth of northern pike in two southern Ohio impoundments', *Trans. Am. Fish. Soc.*, **122**, 228–236.

Healey, F. P. and Hendzel, L. L. 1980. 'Physiological indicators of nutrient deficiency in lake phytoplankton', *Can. J. Fish. Aquat. Sci.*, **37**, 442–453.

Heath, S., Bennett, W. A., Kennedy, J., and Beitinger, T. L. 1994. 'Heat and cold tolerance of the fathead minnow, *Pimephales promelas*, exposed to the synthetic pyrethroid cyfluthrin', *Can. J. Fish. Aquat. Sci.*, **51**, 437–440.

Hecky, R. E. and Hesslein, R. H. 1996. 'Contributions of benthic algae to lake food webs as revealed by stable isotope analysis', *J. North Am. Benthol. Soc.*, **14**, 582–598.

Hecky, R. E. and Kilham, P. 1988. 'Nutrient limitation of phytoplankton in freshwater and marine environments: a review of recent evidence on the effects of enrichment', *Limnol. Oceanogr.*, **33**, 796–822.

Hemond, H. F. 1990. 'Wetlands as the source of dissolved organic carbon to surface waters', in Perdue, E. M. and Gjessing, E. T. (Eds), *Organic Acids in Aquatic Ecosystems*. John Wiley & Sons, New York. pp. 301–313.

Hill, D. K. and Magnuson, J. J. 1990. 'Potential effects of global climate warming on the growth and prey consumption of Great Lakes fish', *Trans. Am. Fish. Soc.*, **119**, 265–275.

Hokanson, K. E. F. 1977. 'Temperature requirements of some percids and adaptations to the seasonal temperature cycle', *J. Fish. Res. Bd. Can.*, **34**, 1524–1550.

Holmes, J. A. 1990. 'Sea lamprey as an early responder to climate change in the Great lakes Basin', *Trans. Am. Fish. Soc.*, **119**, 292–300.

Hondzo, M. and Stefan, H. G. 1991. 'Three case studies of lake temperature and stratification response to warmer climate', *Wat. Resour. Res.*, **27**, 1837–1846.

Hondzo, M. and Stefan, H. G. 1993. 'Regional water temperature characteristics of lakes subjected to climate change', *Climatic Change*, **24**, 187–211.

Houghton, J. J., Meiro Filho, L. G., Callander, B. A., Harris, N., Kattenberg, A., and Maskell, K. (Eds) 1996. *Climate Change 1995: The Science of Climate Change*, Contribution of Working Group I to the Second Assessment Report of the Intergovernmental Panel on Climate Change. Cambridge University Press, New York. 584 pp.

Hynes, H. B. N. 1975. 'The stream and its valley', *Int. Verein. Theor. Ang. Limnol., Verhand.*, **19**, 1–15.
Johnson, T. B. and Evans, D. O. 1990. 'Size dependent winter mortality of young-of-the-year white perch: climate warming and the invasion of the Laurentian Great Lakes', *Trans. Am. Fish. Soc.*, **119**, 779–785.
Jones, B. R., Hokanson, K. E. F., and McCormick, J. H. 1972. 'Winter temperature requirements for maturation and spawning of yellow perch, *Perca flavenscens* (Mitchell)', in Marois, M. (Ed.), *Proceedings, World Conference Towards a Plan of Action for Mankind*, Vol. 3, Biological Balance and Thermal Modifications. Pergamon Press, New York. pp. 189–192.
Kasian, S. E. M., Findlay, D. L., Regehr, G., Beaty, K. G., Cruikshank, D. R., Shearer, J. A., in press. Natural variability of phytoplankton composition over decadal time scales in the Experimental Lakes Area. *Can. J. Fish. Aguat. Sci.*
Kasischke, E. S., Christensen, N. L., Jr., and Stocks, B. J. 1995. 'Fire, global warming, and the carbon balance of boreal forests', *Ecol. Appl.*, **5**, 437–451.
Keith, V. F., DeAvila, C., and Willis, R. M. 1989. 'Effect of climatic change on shipping within Lake Superior and Lake Erie', in *The Potential Effects of Global Climate Change on the United States.* US Environmental Protection Agency, Washington, D.C. App. H, Infrastructure.
Krabbenhoft, D. P. and Webster, K. E. 1995. 'Transient hydrogeological controls on the chemistry of a seepage lake', *Wat. Resour. Res.*, **31**, 2295–2305.
Kratz, T. K., Webster, K. E., Bowser, C. J., Magnuson, J. J., and Benson, B. J. 1997. The influence of landscape position on northern Wisconsin lakes. *Freshwat. Biol.*, **37**, 209–217.
Kunkel, K. E., Mortsch, L. D., and Lewis, P. 1993. 'The climate of the Great Lakes — St. Lawrence Basin', in Lee, D. H. (Ed.), *Climate, Climate Change, Water Level Forecasting and Frequency Analysis*, Final Report of Task Group 2 Working Committee 3 for the IJC Levels Reference Study, Phase II, Vol. 1. pp. 5–14.
Lazerte, B. D. 1993. 'The impact of drought and acidification on the chemical exports from a minerotrophic conifer swamp', *Biogeochemistry*, **18**, 153–175.
Leino, R. L. and McCormick, J. H. 1993. 'Responses of juvenile largemouth bass to different pH and aluminum levels at overwintering temperatures: effects on gill morphology, electrolyte balance, scale calcium, liver glycogen, and depot fat', *Can. J. Zool.*, **71**, 531–543.
Lenihan, J. M. and Neilson, R. P. 1994. 'A rule-based vegetation formation model for Canada', *J. Biogeogr.*, **20**, 615–628.
Lewis, C. F. M. and Anderson, T. W. 1989. 'Oscillations of levels and cool phases of the Laurentian Great Lakes caused by inflows from glacial lakes Agassiz and Barlow-Ojibway', *J. Paleolimnol.*, **2**, 99–146.
Lewis, C. F. M. and Anderson, T. W. 1992. 'Stable isotope (O and C) and pollen trends in eastern Lake Erie, evidence for a locally induced climate reversal of Younger Dryas age in the Great Lakes basin', *Climate Dynam.*, **6**, 241–250.
Lin, P. 1995. 'Adaptations to temperature in fish: salmonids, centrarchids and percids', *PhD Thesis*, University of Toronto, Canada. p. 170.
Liu, K. B. 1990. 'Holocene paleoecology of the boreal forest and Great Lakes — St. Lawrence forest in Northern Ontario', *Ecol. Monogr.*, **60**, 179–212.
Lodge, D. M. 1993. 'Species invasions and deletions: community effects and responses to climate and habitat change', in Kareiva, P. M., Kingsolver, J. G., and Huey, R. B. (Eds), *Biotic Interactions and Global Change*, Sinauer Associates Inc., Sunderland, MA, pp. 367–387.
Magnuson, J. J. 1976. 'Managing with exotics — a game of chance'. *Trans. Am. Fish. Soc.*, **105**, 1–9.
Magnuson, J. J. and DeStasio, B. T. 1996. Thermal niche of fishes and global warming', in Wood, C. M. and McDonald, D. G. (Eds), *Global Warming — Implications for Freshwater and Marine Fish*, SEB Seminar Series. Cambridge University Press, Cambridge. pp. 377–408.
Magnuson, J. J., Crowder, L. B., and Medvick, P. A. 1979. 'Temperature as an ecological resource', *Am. Zool.*, **19**, 331–343.
Magnuson, J. J., Regier, D. K., Holmes, H. A., Meisner, J. D., and Shuter, B. J. 1989. 'Potential responses of Great Lakes fishes and their habitat to global climate warming', in Smith, J. B. and Tirpak, D. A. (Eds), *The Potential Effects of Global Climate Change on the United States.* Environmental Protection Agency, EPA-230-05-89-055, Washington, D.C. App. E, Aquatic Resources.
Magnuson, J. J., Meisner, J. D., and Hill, D. K. 1990. 'Potential changes in the thermal habitat of Great Lakes fish after global climate warming', *Trans. Am. Fish. Soc.*, **119**, 254–264.
Manabe, S. and Stouffer, R. J. 1980. 'Sensitivity of a global climate model to an increase of CO_2 concentration in the atmosphere', *J. Geophys. Res.*, **85**, 5529–5554.
Mandrak, N. E. 1989. 'Potential invasion of the Great Lakes by fish species associated with climate warming', *J. Great Lakes Res.*, **15**, 306–316.
Maykut, G. A. and Untersteiner, N. 1971. 'Some results from a time dependent thermodynamic model of sea ice', *J. Geophys. Res.*, **76**, 1550–1575.
McAndrews, J. H. and Campbell, I. D. 1993. 'The mean July temperature in Eastern Canada from Bartlein and Webb's (1985) pollen transfer functions: comments and illustrations', in Tekla, A. (Ed.), *Proxy Climate Data and Models of the Six Thousand Years Before Present Time Interval: The Canadian Perspective*, Canadian Global Climate Change Program Incidental Report Series No. IR93-3. The Royal Society of Canada. pp. 22–25.
McCormick, M. J. 1990. 'Potential changes in thermal structure and cycle of Lake Michigan due to global warming', *Trans. Am. Fish. Soc.*, **119**, 183–195.
McFadden, J. D. 1965. 'The interrelationship of lake ice and climate in central Canada', *Technical Report Number 20*. Department of Meteorology, University of Wisconsin–Madison. 115 pp.
McFarlane, N. A. 1991. 'The Canadian Climate Centre second-generation global circulation model', in Sanderson, M. (Ed.), *Proceedings, Symposium on Climate Scenarios.* University of Waterloo, Waterloo. pp. 3–15.
McFarlane, N. A., Boer, G. J., Blanchet, J.-P., and Lazare, M. 1992. 'The Canadian Climate Centre second generation general circulation model and its equilibrium climate', *J. Climate*, **5**, 1013.
McIver, D. L. and Isaac, J. L. 1989. *Bioclimatic profiles for Canada, 1951–80.* Canadian Climate Centre, Environment Canada.

McLain, A. S., Magnuson, J. J., and Hill, D. K. 1994. 'Latitudinal and longitudinal differences in thermal habitat for fishes influenced by climate warming: expectations from simulations', *Verh. Int. Verein. Limnol.*, **25**, 2080–2085.

Meisner, J. D. 1990. 'Potential loss of thermal habitat for Brook Trout, due to climatic warming, in two southern Ontario streams', *Trans. Am. Fish. Soc.*, **119**, 282–291.

Meisner, J. D., Goodier, J. L., Regier, H. A., Shuter, B. J., and Cristie, W. J. 1987. 'Lab assessment of climate warming on Great Lakes basin fishes', *J. Great Lakes Res.*, **13**, 340–352.

Meisner, J. D., Rosenfeld, J. S., and Regier, H. A. 1988. 'The role of groundwater in the impact of climate warming on stream salmonines', *Fisheries*, **13**, 2–8.

Mekis, E. and Hogg, W. D. 1997. 'Rehabilitation and analysis of Canadian daily precipitation time series', *J. Climate*, in press.

Melo, O. T. 1989. 'Electric supply and demand in Ontario', in *Report of the First U.S.-Canada Symposium on Impacts of Climate Change on the Great Lakes Basin*, Sept. 27–29, 1988. Oak Brook, Illinois. pp. 134–143.

Metcalfe, J. R., Ishida, S., and Goodison, B. E. 1994. 'A corrected precipitation archive for the Northwest Territories', *Environment Canada–Mackenzie Basin River Study, Interim Report #2*. 110–117.

Meyer, J. L. and Pulliam, W. M. 1992. 'Modification of terrestrial-aquatic interactions by a changing climate', in Firth, P. and Fisher, S. G. (Eds), *Global Climate Change and Freshwater Ecosystems*. Springer-Verlag, New York. pp. 177–191.

Mierle, B. and Ingram, R. 1991. 'The role of humic substances in the mobilization of mercury from watersheds', *Water Air Soil Pollut.*, **56**, 349–357.

Minns, C. K. and Moore, J. E. 1995. 'Factors limiting the distributions of Ontario's freshwater fishes: the role of climate and other variables, and the potential impacts of climate change', in Beamish, R. J. (Ed.), *Climate Change and Northern Fish Populations. Can. Spec. Publ. Fish Aquat. Sci.*, **121**, 137–160.

Minns, C. K., Randall, R. G., Chadwick, E. M. P., Moore, J. E., and Green, R. 1995. 'Potential impact of climate change on the habitat and population dynamics of juvenile Atlantic salmon (*Salmon salar*) in eastern Canada', in Beamish, R. J. (Ed.), *Climate Change and Northern Fish Populations. Can. Spec. Publ. Fish Aguat. Sci.*, **121**, 699–708.

Moore, M. V., Folt, C. L., and Stemberger, R. S. 1995. 'Consequences of elevated temperatures for zooplankton assemblages in temperate lakes', *Arch. Hydrobiol.*, **268**, 1–31.

Moore, M. V., Pace, M. L., Mather, J. R., Murdoch, P. S., Howarth, R. W., Folt, C. L., Chen, C. Y., Hemond, H. F., Flebbe, P. A., and Driscoll, C. T. 1997. Potential effects of climate change on freshwater ecosystems of the New England/Mid Atlantic region', *Hydrol. Process.*, **11**, 925–947.

Mortsch, L. D. (Ed.) (1990) *Eastern Canadian Boreal and Sub-Arctic Wetlands: A Resource Document*. Atmospheric Environment Service, Environment Canada, Climatological Studies No. 42. 169 pp.

Mortsch, L. D. and Quinn, F. H. 1996. 'Climate change scenarios for the Great Lakes Basin ecosystem studies', *Limnol. Oceanogr.*, **41**, 903–911.

Mott, R. J., Grant, D. R., Stea, R. R., and Occhietti, S. 1986. 'A late glacial climatic oscillation in Atlantic Canada — an Allerod/Younger Dryas equivalent', *Nature*, **323**, 247–250.

Overpeck, J. T., Rind, D., and Goldberg, R. 1990. 'Climate-induced changes in forest disturbance and vegetation', *Nature*, **343**, 51–53.

Parkinson, C. L. and Washington, W. M. 1979. 'A large-scale numerical model of sea ice', *J. Geophys. Res.* **84**, 311–337.

Pereira, D. L., Anderson, C. S., Radomski, P. J., Sak, T., and Stevens, A. 1995a. 'Potential to index climate with growth and recruitment of temperate fish', in Beamish, R. J. (Ed.), *Climate Change and Northern Fish Populations. Can. Spec. Publ. Fish. Aquat. Sci.*, **121**, 709–717.

Pereira, D. L., Bingham, C., Spangler, G. R., Cohen, Y., Conner, D. J., and Cunningham, P. K. 1995b. 'Growth and recruitment of freshwater drum (*Aplodinotus grunniens*) as related to long-term temperature patterns', in Beamish, R. J. (Ed.), *Climate Change and Northern Fish Populations. Can. Spec. Publ. Fish. Aquat. Sci.*, **121**, 617–629.

Peters, R. L. 1990. 'Effects of global warming on forests', *Forest Ecol. Manage.*, **35**, 13–33.

Phillips, D. W. 1990. *The Climates of Canada*. Supply and Services Canada. Environment Canada 176 pp.

Phillips, D. W. and McCulloch, J. A. W. 1972. *The Climate of the Great Lakes Basin* Climatological Studies No. 20. Atmospheric Environment Service, Downsview. p. 40.

Pickering, Q. H. 1968. 'Some effects of dissolved oxygen concentrations upon the toxicity of zinc to the bluegill, *Lepomis macrochirus* RAF', *Water Res.*, **2**, 187–194.

Poff, J. L. 1992. 'Regional hydrologic response to climate change: an ecological perspective', in Firth, P. and Fisher, S. G. (Eds), *Global Climate Change and Freshwater Ecosystems*. Springer-Verlag, New York. pp. 88–115.

Post, W. M., Pastor, J., King, A. W., and Emanuel, W. R. 1992. 'Aspects of the interaction between vegetation and soil under global change', *Water Air Soil Pollut.*, **64**, 345–363.

Quinn, F. H. 1981. 'Secular changes in annual and seasonal Great Lakes precipitation, 1854–1979, and their implications for Great Lakes water resource studies', *Wat. Resour. Res.*, **17**, 1619–1624.

Quinn, F. H. and Norton, D. C. 1982. 'Great Lakes precipitation by months, 1900–1980', *NOAA Data Report ERL GLERL-20*. p. 29.

Rastetter, E. B., McKane, R. B., Shaver, G. R., and Melillo, J. M. 1992. 'Changes in C storage by terrestrial ecosystems: how C–N interactions restrict responses to CO_2 and temperature', *Water Air Soil Pollut.*, **64**, 327–344.

Regier, H. A., Holmes, J. A., and Pauly, D. 1990a. 'Influence of temperature changes on aquatic ecosystems: an interpretation of empirical data', *Trans. Am. Fish. Soc.*, **119**, 374–389.

Regier, H. A., Magnuson, J. J., and Coutant, C. C. (Convenors), 1990b. 'Symposium on effects of climate change on fish, Proceedings', *Trans. Am. Fish. Soc.*, **119**, 173–389.

Regier, H. A., Lin, P., Ing, K. K., and Wichert, G. A. 1996. 'Likely responses to climate change of fish associations in the Laurentian Great Lakes: concepts, methods and findings', *Boreal Env. Res.*, **1**, 1–15.

Reinert, R. E., Stone, L. J., and Willford, W. A., 1974. 'Effect of temperature on accumulation of methylmecuric chloride and p,p′ DDT by rainbow trout (Salmo gairdneri)', *J. Fish. Res. Bd Can.*, **31**, 1649–1652.

Reycraft, J. and Skinner, W., 1993. 'Canadian lake ice conditions: an indicator of climate variability', *Climatic Persp.*, **15**, 9–15.

Rizzo, B. W. and Wilken, E. 1992. 'Assessing the sensitivity of Canada's ecosystems to climatic change', *Climatic Change*, **21**, 37–55.

Robertson, D. M., Ragotzkie, R. A., and Magnuson, J. J. 1992. 'Lake ice records used to detect historical and future climate changes', *Climatic Change*, **21**, 407–427.

Rosenberg, N. J., Kimball, B. A., Martin, P., and Cooper, C. F. 1990. 'From Climate and CO_2 enrichment to evapotranspiration', in Waggoner, P. E. (Ed.), *Climate Change and U.S. Water Resources*. Wiley, New York. pp. 151–175.

Rowe, J. S. 1977. *Forest Regions of Canada*. Department of Fisheries and the Environment, Canadian Forestry Service, Publication 1300. p. 172.

St Louis, V. L., Rudd, J. W. M., Kelley, C. A., Beaty, K. G., Bloom, N. S., and Flett, R. J. 1994. 'Importance of wetlands as sources of methyl mercury to boreal forest ecosystems', *Can. J. Fish. Aquat. Sci.*, **51**, 1065–1076.

Schindler, D. W. 1977. 'Evolution of phosphorus limitation in lakes: natural mechanisms compensate for deficiencies of nitrogen and carbon in eutrophied lakes', *Science*, **195**, 260–262.

Schindler, D. W. 1997. 'Widespread effects of climatic warming on freshwater ecosystems', *Hydrol. Process.*, **11**, 1043–1067.

Schindler, D. W. and Fee, E. J. 1973. 'Diurnal variation of dissolved inorganic carbon and its use in estimating primary production and CO_2 invasion in Lake 227', *J. Fish. Res. Bd Can.*, **30**, 1501–1510.

Schindler, D. W., Frost, V. E., and Schmidt, R. V. 1973. 'Production of epilithiphyton in two lakes of the Experimental Lakes Area, northwestern Ontario.' *J. Fish. Res. Bd Can.*, **30**, 1511–1524.

Schindler, D. W., Beaty, K. G., Fee, E. J., Cruikshank, D. R., DeBruyn, E. R., Findlay, D. L., Linsey, G. A., Shearer, J. A., Stainton, M. P., and Turner, M. A. 1990. 'Effects of climatic warming on lakes of the central boreal forest', *Science*, **250**, 967–970.

Schindler, D. W., Bayley, S. E., Parker, B. R., Beaty, K. G., Cruikshank, D. R., Fee, E. J., Schindler, E. U., and Stainton, M. P. 1996a. 'The effects of climatic warming on the properties of boreal lakes and streams at the Experimental Lakes Area, northwestern Ontario', *Limnol. Oceanogr.*, **41**, 1004–1017.

Schindler, D. W., Curtis, P. J., Parker, B. R., and Stainton, M. P. 1996b. 'Consequences of climate warming and lake acidification for UV-B penetration in North American boreal lakes', *Nature*, **379**, 705–708.

Schlesinger, M. and Zhao, Z. 1988. 'Seasonal climate changes induced by doubled CO_2 as simulated by the OSU atmospheric GCM/mixed-layer ocean model', *Climate Research Institute Report*. Oregon State University, Corvallis.

Scully, N. M. and Lean, D. R. S. 1994. 'The attenuation of ultraviolet radiation in temperate lakes', *Arch. Hydrobiol. Beih. Ergebn. Limnol.*, **43**, 135–144.

Sedell, J. R. and Dahm, C. N. 1990. 'Spatial and temporal scales of dissolved organic carbon in streams and rivers', in Perdue, E. M. and Gjessing, E. T. (Eds), *Organic Acids in Aquatic Ecosystems*. John Wiley & Sons, New York. pp. 261–279.

Shearer, J. A., Fee, E. J., DeBruyn, E. R., and DeClercq, D. R. 1987. 'Phytoplankton primary production and light attenuation response to the experimental acidification of a small Canadian Shield lake', *Can. J. Fish. Aquat. Sci.*, **44** (Suppl. 1), 83–90.

Shuter, B. J. and Post, J. R. 1990. 'Climate, population viability, and the zoogeography of temperature fishes', *Trans. Am. Fish. Soc*, **119** 314–336.

Siegel, D. I., Reeve, A. S., Glaser, P. H., and Romanowicz, E. A. 1995. 'Climate-driven flushing of pore water in peatlands', *Nature*, **374**, 531–533.

Sinokrot, B. A. and Stefan, H. G. 1993. 'Stream temperature dynamics: measurements and modeling', *Wat. Resour. Res.*, **29**, 2299–2312.

Sinokrot, B. A., Stefan, H. G., McCormick, J. H., and Eaton, J. G. 1995. 'Modeling of climate change effects on stream temperatures and fish habitats below dams and near groundwater inputs', *Climatic Change*, **30**, 181–200.

Skinner, W. R. and Gullett, D. W. 1993. 'Trends of daily maximum and minimum temperature in Canada during the past century', *Climatol. Bull.*, **27**, 63–77.

Slivitsky, M. and Mathier, L. 1993. 'Climate changes during the 20th century on the Laurentian Great Lakes and their impacts on hydrologic regime', in Duckstein, L. and Parent, E. (Eds), NATO Advanced Studies Institute Series, Series E-Vol. 275, Kluwer, pp. 235–251.

Smayda, T. J. 1980. 'Phytoplankton species succession', in Morris, I. (Ed.), *The Physiological Ecology of Phytoplankton*. University of California Press, Berkeley. pp. 493–570.

Smith, J. B. 1991. 'The potential impacts of climate change on the Great Lakes', *Bull. Am. Meteorol. Soc.*, **72**, 21–28.

Smith, V. H. 1991. 'Competition between consumers', *Limnol. Oceanogr.*, **23**, 707–712.

Smol, J. P. 1982. 'Postglacial changes in fossil algal assemblages from three Canadian lakes', *PhD Thesis*, Queen's University, Kingston, Ontario, Canada. p. 151.

Smol, J. P. and Boucherle, M. M. 1985. 'Postglacial changes in algal and cladoceran assemblages in Little Round Lake, Ontario', *Arch. Hydrobiol.*, **103**, 25–49.

Sreenivasa, M. R. and Duthie, H. C. 1973, 'The postglacial diatom history of Sunfish Lake, southwestern Ontario', *Can. J. Bot.*, **51**, 1599–1609.

Stefan, H. G. and Fang, X. 1994. 'Model simulations of dissolved oxygen characteristics of Minnesota lakes: past and future', *Environ. Manage.*, **18**, 73–92.

Stefan, H. G. and Fang, X. 1995. 'A methodology to estimate year-round effects of climate change on water temperature, ice and dissolved oxygen characteristics of temperate zone lakes with application to Minnesota', *Project Report 377*. St. Anthony Falls Lab, Univ. of Minnesota, Minneapolis.

Stefan, H. G. and Preud'homme, E. B. 1993. 'Stream temperature estimation from air temperature', *Wat. Res. Bull.*, **29**, 27–45.

Stefan, H. G. and Sinokrot, B. A. 1993. 'Projected global climate change impact on water temperatures in five north central U.S. streams', *Climatic Change*, **24**, 353–381.

Stefan, H. G., Hondzo, M., and Fang, X. 1993a. 'Lake water quality modeling for projected future climate scenarios', *J. Environ. Qual.*, **22**, 417–431.

Stefan, H. G., Eaton, J. G., Hondzo, M., Goodno, B. E., Fang, X., Hokanson, K. E. F., and McCormick, J. H. 1993b. 'A methodology to estimate global climate-change impacts on lake waters and fisheries in Minnesota', in Ballentine, J. M. and Stakhiv, E. Z. (Eds), *Proc. First Natl. Conf. on Climate Change and Water Resources Mgt.* US Army Corps Eng., IWR Report 93-R-17. p. II.177–192.

Stefan, H. G., Hondzo, M., Eaton, J. G., and McCormick, J. H. 1995. 'Predicted effects of global climate on fishes in Minnesota lakes', in Beamish, R. J. (Ed.), *Climate Change and Northern Fish Populations. Can. Spec. Publ. Fish. Aquat. Sci.*, **121**, pp. 57–72.

Stefan, H. G., Hondzo, M., Fang, X., Eaton, J. G., and McCormick, J. H. 1996. 'Simulated long-term temperature and dissolved oxygen characteristics of lakes in the north-central U.S. and associated fish habitat limits', *Limnol. Oceanogr.*, **41**, 1124–1135.

Stemberger, R., Herlihy, A. T., Kugler, D. L. and Paulsen, S. G. 1996. 'Climatic forcing on Zooplankton richness in lakes of the northeastern United States', *Limnol. Oceanogr.*, **41**, 1093–1101.

Sterner, R. W., Hagemeier, D. D., Smith, W. L., and Smith, R. F. 1993. 'Phytoplankton nutrient limitation and food quality for *Daphnia*'. *Limnol. Oceanogr.*, **38**, 857–871.

Stirling, G. and McQueen, D. J. 1986. 'The influence of changing temperature on the life history of *Daphniopsis ephemeralis*', *J. Plankton Res.*, **8**, 583–595.

Swackhamer, D. L. and Skoglund, R. S. 1991. 'The role of phytoplankton in the partitioning of hydrophobic organic contaminants in water', in Baker, R. A. (Ed.), *Organic Substances and Sediments in Water*. Lewis Publishers, Chelsea, Michigan. pp. 91–105.

Taylor, W. W., Smale, M. A., and Freeberg, M. H. 1987. 'Biotic and abiotic determinants of lake whitefish (*Coregons clupeaformis*) recruitment in northeastern Lake Michigan', *Can. J. Fish. Aquat. Sci.*, **44**, 313–323.

Teller, J. T. 1985. 'Glacial Lake Agassiz and its influence on the Great Lakes', in Karrow, P. F. and Calkin, P. E. (Eds), *Quaternary Evolution of the Great Lakes*. Geological Association of Canada, Special Paper 30, pp. 1–16.

Teller, J. T. and Thorleifson, L. H. 1983. 'The Lake Agassiz–Lake Superior connection', in Teller, J. T. and Clayton, L. (Eds), *Glacial Lake Agassiz*. Geological Association of Canada, Special Paper 26, pp. 261–290.

Terasmae, J. 1961. 'Notes on late Quarernary climatic changes in Canada', *Annales New York Acad. Sci.* **95**, 658–675.

Tonn, W. M. 1990. 'Climate change and fish communities: a conceptual framework', *Trans. Am. Fish. Soc.*, **119**, 337–352.

Tonn, W. M., Magnuson, J. J., Rask, M., and Toivonen, J. 1990. 'Intercontinental comparison of small-lake fish assemblages: the balance between local and regional processes', *Am. Nat.*, **136**, 345–375.

Turner, M. A., Howell, E. T., Robinson, G. G. C., Campbell, P., Hecky, R. E., and Schindler, E. U. 1994. 'Roles of nutrients in controlling growth of epilithon in oligotrophic lakes of low alkalinity', *Can. J. Fish. Aquat. Sci.*, **51**, 2784–2793.

Twiss, M. R. and Nalewajko, C. 1992. 'Influence of phosphorus nutrition on copper toxicity to three strains of *Scenedesmus acutus* (Chlorophyceae)', *J. Phycol.*, **28**, 291–298.

Vanderploeg, H. A., Bolsenga, S. J., Fahnensteil, G. L., Liebig, R. J., and Gardner, W. S. 1992. 'Plankton ecology in an ice-covered bay of Lake Michigan: utilization of a winter phytoplankton bloom by reproducing copepods', *Hydrobiologia*, **243/244**, 175–183.

Vavrus, S. J., Wynne, R. H., and Foley, J. A. 1996. 'Measuring the sensitivity of southern Wisconsin lake ice to climate variations and lake depth using a numerical model', *Limnol. Oceanogr.*, **41**, 822–831.

Watras, C. J., Morrison, K. A., and Host, J. S. 1995. 'Concentration of mercury species in relationship to other site-specific factors in the surface waters of northern Wisconsin lakes', *Limnol. Oceanogr.*, **40**, 556–565.

Webb, J., III, Bartlein, P. J., and Kutzbach, J. E. 1987. 'Climatic change in eastern North America during the past 18,000 years: comparisons of pollen data with model results', in Ruddiman, W. F. and Wright, H. E., Jr. (Eds), *North America and Adjacent Oceans during the Last Deglaciation. Geology of North America*, Vol. K-3. Geological Society of America, Boulder, Colorado. pp. 142–165.

Webster, K. E. and Brezonik, P. L. 1995. 'Climate confounds detection of chemical trends related to acid deposition in upper midwest lakes in the U.S.A.', *Water Air Pollut.*, **85**, 1575–1580.

Webster, K. E., Newell, A. D., Baker, L. A., and Brezonik, P. L. 1990. 'Climatically induced rapid acidification of a softwater seepage lake', *Nature*, **347**, 374–376.

Webster, K. E., Kratz, T. K., Bowser, C. J., Magnuson, J. J., and Rose, W. J. 1996. 'The influence of landscape position on lake chemical responses to drought in northern Wisconsin', *Limnol. Oceanogr.*, **41**, 977–984.

Wentz, D. A., Rose, W. J., and Webster, K. E. 1995. 'Long-term hydrologic and biogeochemical responses of a softwater seepage lake in north central Wisconsin', *Wat. Resour. Res.* **31**, 199–212.

Wetzel, R. G. 1992. 'Gradient-dominated ecosystems: sources and regulatory functions of dissolved organic matter in freshwater ecosystems', *Hydrobiology*, **229**, 181–198.

Wilson, S. E., Walker, I. R., Mott, R. J., and Smol, J. P. 1993. 'Climatic and limnological changes associated with the Younger Dryas in Atlantic Canada', *Climate Dynam.*, **8**, 177–187.

Wisconsin Department of Natural Resources, 1995. 'Wisconsin's biodiversity as a management issue', *A Report to Department of Natural Resources Managers May 1995*, Madison, Wisconsin. p. 240.

Wong, S. C. 1979. 'Elevated atmospheric partial pressure of CO_2 and plant growth', *Oecologia*, **44**, 68–74.

Wood, C. M. and McDonald, D. G. (Eds.) 1996. *Global Warming — Implications for Freshwater and Marine Fish*. SEB Seminar Series, Cambridge University Press, Cambridge.

Wynne, R. H. and Lillesand, T. M. 1993. 'Monitoring phenological changes in lake ice as a robust indicator of regional climate change using the AVHRR', *Bull. Ecol. Soc. Am., Suppl.*, **74**, 495–496.

Wynne, R. H., Magnuson, J. J., Clayton, M. K., Lillesand, T. M., and Rodman, D. C. 1996. 'Determinants of temporal coherence in the satellite-derived 1987–1994 ice breakup dates of lakes on the Laurentian Shield', *Limnol. Oceangr.*, **41**, 832–838.

Yan, N. D., Keller, W., Scully, N. M., Lean, D. R. S., and Dillon, P. J. 1996. 'Increased UV-B penetration in a lake owing to drought-induced acidification', *Nature*, **381**, 141–143.

Yu, Z. and McAndrews, J. H. 1994. 'Holocene water levels at Rice Lake, Ontario, Canada: sediment, pollen and plant-macrofossil evidence', *The Holocene*, **4**, 141–152.

3

EFFECTS OF CLIMATE CHANGE ON THE FRESHWATERS OF ARCTIC AND SUBARCTIC NORTH AMERICA

WAYNE R. ROUSE,[1] MARIANNE S. V. DOUGLAS,[2] ROBERT E. HECKY,[3] ANNE E. HERSHEY,[4] GEORGE W. KLING,[5] LANCE LESACK,[6] PHILIP MARSH,[7] MICHAEL MCDONALD,[8] BARBARA J. NICHOLSON,[9] NIGEL T. ROULET[10] AND JOHN P. SMOL[11]

[1] *Department of Geography, McMaster University, Hamilton, ON L8S 1V2, Canada*
[2] *Department of Geology, University of Toronto, Toronto, ON M5S 3B1, Canada*
[3] *Freshwater Institute, Fisheries and Oceans Canada, 501 University Cres., Winnipeg, MN R3T 2N6, Canada*
[4] *Department of Biology, University of Minnesota, Duluth, MN 55812, USA*
[5] *Department of Biology, University of Michigan, Ann Arbor, MI 48109, USA*
[6] *Departments of Geography and Biological Sciences, Simon Fraser University, Burnaby, BC V5A 1S6, Canada*
[7] *National Hydrology Research Institute, 11 Innovation Blvd., Saskatoon, SK S7N 3H5, Canada*
[8] *Department of Chemical Engineering, 231 Engineering Bldg., University of Minnesota, Duluth, MN 55812, USA*
[9] *Department of Biological Sciences, Central Connecticut State University, New Britain, CT 06050-4010, USA*
[10] *Department of Geography and the Centre for Climate and Global Change Research, McGill University, Montreal, QC H3A 2K6, Canada*
[11] *Departmental of Biology, Queen's University, Kingston, ON K7L 3N6, Canada*

ABSTRACT

Region 2 comprises arctic and subarctic North America and is underlain by continuous or discontinuous permafrost. Its freshwater systems are dominated by a low energy environment and cold region processes. Central northern areas are almost totally influenced by arctic air masses while Pacific air becomes more prominent in the west, Atlantic air in the east and southern air masses at the lower latitudes. Air mass changes will play an important role in precipitation changes associated with climate warming. The snow season in the region is prolonged resulting in long-term storage of water so that the spring flood is often the major hydrological event of the year, even though, annual rainfall usually exceeds annual snowfall. The unique character of ponds and lakes is a result of the long frozen period, which affects nutrient status and gas exchange during the cold season and during thaw. GCM models are in close agreement for this region and predict temperature increases as large as 4°C in summer and 9°C in winter for a 2 × CO_2 scenario. Palaeoclimate indicators support the probability that substantial temperature increases have occurred previously during the Holocene. The historical record indicates a temperature increase of >1°C in parts of the region during the last century. GCM predictions of precipitation change indicate an increase, but there is little agreement amongst the various models on regional disposition or magnitude. Precipitation change is as important as temperature change in determining the water balance. The water balance is critical to every aspect of hydrology and limnology in the far north. Permafrost close to the surface plays a major role in freshwater systems because it often maintains lakes and wetlands above an impermeable frost table, which limits the water storage capabilities of the subsurface. Thawing associated with climate change would, particularly in areas of massive ice, stimulate landscape changes, which can affect every aspect of the environment. The normal spring flooding of ice-jammed north-flowing rivers, such as the Mackenzie, is a major event, which renews the water supply of lakes in delta regions and which determines the availability of habitat for aquatic organisms. Climate warming or river damming and diversion would probably lead to the complete drying of many delta lakes. Climate warming would also change the characteristics of ponds that presently freeze to the bottom and result in fundamental changes in their limnological characteristics. At present, the food chain is rather simple usually culminating in lake trout or arctic char. A lengthening of the growing season and warmer water temperature would affect the chemical, mineral and nutrient status of lakes and most likely have deleterious effects on the food chain. Peatlands are extensive in region 2. They would move northwards at their southern boundaries, and, with sustained drying, many would change form or become inactive. Extensive wetlands and peatlands are an important component of the global carbon budget, and warmer and drier conditions would most likely change them from a sink to a source for atmospheric carbon. There is some evidence that this may be occurring already. Region 2 is very vulnerable to global warming. Its freshwater systems are probably the least studied and most poorly understood in North America. There are clear needs to improve our current knowledge of temperature and precipitation patterns; to model the thermal behaviour of wetlands, lakes and rivers; to understand better the interrelationships of cold region rivers with their basins; to begin studies on the very large lakes in the region; to obtain a firm grasp of the role of northern peatlands in the global carbon cycle; and to link the terrestrial water balance to the thermal and hydrological regime of the polar sea. Overall, there is a strong need for basic research and long-term monitoring.

INTRODUCTION

Region 2 (Figure 1; Figure 1 in Leavesley *et al.*, 1997) extends across 30° of latitude (52°N to 82°N) and approximately 80° of longitude (62°W to 142°W). This large area, of approximately six million km^2, includes the Mackenzie River Basin, the Yukon River Basin and north slope of Alaska, Keewatin, the Hudson Bay Lowland, northern Quebec and Labrador and the Canadian Arctic Islands. Lakes and deltas are abundant and are an important resource for indigenous people. The surface area of freshwater is about 350 000 km^2. Some of the largest freshwater lakes on the earth lie within the Mackenzie Basin. These are Great Bear and Great Slave Lakes and Lake Athabasca. There are innumerable smaller lakes and ponds. Major river deltas include the Mackenzie Delta (25 000 lakes within its approximately 12 000 km^2 area alone), Peace–Athabasca Delta and the Yukon Delta. Other smaller deltas are formed by most north-flowing rivers. The freshwater ecosystems include approximately 450 000 km^2 of peatlands. Of these, 280 000 km^2 are in the Hudson Bay Lowland, which is the second largest peatland in the world. The entire region is underlain by continuous or discontinuous permafrost (Figure 1a). The covering vegetation is boreal forest, subarctic open forest (taiga) and tundra.

This paper outlines the general climatic and hydrological characteristics of region 2. It then examines model predictions of the magnitude of climate warming and precipitation change for the region and places this in context using historical and palaeoclimatic evidence. Following a discussion of the likely broad-scale effects of climate change on freshwaters, the paper focuses in more detail, on potential effects on small ponds and lakes, rivers and northern peatlands.

CLIMATIC CHARACTERISTICS

As would be expected in a region with such geographic diversity, climate is highly variable. There are very few meteorological stations in region 2 so the generalizations stated below should be accepted with reservations. Exclusive of the ice cap areas in the eastern Queen Elizabeth Islands, annual mean temperature spans 21°C (−18 to +3°C), annual precipitation ranges from 60 to 460 mm, the frost-free period from 10 to <125 days, the median snow-free period from 80 to 245 days and the average global solar radiation from 90 to 160 W/m^2 (Hare and Thomas, 1979). Average net radiation at the surface varies from 3 to 53 W/m^2 (Rouse, 1993). In the most southerly parts about 33% of solar radiation is realized as effective surface net radiant energy. This decreases to about 3% in the most northerly parts. There is a marked seasonality in the climate. Typically, air temperature in the summer months (JJA) is between 34 and 41°C warmer than in the winter months (DJF).

Circulation Patterns, Air Masses and Winds

In both summer and winter, the regional winds are westerly in the southern portions of region 2, but in the north, the flow is north-westerly. In the Arctic Islands the flow is part of a circumpolar vortex that is particularly well developed in winter and centred over northern Baffin and Cornwallis Islands (Hare and Thomas, 1979; Maxwell, 1980). Surface winds are heavily influenced by local and regional factors. For example, in the Arctic Islands at Resolute (latitude 74° 40′) the measured annual mean wind speed is 6·3 m/s,

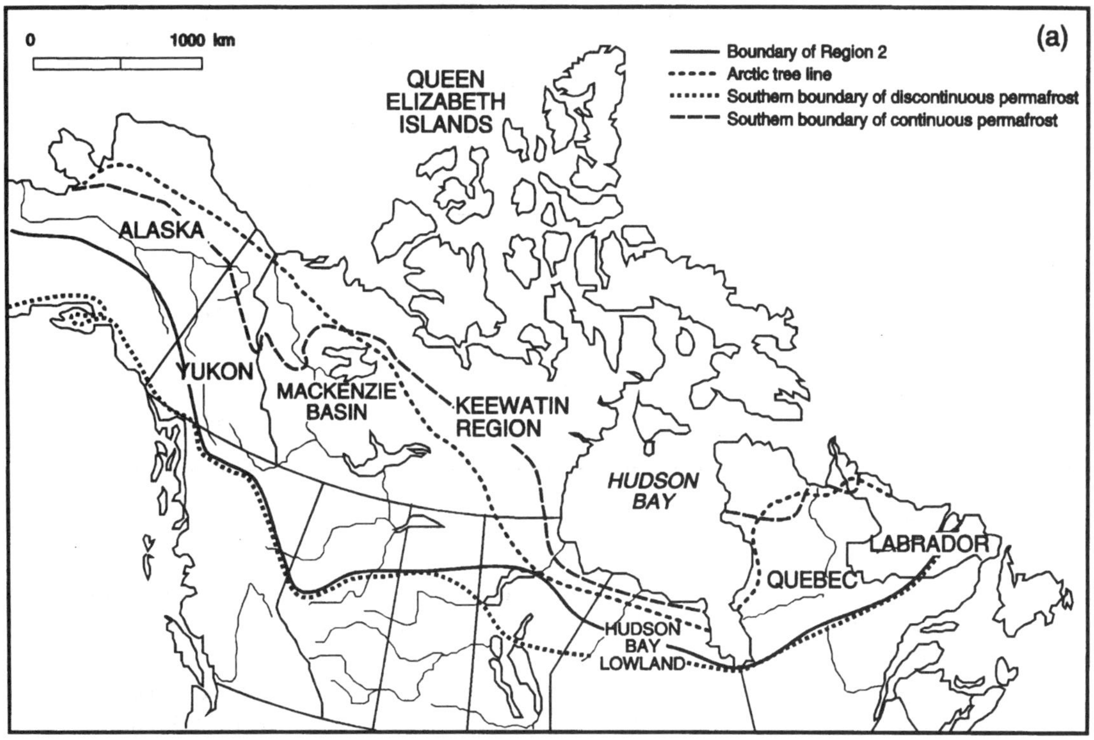

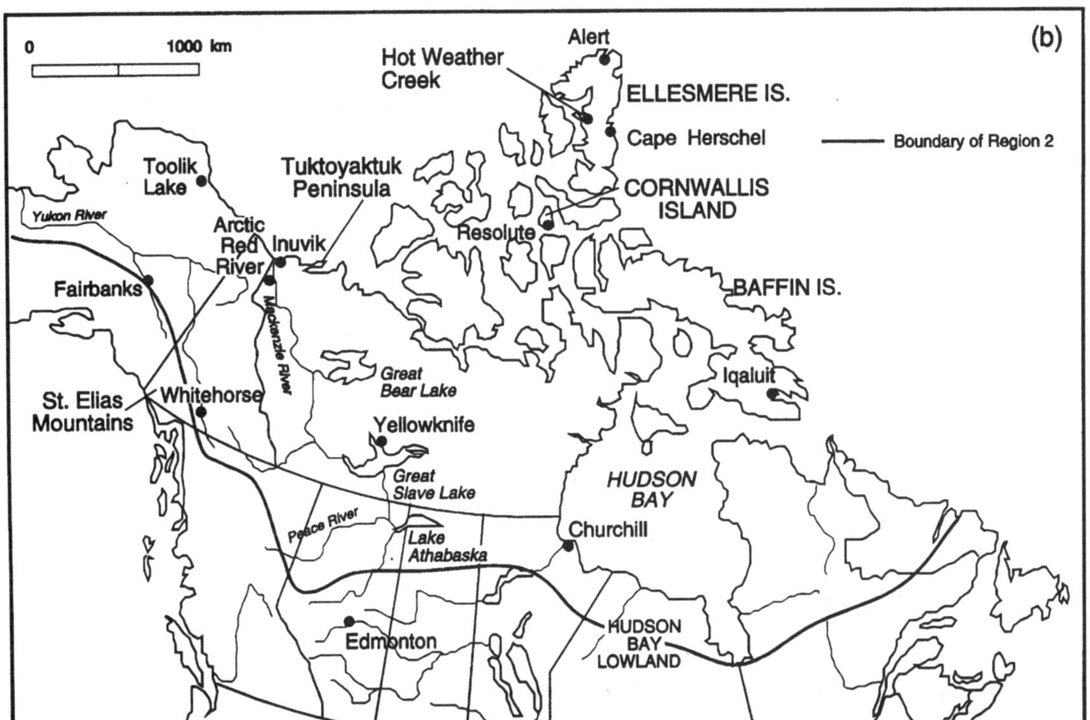

Figure 1. Map of region 2 showing (a) the location of continuous and discontinuous permafrost, Arctic tree-line and specific study areas, and (b) place names referred to in this paper

Table I. Air mass frequency (%) by season. W — winter (DJF); SP — spring (MAM); S — summer (JJA); F — autumn (SON). Estimates are derived from Bryson (1966) and Maxwell (1980)

	Air mass source*																			
	Arctic				Pacific				Atlantic				Southern				Hudson Bay			
Station	W	SP	S	F	W	SP	S	F	W	SP	S	F	W	SP	S	F	W	SP	S	F
Alert	**100**	**100**	**100**	**100**																
Resolution	**100**	**100**	**100**	**100**																
Aqaluit	**100**	**100**	**66**	**100**							**34**									
Baker Lake	**100**	**100**	**70**	**100**											**30**					
Fairbanks†	**100**	**66**	**30**	**45**		**34**	**70**	**55**												
Whitehorse	**100**	**65**	25	**45**		**35**	**75**	**55**												
Inuvik	**100**	**85**	**55**	**88**		15	**45**	12												
Churchill	**100**	**90**	**48**	**85**										10	**30**	3			22	12
Schefferville	**100**	**85**	**45**	**88**						15	**30**	10			20				5	2

* All frequencies ⩾30% are in bold.
† Fairbanks data are estimated by extrapolation.

whereas at Alert (latitude 82° 30′) it is 3·0 m/s (Maxwell, 1980). These differences are the result of local topographical and instrument-siting factors. In general terms, wind speeds increase from south to north and are about 50% higher in winter than in summer (Rouse, 1993). The biggest increase in surface wind accompanies emergence from the boreal forest into open tundra and is a result of the reduced surface drag.

Annual air mass dominance varies widely within the region (Table I). For the high Arctic Islands, arctic air masses dominate for most of the year. As one moves south-eastwards, Atlantic air achieves more importance, whereas south-westwards, Pacific air becomes increasingly common. In south-central parts of region 2, such as the Hudson Bay Lowland, while arctic air is dominant annually, southern air and air from Hudson Bay achieve some importance in summer and autumn. Since arctic air is much drier than Atlantic, Pacific and especially southerly air masses, in scenarios of climate warming, the changing air mass dominance patterns become important, particularly to precipitation. Substantial changes to less domination by arctic air masses can be anticipated for northern Alaska and Yukon, the Mackenzie Basin, Keewatin, the Hudson Bay Lowland, northern Quebec and Labrador and the more southern islands of the Arctic Archipelago.

HYDROLOGICAL CHARACTERISTICS

Extent of Snow, Ice and Permafrost

For most of region 2, snow accounts for 40–80% of the annual precipitation, the majority of which is stored on the ground for 6–9 months of the year. Actual snowfall can be up to two to three times greater than that measured by standard snow collectors at weather stations. This is because of under catch and large numbers of trace precipitation events in the wind-swept tundra (Goodison, 1981; Woo *et al.*, 1983). The hydraulic conductivity of permafrost soils is significantly lower than for unfrozen soils, thus limiting groundwater flows. Consequently, the occurrence of permafrost is important in controlling the areal and spatial distribution of wetlands in many areas of the region.

Snow Redistribution Processes

Intercepted snow in forest canopies and blowing snow on tundra enhance sublimation of the snow and reduce the total snow on the ground at spring melt. In regions of low precipitation, such as most of the tundra and the drier northern regions of the boreal forest, sublimation limits the water available for runoff. In dense coniferous canopies, interception can result in up to 40% of the snow sublimating, while in open or deciduous forests it may be less than 10% (Pomeroy and Gray, 1994). Snow becomes entrained into the atmosphere during high wind events, which both redistribute snow on the ground and result in considerable sublimation. At the northern edge of the prairies, sublimation can comprise up to 40% of the snowfall (Pomeroy and Gray, 1994) and in tundra, over 10% (Marsh *et al.*, 1996).

Cold, high latitude snow covers behave differently to their temperate counterparts (Marsh, 1991a). Beneath temperate snow covers the ground heat flux is seldom important and a small flux into the basal snow helps to increase melt. In northern permafrost soils, heat flows from the snow to the ground. This increases the energy required to melt the snowpack, and delays melt runoff. Significant amounts of meltwater refreeze within the snow, soil and as basal ice. This also delays the start of melt runoff and extends the snowmelt period. Snowmelt infiltration into frozen soils plays an important part in controlling runoff. Depending on the magnitude of autumn rains and soil moisture content at freeze-up, snowmelt infiltration varies from zero to greater than the entire snowpack water equivalent. In addition, soil temperature controls the portion of the frozen soil infiltration that refreezes and will not be available for immediate runoff.

Effects of River and Lake Ice

Seasonal ice cover affects the characteristics of aquatic ecosystems. It controls the amount of light reaching the unfrozen water beneath the ice cover (Prowse and Stephenson, 1986). The high surface albedo of snow atop the ice means that any significant accumulation of snow reflects most of the incident radiation. Beneath the snowpack, black ice transmits most of the solar radiation through to the unfrozen waters, whereas white ice is strongly reflective. The duration of lake ice varies considerably over the study area. For example, Lake Athabasca (58°N) has a mean duration of ice cover from early November to mid-May, and an average maximum thickness of about 1 m. In contrast, lake ice cover near Alert, at 82°N, extends on average from late August to mid-July, with an average maximum thickness of about 2 m. Meanwhile, some high arctic lakes maintain their ice cover throughout the summer (Smol, 1988).

Ice prevents gas exchange between underlying waters and the atmosphere and may commonly lead to depletion of dissolved O_2 and the build up of reduced gases such as CO_2, CH_4 and H_2S. In shallow lakes, a substantial portion of the water column and total lake volume is incorporated into the ice cover, while a smaller portion remains unfrozen. Because solutes are excluded from the ice during freezing, the chemistry of the underlying water changes dramatically over the winter (Schindler *et al.*, 1974; Hobbie, 1980; Welch and Bergmann, 1985; Welch and Legault, 1986; Lesack *et al.*, 1991). Finally, the processes accompanying ice formation during freeze-up and break-up have a wide range of effects on the bed and banks of lakes and rivers. These include frazil ice impact on fish, anchor ice growth, elevated water levels, channel blockage and increased scouring of river banks and beds (Prowse, 1994). These processes have a large effect on the riparian geomorphology and vegetation (Scrimgeour *et al.*, 1994).

River Hydrology

The complexity of runoff magnitude can be illustrated for the Mackenzie Basin (Figure 2). Runoff varies from 100 mm/yr in the northern portions of the basin, to over 1000 mm/yr in the southern, mountainous parts. Mean annual discharge of 173 mm/yr averaged over the whole basin is approximately 42% of total precipitation input to the basin, with the remaining 58% being evapotranspired (Marsh and Prowse, 1993).

On an annual basis, it is often assumed that changes in storage are zero. However, that is not always the case and long-term fluctuations in groundwater, lake and glacial storage can occur. The hydrological effect of glaciers in a watershed is to store water during cold or wet periods, and to release water during warm periods. Lakes and wetlands also store water, although on shorter time scales.

Seasonal variations in flow are important to the freshwater ecosystems in river basins. Four regimes common to northern regions can be identified (Church, 1974; Woo, 1986). Arctic nival regimes occur in areas of continuous permafrost. Spring snowmelt dominates the hydrograph, and runoff from summer rain is generally small since precipitation is low. Winter stream flow is usually very low owing to limited groundwater contribution. Subarctic nival regimes are dominated by spring snowmelt, but summer rain

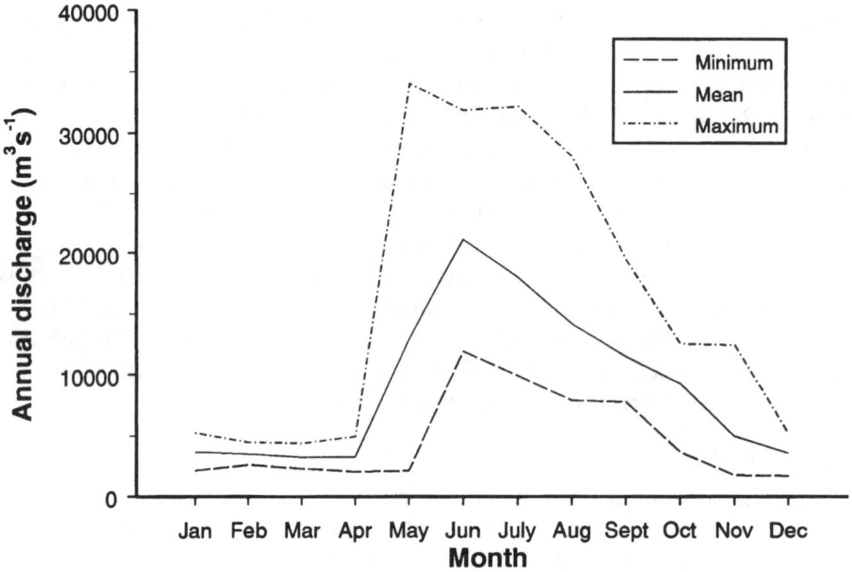

Figure 2. The average monthly discharge for the Mackenzie River at Arctic Red River. The basin at this point is $1{\cdot}66 \times 10^6$ km^2

storms may produce peak flow of similar size to the spring snowmelt (Watt, 1989). Discontinuous permafrost increases groundwater contributions and winter flow may be significant. Wetland regimes occur where drainage is poor owing to low relief or an impermeable substrate. Wetland basins, with low energy gradients, have significant temporary storage capacity, and therefore they tend to attenuate peak flows but still retain high yields. Finally, in proglacial streams, snowmelt runoff is important, but instead of a brief spring peak, flow increases throughout the spring snowmelt period and then continues to increase as meltwater is contributed from higher regions of the glaciated basin.

The Mackenzie River integrates the runoff from major sub-basins that display the characteristics of all four regimes. The main stream flow can be described as subarctic nival, with a dominant spring flood, significant summer peaks owing to rainfall runoff and discharge that continues throughout the winter (Figure 2).

The dramatic effects of ice on river flows are evident in the Mackenzie Basin. In northern portions, channels begin to freeze over in October, with a complete ice cover forming by January. Freeze-up is an important event because the flow conveyance of the channel is reduced as downstream ice cover is formed (Gray and Prowse, 1992). This is mainly a response to the higher roughness of ice-covered channels. As a result, upstream water levels rise as water is placed into storage behind the accumulating ice cover and a period of low discharge prevails downstream. The lowest discharge of the year may occur during freeze-up, not later in the winter when water delivery to the channel is at a minimum. In some cases, river icings can lead to a complete blockage of the channel system (Van Everdingen, 1974).

The Mackenzie, like most rivers in region 2, flows northwards from areas that are warmer to those that are colder. Thus, melt progresses from upstream to downstream. The resulting flood wave often progresses downstream more rapidly than the melt proceeds, and the flood wave encounters thick, resistant ice covers, resulting in very large ice jams (e.g. Andres and Doyle, 1984; Gray and Prowse, 1992). Ice jams make it impossible to measure discharge during what is often the major flood event of the year.

Lake Hydrology

In cold regions, lake ice plays an important hydrological role during the spring break-up. If a significant portion of the snowmelt runoff occurs before the lake ice has melted, floating ice may prevent wind-induced

Table II. Scenarios of temperature change (°C) for a 2 × CO_2 scenario. GCM models: A, Atmospheric Environment Service, Canada; B, Geofluids Dynamics Laboratory, USA; C, United Kingdom Meteorological Service, UK (adapted from Mitchell *et al.*, 1990)

	Alaska–Yukon	Mackenzie	Keewatin, N. Manitoba, N. Ontario	N. Quebec Labrador	High Arctic
Winter					
A	2–8	4–8	4–12	8–12	8–12
B	2–8	0–4	2–8	6–12	8–12
C	2–8	2–6	4–8	6–8	4–8
Summer					
A	2–6	2–4	4–6	4–6	4–6
B	2–6	2–6	2–6	2–6	0–4
C	2–6	2–6	2–4	2–6	2–4

intermixing of the runoff and lake water. As previously mentioned, solute exclusion during the growth of the lake ice can increase solute concentrations substantially in the unfrozen water beneath the ice. The resulting differences in salinity between the unfrozen lake water and inflowing runoff can prevent intermixing while the lake remains frozen (Lesack *et al.*, 1991). In other small lakes with large catchments, a substantial portion of snowmelt can pass through the lake without mixing and this can reduce the utilization of the snowmelt nutrient load by the lake ecosystem (Welch and Bergman, 1985).

CLIMATE CHANGE

GCM Scenarios

A general idea of the potential magnitude of climate change in region 2 can be obtained from GCM (general circulation model) scenarios for 2 × CO_2 warming (Table II). In both summer and winter, composite GCM results indicate that Alaska, Yukon and the Mackenzie Basin could undergo increases of about 4°C; in central parts of region 2, this changes to 4 and 7°C for summer and winter, respectively, and, in the east and the Canadian Arctic Archipelago, increases of the order of 4 and 9°C are suggested. Models generally concur that precipitation will increase in summer and winter, but little agreement exists on the magnitude and regionality (Mitchell *et al.*, 1990). Given that present day precipitation in region 2 is poorly known, there can be little confidence in 2 × CO_2 projections for this variable. Cohen (1993) reports on a composite scenario for the Mackenzie Basin that is derived from a GCM model, contemporary station data and historical and palaeoclimatic information to predict an increase of 3°C in annual temperature and a 10% increase in precipitation. From a selected suite of different GCMs, Maxwell (1992) also provides averages for the lower Mackenzie that indicate substantial temperature and precipitation increases. Models, thus, generally agree in indicating moderately large temperature increases throughout region 2 in summer, and, in winter, large to very large temperature increases in the central and northern portions of the region.

Historical Variability and Palaeoclimatic Reconstructions

The instrumental record indicates a temperature increase of 1·7°C in the forested Mackenzie Basin between 1895 and 1991, and about 0·6°C in areas of tundra region between 1922 and 1991 (Skinner and Gullet, 1993). No cause or effect can be ascribed to these short-term changes and patterns since they are well within the natural variability of the temperature record. Also, regional consistency in temperature change in region 2 has not been demonstrated.

For a 30-year period, Isaac and Stuart (1992) demonstrated a close correspondence between higher than normal precipitation and above-average daily temperatures. This suggests that the amelioration of very cold temperatures in these regions has the potential to increase precipitation, particularly in winter, and especially in the central, eastern and northern areas. This is encouraging because without increased precipitation many systems would dry up under warmer temperatures.

Palaeoclimatic reconstructions during the Holocene cast some light on the magnitude of climate change or variability. From dendroclimatology analysis, Szeicz and MacDonald (1995) reconstructed June–July temperatures for the central Yukon and northern and middle Mackenzie Basin for the period 1638 to 1988. Szeicz's record generally agrees with that of Bradley and Jones (1993), which uses composite tree ring and Greenland ice core proxy data. Both records show a fairly steady increase in temperatures in the last 150 years, of the magnitude of 2°C, and also indicate that late 20th century summer temperatures are as warm or warmer than any in the period 1638–1980. There is also general agreement with the dendroclimatic reconstruction of Jacoby *et al.* (1985) for Alaska and the central Yukon, except that this latter study indicates a post-1950 temperature decrease. All summertime dendroclimatic records indicate that medium-term temperature variations of a similar but lesser magnitude to those suggested by GCM results have been experienced in the recent Holocene. Evidence from eastern parts of region 2 is less easily interpreted because of the dominance of fire in the northern forest dynamics during the last 3000 years (Payette and Morneau, 1993). Palynological evidence suggests that on the Tuktoyaktuk Peninsula near the Mackenzie River Delta, forest limits were at least 70 km polewards of the current tree-line between 9500 and 5000 BP (Ritchie, 1984; Spear, 1993). This evidence is important because it illustrates that the boreal forest has migrated to the Arctic coast during the Holocene (Roots, 1989) and that high latitudes respond strongly to global climate warming of the magnitude indicated by GCM scenarios.

Based on stratigraphic evidence in the central and southern Yukon, Burn (1993) argues that the Holocene has been the only non-glacial period with widespread permafrost. The temperature of the permafrost in these regions is close to 0°C and with small warming could undergo substantial thaw. Burn (1993) notes particularly that cold surface temperatures and thin snow cover caused by the blocking effect of the St Elias Mountains may be responsible for maintaining the permafrost, and that these blocking effects are not simulated by output of GCMs for $1 \times CO_2$. The implication is that $2 \times CO_2$ scenarios probably do not model that influence either.

Koerner's (1991) analysis of stable isotope ratios in glacier ice cores features the following variability. Whereas at higher elevations in Antarctica and Greenland no melting occurs at all, in the Canadian Arctic, melting over the entire ice cap surface occurs nine years out of ten. From 1200 AD to the 1970s there have been several periods in which strong melting persisted over five or more decades, but the period with the strongest magnitudes of melt has occurred during the last 100 years. Koerner cautions, however, that this warming is not unusual in the longer record, and much larger changes are apparent in the last one hundred thousand years. He also notes that the occurrence of the coldest period of the Holocene just 200 years ago (Little Ice Age) tends to highlight the present warm period.

Douglas *et al.* (1994), using diatom indicators and palaeolimnological techniques in the High Arctic, found relatively stable diatom populations over the last 8000 years, but then noted striking successional changes over the last 200 years. They attribute climate warming as the most likely cause of these successional changes.

Overall, there appears to be a congruence in the various indicators that suggests a substantial 20th century warming trend in much of region 2. There is convincing evidence that high latitudes do give a unique high magnitude response to global climate change.

BROAD-SCALE EFFECTS OF CLIMATE WARMING

The permafrost that underlies region 2 varies in thickness along an air temperature and general latitudinal gradient (Table III). It is in those regions where the annual mean temperature is higher than −6°C, or the annual mean ground temperature is hovering around 0°C, that permafrost could disappear completely. By analogy, with current climate–permafrost distribution, the long-term effects of warming scenarios produced by GCM scenarios would be to move the permafrost boundaries polewards by about 500 km or so

Table III. Permafrost thicknesses (m), sample ground temperatures (°C) and mean air temperatures (°C) for select Canadian stations (after Brown, 1970). Stations are ordered according to decreasing daily mean air temperatures

Location	Latitude Longitude	Thickness (m)	Ground temp (°C), Depth (m)	Air temp. (°C)
Thompson, Manitoba	55°45′ N 97°54′ W	15	0, 8	−4·0
Uranium City, Saskatchewan	59°32′ N 108°43′ W	9	0, 9	−4·0
Schefferville, Quebec	54°50′ N 67°00′ W	>75	−0·2, 60	−4·5
Yellowknife, NWT	62°28′ N 114°27′ W	60–90	0, 12	−5·5
Normal Wells, NWT	65°19′ N 126°46′ W	45–60	−2·8, 50	−6·2
Churchill, Manitoba	58°45′ N 94°04′ W	30–60	−2·2, 10	−7·2
Inuvik, NWT	68°18′ N 133°29′ W	>90	−3·9, 30	−9·1
Rankin Inlet, NWT	62°45′ N 92°05′ W	300	−9·0, 30	−11·6
Winter Harbour, NWT	74°50′ N 110°46′ W	460	ND	−16·0
Resolute, NWT	74°43′ N 54°59′ W	400	−13·0, 30	−1·6

(Woo *et al.*, 1992). This would reduce the area of permafrost to less than 80% of its present coverage. Probably the most profound effect from warming would be the melting of widespread massive ground ice (Figure 3a) which is concentrated in the top few metres of permafrost terrain, and which is particularly susceptible to climate warming. This would induce massive terrain slumping (thermokarst) with effects on all surface features, including drainage patterns and sediment loads in rivers and lakes, thus affecting aquatic ecosystems.

Groundwater in permafrost areas may be found above (suprapermafrost), within (intrapermafrost) or below (subpermafrost) the frozen layer (Woo *et al.*, 1992). The suprapermafrost zone provides little water storage in winter and in early spring when the active layer is frozen. In summer its storage capacity changes with thawing and refreezing. Occurrence of intrapermafrost groundwater is governed by the vertical distribution of permafrost. The subpermafrost groundwater often has higher chemical concentrations because it has a longer residence time to be in contact with earth materials. Warming of the permafrost will extend the thaw season, allowing a longer period of water infiltration and deeper active layer development (Hinzman and Kane, 1992; Kane *et al.*, 1992), and a concomitant increase in hydrological activities in the suprapermafrost zone (Woo *et al.*, 1992).

A deeper active layer and a longer unfrozen period will reduce overland flow as both infiltration and active layer water storage capacity increase. Patchy arctic wetlands, currently supported by surface flow from late lying snowbanks, would disappear. Increasing the depth of the active layer will threaten the survival of many arctic wetlands which exist because impermeable permafrost prohibits deep percolation and impedes drainage of the active layer. Similarly, some small lakes exist because the permafrost isolates them from the regional groundwater system or from surface flow. Permafrost degradation will open subsurface flow connections, thereby draining these lakes (Mackay, 1992; Woo *et al.*, 1992).

The spring melt period should be earlier and possibly stretch over a longer time period because of lower radiation receipts occurring earlier in the year (Kane *et al.*, 1992; Figure 3b). Because of increased active layer storage capacity, peak flows associated with both snowmelt and rainfall events would be similar or lower than at present. These expectations are consistent with the analogue of northern basins. Those with

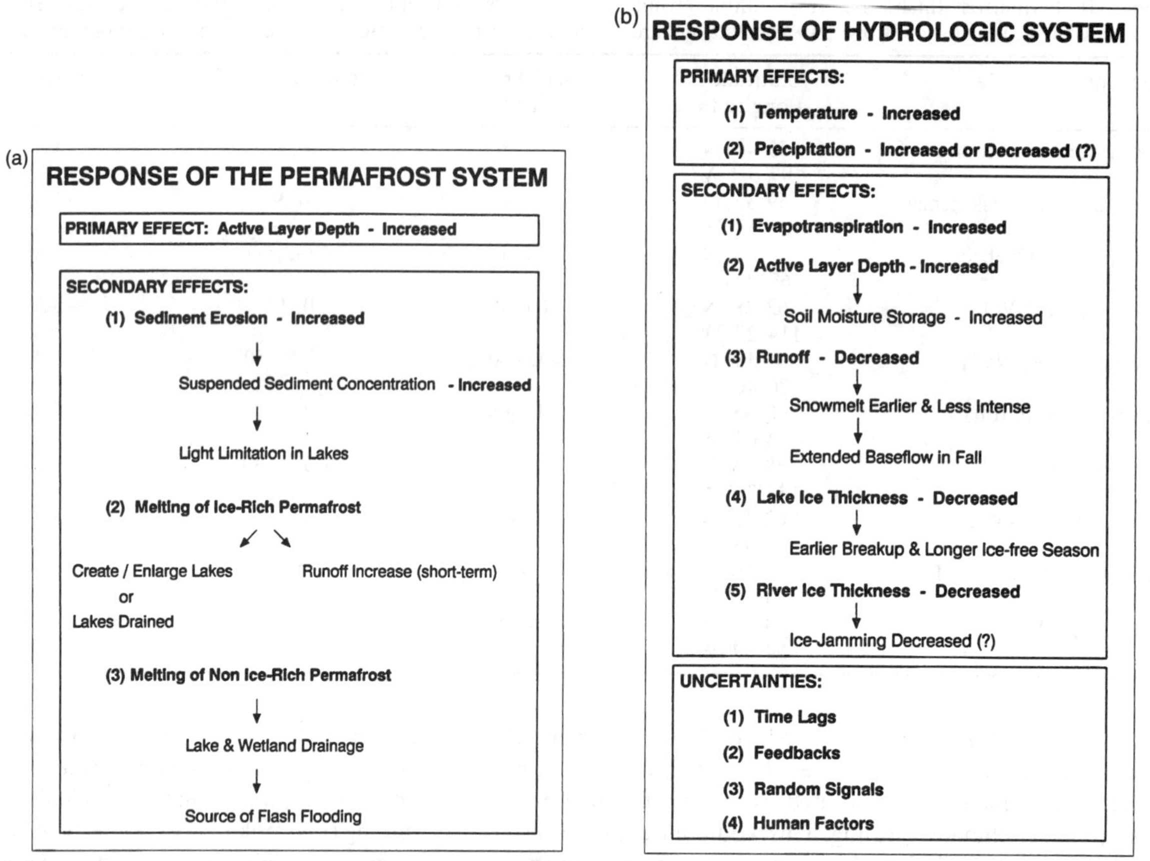

Figure 3. (a) A conceptual summary of the potential response of the permafrost system to climate change. Major primary and secondary effects are listed and potential higher order effects are indicated with arrows. (b) A conceptual summary of the potential response of the hydrological system to climate change. Major primary and secondary effects are listed and potential higher order effects are indicated with arrows

less permafrost, but receiving comparable amounts of precipitation, produce lesser amounts of runoff and smaller ranges in flow volumes.

Lake ice covers will be thinner, will break up earlier and freeze up later. Thus, there will be a longer ice-free season. For large north-flowing rivers, thinner ice covers, smaller temperature gradients between southern and northern portions of the basins and reduced spring flood peaks could lead to less frequent and less severe ice-jamming during the spring break-up.

In the snow-free season, evapotranspiration often exceeds precipitation giving a negative water balance (Woo *et al.*, 1992). An increase in the length of the snow-free period and in summer temperature will increase evaporation and transpiration. If precipitation does not change, or decreases, water balances will become more strongly negative. If there is an increase in precipitation that matches that of evaporation the water balance will not change. Only an increase in rainfall, which exceeds that of evaporation, will give a positive summer water balance to increase lake levels, recharge ground moisture and increase runoff and river flow. The critical equation with respect to the effect of greenhouse warming on the water balance thus involves both the magnitude of the warming effect and the responses of the precipitation regime. Changes in the water balance are particularly important to the surface moisture regime of the Arctic Islands, where, at present, in unglaciated basins, surface runoff and stream flow occur only during spring snowmelt. After melt the terrain becomes hydrologically inactive until the following early summer period. Wildlife is sustained through stored water in localized ecological niches such as Truelove Lowland on Devon Island (Bliss, 1977) and Hot Weather Creek on Ellesmere Island (Figure 1b). Such niches will be strongly influenced by changes in the water balance.

Contemporary evidence suggests that hot and dry summer conditions will increase overall evapotranspiration from wetlands even when they develop a substantial moisture deficiency. This arises because of the large stomatal conductance of transpiring plants, and from the ability of porous peat soils to supply water from wet subsurface layers when atmospheric demand is large (Rouse *et al.*, 1992). Thus, where vascular plants are dominant, they will continue to transpire until the water table drops well below the surface. Where non-transpiring vegetation, such as *Sphagnum* moss or lichen, is dominant, however, evaporation is inhibited after surface drying. A hot, dry year promotes deeper thaw of the active layer than does a comparable wet year (Rouse *et al.*, 1992). At tree-line, with the same ambient conditions, wet tundra is more sensitive to changes in climate than is forest at tree-line and the differences in sensitivities are a function of both physical and physiological responses within the ecosystems (Lafleur and Rouse, 1995). Individual ecosystems in continuous permafrost near tree-line all have differing sensitivities to seasonal changes in temperature and precipitation. For example, the response in evaporative water loss to increases or decreases in air temperature is greatest for dwarf willow–birch forest, and decreases through sedge fens, open forest, upland lichen heath and shallow tundra lakes (W. R. Rouse, unpublished data).

Flora are more or less in equilibrium with the climate, including soil climate. Summer temperature, the length of the summer growing season and intensity of summer warmth show the greatest correlation with vegetation distribution and species diversity (Edlund and Alt, 1989). Seasonal snow cover and soil moisture availability also influence the distribution of species and communities. Permafrost can present a barrier to the downward development of roots. It also ties up water in the frozen state that is unavailable for transpiration. In addition, frost heaving dislocates and damages plant roots, and the maintenance of cold rooting zones and cold soil water add stress factors for all vascular plants. By maintaining a high water table, permafrost can promote anaerobic conditions within rooting zones. Such conditions favour the development of non-vascular plants and contribute to the poor growth of vascular plants, especially boreal forest species near tree-line (Woo *et al.*, 1992). Under a warmer and wetter climate change scenario, there is some consensus that contemporary vegetation belts would move polewards, although there would be a lag between temperature increase and migration. The palaeobotanical record supports this response. Under a warmer and drier scenario, there would also be a northwards migration, but this would favour xerophytic vegetation with less biomass.

GENERAL RESPONSES OF AQUATIC ECOSYSTEMS TO CLIMATE CHANGE

Aquatic Habitat and Primary Production

The general potential responses of aquatic ecosystems to increased temperature and decreased runoff are summarized in Figure 4. The response to increased temperature will be a longer ice-free season for lakes, which will increase the length of the stratified season and could increase the depth of mixing. The longer stratified season should lead to lower oxygen concentrations in the hypolimnion of lakes and an increased stress on cold water organisms.

A longer ice-free season should also enhance the underwater light climate for primary producers and increase the length of the growing season. However, this may be offset, potentially, by a deterioration in light availability because of two different mechanisms. First, in regions where extensive melting of permafrost may occur, land subsidence and formation of new drainage systems could lead to extensive erosion and increased turbidity in streams and lakes. In Southern Indian Lake (northern Manitoba), analogous factors decreased light penetration sufficiently to cause a switch from nutrient limitation to light limitation of primary production, which more commonly controls the primary production of northern lakes (Hecky and Guildford, 1984). Secondly, increases in average DOC concentrations and associated water darkening may also occur among lakes and streams.

RESPONSE OF AQUATIC ECOSYSTEMS

PRIMARY EFFECTS:

(1) Temperature - Increased

(2) Runoff - Decreased

SECONDARY EFFECTS (PHYSICAL-CHEMICAL):

(1) Ice-Free Season - Increased

↓

Prolonged Stratification in Lakes

↓

Lower Oxygen in Hypolimnia

(2) Decomposition & Weathering - Increased

↓

Nutrient Supply - Increased

(3) Water Residence Times - Increased

↓

Dilution of Nutrients - Decreased

HIGHER-ORDER EFFECTS (BIOTA):

(1) Primary Production - Autotrophic Growth Increased
- Imported POM Increased

↑

Longer Growing Season
Enhanced Nutrient Supply
Underwater Light - Variably Decreased
Terrestrial Primary Production - Increased
North-Moving Treeline - Riparian Litter Enhanced

(2) Secondary Production - Variable Response

↑

Temperatures Favor Increased Production
Food Webs & Food Supplies - Ecosystem Specific
Organism Dispersal Abilities - Variable

Figure 4. Potential response of aquatic ecosystems to a temperature increase and a runoff decrease

In northern aquatic ecosystems, both *in situ* primary production and imported organic matter from terrestrial ecosystems can be important sources of energy to fuel aquatic food chains. A general warming through the region should increase terrestrial primary production. It is expected that primary production would also increase because aquatic autotrophs tend to be limited by nutrients. The microbial degradation of organic matter is strongly temperature dependent. Consequently, both the amount of organic matter imported from terrestrial ecosystems and the rate at which the matter is processed should increase. Increased rates of nutrient remineralization derived from microbial decomposition should lead to an enhanced nutrient supply in northern aquatic ecosystems. A general decrease in the amount of runoff would lead to lower flushing rates and the decreased dilution of nutrients and longer contact times with rock minerals within their catchments. In addition, increased temperatures would, at least modestly, increase rates of rock weathering. The net effect of increased weathering plus decreased dilution would be to further enhance the nutrient supply. If an enhanced nutrient supply stimulates the growth rates of aquatic autotrophs, and this is coupled with a longer growing season, there should be an overall increase in annual primary production.

Secondary Production

Secondary production represents a combination of complex microbial and animal communities that process autotrophic production and imported organic matter for their own growth, maintenance and reproduction. Food supply and temperature represent primary regulators on the growth of consumer populations (Benke, 1984; Sweeney, 1984). While the effect of warmer temperatures should generally be favourable throughout the region, the food supply available for consumers would be ecosystem specific, thus giving a variable response of secondary production to climate warming.

The dispersal ability of the organisms may control overall changes in secondary production. Most of the fauna contributing to secondary production, except fishes, have high dispersal ability and should quickly take advantage of warmer temperature regimes. However, there are the glacial relict fauna, which represent animals distributed by and surviving in large proglacial lakes during the last deglaciation. Some of these, such as *Pontoporeia affinis* and *Mysis relicta*, are important fish food in many northern lakes. The potential loss of this fauna under warmer, less oxic conditions, could cause short-term (years to decades) disruptions in the food chain until a warmer water fauna became established.

Climate warming would probably initiate a complex series of immigrations and extinctions of fish species over broad areas. The present distribution originated from a few major refugia that existed during the Pleistocene. As the glaciers waned, the proglacial lakes at the southern boundary of the ice receded northwards and provided ready access for fishes into what became modern lakes. These fishes acclimatized (Legendre and Legendre, 1984) and many species have never before been exposed to temperatures as high as those forecast by GCM models. If these lakes become warmer and less oxygen rich, at least some of these species could be extirpated in southern lakes and in shallower northern lakes in the region. Warmer water species could invade and the short-term effects of the potential invasion of numerous species is largely unknown. However, the destabilizing effects of past introductions of species into a variety of aquatic ecosystems is well documented.

There is some evidence that the total biomass of fish communities may not change significantly in response to climate warming. For example, there is no relation between temperature and yield in Canadian lakes (Schlesinger and Regier, 1982). Schlesinger and Regier (1983) suggest that this is a result of lakes having fish species with different thermal optima. Hence, although the yield of individual species does vary with temperature, there are trade-offs so that the aggregate is independent of temperature. This inference is based on stable fish communities under present temperature conditions, and it is unknown whether this will be maintained after climate warming occurs and the fish communities become destabilized.

Stream Communities

The break-up of river ice represents a major annual disturbance even in high latitude rivers. It is characterized by large increases in current velocity, stage, water temperature, concentrations of suspended materials and substrate scouring (Prowse, 1994). Scrimgeour *et al.* (1994) have argued that if the intermediate disturbance hypothesis (Connell, 1978) is applicable to streams, a shift from intense, dynamic break-ups to less severe, thermal events could potentially alter community richness. Such a shift could well occur in north-flowing rivers as a consequence of climatic warming. Moreover, alteration to flow regimes as a result of climate warming could also alter the structure of aquatic and riparian communities in rivers by affecting ice break-up intensity. Despite the potential importance of river ice break-up, there are few empirical investigations thus far that have addressed the effects on river communities.

Small streams represent shallow-water habitat that has close linkages to riparian vegetation. Because stream communities can be highly dependent on organic matter derived from riparian vegetation, temperature effects can be important. Microbial processing of organic matter and organismal growth and survival may be strongly linked to temperature. Oswood *et al.* (1992) have argued that a 4°C increase in water temperature over a four-month ice-free season can represent an increase of about 500 degree days, or roughly a 50–100% increase in yearly cumulative warmth among Alaskan rivers and streams. An increase in the length of the ice-free season and decreased freezing to the substrate in shallow areas would, potentially, be beneficial to invertebrates and fish (Oswood *et al.*, 1992). However, the potential effect of increased temperature on net processing of organic matter by invertebrate shredders and microbes in concert is less clear. Irons *et al.* (1994) examined litter processing rates along a latitudinal gradient involving streams in Costa Rica, Michigan and Alaska. Surprisingly, the processing rates were similar at the Michigan and Alaskan sites, which should not be the case if temperature were the principal factor controlling the breakdown rate. The authors suggest that the relative importance of invertebrate versus microbial processing may change on a latitudinal gradient, with invertebrates being more important in high latitude systems. A potential consequence is that processing rates may not increase in association with climate warming.

RESPONSES TO CLIMATE WARMING IN SELECTED ARCTIC ECOSYSTEMS

High Arctic Ponds

Arctic ponds are standing water bodies less than 2 m deep that freeze to the bottom, whereas lakes are sufficiently deep that water remains unfrozen under the ice cover (Sheath, 1986). As an example of high arctic ponds, this paper focuses on a suite of 36 ponds from Cape Herschel, Ellesmere Island (Figure 1b), which have been studied extensively over a number of years. Cape Herschel (78°37′ N, 74°42′ W) is a rugged peninsula (2 × 5 km^2) of high relief (0–285 m asl) which lies close to the northern extent of the North Water, a polynya that remains ice free most of the year.

The physical and chemical characteristics of the study ponds have been summarized in Douglas and Smol (1994). The ponds are small (the largest being about 1 km long) and shallow (typically <1 m deep, and usually $<0{\cdot}5$ m deep). These clear, oligotrophic ponds are completely frozen for at least 10 months of the year. During the short summers, water temperature warms substantially to a recorded maximum of 17°C. Temperatures, however, are typically less than 10°C. The majority of ponds are alkaline, reflecting the presence of calcareous tills overlying the granitic bedrock. Major ion concentrations are similar at most sites, and change over the summer owing to the combined effects of cryoconcentration, snowmelt dilution, evaporation and other factors (Douglas and Smol, 1994). Because the ponds are frozen most of the year, vertebrate predators are generally absent, with the possible exception of some migratory birds. As a result, the ponds are teeming with *Daphnia*, copepods, chironomids, rotifers and other invertebrates (Nogrady and Smol, 1989).

Small, shallow water bodies, such as the Cape Herschel ponds, may be especially sensitive to climate warming (Smol *et al.*, 1991; Douglas *et al.*, 1994), but will be dependent on the amount and timing of precipitation that the area might receive. However, assuming air temperature will increase, the following scenarios are proposed.

With increased temperatures and a longer growing season, many limnological changes would be expected. Presently, the ponds are only free of ice from late June or early July, and may refreeze by August (Douglas and Smol, 1994). The length of the ice-free period, and hence growing season, would significantly increase with small warming (Figure 3b). Total annual primary production may increase in the ponds, and more complex periphytic communities could also develop (Douglas and Smol, 1993). Decomposition processes would accelerate, with concomitant increases in nutrient cycling. Assuming that precipitation does not also increase, evaporation and concentration in the ponds would be enhanced, resulting in elevated conductivities. Since many of the ponds are presently very shallow, some could desiccate completely.

Tree-line Lakes

Strong climatic and limnological gradients exist across north–south transects through the Yukon and North-west Territories (Pienitz, 1993; Pienitz and Smol, 1993; Pienitz *et al.*, 1995). For example, a typical transect that included 59 lakes from Whitehorse to north of Tuktoyaktuk showed most lakes to be dilute (mean conductivity = 160 μS/cm) and slightly acidic to alkaline (pH = 5·9–9·3). Although several limnological variables could change polewards, dissolved organic carbon (DOC) and dissolved inorganic carbon (DIC) concentrations clearly change. DOC and DIC also appear to be important chemical variables influencing diatom species assemblages in lakes and some streams (Miller *et al.*, 1992; Pienitz and Smol, 1993). Water depth, surface water temperatures, ice and snow cover are also variables that influence the diatom assemblages (Pienitz *et al.*, 1995). For lakes located near the arctic tree-line, changes in DOC may be especially important. Of particular significance to northern lakes, DOC is closely related to the composition and density of terrestrial vegetation (e.g. spruce trees) in the catchment. Consequently, any climate-induced changes in the position of tree-line will affect the limnology of lakes at this ecotonal boundary. Forsberg (1992) estimates that, in Fennoscandinavia, increased warming would promote higher DOC levels in lakes. In areas north of tree-line, DOC concentrations in groundwater vary with overlying vegetation type (Kling, 1995), and DOC concentrations may change if climate alters the vegetation composition.

DOC has an important influence on, for example, arctic diatom species assemblages (Pienitz and Smol, 1993), although the specific causative factors are not yet determined. In addition, DOC concentrations control the activity of decomposition by bacteria in some lakes on the Alaskan North Slope (Kling, 1995). Lakes with the smallest lake volume:catchment area ratios are likely to be most affected. As with all these regions, however, lake water DOC is only one of several variables that will be changing with a warming climate.

Changes in DOC and related chemical changes would be only one limnological effect of trees migrating to a lake's catchment. For example, evapotranspiration rates will change, with resulting hydrological effects. The amount of wind mixing that a lake would be subjected to would decrease, thus affecting the physical limnology of the system. Many related limnological variables would also be affected.

A few saline lakes are found in the central Yukon (Pienitz *et al.*, 1992). Preliminary data indicate that changes in these high latitude saline lakes are similar to those recorded in more southern regions (Pienitz *et al.*, 1992; Veres *et al.*, 1995). If, with greenhouse warming, evaporation rates are higher, and precipitation does not increase concurrently, these athalassic lakes would increase in salinity, with accompanying dramatic effects on the aquatic biota. Moreover, because these lakes are situated in low precipitation areas, possible further decreases in snow and rainfall, as well as increased evaporation rates, may greatly decrease their size or they may disappear completely.

Shallow Tundra Lakes and the Food-web

Shallow arctic tundra lakes can be surprisingly productive. Some in the western Arctic have annual productivities higher than temperate lakes much further south (Ramlal *et al.*, 1994). Myriad such lakes on the Tuktoyaktuk Peninsula support the feeding activities of abundant migratory and non-migratory populations of broad whitefish and northern pike (Hesslein *et al.*, 1991). A detailed lake carbon budget demonstrated that benthic photosynthesis provided 50% of the organic carbon available to the food-web of the lake, while phytoplankton (20%) and allochthonous carbon (30%) provided lesser amounts (Ramlal *et al.*, 1994). Benthic photosynthetics were most likely carbon limited as in many lakes (Hecky and Hesslein, 1995). Carbon dioxide limitation of benthic photosynthesis occurs within a diffusion-limited boundary layer overlying benthic algae. A $2 \times CO_2$ atmosphere would retard lake degassing and raise PCO_2 in bulk solution. Consequently, benthic photosynthesis could be enhanced if the lakes maintain a positive hydrological balance, and benthic carbon production would provide an even larger share of total organic

MACKENZIE DELTA LAKES - CASE STUDY

OBJECTIVE:

SIMULATION MODEL OF INFREQUENTLY FLOODED LAKES

INPUT = 12 yr Record of:

a) Temperature
b) Solar Radiation
c) Precipitation (Rain & Snow)
d) Water Levels in the Mackenzie River

OUTPUT = Lake Water Level / Water Balance

RESULT:

SCENARIO (1) - Present Conditions
↓
Negative Water Balance If Not Flooded by River

SCENARIO (2) - CCC Model, 2x CO2, +12% Precip
↓
Water Levels Decline More Rapidly Between Flooding Events

SCENARIO (3) - CCC Model, 2x CO2, -10% Precip
↓
Water Levels Decline Much Faster
↓
Many Lakes Could Disappear Within 10 yr If Not Flooded

CONCLUDE:

ICE-JAM FLOODING MAY BE CRITICALLY IMPORTANT

Figure 5. A conceptual summary of the results of a simulation model (Marsh and Lesack, 1996) to evaluate the potential effects of climate change on lakes in the Mackenzie River, Delta. The study focused on lakes that are not necessarily flooded by the river each year (i.e. infrequently flooded lakes) and utilized climate change scenarios forecast by the GCM of the Atmospheric Environment Service, Environment Canada

carbon flow in these shallow lakes. Warmer atmospheric temperatures and longer open water periods would also contribute to higher productivity.

Mackenzie Delta Lakes: an Example of River–Lake Synergism

Lakes in the Mackenzie River Delta may be particularly sensitive to climate change. Water levels are controlled primarily by Mackenzie and Peel River discharge to the delta, growth of river ice, ice break-up and jamming, and changes in sea level (Marsh and Schmidt, 1993). The resulting variations in channel water levels control the flooding of approximately 25 000 delta lakes (Marsh and Hey, 1989). Since these delta lakes tend to have a negative water balance between flooding events (Marsh, 1991b), and changes in the hydrological regime of the Mackenzie Basin could have significant environmental effects (Marsh and Ommanney, 1991).

Figure 5 shows a model that simulates the water balance and lake water levels for Mackenzie Delta lakes that are infrequently flooded (Marsh and Lesack, 1996). A 12-year record of daily air temperature, solar radiation, rain and snowfall, and river levels is used as model input to simulate ice growth rate, maximum ice thickness, melting of the ice cover, duration of the ice-free period and evaporation and lake water levels during the ice-free period (Figure 5). It successfully predicts the major hydrological dynamics of a representative lake over the period of record.

Under present climatic conditions, the 'model lake' consistently shows a slight negative water balance if not flooded by the river. For the Mackenzie Delta region, the GCM of the Canadian Climate Centre (Boer *et al.*, 1990) indicates an increase in air temperature of 3°C for JJA, and 5°C for DJF, and precipitation increases of 15 and 10%, respectively, are forecast. Under this 2 × CO_2 scenario, water levels in the 'model lake' decline more rapidly between flooding events than under the present climate. Since it is generally recognized that forecasts of precipitation by GCMs are not as reliable as temperature, the effect of a 10% precipitation decrease was also modelled. Under this scenario, water levels in the 'model lake' decline much faster between flooding events than at present and many lakes within the delta would disappear within 10 years. In both climate warming scenarios the lake ice cover is generally thinner and the ice-covered period is consistently shorter.

The above evidence indicates that, despite potential increases in precipitation, evapotranspiration may increase sufficiently so that less water will be available to lakes than under the present climate. Also, the model demonstrates the potential importance of ice-jam flooding in maintaining the water balance of delta lakes in north-flowing river systems. Significant changes in the frequency or severity of ice-jam flooding by itself would have a major effect on these lakes.

Indirect evidence of the effects of flooding is provided by the hydrological regime of the Peace–Athabasca Delta, which has been changed significantly by the Bennet Dam on the Peace River. Simulations of natural and regulated water levels in the delta (Farley and Cheng, 1986) demonstrated that peak water levels in the delta have been significantly lower than under natural conditions. Less frequent flooding of perched lake basins within the delta has been evident with significant effects on the delta ecosystem (Peace–Athabasca Delta Implementation Committee, 1987). A number of perched lakes last received water in 1974, and even during 1990, when the highest flows on record occurred, the perched lakes did not receive any floodwater. Ice-jams represent a critical process that, historically, had raised floodwaters sufficiently to flood these lakes (Prowse and Lalonde, 1996).

Alaska Lakes and Rivers, Nutrient Cycling and the Food-web

Global climate change may affect arctic aquatic ecosystems directly through increased temperatures or through increases or decreases in nutrients to these systems. Nutrient input to Toolik Lake is primarily through the inlet streams and occurs during spring snowmelt, with up to 30% occurring in the first 10 days of stream flow (Whalen and Cornwell, 1985). The primary productivity in lakes and ponds (Miller *et al.*, 1986; Kling *et al.*, 1992) and rivers (Petersen *et al.*, 1985) on the North Slope of Alaska is very low owing to low levels of phosphorus or colimitation of phosphorus and nitrogen. The range of chlorophyll in lakes and ponds is 0·3 to 148 μg/l, and in rivers it is 0·1–9·4 $\mu g/cm^2$ (Kling *et al.*, 1992).

Understanding of these effects is often complicated by complex food-web interactions. Arctic lake ecosystems near Toolik Lake, Alaska, are trophically simple (Kling *et al.*, 1992; McDonald *et al.*, 1996; Figure 6). Lakes may have some or all of the trophic linkages shown; food-web complexity is determined by the fish species present, landscape position, geological history and lake depth. Typically, ponds in the area have no fish. Rivers in the area have grayling in them, and in some cases anadromous arctic char.

In a survey of lakes in the Alaskan arctic coastal plain, no significant correlations of zooplankton species and chlorophyll *a* concentration were found (Kling *et al.*, 1992). O'Brien *et al.* (1992) found that with a 12-fold experimental increase in primary production, grazing zooplankton increased less than twofold. Although microplankton densities appear to be correlated with increased primary production (Rublee, 1992), fish predation plays an important role in structuring the macrozooplankton community.

Previous work has shown that arctic lakes and rivers near Toolik have strong benthic–pelagic coupling through the food-web. In lakes, the distribution of the large snail *Lymnea elodes* is controlled by lake trout

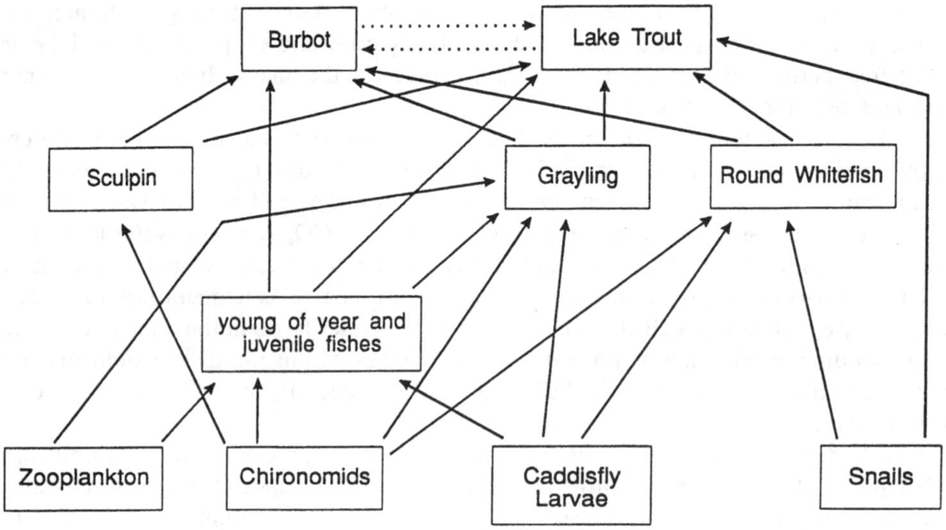

Figure 6. Lake trout are the dominant predator in the Toolik Lake food-web in arctic Alaska because of their numeric superiority and large adult size (based on McDonald *et al.*, 1996)

(*Salvelinus namaycush*) (Merrick *et al.*, 1992). In the absence of lake trout, algal biomass limits snail growth and abundance (Hershey, 1992a). Chironomids dominate the benthic insect fauna found in lakes; their abundance, biomass and community structure are controlled by slimy sculpin (*Cottus cognatus*), invertebrate predation and habitat complexity (Goyke and Hershey, 1992). In rivers, macroinvertebrate consumers are food limited (Hiltner and Hershey 1992; Peterson *et al.*, 1993).

Lake trout are the dominant fish in most large lakes in the Toolik Lake area (Figure 6), but arctic char (*Salvelinus alpinus*) hold a similar position in the food chain in some lakes. Lake trout control the distribution and abundance of large snails (Hershey, 1992b; Merrick *et al.*, 1991), slimy sculpin (McDonald and Hershey, 1992; Hanson *et al.*, 1992), and most likely limit recruitment and in-lake distribution of all salmonines (McDonald *et al.*, 1992). Arctic grayling (*Thymallus arcticus*), present in lakes, feed primarily on zooplankton when small (O'Brien *et al.*, 1979), and switch to emerging and ovipositing insects as they get larger (Merrick *et al.*, 1992). Grayling also occur in rivers, and feed primarily on drifting insects (Hobbie *et al.*,1995). Round whitefish (*Prosoprium cylinbdraceum*) feed on small-bodied snails (Merrick *et al.*, 1992). Burbot (*Lota lota*) feed primarily on sculpin.

Lake trout act as a keystone predator in these systems. However, because of the low food supply and low temperatures, lake trout are near their physiological limits for survival, and may be particularly sensitive to changes in temperature and food supply (McDonald *et al.*, 1996). When the temperature rises, the fish's metabolic rate increases, and if food availability decreases, stays the same or increases only slightly, then the growth rate of the fishes will decline (Brett and Groves, 1979). Because the fishes in the Toolik Lake area are already food limited, subsequent increases in temperature or decreases in food could have deleterious effects on the populations. Modelling of young lake trout in the epilimnion of Toolik Lake has suggested that these small fish will not survive with the current availability of food and a 3°C increase in water temperature (McDonald *et al.*, 1996). Other fishes, such as arctic char, sculpin and grayling, that use the epilimnetic areas extensively, may be similarly affected. However, sculpin and grayling, in the absence of large lake trout predators, may change their use of habitats and resources (McDonald and Hershey, 1992; McDonald *et al.*, 1992). Changes in the growth rates of fishes, especially predators, can greatly affect the survival of the population, and may result in cascading effects through the entire trophic structure (Carpenter *et al.*, 1985).

In addition to changing temperature, changing precipitation will affect these aquatic ecosystems. Current precipitation levels in the Toolik area are very low, and if precipitation decreases, nutrient inputs into the Toolik region could be further reduced. Resulting decreases in primary production could cause cascading effects and reduce biomass at higher trophic levels (Sommer, 1979). Alteratively, precipitation could increase, or melting permafrost could increase the nutrient loading to the lakes. Primary productivity will respond to an increase in nutrients (Hobbie *et al.*, 1995), but under the extremely oligotrophic conditions found in arctic Alaskan lakes, primary consumers appear only weakly coupled to planktonic primary productivity. Even after three years of a 10-fold experimental nutrient addition to limnocorrals in Toolik Lake, increases in zooplankton abundance and biomass of only two- to three-fold occurred (O'Brien *et al.*, 1992). It is unlikely that a 10-fold increase in nutrients to Toolik Lake from rainfall or melting of permafrost in the watershed, needed to produce a doubling of the available zooplankton, will occur. In the Kuparuk River near Toolik Lake, primary productivity and algal biomass increased with experimental nutrient addition (Peterson *et al.*, 1993). The invertebrates response also showed them to be food limited (Hershey *et al.*, 1988). Growth of adult grayling has typically increased with fertilization (Deegan and Peterson, 1992), but growth is also highly correlated with stream flow (Hobbie *et al.*, 1995). At low flows, low growth of adult fish occurs in the Kuparuk River (Hobbie *et al.*, 1995), but young-of-year grayling grow better with lower flow (L. Deegan and B. Peterson, unpublished data). Temperature increases in the rivers naturally coincide with low flow conditions, but with the increased temperatures associated with climate change the adult grayling metabolic demands may further reduce their growth at low flows. As grayling in these systems are food limited, the increased metabolic costs may result in decreased survival of riverine populations. Temperature increases may also alter the biogeography of stream insects and fish through an increase in degree–days and the amelioration of colder thermal regimes, allowing organisms to migrate northwards (Oswood *et al.*, 1992).

It appears that global warming may have the most significant effect on the predators of these, arctic Alaskan aquatic ecosystems. The predators in these systems are long lived, and population changes owing to recruitment failure may not be reflected in the adult populations for many years. However, the eventual loss of top predators from these systems will most likely cascade through the food-web, affecting the structure and function of both the benthic and planktonic communities (Hershey, 1990; Goyke and Hershey, 1992; Hanson *et al.*, 1992; O'Brien *et al.*, 1992).

Northern Peatlands

Most of the wetlands of discontinuous and continuous permafrost regions are peatlands (National Wetlands Working Group, 1988). The location of a peatland is related to regional climate controls on precipitation and evapotranspiration and is thus controlled by the water balance. The most important variable is the position of the water table because it controls the size of the aerobic and anaerobic layers in a peatland, and thus biomass productivity and decomposition, and ultimately peat accumulation. The degree of anaerobicity also determines, in part, the distribution of plants, and the magnitude of the exchange of carbon gas with the atmosphere and the export of dissolved organic carbon. The water table is also a key variable in determining the rate and magnitude of runoff. A secondary climate variable that is critical to peatland function is temperature. Productivity and decomposition rates are related to temperature once the general level of saturation of the peatland has been accounted for.

The wetlands in the interior of region 2 are largely bryophyte-dominated peatlands. These are a result of cool temperatures and an excess of precipitation over evapotranspiration (Gignac and Vitt, 1994). Along the southern boundary of region 2, in the grassland and low boreal forest, seasonal drought limits wetlands to *Typha* marshes and *Salix* swamps. Progressing northwards, a cooler, moister climate limits the frequency and severity of seasonal droughts and the wetlands become sedge- and bryophyte-dominated fens and bogs.

Climate produces variations in the peatland form (Belland and Vitt, 1995). In the southern areas of region 2, peatlands are spatially restricted and the predominant peatland form is the basin bog, basin fen or channel fen. Extensively developed peatlands are a northern feature (Nicholson, 1994). In Canada and Finland, peatland forms have been demonstrated to occur in distinct geographical bands related to climate

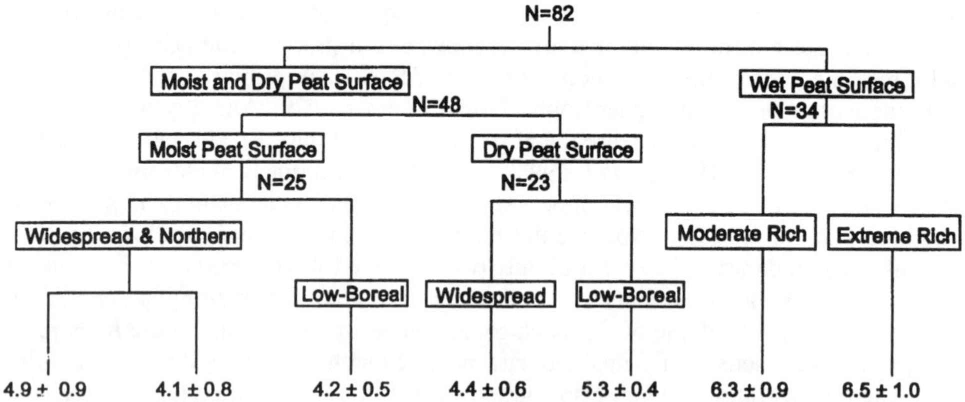

pH	4.9 ± 0.9	4.1 ± 0.8	4.2 ± 0.5	4.4 ± 0.6	5.3 ± 0.4	6.3 ± 0.9	6.5 ± 1.0
Ca (mg/l)	8.7 ± 12.3	5.3 ± 2.4	1.7 ± 0.8	5.1 ± 2.8	18.5 ± 17.7	19.6 ± 17.2	57.0 ± 47.5
Mg (mg/l)	2.5 ± 3.2	2.2 ± 1.8	0.4 ± 0.3	1.7 ± 1.3	4.0 ± 2.7	4.6 ± 3.6	11.7 ± 7.7
Na (mg/l)	4.1 ± 5.8	2.8 ± 2.0	5.1 ± 2.6	3.9 ± 2.6	8.6 ± 5.5	7.3 ± 11.5	7.5 ± 2.3
Gs (days)	169 ± 18	170 ± 12	191 ± 3	174 ± 23	187 ± 6	186 ± 9	179 ± 21
Temp (oC)	-2.3 ± 2.8	-2.5 ± 1.5	1.1 ± 0.6	-1.6 ± 3.6	0.5 ± 0.6	0.1 ± 1.4	-0.7 ± 3.1
Tspp (mm)	248 ± 82	235 ± 54	343 ± 18	267 ± 86	346 ± 56	337 ± 54	301 ± 83
Ypp (mm)	398 ± 91	398 ± 49	486 ± 12	417 ± 87	498 ± 72	482 ± 65	453 ± 87
Height (dm)	1.8 ± 1.0	2.1 ± 0.8	2.4 ± 1.0	3.4 ± 1.3	2.5 ± 0.8	1.4 ± 0.7	1.0 ± 0.9
Σ Sphagnum	10.4	14.0	11.5	11.4	4.1	1.7	0.7
Σ Amblystegiaceae	2.4	2.0	1.1	1.3	1.1	6.0	7.4
Sample size (N)	15	3	7	14	9	27	7
Community group	1	2	3	4	5	6	7
Peatland Type	Widespread Poor Fens	Peat Plateaus with Thermokarst pools	Low-Boreal Bogs	Bogs & Peat Plateaus without Thermokarst pools	Low-Boreal Dry Poor Fens	Moderate Rich Fens	Extreme Rich Fens

Figure 7. TWINSPAN dendrogram outlining the division of 82 peatlands in the Mackenzie River basin into seven TWINSPAN bryophyte stand groups and seven peatland types. Values are the means and standard deviations of climatic and environmental gradients significant on the first DCCA axis for each TWINSPAN stand group

(Ruuhijärvi, 1983; Wells and Hirvonen, 1988). Permafrost supports the development of peat plateaux, thermokarst pools, ice wedge polygons, palsas and peat mounds.

Past climate has had a significant effect on the development of northern peatlands. Basal dates of peatlands in continental western Canada demonstrate that peatland formation was prevented during the early Holocene, south of latitude 54°30′ until after 6000 years BP (Zoltai and Vitt, 1990). North of latitude 54°30′ and in the mountains where it was cooler and moister, peatlands began to form as early as 8000 BP. On the North Slope of Alaska, organic soil accumulation began around 6000–8000 BP (Schell and Zieman, 1983; Eisner, 1991). In general, peatlands in North America have been expanding at a constant rate (Gorham and Janssens, 1992). A few peatlands are older than 10 000 years BP, but most range between 10 000 and 4000 years old.

Figure 7 shows a classification scheme relating peatland type, peatland form and climate. Seven peatland types are identified and it is apparent that there is a large range in floral, pedological, chemical, mineral and environmental characteristics, all of which will be affected by changes in temperature and the water balance.

From a climatological perspective the seven peatland types referred to above can be treated as only two distinct peatland types, based on the physical characteristics of the peat and on the dominant vegetation. The surface of ombrogenic peatlands receives water from precipitation only. In contrast, rheogenic peatlands, such as fens and swamps, may receive additional inputs from surface and subsurface flow systems. The sources of water to a peatland's surface determines how responsive the peatland water storage is to atmospheric forcings. The ombrogenic systems are very responsive and storage changes are large and variable, while rheogenic systems are less responsive and change less. Ombrogenic peatlands are more susceptible to drought than peatlands that receive some groundwater (Siegel *et al.*, 1995). However, since the rheogenic system is dependent on external sources of water, any climate forcings that influence the basin in which the rheogenic peatland is located will result in a change in storage in the peatland.

It is commonly believed that peatlands evapotranspire at rates similar to those of open water surfaces (e.g. Verry, 1988), but recent studies have found peatlands usually evaporate less than lakes (Roulet *et al.*, 1997). If evaporative demand increases, the mechanisms that supply moisture to the surface will ultimately control the rate and magnitude of water loss. Bogs are dominated by mosses that have no active internal system that conducts water. Fens, dominated by vascular plants, are more effectively coupled to the atmosphere. This coupling is physiologically controlled by stomatal and root conductance (Lafleur, 1988). Desjardins *et al.* (1994) measured the regional sensible and latent heat fluxes from the continuous peatland of the Hudson Bay Lowland and observed that the latent heat flux dominated in the fens, but the latent and sensible heat flux were of equal magnitude in areas where bogs were more frequent.

The thermal regime is closely related to water storage. The drier surface layer of the peatland is an efficient insulator since air has a much lower thermal conductivity than water, while the saturated zone has a greater heat capacity (Roulet *et al.*, 1997). This results in a heat storage component similar to that of lakes, wherein substantial energy exchange is required to heat or cool a peatland. However, unlike a lake, the subsurface thermal exchange process is limited by a much smaller thermal conductivity and lack of convection.

When peatlands are frozen, they have a very limited storage capacity and many act as flow-through systems (Roulet and Woo, 1988), but when the frost table is at its lowest during the summer, the hydraulic properties of the peat itself control the storage and flow dynamics of the peatland (Roulet and Woo, 1986). Permafrost confines the water flow in peatlands to a thin 'active' layer, thus restricting the interaction of surface and groundwater in peatlands. As a result, northern peatlands will be more affected by climate change than their southern counterparts. An equally critical issue is the stability of permafrost in, and adjacent to, peatlands. Because of the unique thermal properties of peat, peatlands can contain permafrost in areas that are, in general, free of permafrost. However, permafrost degradation adjacent to peatlands could have a catastrophic effect on the exchange of water, and lead to large-scale mass wasting such as is observed in the collapse scars in peat plateaux and palsas.

The results from two simple, physically based hydrology and climate models to assess changes in water storage in northern fens owing to changes in precipitation and temperature for $2 \times CO_2$ climate scenarios are presented in Figure 8. Roulet *et al.* (1992) assumed a 3·5°C increase in temperature and 1 mm d^{-1} increase in precipitation, while Lafleur (1993) did several model runs for temperature increases of 2, 4 and 8°C for June, July and August. The results from both studies indicate a substantial negative change in water storage in fens. Based on these results, it is reasonable to assume that an increase in temperature could reduce water storage in northern peatlands, even with a persistent, but small, increase in precipitation. Depending on the soil moisture characteristics and the physical properties of the peat, the estimated changes in storage could be an increase in depth to the water table of 10–20 cm during the summer months. Few studies have examined the change in thermal regime in northern peatlands, but Roulet *et al.* (1992) modelled a 0·8°C increase in the near-surface summer temperatures in peat for a 3·5°C increase in air temperature.

Table IV outlines the anticipated major effects of climate warming on the distribution, water relations and carbon cycle in peatlands. As permafrost melts, new drainage channels could form, rapidly removing the excess water, resulting in improved drainage and a drawdown in local water tables. Alternatively, local topography and fine-textured soils could result in the ponding and retention of much of the water currently frozen, creating many new ponds and lakes. Integrated drainage patterns appear on air photographs in permafrost-dominated peat plateaux, soon after forest fires. Forest fires have the effect of removing tree and lichen cover. This increases the amount of snow cover sitting on the peatland surface, slightly insulating the peat over the winter months. During the following summer, more heat is absorbed by the dark-hued ash and dead, standing trees. A greater amount of heat is transmitted into the peat, resulting in large-scale melting of

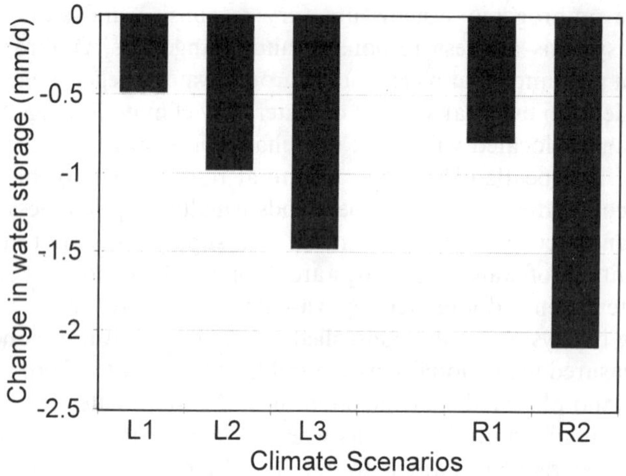

Figure 8. Change in the daily summer (JJA) moisture storage in northern peatlands based on two different climate change modelling scenarios. Cases L1, L2 and L3 (Lafleur, 1993) assume reference climate temperature, wind speed, net radiation and precipitation, and then perturbate the reference temperature by +2°C (L1), +4°C (L2) and +8°C (L3). Cases R1 and R2 (Roulet *et al.*, 1992) assume reference climate temperature, wind speed, net radiation and precipitation, and then perturbate temperature by +0·8°C at −10 cm depth of peat and increase precipitation +1 mm/d. R1 is the loss of water from a non-floating surface peatland and R2 is the loss from a floating surface peatland

the permafrost, collapse of the plateau surface and development of integrated drainage patterns. Under present climatic conditions, a forest fire in a permafrost peatland results in a temporarily wetter landscape. Over time the melting ceases, drainage subsides and the peat plateau is regenerated with the re-establishment of *Sphagnum* hummocks (Zoltai, 1993). This cycle has been documented to take anywhere from 600 to 2000 years to complete. Because of the uncertainties associated with precipitation predictions and the net effect

Table IV. Potential effects of a warmer climate on northern peatlands. ↑ and ↓ indicate an increase or a decrease in the various categories/effect

Regional effects		
Area of peatlands	↓	↓
Change in peatland form	↑	↑
Change in plant communities	↑	↑
Fire frequency	↑	↑
Site-specific effects		
Water relations in peatlands		
Evaporation	↑	↓
Runoff to streams	↑	↓
Growth of vegetation		
Trees	↓	↑
Mosses	↑	↓
Primary production	↑	↑
Decomposition	↓	↑
Carbon cycling		
Peat accumulation	↑	↓
Nutrients available for growth	↓	↑
CO_2 fixation	↑	↑
CO_2 release	↓	↑
CH_4 release	↑	↓
DOC release	↑	↑

that permafrost melting will have on local peatland water tables, we will discuss the effects of climate change on peatlands under two scenarios; a net rise versus a net fall in the water table.

An increase in mean annual temperature will shift the zone where peatlands commonly occur. Peatland losses are anticipated to occur along the southern boundary, where a drop in the water table will increase decomposition of the accumulated peat, and may convert marginal peatlands into *Salix*- or *Picea mariana*-dominated swamps. In some cases, wetlands will cease to exist because of drought, and an increase in forest fires will serve to enhance the rate of peatland loss by removal caused by burning (Wein, 1983). Predictions from a climate change model (2°C increase) put forth by Gignac and Vitt (1994) indicate that the southern boundary of peatlands in northern Alberta will migrate to approximately 200–300 km north of its present position, and that only bryophytes normally found in upland boreal forest communities will exist. Predictions on the impact of a 2 × CO_2 climate change on peatland communities in the Mackenzie River Basin indicate that peatlands are likely to disappear from all of Alberta, and the southern limit will lie just beyond the border between Alberta and the North-west Territories (Nicholson and Gignac, 1996). Widespread poor fens, bogs with peat plateaux, moderate-rich fens and extreme-rich fens (Figure 6) will stretch north of the border to Inuvik. Peat plateaux with thermokarst pools will be pushed northwards to Inuvik, while low boreal bogs, the most southern group, will lie at the southern end of Great Slave Lake.

Gorham (1995) estimates that peatlands are growing in height at a rate of 0.5 mm/yr. Peat accumulation is dependant on the balance of primary production and decomposition. Primary productivity is controlled by nutrients and temperature, whereas decomposition is dependent upon temperature, nutrient levels, depth to the water table and nature of the material. Climate warming is expected to increase decomposition in peatlands owing to increased fungal and bacterial metabolic rates, and an increase in the aerobic zone. Recently, however, it has been demonstrated that in subsurface peats (30–40 cm), an increase in temperature and aeration does not lead to an increase in decomposition (Hogg *et al.*, 1992). Decay resistance of the subsurface peat was attributed to the recalcitrant nature of the material. Furthermore, studies of peat decomposition near Toolik Lake showed that increases in temperatures of less than 9°C had no effect on organic carbon respiration and nitrogen mineralization (Nadelhoffer *et al.*, 1991). Higher water tables have the effect of increasing the rate of peat accumulation by reducing the depth of the aerobic zone, and reducing the exposure of the surface material to aerobic decomposition by fungi and bacteria. Thus, temperature increases on wet peatlands are anticipated to have a negligible effect on the decomposition of the peat, if the water table stays constant or rises.

Exports of carbon from peatlands include particulate, dissolved inorganic and dissolved organic carbon from stream flow, carbon loss through deep seepage, methane efflux and CO_2 efflux. The magnitude of the fluxes of carbon in and out of a peatland is very small and this makes it extremely difficult to assess the contemporary carbon dynamics of northern peatlands, and estimating the future changes in carbon exchange is impossible at present.

A variety of different studies provide estimates of the sensitivity of the carbon balance to climate variability and change. Evidence suggests that the soil CO_2 efflux increases linearly with reduced water table, while the flux of CH_4 decreases exponentially (Moore and Knowles, 1989). The drainage of peatlands can convert them from a net CO_2 sink to a source (Armentano and Menges, 1986). A peatland can be a large sink for CO_2 one year and a large source the next year (Shurapli *et al.*, 1995). A change in water table position of only −10 cm is sufficient to convert a northern forested peatland from a source to a sink for atmospheric CH_4 (Moore and Roulet, 1993). In Finland, the annual soil efflux of CO_2 increased when bogs and fens were drained by about 40 and 90% respectively (Martikainen *et al.*, 1994). Laine *et al.* (1997) found that drained peatlands underwent a net carbon loss. Model results indicate that increased temperature can stimulate plant and soil respiration, but, providing conditions are moist, the losses of biomass are more than compensated for by increased photosynthesis, which results from the redistribution of nitrogen from soil to plants. However, when warming is accompanied by decreases in soil moisture, increased photosynthesis only partially compensates for increased respiration. Under this latter condition there is substantial loss of nitrogen from the ecosystem (McKane *et al.*, in press). Such loss may affect the receiving water bodies

(Kling, 1995). Conflicting field observations suggest that present day wet tundra ecosystems may be either a source of CO_2 (Oechel *et al.*, 1993) or a sink of CO_2 (Shaver *et al.*, 1992).

The plant community structure would likely change with changes in groundwater storage (Bubier and Moore, 1994). In addition, it is not clear how ecosystem productivity might change in a CO_2-enhanced environment. For example, Alaskan ecosystems respond little to CO_2 enhancement without an addition of nutrients or a lowering of the water table (Oechel and Vourlitis, 1994; McKane *et al.*, in press). It is also important to consider the role of changing plant and soil conditions and processes on surface waters. Some northern peatland and wetland areas contain as much as 40–80% water surface (Kling *et al.*, 1991; Roulet *et al.*, 1994), which yield very large fluxes of CO_2 and CH_4 (Kling *et al.*, 1991; Hamilton *et al.*, 1994). Changes in moisture supply and thermal regime could lead to topographical and vegetation changes that could alter the water surfaces of northern peatlands and thus alter the delivery of CO_2 and CH_4 from surface waters to the atmosphere. Oswood *et al.* (1992) provide a thorough summary of how changes in thaw depth could increase DOC loading from land to streams, and how changes in riparian vegetation may affect the quality of water.

SUMMARY AND RECOMMENDATIONS

Region 2 is clearly rich in lakes, ponds, rivers and wetland habitats and represents a major water resource in North America. It is also, arguably, the most susceptible to climate warming because of the high magnitude temperature response to global warming and because of the prevalence of permafrost. This response will have direct effects on habitat and equally important indirect effects through the water balance. Region 2 also has the sparsest research base because of its size and small human population. Thus, many fundamental processes are little understood. These involve both the need for basic research and for long-term monitoring.

Climate

The present day climate in much of region 2 is not well known. Temperatures are measured primarily at settlements, most of which are on coasts, rivers or lakes. They are not representative of the hinterland of the Arctic Islands or much of the terrestrial mainland. Because so much precipitation involves snow, which in tundra areas is blowing snow, precipitation measurements and estimates are unreliable. Thus scenarios involving $2 \times CO_2 - 1 \times CO_2$ are more than normally speculative because $1 \times CO_2$ is, at present, unreliable, although the accuracy of estimations is improving rapidly. At present, the error in water balance calculations employing GCM scenarios could easily be 100% or more.

The solution to these problems of estimation is to increase the number of long-term monitoring stations in unpopulated locations. This is contrary to the present trend, which is to reduce the number. Other needs are to better understand snow depths, densities and sublimation rates, and general precipitation patterns from storms. These can make use of contemporary technology employing remote sensing in the form of passive microwave sensing and radarsat and weather radar. There are promising developments in this direction but the need for standard long-term monitoring still remains. The overall goal should be to develop reliable methods of extrapolation and interpolation from a spatially limited climate network. Palaeoclimatic material should be exploited where possible to help generate some of the long-term data that do not exist for the region.

Hydrological

The need for accurate water balances for lakes, and particularly for wetland systems, has been reiterated several times in this paper. This involves a number of required inputs. Simple, reliable models for determining rates of evaporation from different wetland systems need further development. Water-holding capacities of reservoirs (soil, lake and river) under different permafrost and pedological conditions need documentation.

The thermal behaviour of wetland systems needs to be modelled. In lakes and rivers, this is especially closely related to dates of freeze and thaw and depths of freezing. For example, in a given environment, at what depth does a pond become a lake? How does this relate to the thermal regime and the water balance? Basic information like this is essential for modelling the effect of a warmer world and it impinges on all aspects of the limnology of ponds and lakes. In peatlands the thermal regime operates in influencing permafrost active layer development, evapotranspiration and the water balance and the carbon budget.

The interrelationships between rivers with their basins through cold region processes is little understood. The example of the synergism of the Mackenzie Delta lakes with the river has been developed and this gives an example of the complexities of cold region hydrological systems. It is especially important with reference to the increasing demands to harness and divert north-flowing rivers for hydroelectric and other purposes. Even our limited contemporary knowledge indicates that the effects of diversions are major and mostly negative. They can also be compounded by climate change.

Limnological

Given the very large range in freshwater systems in region 2, our knowledge of the full limnological systems is very limited in spite of excellent local contemporary efforts. Two of the largest freshwater lakes in the world, Great Bear and Great Slave, are largely undocumented, in comparison to the Laurential Great Lakes, for example, although palaeolimnological studies indicate some interesting trends (Stoermer *et al.*, 1990). They represent lakes dominated by cold region processes. Not only are they major ecosystems in their own right, but they also play a major role in the entire lower Mackenzie Basin, including the delta. Smaller lakes and ponds have been somewhat better documented in certain environments but freshwater systems in most of the regions have not been studied.

The carbon budget of freshwater systems is important on a global basis but is largely unknown. The obvious importance of peatlands is through the large carbon reservoir that they represent and the likelihood that they can readily change from sinks to sources of CO_2 and other forms of carbon, if this is not happening already. In this fashion, region 2 and other circumpolar cold lands can interact with the global atmosphere by enhancing the greenhouse effect in a positive feedback.

Multidisciplinary

Another way in which freshwater systems can influence the global climate system is through their freshwater input to the Arctic Ocean. The longevity of sea ice is intimately linked to the freshwater input from north-flowing rivers and responds to any changes in the volume of this input. The Arctic Ocean is, in turn, strongly linked to temperate oceans through deep oceanic circulations. The polar sea is also closely linked to the sun and the surface energy balance through its surface albedo. Thus, the water balance of freshwater systems in region 2 has effects that extend well beyond the immediate environments. There are basic research needs to measure, understand and model this much larger water and heat balance system. In such avenues the needs of the limnological, hydrological, atmospheric and marine scientific communities to understand more fully high latitude processes have become intertwined.

Basic long-term monitoring and interdisciplinary research is necessary in the little-studied environments of region 2. Only then can reliable models be developed and their validity on the regional scale be tested.

REFERENCES

Andres, D. and Doyle, P. F. 1984. 'Analysis of breakup and ice jams on the Athabasca River at Fort McMurray, Alberta', *Can. J. Civil Eng.*, **11**, 444–458.

Armentano, T. V. and Menges, E. S. 1986. 'Patterns of change in the carbon balance of organic soil-wetlands of the temperate zone', *J. Ecol.*, **74**, 755–774.

Belland, R. J. and Vitt., D. H. 1995. 'Bryophyte vegetation patterns along environmental gradients in continental bogs', *Ecoscience*, **2**, 395–407.

Benke, A. C. 1984. 'Secondary production of aquatic insects', in Resh, V. H. and Roserber, D. M. (Eds), *The Ecology of Aquatic Insects*. Praeger, New York. pp. 289–322.

Bliss, L. C. (Ed). 1977. *Truelove Lowland, Devon Island, Canada: A High Arctic Ecosystem*. University of Alberta Press, Edmonton. p. 714.

Boer, G. J., McFarlane, N., Blanchet, J.-P., and Lazare, M. 1990. *Greenhouse gas induced climatic change simulated with the CCC second generation GCM*. Canadian Climate Centre, Atmospheric Environment Centre, Downsview, Ontario. 24 pp.

Bradley, R. S. and Jones, P. D. 1993. 'Little Ice Age summer temperature variations, their nature and relevance to recent global warming trends', *The Holocene*, **3**, 367–376.

Brett, J. R. and Groves, T. D. D. 1979. 'Physiological energetics', in Hoar, W. S., Randall, D. J., and J. R. Brett, J. S. (Eds), *Fish Physiology*, Vol. 8. Academic Press, New York. pp. 599–667.

Brown, R. J. E. 1970. *Permafrost in Canada — its Influence on Northern Development*. University of Toronto Press, Toronto.

Bryson, R. A. 1966. 'Air masses, streamlines and the boreal forest', *Geogr. Bull.*, **8**, 228–269.

Bubier, J. L. and Moore, T. R. 1994. 'An ecological perspective on methane emissions from northern wetlands', *TREE*, **9**, 460–464.

Burn, C. R. 1993. 'Permafrost, tectonics and past and future regional climate change, Yukon and adjacent Northwest Territories', *Can. J. Earth Sci.*, **31**, 182–191.

Carpenter, S. R., Kitchell, J. F., and Hodgson, J. R. 1985. 'Cascading trophic interactions and lake productivity', *BioScience*, **35**, 634–639.

Church, M. 1974. 'Hydrology and permafrost with reference to northern North America', in *Permafrost Hydrology*, Proceedings of Workshop Seminar 1974. Canadian National Committee for the International Hydrological Decade, Ottawa, Ontario. pp. 7–20.

Cohen, S. J. 1993. *Mackenzie Basin Impact Study, Interim Report 1*. Environment Canada, Atmospheric Environment Service. Downsview, Ontario, Canada. 163 pp.

Connell, J. H. 1978. 'Diversity in tropical rainforests and coral reefs', *Science*, **199**, 1302–1310.

Deegan, L. A. and Peterson, B. J. 1992. 'Whole river fertilization stimulates fish production in an arctic tundra river', *Can. J. Fish. Aquat. Sci.*, **49**, 1890–1901.

Desjardins, R. L., MacPherson, J. I., Schuepp, P. H., and Hayhoe, H. N. 1994. 'Airborne flux measurements of CO_2, sensible, and latent heat over the Hudson Bay Lowland', *J. Geophys. Res.*, **99**(D1), 1551–1562.

Douglas, M. S. V. and Smol, J. P. 1993. 'Freshwater diatoms from high arctic ponds Cape Herschel, Ellesmere Island, N.W.T.', *Nova Hedwigia*, **57**, 511–552.

Douglas, M. S. V. and Smol, J. P. 1994. 'Limnology of high arctic ponds Cape Herschel, Ellesmere Island, N.W.T.', *Arch. Hydrobiol.*, **131**, 401–434.

Douglas, M. S. V., Smol, J. P., and Blake, W., Jr. 1994. 'Marked post-18th century environmental change in high Arctic ecosystems', *Science*, **266**, 416–419.

Edlund, S. A. and Alt, B. T. 1989. 'Regional congruence of vegetation and summer climate patterns in the Queen Elizabeth Islands, Northwest Territories, Canada', *Arctic*, **42**, 3–23.

Eisner, W. R. 1991. 'Palynological analysis of a peat core from Imnaviat Creek, the North Slope, Alaska', *Arctic*, **44**, 279–282.

Farley, D. W. and H. Cheng. 1986. 'Hydraulic impact of flow regulation on the Peace–Athabasca Delta', *Can. Water Res. J.*, **11**, 26–42

Forsberg, C. 1992. 'Will an increased greenhouse impact in Fennoscandia give rise to more humic and coloured lakes?', *Hydrobiologia*, **229**, 51–58.

Gignac, L. D. and Vitt, D. H. 1994. 'Responses of northern peatlands to climatic change, effects on bryophytes', *J. Hattori Bot. Lab.*, **75**, 119–132.

Goodison, B. E. 1981. 'Compatibility of Canadian snowfall and snow cover data', *Wat. Resour. Res.*, **17**, 893–900.

Gorham, E. 1995. 'The biogeochemistry of northern peatlands and its possible responses to global warming', in Woodwell, G. M. and Mackenzie, F. T. (Eds), *Biotic Feedbacks in the Global Climate System: Will the Warming Feed the Warming*? Oxford University Press, New York. pp. 169–187.

Gorham, E. and Janssens, J. 1992. 'The paleorecord of geochemist and hydrology in northern peatlands and its relation to global change', *Suo*, **43**, 117–126.

Goyke, A. P. and Hershey, A. E. 1992. 'Effects of fish predation on larval chironomid *Diptera, Chironomidae* communities in an arctic ecosystem' *Hydrobiologia*, **240**, 203–211.

Gray, D. M. and Prowse, T. D. 1992. 'Snow and floating ice', in Maidment D. R. (Editor-in-Chief), *The Handbook of Hydrology*. McGraw-Hill Publishing Co, N. Y., 631–653.

Hamilton, J. D., Kelly, C. A., Rudd, J. W. M., Hesslein, R. H., and Roulet, N. T. 1994. 'Flux to the atmosphere of CH_4 and CO_2 from wetland ponds on the Hudson Bay Lowland', *J. Geophys. Res.*, **99**(D1), 1494–1510.

Hanson, K. L., Hershey, A. E., and McDonald, M. E. 1992. 'A comparison of slimy sculpin *Cottus cognatus* populations in arctic lakes with and without piscivorous predators', *Hydrobiologia*, **240**, 189–201.

Hare, F. K. and Thomas, M. K. 1979. *Climate Canada*, 2nd edn. John Wiley and Sons, Toronto. 230 pp.

Hecky, R. E. and Guildford. S. J. 1984. 'The primary productivity of Southern Indian Lake before, during and after impoundment and Churchill River diversion', *Can. J. Fish. Aquat. Sci.*, **41**, 591–604.

Hecky, R. E. and Hesslein, R. H. 1995 'Contributions of benthic algae to lake food webs as revealed by stable isotope analysis', *J. North Am. Benth. Soc.*, **14**, 631–653.

Hershey, A. E. 1990. 'Snail populations in arctic lakes: competition mediated by predation?', *Oecologia*, **82**, 26–32.

Hershey, A. E. 1992a. 'Effects of experimental fertilization on the benthic microinvertebrate community of an arctic lake', *J. North Am. Benthol. Soc.*, **11**, 204–217.

Hershey, A. E. 1992b. 'Effects of predatory sculpin on the chironomid communities of an arctic lake', *Ecology*, **66**, 1131–1138.

Hershey, A. E., Hiltner, A. L., Hullar, M. A. J., Miller, M. C., Vestal, J. R., Lock, M. A., Rundle, S., and Peterson, B. J. 1988. 'Nutrient influence on a stream grazer: Orthocladius microcommunities track nutrient input', *Ecology*, **69**, 1383–1392.

Hesslein, R. H., Capel, M. J., Fox, D. E., and Hallard, K. A. 1991. 'Stable isotopes of sulfur, carbon and nitrogen as indicators of trophic level and fish migration in the lower Mackenzie River basin, Canada', *Can. J. Fish. Aquat. Sci.*, **48**, 2258–2265.

Hiltner, A. L. and Hershey, A. E. 1992. 'Black fly response to phosphorus enrichment of an arctic tundra stream', *Hydrobiologia*, **240**, 259–265.

Hinzman, L. D. and Kane, D. L. 1992. 'Potential response of an arctic watershed during a period of global warming', *J. Geophys. Res.*, **97**, 2811–2820.

Hobbie, J. E. (Ed.) 1980. *The Limnology of Tundra Pond.* Dowden, Hutchinson and Ross, Stroudsburg, PA. 515 pp.

Hobbie, J. E., Deegan, L. A., Peterson, B. J., Rastetter, E. B., Shaver, G. R., Kling, G. W., O'Brien, W. J., Chapin, F. S. T., Miller, M. C., Kipphut, G. W., Bowden, W. B., Hershey, A. E., and McDonald, M. E. 1995. 'Long-term measurements at the arctic LTER site', in Powell, T. M. and Steele, J. H. (Eds), *Ecological Time Series.* Chapman Hall Publ., New York. pp. 391–409.

Hogg, E. H., Leiffers, V. J., and Wein, R. W. 1992. 'Potential carbon losses from peat profiles, effects of temperature, drought cycles and fire', *Ecol. Appl.*, **2**, 298–306.

Irons, J. G., Oswood, M. W., Stout, R. J., and Pringle, C. M. 1994. 'Latitudinal patterns in leaf litter breakdown: is temperature really important?', *Freshwat. Biol.*, **32**, 401–411.

Isaac, G. A. and Stuart, R. A. 1992. 'Temperature-precipitation relationships for Canadian Stations', *J. Climate*, **5**, 822–830.

Jacoby, G. C., Jr., Cook, E. R., and Ulan, L. D. 1985. 'Reconstructed summer degree days in central Alaska and northwestern Canada since 1524', *Quat. Res.*, **23**, 18–26.

Kane, D. L., Hinzman, L. D., Woo, M. K., and Everett, K. R. 1992. 'Arctic hydrology and climate change' in Chapin, F. S., III, Jeffries, R. L., Reynolds, J. F., Shaver, G. R., Svoboda, J., and Chu, E. W. (Eds), *Arctic Ecosystems in a Changing Climate: an Ecophysiological Perspective.* Academic Press, San Diego. pp. 35–57.

Kling, G. W. 1995. 'Land-water linkages: the influence of terrestrial diversity on aquatic systems', in Chapin, F. S. and Korner, C. (Eds), *Arctic and Alpine Diversity.* Springer-Verlag, Berlin. pp. 295–308.

Kling, G. W., Kipphut, G. W., and Miller, M. C. 1991. 'Arctic lakes and streams as gas conduits to the atmosphere: implications for tundra carbon budgets', *Science*, **251**, 298–301.

Kling, G. W., O'Brien, W. J., Miller, M. C., and Hershey, A. E. 1992. 'The biogeochemistry and zoogeography of lakes and rivers in arctic Alaska', *Hydrobiologia*, **240**, 1–14.

Koerner, R. M. 1991. 'Past climate changes as deduced from Canadian ice cores', in Woo, M. K. and Gregor, D. J. (Eds), *Arctic Environment, Past, Present and Future.* McMaster University, Hamilton. pp. 61–70.

Lafleur, P. 1988. 'Leaf conductance of four species growing in a subarctic marsh', *Can. J. Bot.*, **66**, 1367–1375.

Lafleur, P. M. 1993. 'Potential water balance response to climatic warming, the case of a coastal wetland ecosystem of the James Bay Lowland', *Wetland*, **13**, 270–276.

Lafleur, P. M. and Rouse, W. R. 1995. 'Energy partitioning at treeline forest and tundra sites and its sensitivity to climatic change', *Atmos.-Ocean*, **33**, 121–133.

Laine, J., Minkkinen, K., Sinisalo, J., Savolainen, I., and Martikainen, P. J. 1997. 'Greenhouse impact of a mire after drainage for forestry', in Trettin, C. C., Jurgensen, M. F., Grigal, D. F., Gale, M. R., and Jeglum, J. K. (Eds), *Northern Forested Wetlands: Ecology and Management*, CRC Press.

Leavesley, G. H., Turner, K., D'Agnese, F. A., and McKnight, D. 1997. 'Regional delineation of North America for the assessment of freshwater ecosystems and climate change', *Hydrol. Process.*, **11**, this issue.

Legendre, P. and Legendre, V. 1984. 'Postglacial dispersal of freshwater fishes in the Quebec peninsula', *Can. J. Fish. Aquat. Sci.*, **41**, 1781–1802.

Lesack, L. F. W., Marsh, P., and Hecky, R. E. 1991. 'Ice cover growth and freeze-out of solutes in a Mackenzie Delta lake', in Prowse, T. D. and Ommanney, C. S. L. (Eds), *Northern Hydrology, Selected Perspectives.* Environ. Can., NHRI Symp. No. 6, July 1990, Saskatoon, pp. 219–236.

Mackay, J. R. 1992. 'Lake stability in an ice-rich permafrost environment: examples from the western arctic coast', in Roberts, R. D. and Bothwell, M. L. (Eds), *Aquatic Ecosytems in Semi-Arid Regions: Implications for Resource Management.* NHRI Symp. Ser., No. 7, Saskatoon. pp. 1–26.

Marsh, P. 1991a. 'Water flux in melting snow covers', in Corapcioglu, M. Y. (Ed.), *Advances in Porous Media*, Vol. 1, Elsevier, Amsterdam. pp. 61–124.

Marsh, P. 1991b. 'Evaporation and ice growth in Mackenzie Delta lakes', in *Hydrology of Natural and Manmade Lakes, Proc., Vienna Symposium, IAHS Publ.*, **206**, 257–266.

Marsh, P. and Hey, M. 1989. 'The flooding hydrology of Mackenzie Delta lakes near Inuvik, N.W.T., Canada', *Arctic*, **42**, 41–49.

Marsh, P. and Lesack, L. F. W. 1996. 'Climate change and the hydrologic regime of lakes in the Mackenzie Delta', *Limnol. Oceanogr., Spec. Iss.*, **41**, 849–856.

Marsh, P. and Ommanney, C. S. L. (Eds), 1991. *Mackenzie Delta, Environmental Interactions and Implications of Development.* Proc., Workshop on the Mackenzie Delta. National Hydrology Research Institute Symposium No. 4, Saskatoon. 195 pp.

Marsh, P. and Prowse, T. D. 1993. 'Hydrology: hydrologic regime of the Mackenzie Basin, potential modelling approaches, and future research needs for addressing climate change issues', in *Mackenzie Basin Impact Study, Interim Report #1*, Cohen, S. J. (Ed.) Environment Canada, Downsview, Ontario. Canada. pp. 37–49.

Marsh, P. and Schmidt, T. 1993. 'Influence of a Beaufort Sea storm surge on channel levels in the Mackenzie Delta', *Arctic*, **46**, 35–41.

Marsh, P., Quinton, B., and Pomeroy, J. W. 1996. 'Hydrological processes and runoff at the Arctic treeline in northwestern Canada', *Proc. 10th Northern Research Basins.* IHP, Norway. 368–397.

Martikainen, P. J., Nykanen, H., Alm, J., and Silvola, J. 1994. 'Change in fluxes of carbon dioxide, methane and nitrous oxide due to forest drainage of mire sites of different trophy', *Plant Soil*, in press.

Maxwell, J. B. 1980. 'The climate of the Canadian Arctic Islands and adjacent waters', *Clim. Ser., No. 30*, Vol. 1. Environment Canada, Atm. Environ. Serv., Downsview, Canada. 531 pp.

Maxwell, J. B. 1992. 'Arctic climate, potential for change under global warming', in Chapin, F. S., Jeffries, R. L., Reynolds, J. F., Shaver, G. R., and Svoboda, J. (Eds), *Arctic Ecosystems in a Changing Climate.* Academic Press. N. Y., pp. 11–34.

McDonald, M. E. and Hershey, A. E. 1992. 'Shifts in abundance and growth of slimy sculpin in response to changes in the predator population in an arctic Alaskan lake', *Hydrobiologia*, **240**, 219–223.

McDonald, M. E., Hershey, A. E., and O'Brien, W. J. 1992. 'Cost of predation avoidance in young-of-year lake trout (*Salvelinus namaycush*): growth differential in sub-optimal habitats', *Hydrobiologia*, **240**, 213–218.

McDonald, M. E., Hershey, A. E., and Miller, M. C. 1996. 'Global warming impacts on trophic structure in arctic lakes', *Limnol. Oceanogr.*, **41**, 1102–1108.

McKane, R. B., Rastetter, E. B., Shaver, G. R., Nadelhoffer, K. J., Giblin, A. E., Laundre, J. A., and Chapin, F. S., III. 'Effects of changes in CO_2 and climate on carbon storage in arctic tundra', *Ecology*, in press.

Merrick, G. W., Hershey, A. E., and McDonald, M. E. 1991. 'Lake trout (*Salvelinus namaycush*) control of snail density and size distribution in an arctic lake', *Can. J. Fish. Aquat. Sci.*, **48**, 498–502.

Merrick, G. W., Hershey, A. E., and McDonald, M. E. 1992. 'Salmonid diet and the size, distribution, and density of benthic invertebrates', *Hydrobiologia*, **240**, 225–233.

Miller, M. C., Hater, G. R., Spatt, P., Westlake, P., and Yeakel, D. 1986. 'Primary production and its control in Toolik Lake, Alaska', *Arch. Hydrobiol. Suppl.*, **74**, 97–131.

Miller, M. C., DeOliveira, P., and Gibeau, G. G. 1992. 'Epilithic diatom community response to years of PO_4 fertilization: Kaparuk River, Alaska', *Hydrologia*, **240**, 103–120.

Mitchell, J. F. B., Manabe, S., Meleshko, V., and Tolioka, T. 1990. 'Equilibrium climate change and its implications for the future', in Houghton, J. T., Jenkins, G. J., and Ephraums, J. J. (Eds), *Climate Change. The IPCC Scientific Assessment*. Cambridge University Press, New York, pp. 135–164.

Moore, T. R. and Kowles, R. 1989. 'The influence of water table levels on methane and carbon dioxide emissions from peatland soils', *Can. J. Soil Sci.*, **67**, 77–81.

Moore, T. R. and Roulet, N. T. 1993. 'Methane flux, water table relations in northern wetlands', *Geophys. Res. Lett.*, **20**, 587–590.

Nadelhoffer, K. J., Giblin, A. E., Shaver, G. R., and Laundre, J. A. 1991. 'Effects of temperature and organic matter quality on C, N, and P mineralization in soils from six arctic ecosystems', *Ecology*, **72**, 242–253.

National Wetlands Working Group, 1988. *Wetlands of Canada. Ecological Land Classification Series*, No. 24. Sustainable Development Branch, Quebec, Environment Canada, Ottawa, Ontario, and Polyscience Publ. Inc., Montreal. 452 pp.

Nicholson, B. J. 1994. 'Wetland development at Elk Island National Park, Alberta, Canada', *J. Paleolimnol.*, **12**, 19–34.

Nicholson, B. J. and Gignac, L. D. 1996. 'Niche dimensions of peatland bryophyte indicator species along environmental and climatic gradients in the Mackenzie River Basin', *The Bryologist*, **98**, 437–451.

Nogrady, T. and Smol, J. P. 1989. 'Rotifers from five high arctic ponds, Cape Herschel, Ellesmere Island', *Hydobiologia*, **173**, 231–242.

O'Brien, W. J., Buchanan, C., and Haney, J. F. 1979. 'Arctic zooplankton community structure: exceptions to some general rules', *Arctic*, **32**, 237–247.

O'Brien, W. J., Hershey, A. E., Hobbie, J. E., Hullar, M. A., Kipphut, G. W., Miller, M. C., Moller, B., and Vestal, J. R. 1992. 'Control mechanisms of arctic lake ecosystems: a limnocorral experiment', *Hydrobiologia*, **240**, 143–188.

Oechel, W. C. and Vourlitis, G. L. 1994. 'The effects of climate change on land–atmosphere feedbacks in arctic tundra regions', *TREE*, **9**, 324–329.

Oechel, W. C., Hastings, S. J., Vourlitis, G., Jenkins, M., Riechers, G., and Grulke, N. 1993. 'Recent change of arctic tundra ecosystems from a net carbon dioxide sink to a source', *Nature*, **361**, 520–525.

Oswood, M. W., Milner, A. M., and Irons, J. G., III. 1992. 'Climate change and Alaskan rivers and streams', in Firth, P. and Fisher, S. G. (Eds), *Global Change and Freshwater Ecosystems*. Springer-Verlag, New York. pp. 192–210.

Payette, S. and Morneau, C. 1993. 'Holocene relict woodlands at the eastern Canadian treeline', *Quart. Res.*, **39**, 84–89.

Peace–Anthabasca Delta Implementation Committee, 1987. *Peace Anthabasca Delta Water Management Works Evaluation*, Final Report. Peace–Anthabasca Delta Implementation Agreement, Canada/Alberta/Saskatchewan. 63 pp.

Peterson, B. J., Hobbie, J. E., Hershey, A. E., Lock, M., Ford, T., Vestal, R., Hullar, M., Ventullo, R., and Volk, G. 1985. 'Transformation of a tundra river from heterotrophy to autotrophy by addition of phosphorus', *Science*, **229**, 1383–1386.

Peterson, B. J., Deegan, L. A., Helfrich, J., Hobbie, J. E., Hullar, M. A. J., Moller, B., Ford, T. E., Hershey, A. E., Hiltner, A. L., Kipphut, G. W., Lock, M. A., Fiebig, D. M., McKinley, V. L., Miller, M. C., Vestal, J. R., Ventullo, R. M., and Volk, G. S. 1993. 'Biological responses of a tundra river to fertilization', *Ecology*, **74**, 653–672.

Pienitz, R. 1993. 'Paleoclimate proxy data inferred from fresh water diatoms for the Yukon and Northwest Territories', *PhD Thesis*, Queen's University, Kingston, Ontario. 220 pp.

Pienitz, R. and Smol, J. P. 1993. 'Diatom assemblages and their relationship to environmental variables in lakes from the boreal forest-tundra ecotone near Yellowknife, Northwest Territories, Canada', *Hydrobiologia*, **269/270**, 391–404.

Pienitz, R., Walker, I. R., Zeeb, B. A., Smol, J. P., and Leavitt, P. R. 1992. 'Biomonitoring past salinity changes in an athalassic sub-arctic lake', *Int. J. Salt Lake Res.*, **12**, 91–123.

Pienitz, R., Smol, J. P., and Birks, H. J. B. 1995. 'Assessment of freshwater diatoms as quantitative indicators of past climatic change in the Yukon and Northwest Territories, Canada', *J. Paleolimnol.*, **13**, 21–49.

Pomeroy, J. W. and Gray, D. M. 1994. 'Sensitivity of snow relocation and sublimation to climate and surface vegetation', in Jones, H. G., Davies, T., Ohmura, A., and Morris, E. M. (Eds), *Snow and Ice Covers, Interactions with the Atmosphere and Ecosystems, IAHS Publ.*, **223**, 213–226.

Prowse, T. D. 1994. Environmental significance of ice to streamflow in cold regions. *Freshwat. Biol.*, **32**, 241–259.

Prowse, T. D. and Lalonde, V. 1996. 'Open, water and ice jam flooding of a northern delta, *Nord. Hydrol.*, **27**(1/2), 85–100.

Prowse, T. D. and Stephenson, R. L. 1986. 'The relationship between winter lake cover, radiation receipts and the oxygen deficit in temperate lakes', *Atmos.-Ocean*, **24**, 386–403.

Ramlal P. S., Hesslein, R. E., Hecky, R. E., Fee, E. J., Rudd, J. W. M., and Guilford, S. J. 1994. 'The organic carbon budget of a shallow, arctic tundra lake on the Tuktoyaktuk Peninsula, N.W.T., Canada'. *Biogeochemistry*, **24**, 145–172.

Ritchie, J. C. 1984. *Past and Present Vegetation of the Far Northwest of Canada*. University of Toronto Press, Toronto.

Roots, E. R. 1989. 'Climate change, high latitude regions', *Climatic Change*, **15**, 223–253.

Roulet, N. T. and Woo, M.-K. 1986. 'Hydrology of a wetland in the continuous permafrost region', *J. Hydrol.*, **89**, 73–91.
Roulet, N. T. and Woo, M.-K. 1988. 'Runoff generation in a low Arctic drainage basin', *J. Hydrol.*, **101**, 213–226.
Roulet, N. T., Moore, T., Bubier, J., and Lafleur, P. 1992. 'Northern fens, methane flux and climatic change', *Tellus*, **44B**, 100–105.
Roulet, N. T., Jano, A., Kelly, C. A., Klinger, L. F., Moore, T. R., Protz, R., Ritter, J. A., and Rouse, W. R. 1994. 'Role of the Hudson Bay Lowland as a source of atmospheric methane', *J. Geophys. Res.*, **99**(D1), 1439–1454.
Roulet, N. T., Munro, S., and Morstch, L. 1997. 'Surface climate of wetlands', in Bailey, W., Rouse, W. R., and Oke, T. R. (Eds), *Surface Climates of Canada.* McGill-Queens Press, Montreal, Canada. In press.
Rouse, W. R. 1993. 'Northern climates', in French, H. and Slaymaker, O. (Eds), *Canada's Cold Environments.* McGill-Queens Press. pp. 65–92.
Rouse, W. R., Carlson, D. W., and Weick, E. J. 1992. 'Impacts of summer warming on the energy and water balance of wet tundra', *Climatic Change*, **22**, 305–326.
Rublee, P. A. 1992. 'Community structure and bottom-up regulation of heterotrophic microplankton in arctic LTER lakes', *Hydrobiologia*, **240**, 133–142.
Ruuhijärvi, R. 1983. 'The Finnish mire types and their regional distribution', in Gore, A. J. P. (Ed.), *Ecosystems of the World 4b. Mires, Swamp, Bog, Fen and Moor. Regional Studies*, Elsevier Science Publishers, New York. pp. 47–67.
Schindler, D. W., Welch, H. E., Kalff, J., Brunskill, G. J., and Kritsch, N. 1974. 'Physical and chemical limnology of Char Lake, Cornwallis Island (75°N Lat.)', *J. Fish. Res. Bd Can.*, **31**, 585–607.
Schell, D. M. and Zieman, P. J. 1983. 'Accumulation of peat carbon in the Alaska Arctic coastal plain and its role in biological productivity', in *Permafrost. Fourth International Conference.* National Academy Press, Washington, D.C. pp. 1105–1110.
Schlesinger, D. A. and Regier, H. A. 1982. 'Climatic and morphoedaphic indices of fish yields from natural lakes', *Trans. Am. Fish. Soc.*, **111**, 141–150.
Schlesinger, D. A. and Regier, H. A. 1983. 'Relationship between environmental temperature and yields of subarctic and temperate zone fish species', *Can. J. Fish. Aquat. Sci.*, **40**, 1829–1837.
Scrimgeour, G. J., Prowse, T. D., Culp, J. M., and Chambers, P. A. 1994. 'Ecological effects of river ice break-up: a review and perspective', *Freshwat. Biol.*, **32**, 261–275.
Shaver, G. R., Billings, W. D., Chapin, F. S. III, Giblin, A. E., Nadelhoffer, K. J. Oechel, W. C., and Rastetter, E. B. 1992. 'Global change and the carbon balance of arctic ecosystems', *Bioscience*, **42**, 433–441.
Sheath, R. G. 1986. 'Seasonality of phytoplankton in northern tundra ponds', *Hydrobiologia*, **138**, 75–83.
Shurapli, N. J., Verma, S. B., Kim, J., and Afkebauer, T. J. 1995. 'Carbon dioxide exchange in a peat ecosystem', *J. Geophys. Res.*, **100**(D1), 14319–14326.
Siegel, D. I., Reeve, A. S., Glaser, P. H., and Romanowicz, E. A. 1995. 'Climate-driven flushing of pore water in peatlands', *Nature*, **374**, 531–533.
Skinner, W. R. and Gullet, D. W. 1993. 'Temperature trends in Canada', *Climatol. Bull.*, **27**, 63–77.
Smol, J. P. 1988. 'Paleoclimate proxy data from freshwater arctic diatoms', *Verh. Int. Verein. Limnol.*, **23**, 837–844.
Smol, J. P., Walker, I. R., and Leavitt, P. R. 1991. 'Paleolimnology and hindcasting climatic trends', *Verh. Int. Verein. Limnol.*, **24**, 1240–1246.
Sommer, M. E. 1979. 'Role of zooplankton grazers in determining the composition and productivity of seston in arctic lakes and ponds', *MSc Thesis*, University of Cincinnati, Ohio.
Spear, R. W. 1993. 'The palynological record of late Quaternary arctic treeline in northwestern Canada', *Rev. Paleobot. Palynol.*, **79**, 99–112.
Stoermer, E. F., Schelske, C. L., and Wolin, J. A. 1990. 'Siliceous microfossil succession in the sediments of McLeod Bay, Great Slave Lake, Northwest Territories', *Can. J. Fish. Aquat. Sci.*, **47**, 1865–1874.
Sweeney, B. W. 1984. 'Factors influencing life-history patterns of aquatic insects', in Resh, V. H. and Rosenberg, D. M. (Eds), *The Ecology of Aquatic Insects.* Praeger, New York. pp. 56–100.
Szeicz, J. M. and MacDonald, G. M. 1995. 'Dendroclimatic reconstruction of summer temperatures in northwestern Canada since A.D. 1638 based on age-dependent modeling', *Quat. Res.*, **44**, 257–266.
Van Everdingen, R. O. 1974. 'Groundwater in permafrost regions of Canada', *Permafrost Hydrol.: Proc. Workshop Seminar, Can. Nat. Comm., The IHD*, pp. 83–93.
Veres, A., Pienitz, R., and Smol, J. P. 1995. 'Lake water salinity and periphytic diatom succession in three subarctic lakes, Yukon Territory, Canada', *Arctic*, **48**, 63–70.
Verry, E. S. 1988. 'The hydrology of wetlands and man's influence on it', in *Int. Symp. Hydrol. Wetlands in Temperate and Cold Regions*, Vol. 2. Suomen Akatemian Julkaisuja, Joensuu, Finland. pp. 41–61.
Watt, W. E. (Editor-in Chief), 1989. *Hydrology of Floods in Canada: a Guide to Planning and Design.* N.R.C., Ottawa, Canada. 245 pp.
Wein, R. W. 1983. 'Fire behavior and ecological effects in organic terrain', in Wein, R. W. and MacLean, D. A. (Eds), *The Role of Fire in Northern Circumpolar Ecosystems.* John Wiley & Sons, Toronto. pp. 81–95.
Welch, H. E. and Bergman, M. A. 1985. 'Water circulation in small arctic lakes in winter', *Can. J. Fish. Aquat. Sci.*, **43**, 506–520.
Welch, H. E. and Legault, J. A. 1986. 'Precipitation chemistry and chemical limnology of fertilized and natural lakes at Saqvagjuak', N.W.T.', *Can. J. Fish. Aquat. Sci.*, **43**, 1104–1134.
Wells, E. D. and Hirvonen, H. E. 1988. 'Wetlands of Atlantic Canada', in *Wetlands of Canada. Ecological Land Classification Series*, No. 24. Sustainable Development Branch Quebec. Environment Canada, Ottawa, Ontario, and Polyscience Publ. Inc., Montreal. pp. 251–303.
Whalen, S. C. and Cornwell, J. C. 1985. 'Nitrogen, phosphorus, and organic carbon cycling in an arctic lake', *Can. J. Fish. Aquat. Sci.*, **42**, 797–808.
Woo, M.-K. 1986. 'Permafrost hydrology in North America', *Atmos.-Ocean*, **24**, 201–234.

Woo, M.-K., Heron, R., Marsh, P., and Steer, P. 1983. 'Comparison of weather station snowfall with winter snow accumulation in high arctic basins', *Atmos.-Ocean*, **21**, 312–325.
Woo, M.-K., Lewkowicz, A. G., and Rouse, W. R. 1992. 'Response of the Canadian permafrost environment to climatic change', *Phys. Geogr.*, **134**, 287–317.
Zoltai, S. C. 1993. 'Cyclic development of permafrost in peatlands of northwestern Alberta, Canada', *Arctic Alp. Res.*, **25**, 240–246.
Zoltai, S. C. and Vitt, D. H. 1990. 'Holocene climatic change and the distribution of peatlands in western interior Canada', *Quart. Res.*, **33**, 231–240.

4

ASSESSMENT OF CLIMATE CHANGE AND FRESHWATER ECOSYSTEMS OF THE ROCKY MOUNTAINS, USA AND CANADA

F. RICHARD HAUER,[1] JILL S. BARON,[2] DONALD H. CAMPBELL,[3] KURT D. FAUSCH,[4] STEVE W. HOSTETLER,[5] GEORGE H. LEAVESLEY,[3] PETER R. LEAVITT,[6] DIANE M. MCKNIGHT[7] AND JACK A. STANFORD[1]

[1]*Flathead Lake Biological Station, The University of Montana, Polson, MT 59860, USA*
[2]*Colorado State University and National Biological Service, Fort Collins, CO 80523, USA*
[3]*US Geological Survey, Denver, CO 80225, USA*
[4]*Colorado State University, Fort Collins, CO 80523, USA*
[5]*US Geological Survey, Corvallis, OR 97333, USA*
[6]*University of Regina, Regina, SK S4S 042, Canada*
[7]*US Geological Survey, Boulder, CO 80303, USA*

ABSTRACT

The Rocky Mountains in the USA and Canada encompass the interior cordillera of western North America, from the southern Yukon to northern New Mexico. Annual weather patterns are cold in winter and mild in summer. Precipitation has high seasonal and interannual variation and may differ by an order of magnitude between geographically close locales, depending on slope, aspect and local climatic and orographic conditions. The region's hydrology is characterized by the accumulation of winter snow, spring snowmelt and autumnal baseflows. During the 2–3-month 'spring runoff' period, rivers frequently discharge >70% of their annual water budget and have instantaneous discharges 10–100 times mean low flow.

Complex weather patterns characterized by high spatial and temporal variability make predictions of future conditions tenuous. However, general patterns are identifiable; northern and western portions of the region are dominated by maritime weather patterns from the North Pacific, central areas and eastern slopes are dominated by continental air masses and southern portions receive seasonally variable atmospheric circulation from the Pacific and the Gulf of Mexico. Significant interannual variations occur in these general patterns, possibly related to ENSO (El Niño–Southern Oscillation) forcing.

Changes in precipitation and temperature regimes or patterns have significant potential effects on the distribution and abundance of plants and animals. For example, elevation of the timber-line is principally a function of temperature. Palaeolimnological investigations have shown significant shifts in phyto- and zoo-plankton populations as alpine lakes shift between being above or below the timber-line. Likewise, streamside vegetation has a significant effect on stream ecosystem structure and function. Changes in stream temperature regimes result in significant changes in community composition as a consequence of bioenergetic factors. Stenothermic species could be extirpated as appropriate thermal criteria disappear. Warming temperatures may geographically isolate cole water stream fishes in increasingly confined headwaters. The heat budgets of large lakes may be affected resulting in a change of state between dimictic and warm monomictic character. Uncertainties associated with prediction are increased by the planting of fish in historically fishless, high mountain lakes and the introduction of non-native species of fishes and invertebrates into often previously simple food-webs of large valley bottom lakes and streams. Many of the streams and rivers suffer from the anthropogenic effects of abstraction and regulation. Likewise, many of the large lakes receive nutrient loads from a growing human population.

We concluded that: (1) regional climate models are required to resolve adequately the complexities of the high gradient landscapes; (2) extensive wilderness preserves and national park lands, so prevalent in the Rocky Mountain Region, provide sensitive areas for differentiation of anthropogenic effects from climate effects; and (3) future research should encompass both short-term intensive studies and long-term monitoring studies developed within comprehensive experimental arrays of streams and lakes specifically designed to address the issue of anthropogenic versus climatic effects.

INTRODUCTION

The purpose of this paper is to address the potential effect of climate change on freshwater ecosystems of the Rocky Mountain Region. We examine current climatic conditions and past climate change and hypothesize on future change, within the context of the spatial and temporal variation of this broad geographical region. We consider responses of individual aquatic ecosystems, examine limitations to evaluating climatic change in light of various anthropogenic factors (e.g. river and lake regulation, water abstraction), make recommendations regarding future research and monitoring needs and evaluate the region's suitability as a global indicator of climate change.

The Rocky Mountains encompass the interior cordillera of western North America from the southern Yukon Territory, Canada (63°N) to northern New Mexico, USA (35°N). Many of North American's great rivers (e.g. Athabasca, Colorado, Columbia, Mackenzie, Missouri, Platte, Rio Grande, Saskatchewan, Yukon) have their headwaters in the Rockies. Multiple mountain ranges generally orientate in a north–north-west to south–south-east direction and are characterized by extreme geological and environmental gradients, often in close geographical proximity. Individual ranges frequently contain peak to valley floor elevation differentials greater than 1500 m and exhibit high variance in climate and vegetation. The landscapes reflect both the dynamic geological processes that formed the region, as well as hydrological and glacial processes. The bedrock of sedimentary and igneous origin provides markedly different source material to weathering processes, development of soil and macro- and micro-nutrients contributed to ground and surface waters.

Climatic patterns vary across a variety of spatial and temporal scales within the Rockies. The southern mountains of New Mexico and Colorado have a continental climate, but exhibit high seasonal and interannual variation in the type, frequency and duration of precipitation events (Dahm and Molles, 1992). The northern ranges are dominated by maritime air masses. Weather throughout the Rocky Mountains is modified by local thermodynamic processes that produce distinctive, local microclimates on different slopes and aspects that affect both water and energy fluxes (Barry, 1992).

The Rocky Mountains contain a diversity of freshwater ecosystems. Headwater streams are generally high gradient environments whose channels are controlled by bedrock. Larger streams and rivers, located in the montane valleys, often contain serial sequences of confined and unconfined reaches (Stanford and Ward, 1993). The most prevalent wetlands are associated with seepage slopes, river floodplains or surficial depressions, or the landforms associated with glacial retreat. Tens of thousands of alpine and subalpine lakes reflect a broad array of climatic and geological conditions and biotic histories. Large lakes (e.g. Flathead, Pend Oreille, Kootenay, McDonald) occupy glaciated valleys and are often >100 m deep.

Most aquatic ecosystems of the Rocky Mountains are strongly influenced by highly variable hydrographical regimes. The region is dominated by the annual accumulation of winter precipitation as snow and elevated stream flows in the late spring and early summer (Poff and Ward, 1989). During the 2–3-month 'spring runoff' period, rivers frequently discharge $>70\%$ of their annual water budget and have instantaneous discharges 10–100 times mean low flow (Stanford and Hauer, 1992). Changes in the fundamental climatic variables of temperature and precipitation may affect all aspects of the hydrological regime, including water temperature, discharge, nutrient flux, groundwater and lake levels, sediment transport, ice and snow cover, biological production, food-web dynamics and biogeography (Firth and Fisher, 1992).

REGIONAL CLIMATE

Current climate

Large-scale circulation patterns and their interactions with the Rocky Mountains are well studied. In the northern ranges, weather patterns throughout the year are primarily Pacific maritime, west of the continental divide. Precipitation is closely correlated with cold fronts moving east from the North Pacific and Gulf of

Alaska. Along the east front ranges of western Montana and Alberta continental air masses predominate, particularly during winter, as Arctic fronts with high barometric pressure and cold temperatures move from northern Canada to the Great Plains. Midwinter temperature differentials between maritime and Arctic air masses are often >40°C. Thus, during winter, locations are alternatively dominated by different atmospheric sources, and temperature can vary greatly over short time periods. Winter precipitation throughout the Rocky Mountains occurs primarily as snow. Winter snowpacks occur at high elevations that vary with latitude (e.g. *ca.* 2200 m in Colorado, *ca.* 1400 m in Montana). Spring snowmelt increases soil moisture recharge, runoff, groundwater recharge and stream flow. Analysis of spring runoff patterns, precipitation and temperature in western Montana shows that the greatest floods (1911–1990) have occurred in late May or early June following a cool, wet spring and were the result of warm, rain-on-snow events at high elevation (Hauer, 1991).

In more southern ranges, such as the Front Range in Colorado, sources of moisture vary from predominantly Pacific origin during the winter, to a southerly and south-easterly source in the summer (Barry, 1973). Spring precipitation maxima are apparent at low elevations, while autumn and winter maxima are observed near the continental divide. Summer precipitation arises from advective cooling and can occur either as rain or snow at the higher elevations. Severe summer thunderstorms can cause floods up to elevations of *ca.* 2300 m in the Colorado Front Range (Jarrett, 1990). Analysis of flow data over a 60- and 68-year period from two New Mexico rivers shows that spring flows during snowmelt were significantly increased during El Niño years (elevated sea surface temperature and reduced barometric pressure in the eastern tropical Pacific) and significantly reduced during La Niña years (reduced temperature and elevated barometric pressure). This pattern suggests a linkage between accumulations of winter snowpack in the southern Rockies with ENSO (El Niño–Southern Oscillation) phenomena (Molles and Dahm, 1990).

Precipitation and temperature patterns throughout the Rockies exhibit a characteristically high spatial diversity. For example, some west-facing slopes in the Flathead Basin, Montana, have average annual precipitation >150 cm. Other areas only a few km distant average <15 cm per year. The Rocky Mountains are also known for high intra-annual variance in precipitation. This is particularly significant when low soil moisture content during midsummer leads to dry forest conditions and wildfire. An important structuring mechanism in Rocky Mountain forest vegetation is susceptibility to wildfire and the resulting mosaic of vegetation types and ages [for reviews see *BioScience*, 1989, **39**(10)].

Past climate change

Evidence of post-glacial climatic change in the Rocky Mountains has been obtained from three sources; palaeoecological analyses of lake sediments, climatic inferences from various palaeochrono sequence data (e.g. tree-rings, moraines, sediment laminae) and direct instrumental observations. These sources differ in the length of record, temporal resolution, interpretability and sensitivity. Climatic reconstructions from lake sediments record long-term climatic trends over broad time-scales (i.e. the Holocene). However, fossil data are difficult to interpret because identification of the climatic drivers is ambiguous. Furthermore, ancient climatic change may have arisen because of different forcing mechanisms from those that occur today. Consequently, high-resolution palaeoecological and tree-ring analyses over intermediate time-scales (*ca.* 100–1500 years) may provide the greatest insights into the effects of climatic change within montane ecosystems. Such meso-scale analyses are often annually resolved and record regional rather than local environmental changes (Leonard, 1986a). In contrast, instrumental data are the most precise and easily interpreted, but reflect relatively high variance as a result of local phenomena (e.g. cloud cover, localized rainfall). Additionally, these data are comparatively short (<100 years).

Much of what is currently known about Holocene environmental change was first identified by palaeoecological studies in the Rocky Mountains (reviewed in Luckman, 1990). Although those investigations are relatively few, they show remarkable consistency in the pattern of post-glacial climatic change. Recent studies have confirmed that glaciers in the main ranges of the central Canadian Rockies quickly receded between 11 500 and 10 000 years before present (BP) (Reasoner and Rutter, 1988; Reasoner and Hickman, 1989; Reasoner *et al.*, 1994). The onset and rate of glacial recession depended on latitude, elevation, aspect and longitudinal position within the mountain ranges (Luckman, 1990; Hickman and Schweger, 1993). In general, early Holocene climates were considered to be significantly warmer (1–1·5°C) than at present and were maintained for a period of between 1000 and 5000 years, depending on the region (Luckman, 1990). This hypsithermal period was followed by climatic cooling over several millennia, reaching near-present conditions by *ca.* 4000 BP. At least three neoglacial periods have been identified over the last 3000 years, with the most pronounced beginning *ca.* 800 BP and reaching its maximum extent in the mid-eighteenth century and mid- to late-nineteenth century (Little Ice Age). Since 1850, temperatures have appeared to increase, although there is considerable variation between sites (Luckman, 1990).

Climatic change at the start of the Holocene may be an excellent analogue for forecasting a response to global warming. In some cases, glacial retreat occurred over a few decades, rather than centuries as previously supposed (Reasoner *et al.*, 1994). The magnitude and speed of inferred change is similar to projected changes arising from elevated CO_2 concentrations. Recent research suggests that glacial retreat may have consisted of at least two phases, separated by a cold period of up to 1000 years. This cold period is well known from Europe (i.e. the Younger Dryas), and other northern Atlantic regions, and has recently been verified in western North America (Reasoner *et al.*, 1994).

Most evidence of early Holocene climatic warming comes from changes in the species composition of plant macro-fossils, pollen and spores. In particular, physical remains from vascular plants (e.g. boles, needles, seeds) have been useful because they are not subject to long-distance transport like pollen, and because they can be used in accelerator mass spectroscopy (AMS) ^{14}C determinations of sediment age. Transitions from herbaceous, tundra-like communities (e.g. *Artemisia*) to forests of *Pinus* and *Abies* (fir) mark the retreat of glaciers and the stabilization of hillslopes as a result of tree colonization. Lake sediments of this era are often very inorganic because of elevated concentrations of glacial silt ('flour') and carbonates (Reasoner and Hickman, 1989; Reasoner *et al.*, 1994). In fact, the inverse relation between sediment organic content and glacier activity has been proposed as a palaeoclimatic marker of glacial activity (Leonard, 1986a, b).

Temperatures may have continued to rise after the initial glacial retreat (*ca.* 10 000 BP) to reach a post-glacial maximum *ca.* 8000 BP. Inferences of warmer temperatures are derived from analyses of fossil pollen and plant macro-fossils, the presence of buried trees in present day alpine meadows and increased organic content of lake sediments. High relative abundance of fossils from presently lower elevation taxa (e.g. *Abies*, *Picea*, *Pinus*) are interpreted as indicating warmer and perhaps drier conditions. When calibrated to current tree-line distribution and composition (Beaudoin, 1987), these pollen ratios can be sensitive indicators of timber-line migration (e.g. Reasoner and Hickman, 1989). In general, tree-line occurs as the 10°C July isotherm in the central Rocky Mountains and migrates approximately 150 vertical metres or 100 latitudinal km for every 1°C change (Luckman, 1990 and references within). However, comparisons of fossil profiles from paired alpine and subalpine lakes suggest that changes in terrestrial vegetation were most dramatic at alpine, rather than subalpine or montane sites (Hickman and Reasoner, 1994).

Climatic cooling in the Rocky Mountain Region after the mid-Holocene temperature maximum was relatively slow in comparison to earlier warming during glacial retreat. Analyses of fossil pollen ratios and macro-fossils suggest declining tree-line elevations (Reasoner and Hickman, 1989). Increased carbonate and mineral contents of lake sediments indicate that the extent of glaciers increased during the same period (Leonard, 1986a, b). The rate and extent of forest change appears to be site specific, depending on the presence of glaciers within the catchment, lake elevation, lake basin aspect and the responsiveness of the vegetation (e.g. herbs vs trees). In general, most reconstructions suggest that the climate cooled, and perhaps became wetter, until *ca.* 4000 BP when an approximately present day climate was established (Luckman, 1990).

Temperatures during the last two or three millennia have probably been colder than at any time since the start of the Holocene. Throughout this period, alpine lakes with glaciers in their catchments are characterized by highly inorganic, carbonate-rich sediments containing few macro-fossils from trees

(Reasoner and Hickman, 1989; Reasoner *et al.*, 1994). Leonard (1986a, b) calibrated annually laminated sediments with glacial meltwater patterns and showed that the deposition of glacial silts was proportional to the extent of glacier cover within the catchment when averaged over a century. Fossil reconstructions suggested that temperature minima were reached during 1200–1300 AD and 1600–1900 AD, and maximum extent of ice occurred during the 'Little Ice Age' in the mid- to late-nineteenth century.

Highly detailed information on climatic change in the last millennium has been gathered from a wide variety of sources. Neoglacial moraines can be aged and provide data of maximum extent of glacial ice within the Holocene. More detailed information can be gained by analyses of annual lake sediment accumulation (see above) or tree growth patterns. Ring width, wood density and isotope content in timber-line trees are all highly correlated to mean summer temperature (Luckman, 1990). Many trees at tree-line can live over 600 years and records can be further extended using downed trees that have been well preserved. In general, analyses from central Rocky Mountains suggest that the climate has been steadily warming since *ca.* 1850 AD, the start of the earliest instrumental records (Luckman, 1990).

Comparison of long-term instrumental data from closely located sites illustrates that, while there are some overall similarities, recent climatic change is dominated by high interannual variability. For example, comparison of mean annual temperatures from five locations in Banff and Jasper National Parks (Alberta, Canada) showed that temperatures were above average during 1920–1945 and through the 1960s and the late 1980s. However, only one site (Banff) showed a statistically significant warming over the last 100 years. It is unclear whether this reflects a stable climate or statistical insensitivity owing to high interannual variation. Interestingly, temperatures at three sites during the late 1980s were near-historical maxima, a pattern also seen immediately to the east in Edmonton, Alberta (Luckman, 1990).

Evaluating future climate change

The preferred method for estimating the effects of climate change on freshwater ecosystems is the use of output from general circulation models (GCMs) to derive input data sets for surface hydrological models (e.g. US EPA, 1989). This approach has been applied in a variety of studies aimed at assessing the sensitivity and response of lakes (e.g. Croley, 1990; McCormick, 1990; Hondzo and Stefan, 1993) or stream flow (Gleick, 1987, 1989; McCabe and Ayers, 1989; Lettenmaier and Gan, 1990) to global warming. GCM output is usually expressed as a climate anomaly, or as a difference between control and hypotheses-oriented simulations (e.g. doubled CO_2) or to observed data (e.g. a 2°C air temperature added to the observed average air temperature values). Modelling studies incorporating GCM output are mainly useful for establishing first-order sensitivities of freshwater ecosystems, rather than as a basis for detailed projections.

A primary limitation to the use of GCMs in evaluating climate change is their coarse spatial resolution. Typically, GCMs differentiate at spatial scales covering several degrees of latitude and longitude. Existing supercomputers are incapable of implementing models at finer resolution for extended climate simulations. Consequently, most GCMs accurately simulate annual and seasonal climate, but are less reliable in simulating regional spatial- and temporal-scale features that are essential for climate change assessment (Mearns *et al.*, 1990; Grotch and MacCracken, 1991; US EPA, 1991; Hostetler, 1994; Bates *et al.*, 1995). This is particularly true for the Rocky Mountains, where the topography is poorly represented and the model output is very different from the true climate of the locale. For example, in one GCM the topography for the western United States is simulated as a gently sloping mountain originating on the Pacific coast and coming to a point over Utah (Giorgi and Mearns, 1991). Furthermore, the output from different GCMs varies significantly (e.g. Grotch and MacCracken, 1991; US EPA, 1991), posing the problem of which GCM to select for ecosystem assessments.

One approach to resolving the scale problems of GCMs has been to nest a higher resolution or regional climate model (RCMs) within a GCM (Dickinson *et al.*, 1989; Giorgi and Bates, 1989; Giorgi *et al.*, 1990, 1993a, b; Bates *et al.*, 1993, 1995). In this configuration, the GCM is run first, and the initial and lateral boundary conditions for the RCM are derived from the output of the GCM. The RCM is then used to provide high resolution simulations of climate over a limited area at horizontal resolutions of tens of kilometres. The use of RCMs appears to hold some promise for attaining simulations of regional climate and surface hydrology that are relevant to the complex topography of the Rocky Mountains (Running and Nemani, 1991; Giorgi *et al.*, 1993a; Hostetler and Giorgi, 1993). However, the quality of a simulation attained from an RCM depends strongly on the ability of the driving GCM to simulate correctly the timing, location and characteristics of synoptic-scale atmospheric features (Hostetler, 1994).

Given the problems inherent in GCMs and nested RCMs, researchers have resorted to alternative approaches to evaluate future climate change. One of the most widely used methods has been to estimate average annual changes in precipitation and temperature using GCMs and then to apply these estimates to adjust historical meteorological time-series data. In the simplest case, the adjustment is made by increasing historical precipitation values by a fixed percentage. Hypothetical scenarios using instrumental estimates of climate change, instead of GCM results, appear to be most promising for the highly variable topographies and microclimates of the Rocky Mountains.

STREAM AND RIVER ECOSYSTEMS

The streams and rivers of the Rocky Mountains are dynamically linked to basin geomorphology, climatic–hydrological regimes and ecological processes and patterns (Hynes, 1975; Minshall, 1988; Reice *et al.*, 1990; Burt, 1992; Stanford and Ward, 1993). Biotic and abiotic stream characteristics change along extreme elevation gradients that are analogous to those seen along latitudinal gradients (see Hauer and Stanford, 1982a; Stanford *et al.*, 1988; Sweeney *et al.*, 1992). Terrestrial ecologists have recognized for over a century the relationship between vegetation composition and elevation within montane regions (Merriam, 1890). Such clinal changes also exist in lotic ecosystems and include changes in thermal regime, bed and channel form, hydrological variability, biogeochemistry and the structure and function of biota along the stream corridor.

Hydrology and temperature regimes

The hydrology of the Rocky Mountain Region is dominated by snow accumulation and melt (Poff and Ward, 1989). Precipitation, in the form of snow, is normally accumulated during winter and released as snowmelt during summer. At mid-continental latitudes, high elevation snowpacks generally begin in mid-autumn and persist until late spring and even into summer (Turk and Spahr, 1989; Baron, 1992). The volume of water stored in snowpack is highly variable but generally increases with increasing elevation. Snowmelt and runoff typically being in March or April, reach a peak in May or June and recede to baseflow conditions in July or August. Convective thunderstorms may produce late spring flooding, particularly when rain on snow conditions occur (Hauer, 1991). During the late summer and autumn, streamflow is predominantly supported by groundwater discharge.

Diurnal variations in stream flow during the snowmelt period are very dynamic (Figure 1). Changes in discharge reflect the variation in meteorological variables that drive the melt process (e.g. solar radiation, air temperature). Daily snowmelt also moves vertically to recharge groundwater, both within and over the unsaturated soil zone. In some areas, the water table reaches the soil surface and groundwater discharge becomes surface runoff (e.g. seepage–slope wetlands and spring brooks) before reaching a stream channel. As stream size and discharge increase, diurnal variation around the mean daily discharge is somewhat reduced.

Mountain streams have several types of flow regimes. For example, snowmelt streams flow throughout the late spring and summer and are supplied by surface snowmelt, groundwater from saturated soils and snowmelt waters moving rapidly through unconsolidated till and talus. Both permanent and ephemeral snowmelt streams are characterized by low concentrations of suspended particulates, dissolved nutrients and dissolved organic compounds (Hauer and Stanford, 1982b). In contrast, glacial outwash streams receive their waters primarily from glacier ablation and have continuous stream flow throughout the summer. Similarly to snowmelt streams, glacial streams also tend to be low in nutrients and dissolved organic

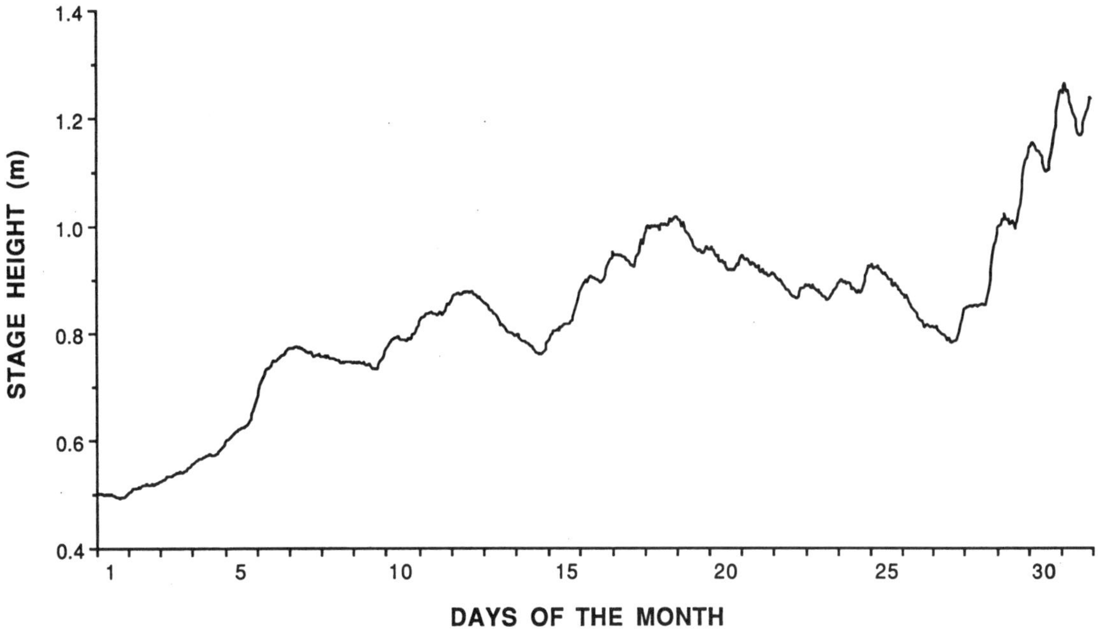

Figure 1. Mean hourly stage height recordings (m) for 1–31 May 1995 in McDonald Creek, Glacier National Park, Montana, USA. Note change in discharge throughout month as well as diel change

compounds, but generally have high concentrations of inorganic particulates (glacial flour). Changes in snow accumulation in winter or changes in the rate of snowmelt in summer would directly affect the hydrological regime of mountain streams, whether they are snowmelt fed (e.g. permanent streams becoming ephemeral with a decrease in snowpack) or glacial fed (e.g. rapid glacier ablation and disappearance).

Changes in temperature are likely to affect montane stream ecosystems through a broad influence upon precipitation state (i.e. snow vs rain), the rate of snowmelt, biogeochemical flux rates and the rate of the metabolic functions of biota. Montane stream temperatures (e.g. annual degree–days, summer mean, summer maximum) are closely correlated with elevation and stream size (e.g. Figure 2), although temperatures may vary significantly between alpine sites. For example, water temperatures may remain near 0°C for north-facing groundwater, surficial snowmelt or glacial sources, whereas temperatures in streams on south-facing slopes may rise as much as 16°C because of direct solar radiation (F. R. Hauer, unpublished data). In contrast, summer temperatures in subalpine streams generally remain <10°C in forested areas, while third- and fourth-order streams generally reach 10–15°C. Large montane valley rivers (fifth to seventh order) experience direct solar radiation and summer maximum temperatures between 18 and 24°C (Hauer and Stanford, 1982b; Stanford *et al.*, 1988).

Climate warming may significantly alter surface and groundwater temperatures. Warmer atmospheric temperatures should particularly affect streams that do not receive direct solar radiation and are more influenced by air temperatures. However, it is unclear whether projected climate change would result in a generalized warming trend throughout the Rocky Mountain Region, or whether change would be strongly regionalized. Furthermore, weather patterns and the influences of storms and cold fronts may have a profound effect on hydrological output (Dahm and Molles, 1992) and thermal regimes, particularly at mid-continental latitudes where maritime influence is most acute.

Fluvial processes, hyporheic habitats and floodplain development

Glaciation has dramatically affected landforms throughout the Rocky Mountain Region (Alt and Hyndman, 1973). Arêtes, glacial cols, cirque headwalls, horns and other erosional landforms of valley

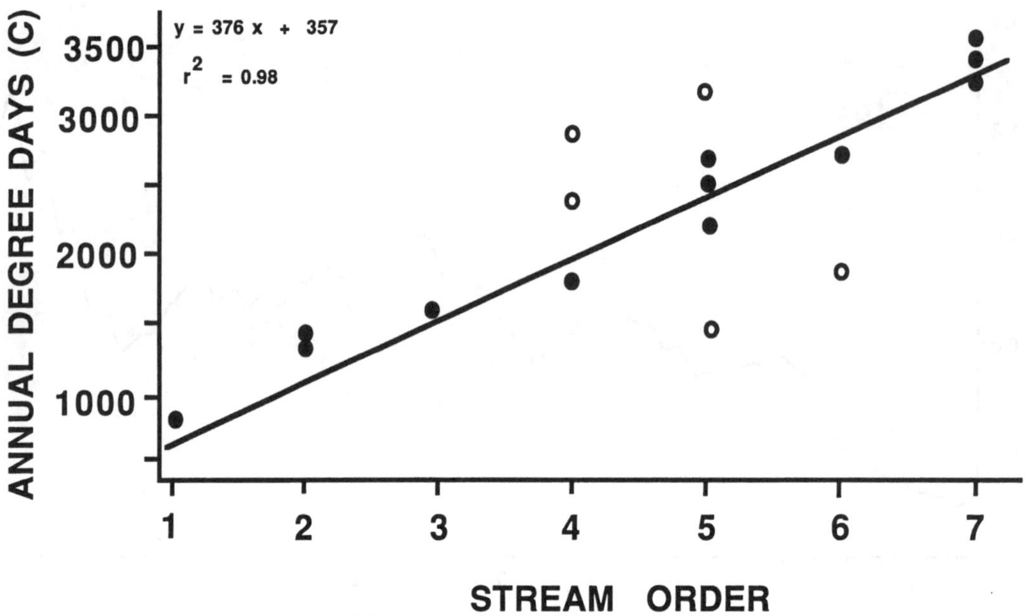

Figure 2. Relationship between temperature [expressed as annual degree days (°C)] and stream order at a series of stream sites in the Flathead Basin, north-western Montana. Solid circles represent sites along the river continuum from high elevation, small streams to low elevation, large rivers. Open circles represent sites below lakes (warmed) or hypolimnion release hydropower dams (cooled). Linear regression is taken only from solid data points. (Modified from Stanford *et al.*, 1988)

glaciers characterize much of the region. Small, high elevation streams generally possess turbulent flow along steep gradients under bedrock control of bed and channel form. Upon reaching the valley floor of the glacial trough, streams frequently descend a series of steps (i.e. glacial stairway) forming confined and unconfined reaches. Low gradient, unconfined reaches are often filled with alluvium. These river segments are characterized by extremely porous substrata saturated by river water that flows interstitially through a network of palaeochannels distributed across the floodplain, creating extensive hyporheic habitats (Stanford and Ward, 1988). Palaeochannels form a vertical lattice of substrata created by the chrono-sequence of laterally abandoned main channels that have become buried by fine sediments deposited by surficial flooding (Stanford and Ward, 1993). Palaeochannels act as conduits of interstitial water as a function of the vertical hydraulic gradient of the flood plain and are often an integral part of the hydrogeomorphology of riparian wetlands. Phreatic (deep) groundwaters may enter the floodplain from unconsolidated glacial till and colluvium along the valley wall and mix into floodplain aquifers. Interstitial flow of water from the main channel, or from tributaries, delimit the hyporheic zone (Stanford and Ward, 1993). Likewise, regular flooding of the river floodplain maintains surface and subsurface connectivity between the main channel, hyporheic habitats and riparian wetlands of the gravel-bed river systems.

Climate change affecting the magnitude and frequency of flooding will directly and indirectly affect these river flood plains. Climatic warming, leading to reduced winter snowpack, would result in a lower total discharge and, generally, a lower peak in spring runoff. Since the annual spring flood is primarily responsible for the redistribution of alluvium as well as stream channel development (i.e. creating new channels and abandoning old channels), reduction in the frequency or intensity of flooding would reduce the long-term complexity of the floodplain. Changes in the timing and magnitude of flooding will directly affect the flux of water, materials (e.g. nutrients, sediment, organic carbon) and biota between the main channel, hyporheic habitats and riparian wetlands. The complexity of these structures and the degree of their interactions directly affect stream corridor biodiversity and ecosystem integrity (*sensu* Angermeier and Karr, 1994).

Biogeochemical flux and cycles

Alpine and subalpine biogeochemical fluxes are strongly controlled by snowmelt. Mass flux through alpine watersheds is dominated by snowmelt runoff, which typically results in $\geqslant 70\%$ of the flux of water the solutes during May–July. Solutes in surface waters are dependent on atmospheric inputs, hydrological flow paths, mineral weathering products, soil chemistry and hydrological temporal dynamics (Campbell *et al.*, 1995). Processes in the terrestrial ecosystem strongly influence inputs to the aquatic ecosystem. Solutes stored in snow from the previous winter are transported over snow and ice, into and through soils and enter surface and ground waters. Vegetation frequently has an important influence on the biogeochemical cycling of carbon, nitrogen and many other nutrients (e.g. Mast and Baron, 1990; Running and Nemani, 1991; Baron, 1992; Arthur and Frahey, 1993).

The potential effects of climate change on biogeochemical fluxes in alpine and subalpine aquatic ecosystems would most likely be caused by shifts in the magnitude and timing of snow accumulation, snowmelt runoff and in the proportion of snow or rain. Biogeochemical reactions in the soil and groundwater have substantial effects on water chemistry only after snowmelt is mostly complete (*ca.* July–August). Yet, by late August, night temperatures frequently fall below freezing, and terrestrial biological activity beings to slow. This makes alpine watersheds sensitive indicators of climate change, because even small changes in temperature or precipitation could cause substantial changes in the duration of the late summer period when conditions are favourable for biogeochemical cycling.

The precise response of biogeochemical processes in lotic systems to climate change will be difficult to predict. For example, the source of stream water during snowmelt includes both direct snowmelt and displacement of 'old' subsurface water. Preferential elution causes early snowmelt water to be acidic and rich in solutes such as nitrate and sulfate that are found in atmospheric deposition. In contrast, the displaced subsurface water is often well buffered and rich in weathering products. An increase in direct snowmelt may favour the acidic components, thus increasing the risk of episodic stream acidification. Alternatively, high elevation systems may become more productive owing to increased maximum summer temperatures, longer growing season, longer water retention times and increased atmospheric loading of essential nutrients.

Lotic biota

A fundamental feature differentiating alpine from lower elevation streams is the character of the streamside vegetation. Alpine streams, being above the timberline, receive direct solar radiation and have few allochthonous inputs from riparian vegetation. Lower elevation streams have woody streamside vegetation that shades the stream, although sources of allochthonous material may be small and highly refractory (Triska and Sedell, 1976). Montane valley streams also contain large woody debris that influence transport and retention of organic matter and the trophic relations of benthic invertebrates (Bilby, 1981; Anderson *et al.*, 1984; Malanson and Butler, 1990; Ralph *et al.*, 1993). Climatic change leading to elevation change in the timberline would have a profound effect on the structure and function of alpine stream systems. For example, energy budgets of high elevation, open canopy streams tend to be primarily autochthonously based. The trophic relationships between the benthic fauna of these streams are dominated by grazers andcollector-gatherers (*sensu* Merritt and Cummins, 1996). Small streams below the timber-line (i.e. closed canopy) are allochthonously based, with benthic fauna dominated by collectors (e.g. net-spinning caddis-flies), collector-gatherers (e.g. ephermerellid mayflies) and predators (e.g. perlid stone-flies).

Many factors influence the distribution and abundance of stream invertebrates along elevational gradients (Hauer and Stanford, 1982a, 1991; Hawkins, 1984; Stanford *et al.*, 1988), especially temperature (Ward and Stanford, 1982). Most lotic invertebrates have well-defined distributions that reflect the close linkage to temperature (Sweeney, 1984). For example, many species of the caddis-fly family, Hydropsychidae, undergo serial replacements along elevation gradients (Figure 3), even though each species performs the same ecological function (filtering seston using a silk net). Thus, changes in stream temperatures are expected to affect species-specific distributions. In high alpine environs, cold stenothermic species may be extirpated if

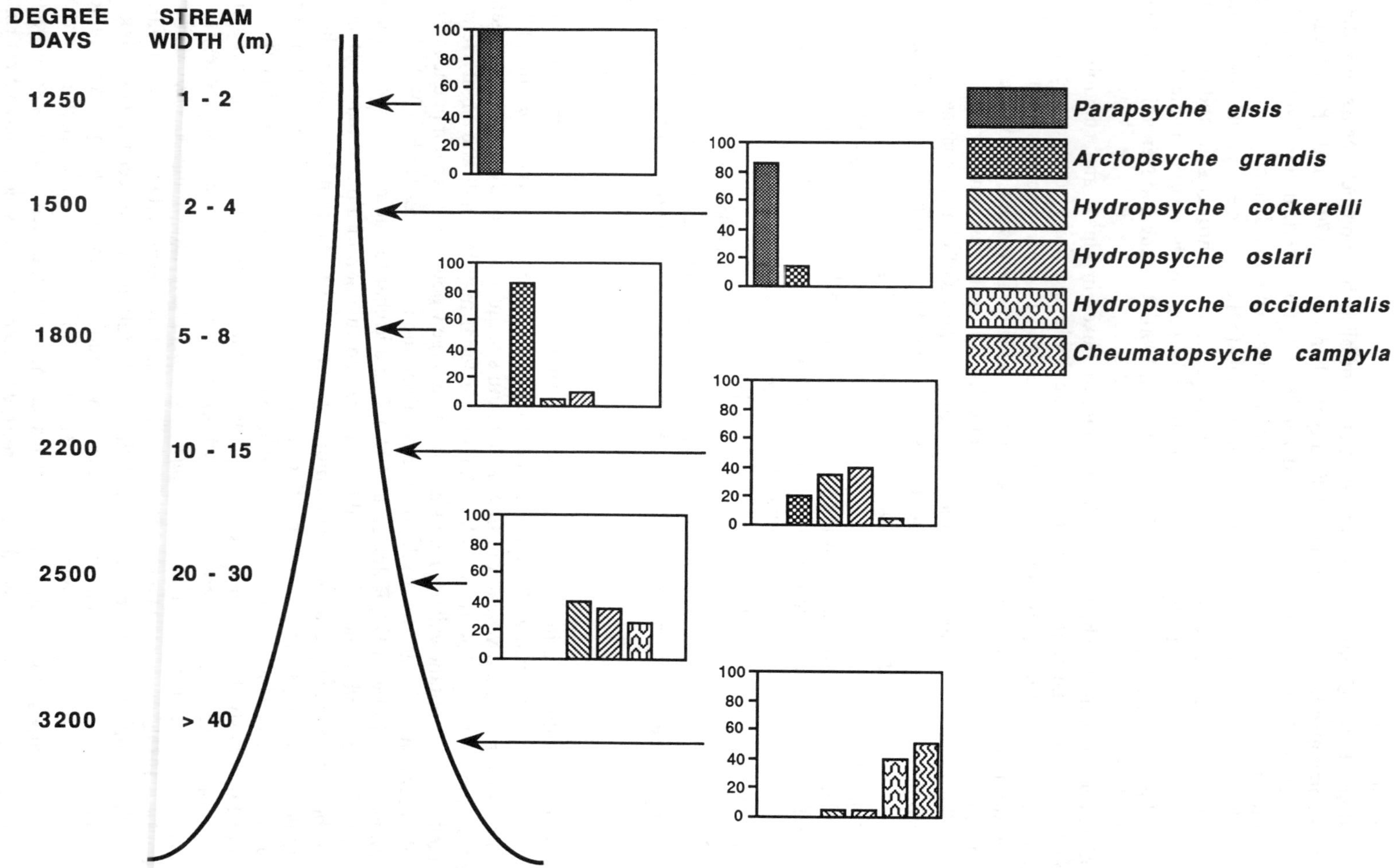

Figure 3. The distribution and relative abundance (percentage of total composition) of the six commonly occurring net-spinning caddis-flies (Trichoptera: Hydropsychidae) as they occur along the river continuum and elevation and temperature gradient in Rocky Mountain streams of southern British Columbia and Alberta, and western Montana. Note the species replacement from *Parapsyche elsis* in the small, cold streams to *Cheumatopsyche campyla* in the large, warmer rivers

temperatures rise beyond tolerable limits or if precipitation changes result in a stream becoming intermittent. Thus, changes in organic matter sources, temperature or hydrological flows and regime would have a profound effect on zoobenthic biodiversity and system integrity.

Native fish populations in cold streams of the Rocky Mountains are, generally, numerically dominated by salmonids. Native salmonids originally consisted of subspecies of cutthroat trout (*Oncorhynchus clarki*; see Behnke, 1992), along with combinations of bull char (*Salvelinus confluentus*), mountain whitefish (*Prosopium williamsoni*) and Arctic grayling (*Thymallus arcticus*) in the northern ranges. In the southern Rocky Mountains of Arizona and New Mexico, Gila trout (*O. gilae gilae*) and Apache trout (*O. g. apache*) are the native salmonids. However, non-native brook trout (*Salvelinus fontinalis*), brown trout (*Salmo trutta*) and rainbow trout (*O. mykiss*) have been introduced throughout the region (e.g. Fausch, 1988, 1989).

Temperature is also an important environmental factor structuring salmonid guilds along altitudinal gradients (Fausch *et al.*, 1994). Modified thermal and hydrological regimes resulting from climatic changes should have both direct and indirect effects on stream salmonds. For example, Keleher and Rahel (1996) estimated that 62% of the geographical area suitable for salmonids in the Rocky Mountain Region would become too warm, and 31% of stream length suitable for salmonids in Wyoming would become unsuitable, with a 4°C warming in air temperature. This is consistent with the results of Meisner (1990, 1993) and Flebbe (1993) regarding the effect of groundwater warming on salmonid distributions in the southern Appalachian Mountains.

Changes in the stream discharge regime may have marked direct effects on salmonid reproduction. Native salmonids are either early summer spawners, adapted to lay eggs and complete juvenile emergence on the descending limb of the snowmelt hydrograph (e.g. cutthroat trout), or autumn spawners that incubate eggs during winter and emerge the following spring before runoff (e.g. bull char). An increase in rain on snow events owing to winter warming would affect incubating eggs of autumn-spawning salmonids (e.g. Seegrist and Gard, 1972; Erman *et al.*, 1989), whereas reduced summer flows would alter recruitment of all species (Northcote, 1992). Similarly, high snowmelt runoff may depress recruitment of recently emerged juveniles (Nehring and Anderson, 1993).

Climate warming may have a number of subtle, indirect effects on stream salmonids. Highly plastic life histories have permitted salmonids to adapt to diverse habitats. For example, fluvial cutthroat trout and bull char spawn and rear in small streams, but move downstream to grow and mature in larger rivers. Adfluvial forms also spawn and rear in small streams, but grow and mature in large lakes (Varley and Gresswell, 1988; Fraley and Shepard, 1989; Rieman and McIntyre, 1993). Both fluvial and adfluvial forms are known to move relatively long distances to complete their life history (Gowan *et al.*, 1994; Young, 1994; Riley and Fausch, 1995; Schlosser and Angermeier, 1995), and are likely to be linked as metapopulations by dispersal (Fausch and Young, 1995). Fish populations in headwater habitats may often be extirpated by stochastic perturbations [e.g. fire, floods (Propst *et al.*, 1992)], but were probably recolonized from downstream populations (Harrison, 1991). However, if climate warming eliminates downstream source populations, headwater subpopulations will most likely collapse (Flebbe, 1993).

Climate warming may also affect stream salmonids by exacerbating biotic interactions with non-native species. In many locales, native cutthroat trout have been excluded from their natural stream habitat by introduced brook trout and brown trout (Griffith, 1988; Fausch, 1989). Recent laboratory experiments showed that brook trout behaviourally dominated cutthroat trout and garnered more food at 20°C, whereas competitive abilities were similar at 10°C (DeStaso and Rahel, 1994). Thus, warming of otherwise appropriate stream habitat could provide non-native trout with a competitive advantage.

LAKE ECOSYSTEMS

Montane lakes may be subject to many of the effects of climate change identified elsewhere in this volume. Possible changes in the physical characteristics of lakes include: temperature structure and mixed layer depth, summer and winter heat budgets, timing of the onset and breakdown of both summer and winter stratification, timing of the onset and break-up of ice cover, river through-flow pattern and drainage basin storage and flushing and ventilation rates. In addition, mountain lakes may have a number of unique responses to warming, which vary both with lake size and elevation.

High elevation lakes are usually small, ultraoligotrophic, poorly stratified and poorly buffered systems that lack natural populations of fish, macrophytes or insect predators (Anderson, 1974a). Waters have low concentrations of dissolved organic compounds (Baron *et al.*, 1991), are transparent to UV radiation and may be either clear, or turbid, depending on the local importance of glaciers. Their planktonic food-webs are dominated by low densities of large (>3 mm) brightly coloured predacious copepods (*Diaptomus, Hesperodiaptomus*), large herbivores (*Daphnia pulex, D. middendorfianna*) and several species of rotifers, flagellates, diatoms and cyanobacteria (McKnight *et al.*, 1991; Stockner, 1991; Thomas *et al.*, 1991). Low maximum summer temperatures (e.g. <12°C) may restrict successful colonization by many temperate species (e.g. Lamontange and Schindler, 1994). Benthic communities frequently support high densities of *Gammarus* (Leavitt *et al.*, 1994) and benthic cyanobacteria, *Chara*, or *Cladophora* (Stanford and Prescott, 1988), but have relatively depauperate chironomid communities (Walker and Mathewes, 1989).

Finger lakes (e.g. McDonald, Waterton) and graben lakes (e.g. Pend Oreille, Flathead) are common among the intermontane valleys of the Rocky Mountain Region. These large lakes are generally oligotrophic with low concentration of nutrients (< 1μg/l soluble reactive P) and primary production (10–20 mg/m^2/yr, <1 mg/l Chl *a* standing crop). Phytoplankton populations are generally dominated by diatoms in spring and green algae in summer. Historically, planktonic food-webs have been dominated by large cladocerans and pelagic fishes; however, introductions of non-native invertebrates and fish have significantly affected native populations and food-web structure (Spencer *et al.*, 1991).

Lake hydrology

Most alpine lakes are small and have highly variable hydrology. Lake flushing rates can exceed 0·71 per day in spring (Baron, 1992), but are generally negligible in winter and intermediate in summer. In many montane watersheds, water retention during spring runoff may be very short and lakes are particularly sensitive to change in water budget or change in biogeochemical processes occurring during snowmelt. Changes in the magnitude or duration of snowmelt from climate change would most likely cause a shift in the relative importance of the in-lake biogeochemical processes. Furthermore, high flushing rates select for phytoplankton species with rapid growth rates (e.g. diatoms), whereas cyanobacteria and flagellates often dominate during relatively quiescent late-summer periods (McKnight *et al.*, 1991). Thus, changes in precipitation patterns may dramatically affect hydrology and hence plankton community structure (Thomas *et al.*, 1991).

The long-term hydrological response of lakes with glaciers may be easier to predict. Over a time-scale of 100 years, the input of suspended inorganic matter to alpine lakes is strongly correlated with glacier size and summer temperature (Leonard, 1986a, b). Long-term fossil analyses show that silt input is greatest during paraglacial periods, when glaciers are extensive, but retreating (e.g. Hickman and Reasoner, 1994). Under these circumstances there is a sufficient volume of water for transport of sediments, and recently exposed substrates are unstable because of sparse vegetation cover. Because the recent 'Little Ice Age' included the coldest period since the start of the Holocene, current climatic warming is occurring during such a paraglacial era and water flow and silt loads may be exceptionally high. For example, estimates from the Premier Range in British Columbia suggest that 23% of glacier cover was lost between *ca.* 1850 and 1970, especially from small glaciers. Rates of recession can exceed 50 m/yr (Leonard, 1986b).

The importance of glaciers in hydrological processes should vary across elevational and latitudinal gradients. There are few glaciers and ice fields among the southern Rocky Mountains. However, glaciers may cover over 15% of mountainous regions in the Canadian Rockies (Luckman, 1990). Similarly, glaciers are less common in rain shadow regions or on south-facing slopes than at other sites. For example, Hickman and Schweger (1993) showed that lakes in the eastern Rocky Mountain foothills and plains were initially saline in the early Holocene because glaciers blocked the flow of moist Pacific air. However, as glaciers melted, present day circulation was established, and lakes freshened. In contrast, mountain and intermontane lakes of the

Rockies were fresh throughout the Holocene. Such variability suggests that there are multiple responses of alpine hydrology to climatic warming.

The water budgets of the large lakes in the mountain and piedmont valleys of the Rocky Mountains are controlled by the large snowmelt rivers that flow into them. Most of these lake basins occupy <5% of their catchments and the calculated retention times are often less than 3–5 years, even though lake depths are often >100 m. Water flux and associated limnological responses have not been well documented over all large lakes. However, in Flathead Lake, Montane, >60% of the annual inflow generally occurs during the 4–8 weeks of the spring runoff (J. Stanford, unpublished data). Generally, river water during this time is 2–5°C warmer and considerably less dense than lake water, thus spring floodwater may not reside in the lake more than a few days as rivers may exhibit overflow with little mixing of the two water masses. Montane lakes usually exhibit dimictic thermal stratification. However, the very largest lakes (e.g. Flathead, Kootenay, Pend Orielle) do not freeze over unless becalmed for several days during exceptionally cold years. Wind-driven mixing usually keeps the pelagic area ice free, but the entire water column may cool to 1–2°C.

Catchment temperature and precipitation patterns are the principal meteorological factors affecting the limnology of large valley-bottom lakes in the Rocky Mountains. Mixing of water and retention of nutrients are determined in large part by temperature differentials of the incoming river water and in-lake water, as well as interactions with wind energy. Climate change resulting in warmer temperatures and a changed hydrological regime would alter the flux of water, heat and nutrients and thereby influence the seasonal dynamics of phytoplankton composition, responses of zooplankton, fish reproduction and overall bio-production.

Biogeochemical flux and cycling

Primary production in Rocky Mountain alpine lakes is usually limited by a combination of low temperatures, rapid flushing and extremely low phosphorus (P) supply. Concentrations of total P (TP) and chlorophyll (Chl) are often less than 2–5 μg/l in alpine lakes, and are usually less than subalpine sites within the same drainage (but see McKnight *et al.*, 1991). Increased temperatures may lead to increased nutrient loading. Longer ice-free seasons and higher temperatures would immediately increase the rates of mineralization, although this relation may be complicated by changes in precipitation patterns and hydrology (see above). Over longer time scales, the introduction of forest vegetation would improve biotic cycling of nutrients within soils and increase nutrient supply to aquatic systems.

Development of forests at alpine sites should increase the concentrations of dissolved organic matter (DOM) and reduce lake transparency to ultraviolet radiation. UV intensities rise by *ca.* 15% for every 1000 m elevation and thus are high in alpine lakes. However, unlike subalpine sites, alpine lakes have low levels (<2 mg/l) of both autochthonous and allochthonous dissolved organic carbon (DOC) (Baron *et al.*, 1991). Because terrestrially derived DOC from wetlands and soils are the main agents attenuating UV radiation (Schindler *et al.*, 1992), UV-A (320–400 nm) radiation may penetrate to >8 m depth in alpine sites (D. Lean, National Water Research Institute, Burlington, Ontario, personal communication). In addition, the products of UV photochemical reactions (radicals, peroxides) can circulate throughout the water column of unstratified alpine lakes. Consequently, the biota of alpine lakes may exhibit high levels of UV-related stress (e.g. Vincent and Roy, 1993; Williamson *et al.*, 1994). Prolonged climatic warming would allow the development of terrestrial vegetation and soils, increase inputs of DOC and reduce UV penetration.

Production in most large, valley bottom lakes appears to be controlled by the availability of labile forms of phosphorus and nitrogen (Dodds *et al.*, 1989; Dodds and Priscu, 1990; Spencer and Ellis, 1990). Since most large Rocky Mountain lakes have experienced significant change in land use within their drainages during the past 4–5 decades (e.g. increased population, timber harvest), trophic dynamics in these lakes reflect a multitude of environmental changes (e.g. sedimentation rates, nutrient flux, primary production, food-web composition) (Stanford and Ward, 1992).

Warming of mountain climes will most likely change the water, heat, nutrient (C, N, P) and sediment flux in large valley bottom lakes, and the influences of these factors will be interactive with increasing human activities occurring in the catchment basins. For example, extremely diverse communities of crysophytes (diatoms) and chlorophytes (green algae) currently dominate the phytoplankton communities. Human-mediated nutrient loads of C, N and P increase microbial production and result in the eventual predominance of nitrogen-fixing cyanophytes. Oxygen depletions in the deep waters associated with increased C, N and P loading may result, and can affect the redox conditions on the lake bottom, dramatically enhancing internal nutrient loading (Wetzel, 1983). Indications of eutrophication, such as oxygen depletion, have already been observed in Flathead and other lakes (J. Stanford, unpublished data).

Lake biota

The food-webs of alpine lakes are often simple both in terms of the number of trophic levels and the degree of functional redundancy (species per trophic level). Over 95% of all alpine lakes in western North America are naturally fishless (Bahls, 1992). In addition, larger alpine lakes are usually too cold to support predacious insects as *Chaoborus* (Lamontange *et al.*, 1994) or Cladocera such as *Leptodora* (Anderson, 1974a). Instead, fishless lakes are dominated by predacious copepods of the subgenus *Hesperodiaptomus* (Anderson, 1974a), large red copepods that greatly reduce the abundance of rotifers and other small zooplankton (Anderson, 1970; Paul and Schindler, 1994). Herbivore communities are also simple and dominated by large-bodied *Daphnia*, rather than small cladocerans (Anderson, 1974a). Perhaps because of their simple nature, alpine food-webs may be especially sensitive to projected environmental changes.

Climatic change may have its greatest impact on primary production in alpine lakes. Analyses of fossil diatoms (Hickman and Reasoner, 1994) and pigments (P. R. Leavitt, M. A. Reasoner and M. Hickman, unpublished data) suggest that both species abundance and diversity increase as forests develop in alpine watersheds, possibly because of increased ice-free seasons. Analyses suggest that planktonic species may benefit particularly from climatic warming (Hickman and Reasoner, 1994). In addition, rooted macrophytes are rare in alpine lakes, but are present in subalpine and low elevation systems and may benefit from increased water temperatures or growing seasons.

Elevated nutrient inputs from developing terrestrial soils would be expected to boost algal biomass in phosphorus-limited alpine lakes. However, minor changes in nutrient loading may also have unexpectedly large effects on algal species composition and vertical distribution. For example, the phytoplankton of many alpine lakes show significant adaptations to high light intensities. Flagellates and picoplankton often avoid inhibiting irradiance by forming deep water populations, even in deep unstratified lakes (Anderson, 1974b; Stockner, 1991; Thomas *et al.*, 1991). These deep populations are highly sensitive to changes in water chemistry and are rapidly lost upon minor lake fertilization (Stockner and Shortreed, 1991).

Recent experiments have demonstrated that natural levels of UV radiation (300–400 nm) at alpine elevations can suppress growth of attached periphyton >50% (R. D. Vinebrooke and P. R. Leavitt, unpublished data). However, natural periphyton communities are often dominated by filamentous cyanobacteria in alpine lakes. These taxa can produce unique pigments that allow them to withstand high UV flux (Garcia-Pichel and Castenholz, 1991; Vincent *et al.*, 1993).

Not all phytoplankton will increase under a scenario of climatic warming. Phytoplankton in shallow, rapidly flushed alpine lakes appear to be controlled by hydrological regimes, rather than by changes in invertebrate community structure, nutrient flux or UV exposure (McKnight *et al.*, 1991; Spaulding *et al.*, 1993). Similarly, production may decline in glacier-fed lakes, as increased silt input favours thinly silicified diatoms or small picoplankton over thick benthic mats of *Chara*, *Cladophora* or cyanobacteria. These mats seem to dominate primary production in the most extreme alpine sites (Stanford and Prescott, 1988). Finally, changes in snowpack distribution may lead to unpredictable changes in primary production. Under-ice algal production can be significant through much of the winter if high winds remove most of the snow cover (Pennak, 1968; Spaulding *et al.*, 1993). Altered precipitation or circulation patterns may lead to an increased depth of snow and reduced light penetration in winter.

It is unclear if projected climatic warning would lead to colonization of alpine lakes by species of temporate fish. Most of the 1464 lakes of the Canadian Rocky Mountain National Parks were fishless prior to the twentieth century, or were inhabited by either mountain whitefish (*Prosopium williamsoni*) or longnose suckers (*Catostomus catostomus*). However, 305 lakes were stocked with non-native trout during this century (Donald, 1987). The persistence and abundance of introduced stocks were directly proportional to either the size of the lake outlet or lake depth, depending on the species introduced. In addition, the presence of mountain whitefish, longnose suckers or lake trout (*Salvelinus namaycush*) reduced the probability of colonization. Consequently, there is no simple prediction regarding the likelihood of successful colonization of alpine lakes by fishes resulting from projected climatic change.

Colonization of alpine lakes by fishes as a result of climatic warming would significantly reconfigure alpine food-webs and alter biogeochemical cycles (Leavitt *et al.*, 1994). Size-selective predation by introduced fish was associated with the elimination of large-bodied invertebrate predators (*Hesperodiaptomus*), herbivores (*Daphnia*) and zoobenthos (*Gammarus*) in many alpine lakes in the northern Rocky Mountains (Anderson, 1972, 1974a, 1980). Loss of the top invertebrates predator almost certainly allowed small herbivorous copepods, rotifers and predacious cyclopoid copepods to increase (Anderson, 1972). In small lakes, fish may switch to less preferred prey (terrestrial insects) after extirpation of large aquatic invertebrates, thereby increasing nutrient recycling between the littoral zone and pelagic waters (Leavitt *et al.*, 1994).

Some invertebrates may colonize alpine lakes in response to increased lake water temperature. Palaeolimnological study of invertebrate communities shows that alpine lakes are often inhabited by cold stenotherms with slow growth rates and poor competitive abilities (Walker and Mathewes, 1989). Many invertebrates, important in structuring the pelagic community, are excluded from alpine lakes because of insufficient degree–days to complete life cycles, poor resource bases or inadequate terrestrial (for adults) or aquatic (immatures, larvae) habitats. For example, *Chaoborus* are important invertebrates predators in many lakes (Lamontagne *et al.*, 1994). Small species (*C. punctipennis*) did not occur in lakes where temperatures were $<20°C$ and large species (*C. americanus*) did not occur in lakes where temperatures were $<15°C$. Interestingly, *Chaoborus* were present in shallow, warm alpine ponds, and were rare in cold, low elevation lakes, suggesting that the terrestrial environment was less important than water temperature in controlling species presence (Lamontange *et al.*, 1994). Colonization of *Chaoborus* in lakes at tree-line might lead to significant changes in the invertebrates predation regime and zooplankton community structure.

Benthic invertebrate communities will also be influenced by climatic warming. Walker and Mathewes (1989) showed that the chironomid communities of alpine lakes are depauperate relative to similar subalpine sites. Taxa characteristic of high elevations (*Heterotrissocladius*, *Parakiefferiella*, *Paraclaudius*) were also common in high Arctic sites, deep cold lakes and early Holocene fossil assemblages, again suggesting that water temperature was the dominant control of community composition. Similarly, Donald and Anderson (1980) found that lentic stone-flies were restricted to alpine sites or large, cold subalpine lakes. Such distinct alpine communities probably result from direct temperature limitation of more temperate species, rather than a preference of alpine taxa for cold waters (Walker and Mathewes, 1989). Consequently, lake water warming and species introduction could lead to loss of alpine fauna via competition or predation.

Invertebrate food-webs may be altered as a result of the indirect effects of climatic warming. In cases where invertebrate predators are resource limited, increased production allows the establishment of additional trophic levels (Neill and Peacock, 1980; Neill, 1988) or alternative invertebrate predators. Similarly, increased water temperatures may allow the development of littoral macrophyte beds that are normally absent from alpine lakes.

The biota of the large, valley bottom lakes will be affected by changes in material flux from their upper watersheds. For example, sediment and nutrients entering Flathead Lake in the spring have a significant effect on summer primary productivity; a low peak in spring runoff delivers less C, N and P from the upper basin than do high runoff years, directly affecting both auto- and heterotrophic production (Ellis and Stanford, 1982, 1988). Microbial production is also closely related to light penetration and the depth of summer mixing in large, oligotrophic lakes (Jassby *et al.*, 1992). Consequently, the effects of changing light conditions (e.g. higher UV-B) on photosynthetic efficiency and species composition may be a factor in these large, low-elevation lakes, as well as in the small, alpine lakes.

In general, the large lakes of the region have a long history of food-web changes mediated by the introduction of non-native species. Non-native species have produced both top-down and bottom-up cascading effects on lake trophic dynamics (e.g. Spencer *et al.*, 1991). Many native species, particularly fishes and amphibians, are currently in rapid decline throughout the Rocky Mountain Region (Gresswell *et al.*, 1995; Nehlsen *et al.*, 1991) as a consequence of habitat degradation and food-web change. Climate warming or eutrophication of the valley bottom lakes will exacerbate problems of biodiversity and ecosystem integrity already associated with destabilized food-web structure and enhancement of non-native species populations. Native species that have evolved complex life histories in the cold oligotrophic environs of the large lakes (e.g. bull trout, Fraley and Shepard, 1989; Yellowstone cutthroat trout, Gresswell *et al.*, 1995) are especially vulnerable.

CONFOUNDING FACTORS

The effects of resource management are felt throughout the Rocky Mountain Region. In many locations these disturbances may obscure or entirely mask the effects of climate change. Deforestation and mining have produced acute, localized perturbations with deleterious effects on downstream freshwater ecosystems. Air pollutants are more diversified, but result in chronic and episodic effects to both terrestrial and aquatic systems, especially in poorly buffered, high elevation habitats (Turk and Spahr, 1991; Baron, 1992; Baron and Denning, 1993). To date, the most pervasive anthropogenic factors affecting freshwater ecosystems within the Rocky Mountain Region are: (1) alterations of land use that increase erosion and nutrient mobilization; (2) construction of dams and diversions that alter hydrological and thermal regimes and interfere with the movement of materials and biota; (3) introduction of non-native species that reconfigure food-webs; and (4) urbanization and the associated nutrient enrichment.

Runoff of sediment and nutrients from agricultural lands or managed forests is a pervasive problem throughout the Rocky Mountains. Removal of trees from mountainous watersheds increases annual stream discharge and changes runoff regimes (Cheng, 1989; Hauer, 1991). Likewise, the short-term flux of nutrients, particularly phosphorus and nitrogen, increases significantly when watersheds are logged (Hauer and Blum, 1991). These ecosystem-level perturbations have affected aquatic system biodiversity even in headwater drainages (Frissell *et al.*, 1993).

Most large river systems of the Rocky Mountain Region are regulated (i.e. dams, diversions, abstraction). Much of this activity has been concentrated in intermontane valleys where river waters are extracted for crop irrigation and municipal water supplies. In the absence of population centres, high-head hydroelectric dams inundate tens of kilometres of river segments and associated wetlands. Biota dependent on riparian habitats for portions of their life histories are invariably affected. Rivers downstream of dams experience significant change in hydrographs, thermal characteristics and alteration of zoobenthic and fish populations (Stanford and Hauer, 1992). Although most dams are located in intermontane valleys, there is increasing interest in micro-hydroelectric generation in headwater areas.

Introduction of non-native fish, invertebrates and macrophytes is widespread in Rocky Mountain freshwater ecosystems. Over 95% of mountain lakes are naturally fishless, yet >55% of the >16 000 US lakes have been stocked (Bahls, 1992; Donald, 1987). In alpine lakes, introduced trout rapidly eliminate large-bodied invertebrates that dominate food-webs (e.g. Anderson, 1972; Anderson and Donald, 1978) and biogeochemical cycles (Leavitt *et al.*, 1994). In many cases, stocked fish die out (Donald, 1987), yet invertebrate communities are slow to recover (Anderson and Donald, 1978; Paul and Schindler, 1994). The high variability in recovery rates between lakes may obscure our ability to detect lake response to climatic change. Mountain lakes that historically contained fish have also been damaged by indiscriminate stocking practices of non-native species and occasional poisoning with non-specific toxins (rotenone, toxaphene) to eliminate 'undesirable' native fish (Miskimmin *et al.*, 1994).

The long-term effects of global change are likely to exacerbate the effects of direct anthropogenic factors that are already detrimental to stream salmonids. For example, all the native species of salmonids, including six endemic subspecies of cutthroat trout, have declined during the last century (Behnke, 1992; Young, 1995). Most are either already listed as threatened under the US Endangered Species Act or are candidates for such a listing. Numerous large, valley bottom lakes (e.g. Flathead, Whitefish, Pend Oreille) inhabited by adfluvial species (e.g. cutthroat trout, bull char) have been further disrupted by introductions of *Mysis relicta*, an invertebrate predator that has decimated zooplankton populations and significantly altered the food-web dynamics (e.g. Spencer *et al.*, 1991). Many salmonid populations are becoming endangered, a problem that will most likely be exacerbated by projected climate change.

The Rocky Mountain Region has experienced rapid human population growth over the past two decades. While the very largest population centres have grown along the piedmont of the east slope (e.g. Calgary, Denver), municipalities in the intermontane valleys have seen explosive growth in the past decade. Rapid population growth often depletes freshwater resources as municipalities use nearby rivers and lakes for domestic water supplies and downstream waters as the recipient of municipal sewage.

RECOMMENDATIONS

Atmospheric increase in greenhouse gases (e.g. CO_2, CH_4, N_2O, CFCs) during the past century, and particularly since 1950, are well documented (reviewed in Levine, 1992). A World Meteorological Organization (1985) study concluded that the predicted equilibrium surface warming as a consequence of these increases for the period 1980–2030 ranges from 0·8 to 4·1°C. However, adequately predicting the effects of climate change at the regional level has been particularly problematic. Global variables affecting climate range from being easily measured (e.g. atmospheric increase in CO_2) to being very difficult to measure (e.g. thermal response of deep ocean currents to solar radiation). Coupling models of ocean response to increased concentration of greenhouse gases and atmospheric response variables at a global scale have proven to be particularly difficult. Likewise, global climate models have not been dynamically linked to regional climate or hydrological models. Although freshwater ecologists have speculated as to the potential consequences of global warming and the subsequent change in global climate, they have been limited on several areas of study; partly because of the fundamental problems of predicting climate change and partly because biological systems do not always respond as anticipated. In spite of these limitations, strong ecological inferences have been forthcoming, including formalization of our current understanding and identification of critical areas of uncertainty that require further investigation.

Both long-term (monitoring) and short-term (experimental) interdisciplinary studies are needed to resolve what remain as critical uncertainties in our understanding of climate change at the regional and local spatial scale and the response variables that underpin freshwater ecosystem integrity (*sensu* Angermeier and Karr, 1994). Research is needed to make accurate predictions of consequences regarding climate change and freshwater ecosystem response; and, as importantly, to formulate informed and appropriate management responses. This research should be conducted at the watershed (catchment) scale within an organizational framework of research nodes of at least 3–4 locales along the Rocky Mountains extending from northern Canada through Montana to Colorado. Two such research nodes are currently conducting watershed-level research specifically focusing on freshwater response to climate change; McDonald Basin, Glacier National Park in Montana, and Loch Vail in Colorado. Similar research nodes should be established in a far northern locale (e.g. Jasper, Canada) and in a central Rockies locale (e.g. Wyoming). Research nodes should be programmatically linked to coordinate methods, databases and information exchange.

Long-term, time-series data need to be systematically collected at research nodes to establish firmly the baseline data of thermal regimes, radiant energy flux, precipitation, hydrology, biogeochemical fluxes and an inventory of biological populations. These data should include remotely sensed areal features such as snow cover, lake optical properties, vegetation and land use. Short-term, experimental studies in the field and laboratory are needed to establish firmly the role of climatic drivers and the response variables of the aquatic ecosystems at different levels of temporal and spatial organization (*sensu* Minshall, 1988). For example, we need to understand, at the biogeochemical level, the source, fate and flux rates of nutrients that control bioproduction, at the species level, the bioenergetics of dominant or keystone species in an array of lake and stream types to determine species-specific response to thermal regimes; at the population level, the dynamics and the effects of habitat fragmentation on metapopulations; and, at the community level, the interactions between temperature and biotic interactions (e.g. competitive exclusion of cutthroat trout by brook trout at higher temperatures). Finally, palaeoecological studies are needed, using glacial geomorphometry, pigments, invertebrate remains, pollen, macro-fossils and geochemistry, to understand past responses of aquatic systems to warming since the 'Little Ice Age' (*ca.* 150–300 BP) and the 'Younger Dryas' (*ca.* 12 000 BP).

Although freshwater ecosystems in the Rocky Mountains have received anthropogenic effects over the past century, near-pristine conditions exist in many hydrologically and geographically diverse drainages throughout the region. Many of the continent's premiere parks (e.g. Jasper, Waterton-Glacier, Yellowstone, Rocky Mountain) and wilderness areas (e.g. Bob Marshall, Frank Church), as well as many of the headwaters of the continent's large river systems, are distributed along the Rocky Mountain spine. The Rockies contain some of the least altered freshwaters among temporate regions of the world. High elevation waters generally have few dissolved ions, poor buffering capacity and relatively simple communities and trophic structures. The biogeochemistry, biological production, food-web dynamics and biogeography of montane lakes, streams and wetlands are greatly influenced by snowpack depth, rate of snowmelt and thermal budget; each of which are directly affected by the climate. Thus, the many pristine aquatic systems distributed across well-defined climatic and elevational gradients throughout the Rocky Mountains may also be among our most sensitive indicators of global climate change and confer important implications for the suitability of the region as a global indicator of change and the effects on aquatic ecosystems. Furthermore, because freshwater resources are fundamental to the human economy, as well as to biodiversity, it is critical that we come to an understanding of the interrelationships of climate change and freshwater ecosystem change in this critically important region.

REFERENCES

Alt, D. D. and Hyndman, D. W. 1973. *Rocks, Ice and Water: The Geology of Waterton-Glacier Park.* Mountain Press Publ. Co., Missoula. 104 pp.

Anderson, R. S. 1970. 'Predator–prey relationships and predation rates for crustacean zooplankters from some lakes in western Canada', *Can. J. Zool.*, **48**, 1229–1240.

Anderson, R. S. 1972. 'Zooplankton composition and change in an alpine lake', *Int. Verein. Theor. Angew. Limnol.*, **18**, 264–268.

Anderson, R. S. 1974a. 'Crustacean plankton communities of 340 lakes and ponds in and near the national parks of the Canadian Rocky Mountains', *J. Fish. Res. Bd Can.*, **31**, 855–869.

Anderson, R. S. 1974b. 'Primary production patterns in seven lakes and ponds in Alberta (Canada)', *Oecologia*, **14**, 1–17.

Anderson, R. S. 1980. 'Relationships between trout and invertebrate species as predators and the structure of the crustacean and rotiferan communities in mountain lakes', in Kerfoot, W. C. (Ed.), *Evolution and Ecology of Zooplankton Communities.* University Press of New England, New Hampshire. pp. 635–641.

Anderson, R. S. and Donald, D. B. 1978. 'Limnological survey of some small lakes in the vicinity of the Cascade Trail, Banff National Park', *Canadian Wildlife Service Ms. Report.* 162 pp.

Anderson, N. H., Steedman, R. J. and Dudley, T. 1984. 'Patterns of exploitation by stream invertebrates of wood debris (xylophagy)', *Verh. Int. Verein. Limnol.*, **22**, 1847–1852.

Angermeier, P. L. and Karr. J. R. 1994. 'Biological integrity versus biological diversity as policy directives', *BioScience*, **44**, 690–697.

Arthur, M. A. and Frahey, T. J. 1993. 'Controls on soil solution chemistry in a subalpine forest in north-central Colorado', *Soil Sci. Soc. Am. J.*, **57**, 1122–1130.

Bahls, P. 1992. 'The status of fish populations and management of high mountain lakes in the western United States', *Northwest Sci.*, **66**, 183–193.

Baron, J. S. 1992. 'Surface waters', in Baron, J. S. (Ed.), *Biogeochemistry of a Subalpine Ecosystem: Loch Vail Watershed*, Ecol. St. 90. Springer-Verlag, New York. pp. 142–186.

Baron, J. S. and Denning, A. S. 1993. 'The influence of mountain meteorology and precipitation chemistry at low and high elevation of the Colorado Front Range, USA', *Atmos. Environ.*, **27A**, 2337–2349.

Baron, J. S., McKnight, D. M. and Denning, A. S. 1991. 'Sources of dissolved and particulate organic material in Loch Vale Watershed, Rocky Mountain National Park, Colorado, USA', *Biogeochemistry*, **15**, 89–110.

Barry, R. G. 1973. 'A climatological transect on the east slope of the Front Range, Colorado', *Artic Alp. Res.*, **5**, 89–110.

Barry, R. G. 1992. *Mountain Weather and Climate.* Methuen & Co., New York. 402 pp.

Bates, G. T., Giorgi, F. and Hostetler, S. W. 1993. 'Toward the simulation of the effects of the Great Lakes on regional climate', *Mon. Weath. Rev.*, **121**, 1373–1387.

Bates, G. T., Hostetler, S. W. and Giorgi, F. 1995. 'Two-year simulation of the Great Lakes region with a coupled modeling system', *Mon. Weath. Rev.*, **123**, 1505–1522.

Beaudoin, A. B. 1987. 'Using *Picea/Pinus* ratios from the Wilcox Pass core, Jasper National Park, Alberta, to investigate Holocene timberline fluctuations', *Geogr. Phys. Quatern.*, **2**, 145–152.

Behnke, R. J. 1992. *Native Trout of Western North America* Monograph 6. American Fisheries Society, Bethesda, Maryland.

Bilby, R. E. 1981. 'Role of organic debris dams in regulating the export of dissolved and particulate matter from a forested catchment', *Ecology*, **62**, 1234–1243.

Burt, T. P. 1992. 'The hydrology of headwater catchments', in Calow, P. C. and Petts, G. E. (Eds), *The Rivers Handbook*, Volume 1. Blackwell Scientific Publ., Oxford. pp. 3–28.

Campbell, D. H., Clow, D. W., Ingersoll, G. P., Mast, M. A., Spahr, N. E. and Turk, J. T. 1995. 'Processes controlling the chemistry of two snowmelt-dominated streams in the Rocky Mountains', *Wat. Resour. Res.*, **31**, 2811–2821.

Cheng, J. D. 1989. 'Streamflow changes after clear-cut logging of a pine beetle-infested watershed in southern British Columbia, Canada', *Wat. Resour. Res.*, **25**, 449–456.

Croley, T. E. 1990. 'Laurentian Great Lakes double CO_2 climate change hydrological impacts', *Climate Change*, **17**, 27–47.

Dahm, C. N. and Molles, M. C. Jr. 1992. 'Streams in semiarid regions as sensitive indicators of global climate change', in Firth, P. and Fisher, S. G. (Eds), *Global Climate Change and Freshwater Ecosystems*. Springer-Verlag, New York. pp. 250–260.

DeStaso, J., III and Rahel, F. J. 1994. 'Influence of water temperature on interactions between juvenile Colorado River cutthroat trout and brook trout in a laboratory stream', *Trans. Am. Fish. Soc.*, **123**, 289–297.

Dickinson, R. E., Errico, R. M., Giorgi, F. and Bates, G. T. 1989. 'A regional climate model for the western United States', *Climatic Change*, **15**, 383–422.

Dodds, W. K. and Priscu, J. C. 1990. 'A comparison of methods for assessment of nutrient deficiency of phytoplankton in a large oligotrophic lake', *Can. J. Fish. Aquat. Sci.*, **47**, 2328–2338.

Dodds, W. K., Johnson, K. R. and Priscu, J. C. 1989. 'Simultaneous nitrogen and phosphorus deficiency in natural phytoplankton assemblages: theory, empirical evidence and implications for lake management', *Lake Reservoir Manage.*, **5**, 21–26.

Donald, D. B. 1987. 'Assessment of the outcome of eight decades of trout stocking in the mountain national parks, Canada', *Am. J. Fish. Manag.*, **7**, 545–553.

Donald, D. B. and Anderson, R. S. 1980. 'The lentic stoneflies (Plecoptera) from the Continental divide region of southwestern Canada', *Can. Entomol.*, **112**, 753–758.

Ellis, B. K. and Stanford, J. A. 1982. 'Comparative photoheterotrophy, chemoheterotrophy and photolithotrophy in a eutrophic reservoir and an oligotrophic lake', *Limnol. Oceanogr.*, **27**, 440–454.

Ellis, B. K. and Stanford, J. A. 1988. 'Phosphorus bioavailability of fluvial sediments determined by algal assays', *Hydrobiologia*, **160**, 9–18.

Erman, D. C., Andrews, E. D. and Yoder-Williams, M. 1989. 'Effects of winter floods on fishes in the Sierra Nevada', *Can. J. Fish. Aquat. Sci.*, **45**, 2195–2200.

Fausch, K. D. 1988. 'Tests of competition between native and introduced salmonids in streams: what have we learned?', *Can. J. Fish. Aquat. Sci.*, **45**, 2238–2246.

Fausch, K. D. 1989. 'Do gradient and temperature affect distributions of, and interactions between, brook charr (*Salvelinus fontinalis*) and other resident salmonids in streams', *Physiol. Ecol. Japan. Special Vol.*, **1**, 303–322.

Fausch, K. D. and Young, M. K. 1995. 'Evolutionarily significant units and movement of resident stream fishes: a cautionary tale', *Am. Fish. Soc. Symp.*, **17**, 360–370.

Fausch, K. D., Nakano, S. and Ishigaki, K. 1994. 'Distribution of two congeneric charrs in streams of Hokkaido Island, Japan: considering multiple factors across scales', *Oecologia*, **100**, 1–12.

Firth, P. and Fisher, S. G. (Eds) 1992. *Global Climate Change and Freshwater Ecosystems*. Springer-Verlag, New York.

Flebbe, P. A. 1993. 'Comment on Meisner (1990): effect of climatic warming on the southern margins of the native range of brook trout, *Salvelinus fontinalis*', *Can. J. Fish. Aquat. Sci.*, **50**, 883–884.

Fraley, J. J. and Shepard, B. B. 1989. 'Life history, ecology and population status of migratory bull trout (*Salvelinus confluentus*) in the Flathead lake and river system, Montana', *Northwest Sci.*, **63**, 133–143.

Frissell, C. A., Liss, W. J. and Bayles, D. 1993. 'An integrated, biophysical strategy for ecological restoration of large watersheds', in Potts, D. E. (Ed.), *Proceedings, Symposium on Changing Roles in Water Resources Management and Policy*. Am. Wat. Res. Assoc., Bethesda. pp. 449–456.

Garcia-Pichel, F. and Catenholz, R. W. 1991. 'Characterization and biological implications of scytenemin, a cyanobacterial sheath pigment', *J. Phycol.*, **27**, 395–401.

Giorgi, F. and Bates, G. T. 1989. The climatological skill of a regional model over complex terrain', *Mon. Weath. Rev.*, **117**, 2325–2347.

Giorgi, F. and Mearns, L. O. 1991. 'Approaches to the simulation of regional climate change: a review', *Rev. Geophys.*, **29**, 191–216.

Giorgi, F., Marinucci, M. R. and Visconti, G. 1990. 'Use of a limited area model nested within a general circulation model for regional climate simulation over Europe', *J. Geophys. Res.*, **95**, 18413–18431.

Giorgi, F., Bates, G. T. and Niemann, S. J. 1993a. 'The multi-year surface climatology of a regional atmospheric model over the western United States', *J. Climatol.*, **6**, 75–95.

Giorgi, F., Hostetler, S. W. and Shields-Brodeur, C. 1993b. 'Analysis of the surface hydrology in a regional climate model', *Q. J. R. Meteorol. Soc.*, **120**, 161–183.

Gleick, P. H. 1987. 'The development and testing of a water balance model for climate impact assessment: modeling the Sacramento Basin', *Wat. Resour. Res.* **23**, 1049–1061.

Gleick, P. H. 1989. 'Climate change and hydrology', *Rev. Geophys.*, **27**, 329–344.

Gowan, C., Young, M. K., Fausch, K. D. and Riley, S. C. 1994. 'Restricted movement in resident stream salmonids: a paradigm lost?', *Can. J. Fish. Aquat. Sci.*, **51**, 2626–2637.

Gresswell, R. E., Liss, W. J. and Larson, G. L. 1995. 'Life-history organization of Yellowstone cutthroat trout (*Oncorhynchus clarki bouvieri*) in Yellowstone Lake', *Can. J. Fish. Aquat. Sci.*, **51**, 298–309.
Griffith, J. S. 1988. 'Review of competition between cutthroat trout and other salmonids', *Am. Fish. Soc. Symp.*, **4**, 134–140.
Grotch, S. L. and MacCracken, M. C. 1991. 'The use of general circulation models to predict regional climatic change', *J. Climatol.*, **4**, 286–303.
Harrison, S. 1991. 'Local extinction in a metapopulation context: an empirical evaluation', *Biol. J. Linn. Soc.*, **42**, 73–88.
Hauer, F. R. 1991. *An Analysis of the Effect of Timber Harvest on Streamflow Quantity and Regime: an Examination of Historical Records*. Flathead Basin Commission, Kalispell. 42 pp.
Hauer, F. R. and Blum, C. O. 1991. *The Effect of Timber Management on Stream Water Quality*. Flathead Basin Commission, Kalispell. 33 pp.
Hauer, F. R. and Stanford, J. A. 1982a. 'Ecological responses of hydropsychid caddisflies to stream regulation', *Can. J. Fish. Aquat. Sci.*, **39**, 1235–1242.
Hauer, F. R. and Stanford, J. A. 1982b. 'Ecology and life histories of three net-spinning caddisfly species (Hydropschidae: Hydropsyche) in the Flathead River, Montana', *Freshwat. Invert. Biol.*, **1**, 18–29.
Hauer, F. R. and Stanford, J. A. 1991. 'Distribution and abundance of Trichoptera in a large regulated river', *Verh. Int. Verein. Limnol.*, **24**, 1636–1639.
Hawkins, C. P. 1984. 'Substrate associations and longitudinal distributions in species of Ephemerellidae (Ephemeroptera: Insecta) from Western Oregon', *Freshwat. Invert. Biol.*, **3**, 181–188.
Hickman, M. and Reasoner, M. A. 1994. 'Diatom response to late Quaternary vegetation and climate change, and to deposition of two tephras in an alpine and a sub-alpine lake in Yoho National Park, British Columbia', *J. Paleolimnol.*, **11**, 173–188.
Hickman, M. and Schweger, C. E. 1993. 'Late glacial–early holocene paleosalinity in Alberta, Canada — climate implications', *J. Paleolimnol.*, **8**, 149–161.
Hondzo, M. and Stefan, H. G. 1993. 'Regional water temperature characteristics of lakes subjected to climate change', *Climatic Change*, **24**, 187–311.
Hostetler, S. W. 1994. 'Hydrologic and atmospheric models — the problem of discordant scales', *Climatic Change*, **27**, 345–350.
Hostetler, S. W. and Giorgi, F. 1993. 'Use of high-resolution climate model data in landscape-scale hydrologic models: a demonstration', *Wat. Resour. Res.*, **29**, 1685–1695.
Hynes, H. B. N. 1975. 'The stream and its valley', *Verh. Int. Verein. Limnol.*, **19**, 1–15.
Jarrett, R. D. 1990. 'Hydrologic and hydraulic research in mountain rivers', *Wat. Resour. Bull.*, **26**, 419–429.
Jassby, A. D., Goldman, C. R. and Powell, T. M. 1992. 'Trend, seasonality, cycle and irregular fluctuations in primary productivity at Lake Tahoe, California–Nevada, USA', *Hydrobiologia*, **246**, 195–203.
Keleher, C. J. and Rahel, F. J. 1996. 'Thermal limits to salmonid distributions in the Rocky Mountain region and potential habitat loss due to global warming: a geographic information system (GIS) approach', *Trans. Am. Fish. Soc.*, **125**, 1–13.
Lamontagne, S. and Schindler, D. W. 1994. 'Historical status of fish populations in Canadian Rocky Mountain lakes inferred from subfossil *Chaoborus* (Diptera: Chaoboridae) mandibles', *Can. J. Fish. Aquat. Sci.*, **51**, 1376–1383.
Lamontagne, S., Donald, D. B. and Schindler, D. W. 1994. 'The distribution of four *Chaoborus* species (Diptera: Chaoboridae) along an elevation gradient in Canadian Rocky Mountain lakes', *Can. J. Zool.*, **72**, 1531–1537.
Leavitt, P. R., Schindler, D. E., Paul, A. J., Hardie, A. K. and Schindler, D. W. 1994. 'Fossil pigment records of phytoplankton in trout-stocked alpine lakes', *Can. J. Fish. Aquat. Sci.*, **51**, 2411–2423.
Leonard, E. M. 1986a. 'Use of lacustrine sedimentary sequences as indicators of Holocene glacial history, Banff National Park, Alberta, Canada', *Quat. Res.*, **26**, 218–231.
Leonard, E. M. 1986b. 'Varve studies at Hector Lake, Alberta, Canada, and the relationship between glacial activity and sedimentation', *Quat. Res.*, **25**, 199–214.
Lettenmaier, D. P. and Gan, T. Y. 1990. 'Hydrologic sensitivities of the Sacramento–San Joaquin River Basin, California, to global warming', *Wat. Resour. Res.*, **26**, 69–86.
Levine, J. S. 1992. 'Global climate change', in Firth, P. and Fisher, S. G. (Eds), *Global Climate Change and Freshwater Ecosystems*. Springer-Verlag, New York. pp. 1–25.
Luckman, B. H. 1990. 'Mountain areas and global change: a view from the Canadian Rockies', *Mountain Res. Devel.*, **10**, 183–195.
Malanson, G. P. and Butler, D. R. 1990. 'Woody debris, sediment, and riparian vegetation of a subalpine river, Montana, USA', *Arctic Alp. Res.*, **22**, 183–194.
Mast, M. A. and Baron, J. S. 1990. 'Chemical weathering in the Loch Vail Watershed, Rocky Mountain National Park, Colorado', *Wat. Resour. Res.*, **26**, 2971–2978.
McCabe, G. J. and Ayers, M. A. 1989. 'Hydrologic effects of climate change in the Delaware River basin', *Wat. Resour. Bull.*, **25**, 1231–1242.
McCormick, M. J. 1990. 'Potential changes in thermal structure and cycle of Lake Michigan due to global warming', *Trans Am. Fish. Soc.*, **119**, 183–194.
McKnight, D. M., Smith, R. L., Bradbury, J. P., Baron, J. S. and Spaulding, S. 1991. 'Phytoplankton dynamics in three Rocky Mountain lakes, Colorado, USA', *Arctic Alp. Res.*, **22**, 264–274.
Mearns, L. O., Schneider, S. H., Thompson, S. L. and McDaniel, L. R. 1990. 'Analysis of climate variability in general circulation models: comparison with observations and changes in variability in 2 × CO_2 experiments', *J. Geophys. Res.*, **95**, 20469–20490.
Meisner, J. D. 1990. 'Effect of climatic warming on the southern margins of the native range of brook trout, *Salvelinus fontinalis*', *Can. J. Fish. Aquat. Sci.*, **47**, 1065–1070.
Meisner, J. D. 1993. 'Reply to Flebbe 1993', *Can. J. Fish. Aquat. Sci.*, **40**, 884.
Merriam, C. H. 1890. 'The geographic distribution of life in North America', *Proc. Biol. Soc. Wash.*, 1–64.

Merritt, R. W. and Cummins, K. W. (Eds) 1996. *An Introduction to the Aquatic Insects of North America*, 3rd edn. Kendall/Hunt, Dubuque, Iowa.

Minshall, G. W. 1988. 'Stream ecosystem theory: a global perspective', *J. North Am. Benthol. Soc.*, **7**, 263–288.

Miskimmin, B. M., Leavitt, P. R. and Schindler, D. W. 1994. 'Fossil record of cladoceran and algal responses to fishery management practises', *Freshwat. Biol.*, **34**, 177–190.

Molles, M. C. Jr. and Dahm, C. N. 1990. 'A perspective on El Niño and La Niña: global implications for stream ecology', *J. North Am. Benthol. Soc.*, **9**, 68–76.

Nehlsen, W., Williams, J. E. and Lichatowich, J. A. 1991. 'Pacific salmon at the crossroads: stocks at risk from California, Oregon, Idaho and Washington', *Fisheries*, **16**, 4–21.

Nehring, R. B. and Anderson, R. M. 1993. 'Determination of population-limiting critical salmonid habitats in Colorado streams using the physical habitat simulation system', *Rivers*, **4**, 1–19.

Neill, W. E. 1988. 'Complex interactions in oligotrophic lake food-webs: responses to nutrient enrichment', in Carpenter, S. R. (Ed.), *Complex Interactions in Lake Communities*. Springer-Verlag, New York. pp. 31–44.

Neill, W. E. and Peacock, A. 1980. 'Breaking the bottleneck: interactions of invertebrate predators and nutrients in oligotrophic lakes', in Kerfoot, W. C. (Ed.), *Evolution and Ecology of Zooplankton Communities*. University Press of New England, New Hampshire. pp. 715–725.

Northcote, T. G. 1992. 'Prediction and assessment of potential effects of global environmental change on freshwater sport fish habitat in British Columbia', *GeoJournal*, **28**(1), 39–49.

Paul, A. J. and Schindler, D. W. 1994. 'Regulation of rotifers by predatory calanoid copepods (Subgenus: *Hesperodiaptomus*) in lakes of the Canadian Rocky Mountains', *Can. J. Fish. Aquat. Sci.*, **51**, 2520–2528.

Pennak, R. W. 1968. 'Field and experimental winter limnology of three Colorado mountain lakes', *Ecology*, **49**, 505–520.

Poff, N. L. and Ward, J. V. 1989. 'Implications of streamflow variability and predictability for lotic community structure: a regional analysis of streamflow patterns', *Can. J. Fish. Aquat. Sci.*, **46**, 1805–1818.

Propst, D. L., Stefferud, J. A. and Turner, P. R. 1992. 'Conservation and status of Gila trout, *Oncorhynchus gilae*', *Southwest. Nat.*, **37**, 117–125.

Ralph, S. C., Poole, G. C., Conquest, L. L. and Naiman, R. J. 1993. 'Steam channel morphology and woody debris in logged and unlogged basins of western Washington', *Can. J. Fish. Aquat. Sci.*, **51**, 37–51.

Reasoner, M. A. and Hickman, M. 1989. 'Late Quaternary environmental change in the Lake O'Hara region, Yoho National Park, British Columbia', *Paleogeog. Paleoclim. Paleoecol.*, **72**, 291–316.

Reasoner, M. A. and Rutter, N. W. 1988. 'Late Quaternary history of the Lake O'Hara region, British Columbia — an evaluation of sedimentation rates and bulk amino acids in lacustrine records', *Can. J. Earth Sci.*, **25**, 1037–1048.

Reasoner, M. A., Osborn, G. and Rutter, N. W. 1994. 'Age of the Crowfoot advance in the Canadian Rocky Mountains: a glacial event coeval with the Younger Dryas oscillation', *Geology*, **22**, 439–442.

Reice, S. R., Wissmar, R. C. and Naiman, R. J. 1990. 'Disturbance regimes, resilience, and recovery of animal communities and habitats in lotic ecosystems', *Environ. Manage.*, **14**, 647–660.

Rieman, B. E. and McIntyre, 1993. 'Demographic and habitat requirements for conservation of bull trout', *General Technical Report INT-302*. US Forest Service, Ogden, Utah. 37 pp.

Riley, S. C. and Fausch, K. D. 1995. 'Trout population response to habitat manipulation in six northern Colorado streams', *Can. J. Fish. Aquat. Sci.*, **52**, 34–53.

Running, S. W. and Nemani, R. R. 1991. 'Regional hydrologic carbon balance responses of forests resulting from potential climate change', *Climatic Change*, **19**, 349–368.

Schindler, D. W., Bayley, S. E., Curtis, P. J., Parker, B. R., Stainton, M. P. and Kelly, C. A. 1992. 'Natural and man-caused factors affecting the abundance and cycling of dissolved organic substances in precambrian shield lakes', *Hydrobiologia*, **229**, 1–21.

Schlosser, I. J. and Angermeier, P. L. 1995. 'Spatial variation in demographic processes of lotic fishes: conceptual models, empirical evidence, and implications for conservation', *Am. Fish. Soc. Symp.*, **17**, 392–401.

Seegrist, D. W. and Gard, R. 1972. 'Effects of floods on trout in Sagehen Creek, California', *Trans. Am. Fish. Soc.*, **101**, 478–482.

Spaulding, S. A., Ward, J. V. and Baron, J. S. 1993. 'Winter phytoplankton dynamics in a subalpine lake, Colorado, USA', *Arch. Hydrobiol.*, **129**, 179–198.

Spencer, C. N. and Ellis, B. K. 1990. 'Co-limitation by phosphorus and nitrogen, and effects of zooplankton mortality, on phytoplankton in Flathead Lake', Montana, U.S.A. *Verh. Internat. Verein. Limnol.*, **24**, 206–209.

Spencer, C. N., McClelland, B. R. and Stanford, J. A. 1991. 'Shrimp stocking, salmon collapse, and eagle displacement', *BioScience*, **41**, 14–21.

Stanford, J. A. and Hauer, F. R. 1992. 'Mitigating the impacts of stream and lake regulation in the Flathead River catchment, Montana, USA: an ecosystem perspective', *Aquat. Conserv. Mar. Freshwat. Ecosys.*, **2**, 35–63.

Stanford, J. A. and Prescott, G. W. 1988. 'Limnological features of a remote alpine lake in Montana, including a new species of *Cladophora* (Chlorophyta)', *J. North Am. Benthol. Soc.*, **7**, 140–151.

Stanford, J. A. and Ward, J. V. 1988. 'The hyporheic habitat of river ecosystems', *Nature*, **335**, 64–66.

Stanford, J. A. and Ward, J. V. 1992. 'Management of aquatic resources in large catchments: recognizing interactions between ecosystem connectivity and environmental disturbance', in Naiman, R. J. (Ed.), *Watershed Management*. Springer-Verlag, New York. pp. 91–124.

Stanford, J. A. and Ward, J. V. 1993. 'An ecosystem perspective of alluvial rivers: connectivity and the hyporheic corridor', *J. North Am. Benthol. Soc.*, **12**, 48–60.

Stanford, J. A., Hauer, F. R. and Ward, J. V. 1988. 'Serial discontinuity in a large river system', *Verh. Int. Verein. Limnol.*, **23**, 1114–1118.

Stockner, J. G. 1991. 'Autotrophic picoplankton in freshwater ecosystems: the view from the summit', *Int. Rev. Ges. Hydrobiol.*, **76**, 483–492.

Stockner, J. G. and Shortreed, K. S. 1991. 'Autotrophic picoplankton: community composition, abundance, and distribution across a gradient of oligotrophic British Columbia and Yukon Territory lakes', *Int. Rev. Ges. Hydrobiol.*, **76**, 581–601.

Sweeney, B. W. 1984. 'Factors influencing life-history patterns of aquatic insects', in Resh, V. H. and Rosenberg, D. M. (eds) *The Ecology of Aquatic Insects*, pp 56–100. Praeger, New York.

Sweeney, B. W., Jackson, J. K., Newbold, J. D. and Funk, D. H. 1992. 'Climate change and the life histories and biogeography of aquatic insects in eastern North America', in Firth, P. and Fisher, S. G. (Eds), *Global Climate Change and Freshwater Ecosystems*. Springer-Verlag, New York. pp. 321.

Thomas, W. H., Cho, B. C. and Azam, F. 1991. 'Phytoplankton and bacterial production and biomass in subalpine Eastern Brook Lake, Sierra Nevada, California. II. Comparison with other high-elevation lakes', *Arctic Alp. Res.*, **23**, 296–302.

Triska, F. J. and Sedell, J. R. 1976. 'Decomposition of four species of leaf litter in response to nitrate manipulation', *Ecology*, **57**, 783–792.

Turk, J. T. and Spahr, N. E. 1989. 'Rocky Mountains: controls on lake chemistry', in Adriano, D. C. and Havas, M. (Eds), *Acidic Deposition and Aquatic Ecosystems: Case Studies*. Springer-Verlag, New York. pp. 181–208.

Turk, J. T. and Spahr, N. E. 1991. 'Rocky Mountains', in Charles, D. (Ed.), *Acidic Deposition in Aquatic Ecosystems: Regional Case Studies*. Springer-Verlag, New York. pp. 471–501.

US EPA (Environmental Protection Agency) 1989. 'The potential effects of global climate change on the United States', in Smith, J. B. and Tirpak, D. (Eds), US EPA Office of Policy, Planning, and Evaluation, Office of Research and Development, Washington, D.C. 413 pp.

US EPA (Environmental Protection Agency) 1991. 'Global comparisons of selected GCM control runs and observed climate data', in Kalkstein, L. S. (Ed.), US EPA Office of Policy, Planning and Evaluation, Climate Change Division, Washington, D.C. 251 pp.

Varley, J. D. and Gresswell, R. E. 1988. 'Ecology, status, and management of the Yellowstone cutthroat trout', *Am. Fish. Soc. Symp.*, **4**, 13–24.

Vincent, W. F. and Roy, S. 1993. 'Solar ultraviolet-B radiation and aquatic primary production: damage, protection, and recovery', *Environ. Rev.*, **1**, 1–12.

Vincent, W. F., Downes, M. T., Castenholz, R. W. and Howard-Williams, C. 1993. 'Community structure and pigment organization of cyanobacteria-dominated microbial mats in Antarctica', *Eur. J. Phycol.*, **28**, 213–221.

Walker, I. R. and Mathewes, R. W. 1989. 'Chironomidae (Diptera) remains in surficial lake sediments from the Canadian Cordillera: analysis of the faunae across an altitudinal gradient', *J. Paleolimnol.*, **2**, 61–80.

Ward, J. V. and Stanford, J. A. 1982. 'Thermal responses in the evolutionary ecology of aquatic insects', *Ann. Rev. Entomol.*, **27**, 97–117.

Wetzel, R. G. 1983. *Limnology*. W. B. Saunders, Philadelphia. 767 pp.

Williamson, C. E., Zagarese, H. E., Schulze, P. C., Hargreaves, B. R. and Seva, J. 1994. 'The impact of short-term exposure to UV-B radiation on zooplankton communities in north temperate lakes', *J. Plankton Res.*, **16**, 205–218.

World Meteorological Organization. 1985. Trace gas effects on climate. Atmospheric Ozone 1985. Global Ozone Research and Monitoring Project. Report No. 16. Vol. III. Chapter 15. p. 821.

Young, M. K. 1994. 'Mobility of brown trout in south-central Wyoming streams', *Can. J. Zool.*, **72**, 2078–2083.

Young, M. K. 1995. 'Conservation assessment for inland cutthroat trout', *USDA Forest Service General Technical Report RM-GTR-256*. Rocky Mountain Forest and Range Experiment Station, Fort Collins, Colorado. 61 pp.

5

POTENTIAL EFFECTS OF CLIMATE CHANGE ON FRESHWATER ECOSYSTEMS OF THE NEW ENGLAND/MID-ATLANTIC REGION

MARIANNE V. MOORE,[1] MICHAEL L. PACE,[2] JOHN R. MATHER,[3] PETER S. MURDOCH,[4] ROBERT W. HOWARTH,[5] CAROL L. FOLT,[6] CELIA Y. CHEN,[6] HAROLD F. HEMOND,[7] PATRICIA A. FLEBBE[8] AND CHARLES T. DRISCOLL[9]

[1]*Department of Biological Sciences, Wellesley College, Wellesley, MA 02181-8283, USA*
[2]*Institute of Ecosystem Studies, Box AB, Millbrook, NY 12545, USA*
[3]*216 Newark Hall, Geography Department, University of Delaware, Newark, DE 19716, USA*
[4]*US Geological Survey, 425 Jordan Road, Troy, NY 12180, USA*
[5]*Section of Ecology & Systematics, Corson Hall, Cornell University, Ithaca, NY 14853, USA*
[6]*Department of Biological Sciences, 6044 Gilman Laboratory, Darmouth College, Hanover, NH 03755-3576, USA*
[7]*Ralph Parsons Laboratory, Department of Civil and Environmental Engineering, Massachusetts Institute of Technology, Cambridge, MA 02139, USA*
[8]*Department of Fisheries and Wildlife, Virginia Tech, Blacksburg, VA 25061-0321, USA*
[9]*Department of Civil & Environmental Engineering, Syracuse University, Syracuse, NY 13244, USA*

ABSTRACT

Numerous freshwater ecosystems, dense concentrations of humans along the eastern seaboard, extensive forests and a history of intensive land use distinguish the New England/Mid-Atlantic Region. Human population densities are forecast to increase in portions of the region at the same time that climate is expected to be changing. Consequently, the effects of humans and climatic change are likely to affect freshwater ecosystems within the region interactively. The general climate, at present, is humid continental, and the region receives abundant precipitation. Climatic projections for a $2 \times CO_2$ atmosphere, however, suggest warmer and drier conditions for much of this region. Annual temperature increases ranging from 3–5°C are projected, with the greatest increases occurring in autumn or winter. According to a water balance model, the projected increase in temperature will result in greater rates of evaporation and evapotranspiration. This could cause a 21 and 31% reduction in annual stream flow in the southern and northern sections of the region, respectively, with greatest reductions occurring in autumn and winter. The amount and duration of snow cover is also projected to decrease across the region, and summer convective thunderstorms are likely to decrease in frequency but increase in intensity.

The dual effects of climate change and direct anthropogenic stress will most likely alter hydrological and biogeochemical processes, and, hence, the floral and faunal communities of the region's freshwater ecosystems. For example, the projected increase in evapotranspiration and evaporation could eliminate most bog ecosystems, and increases in water temperature may increase bioaccumulation, and possibly biomagnification, of organic and inorganic contaminants. Not all change may be adverse. For example, a decrease in runoff may reduce the intensity of ongoing estuarine eutrophication, and acidification of aquatic habitats during the spring snowmelt period may be ameliorated.

Recommendations for future monitoring efforts include: (1) extending and improving data on the distribution, abundance and effect of anthropogenic stressors (non-point pollution) within the region; and (2) improving scientific knowledge regarding the contemporary distribution and abundance of aquatic species. Research recommendations include: (1) establishing a research centre(s) where field studies designed to understand interactions between freshwater ecosystems and climate change can be conducted; (2) projecting the future distribution, activities and direct effects of humans within the region; (3) developing mathematical analyses, experimental designs and aquatic indicators that distinguish between climatic and anthropogenic effects on aquatic systems; (4) developing and refining projections of climate variability such that the magnitude, frequency and seasonal timing of extreme events can be forecast; and (5) describing quantitatively the flux of materials (sediments, nutrients, metals) from watersheds characterized by a mosaic of land uses.

INTRODUCTION

Water-rich, extensively forested and densely populated by humans along the eastern seaboard, this region extends from eastern maritime Canada to Chesapeake Bay and westwards to include much of the Ohio River drainage basin. In this region, past deforestation, present reforestation and expansion of the human population are key factors that impinge upon considerations of climate change and its effect on freshwaters. In this paper, we explore interactions between climate change and direct anthropogenic stress on the freshwater ecosystems of the region. We contend that the dual effect of climate change and direct anthropogenic stress could be as equally transforming as the historical land use alterations that affected freshwater ecosystems within the region during the late eighteenth and nineteenth centuries.

The paper is structured in the following manner. We begin by defining the region's boundaries and its distinctive features including its unique history of intensive land use. The current and projected climates for the region are presented, and a brief description of urban climates and the possibilities they offer for global climate change research then follows. Five examples demonstrating how climate change and direct human effects may interactively affect aquatic ecosystems in the region are presented next, and recommendations for future research and monitoring are discussed.

BOUNDARIES

This region projects beyond the conventional state boundaries frequently used to define the New England/Mid-Atlantic area (Figure 1 in Leavesley *et al.*, this volume). Owing to inclusion of much of the Ohio River drainage system, the region extends south-west to include the states of West Virginia, Kentucky and much of Ohio. Nova Scotia and eastern New Brunswick define the northernmost portion of the region. Unless otherwise noted, regional statistics in this paper were calculated for the entire land area defined by the states listed in Table I. Although regional statistics do not include Nova Scotia and the appropriate portions of New Brunswick, Tennessee, Indiana, Illinois, North Carolina and Virginia, many of the ideas in this paper can be extended to these areas.

Table I. Percentage land cover or use of non-federal land (excluding surface water) per state, region, and US in 1992. Percentages were derived using data from the 1992 National Resources Inventory (US DOA, 1994). Values in parentheses in state and forested columns represent the rankings of the top 12 forested states in the US and % forested of total land (federal plus non-federal land), respectively (Alampi, 1994). 'Developed' refers to urban areas or those cities with more than 2500 individuals. Land use percentages for the US do not sum to 100% because rangeland, absent from the New England/Mid-Atlantic Region, was omitted from the table

State (Rank)	% Developed	% Forested (% of total land)	% Crop	% Pasture	% Minor use
Connecticut (12)	26·7	57·6 (59)	7·5	3·6	4·6
Delaware	16·9	29·1	41·2	2·1	10·7
Kentucky	6·9	43	21·2	24·4	4·5
Maine (1)	3·6	90 (90)	2·3	0·6	3·6
Maryland	18·2	39·2	27·7	9	5·9
Massachusetts (9)	27·1	57·4 (62)	5·6	3·5	6·4
New Hampshire (2)	11·4	79·4 (88·0)	2·9	2	4·4
New Jersey	34·9	38·8	14·3	3·5	8·5
New York (9)	10·1	57·6 (62·0)	18·8	10·1	3·3
Ohio	13·9	25·0	46·5	8·8	5
Pennsylvania (12)	12·3	55·1 (59)	20·1	8·4	4·1
Rhode Island (11)	28·7	59·4 (60)	3·8	3·7	4·5
Vermont (4)	5·9	75 (76)	11·5	6·3	1·4
West Virginia (3)	4·9	74·5 (79)	6·5	11·4	2·8
Region	**11·1**	**55·3**	**19·6**	**9·7**	**4·2**
USA	6·2	26·6	27·5	8·5	6

DISTINCTIVE FEATURES

Freshwater ecosystems in the New England/Mid-Atlantic Region are abundant and diverse. This is a result, in part, of the Wisconsin glaciation, the altitudinal relief of the Appalachian Mountains and the presence of the Atlantic coast. For example, 11 500 lakes and reservoirs (surface area >1 ha) occur in New England, New York and New Jersey alone (Paulsen *et al.*, 1991). Of the natural lakes, most are kettle lakes or of glacial origin. Based on total phosphorus and chlorophyll *a* data, 38% are oligotrophic, 42% mesotrophic and 21% eutrophic to hypereutrophic. In the coastal lowlands, however, where humans are concentrated, a much greater percentage of these lakes (51%) are eutrophic and hypereutrophic (US EPA, 1994). South of the Wisconsin glaciation (all mid-Atlantic and interior states excluding New York), natural lakes are largely absent and are replaced by reservoirs and numerous streams. Lotic habitats range from cold water streams in the Appalachian Mountains to large, turbid rivers along the Atlantic coastal plain. Most streams throughout the region have been classified as perennial runoff streams that flow throughout the year but have low baseflows with little between-year variation in low flow (Poff, 1992). More than 480 estuaries (>2 km^2) also occur here, and this density is second only to that in the south-eastern region of the US (K. Summers, EMAP, personal communication). These estuaries include many large, culturally rich and economically important systems such as Chesapeake Bay, Delaware Bay, the Hudson River, Long Island Sound, Massachusetts Bay and the Bay of Fundy. Finally, at least 6% of the land area is covered by wetlands and, of this, 92% is freshwater wetlands (R. Tiner, USFWS, personal communication). Of the freshwater wetlands, most located south of Maine and east of the Appalachians are red maple swamps, whereas bogs are prevalent in the interior of Maine and the Canadian maritimes. Riparian wetlands predominate in the Ohio River drainage.

More humans reside within the New England/Mid-Atlantic Region than in any other region of North America. Humans, however, are largely concentrated in coastal areas where population density is projected to increase at the same time that climate is expected to change (Culliton *et al.*, 1990). The proportion of urban land in coastal states of the region is 2–5 times greater than the average for the US (Table I). Indeed, eight of the ten most urban states in the USA occur in this region, and all are coastal except Ohio (Morgan *et al.*, 1993). Estuaries are probably affected most by coastal metropolitan areas via urban runoff, combined sewer overflows, municipal and industrial wastes and harbour dredging. These effects are likely to increase (Kennedy, 1990) because the coastal population of the New England/Mid-Atlantic area is expected to increase by four million people, the current combined population of Connecticut and Rhode Island, by the year 2010 (Culliton *et al.*, 1990). Importantly, however, humans are unevenly distributed in the region. Some areas are largely rural. For example, Maine, Vermont and West Virginia are ranked among the most rural states on the basis of land use (Table I; Morgan *et al.*, 1993). Nevertheless, humans currently, and in the past, have greatly modified the landscape and freshwater ecosystems throughout the entire New England/Mid-Atlantic Region, including the most rural areas.

The long history of intensive land use is another signature feature of the New England/Mid-Atlantic Region. Forest clearance, during the late eighteenth and nineteenth centuries, was probably the most obvious alteration to the landscape. At the time of European settlement, the hardwoods forest of eastern North America stretched nearly unbroken from the Atlantic Coast to the Mississippi River (Greeley, 1925; Whitney, 1994). According to the diaries and letters of Europeans, it was 'an ocean of woods', a place where 'a savage might skulk from the Hudson to Lake Erie without once exposing himself to the glare of the sun' (Whitney, 1994). This virgin forest, however, was almost entirely swept off the landscape and converted to farmland or burned for domestic and industrial purposes (Whitney, 1994). The maximum extent of forest clearance was remarkable. Historical records reveal that approximately 80–90% of the wooded area in portions of central Massachusetts and eastern New York had been removed by the mid-1800s (Foster, 1992; reviewed by Whitney, 1994).

In addition to forest clearance, changes were also occurring in freshwater ecosystems. Increased stream flow undoubtedly resulted from forest removal, because forests were no longer transpiring large quantities of water during the growing season. Contemporary deforestation experiments of entire watersheds in the East have increased stream flow from 20–40% (Lee, 1980) or 25 mm for each 10% decrease in forest cover (Bosch and Hewlett, 1982). Thus, during much of the nineteenth century, forest clearance and increased stream flow, together with the continued expansion of agriculture, resulted in soil erosion and massive increases in sedimentation rates within lakes and other aquatic systems. For example, a fourfold increase in sediment deposition occurred at this time at several sites in Chesapeake Bay (Brush, 1986), and sedimentation caused the abandonment of coastal port towns (Whitney, 1994). Also, large-scale drainage of wetlands began in the latter half of the nineteenth century. Thus, just as agriculture was largely responsible for forest clearance, it was also the major driving force behind the destruction of wetlands (Tiner, 1984). Loss of wetlands between pre-settlement times and the 1980s was greatest in the Midwest (e.g. 90% of wetland area, Ohio). Extent of loss, however, was considerable in the Mid-Atlantic states (58%), but much less in New England (25%) (Dahl, 1990).

During the last 100 years, farm abandonment and the decline of rural agriculture resulted in rapid reforestation in much of the eastern United States (Foster, 1995; Whitney, 1994). For example, two million acres of cleared land were reforested from 1880 to 1900 in Vermont, New Hampshire and Maine at the same time that 11 000 farms were abandoned (Robinson, 1988). The pattern, of farm expansion and contraction, occurred in areas to the west, such as Ohio and Pennsylvania, and it continues today along the western margin of Appalachia (Whitney, 1994). Importantly, these major historical shifts in land use resulted more from changing social and economic factors than a decline in soil fertility or a regional climatic shift (Foster, 1995; Whitney, 1994).

Today, nine of the 12 most forested states in the USA occur in the New England/Mid-Atlantic Region, and 59–90% of the total land area of these states is covered in forest (Table I; Figure 1 in Leavesley *et al.*, this volume). In New England and New York state, forest harvesting is proceeding much more slowly than growth (Smith, 1990; Foster, 1995). Consequently, the forest is characterized by continual ageing and growth resulting in a net increase in stored carbon (Wofsy *et al.*, 1993). Since World War II, however, the forest has become increasingly fragmented into individual parcels of relatively small land area (Foster, 1995) as more urban dwellers purchase ‘a piece of the woods’.

Although forests dominate the landscape, farming and mining, and, increasingly, urbanization represent important land uses in localized portions of the region (Figure 1 in Leavesley *et al.*, this volume; Table I) (Alampi, 1994; US DOA, 1994). Suburbanization and urbanization represent a type of land use that has increased substantially since World War II (Figure 1). For example, 2·75 million acres of land in the New England/Mid-Atlantic Region were converted to urban or suburban use between 1982 and 1992 (US DOA, 1994). When expressed as a proportion of non-federal land area, an additional 1·6% of the region was developed during the 10-year period, a rate nearly twice that observed for the entire USA (0·9% per 10 year) over the same time period.

At present, the most pervasive anthropogenic stressor affecting freshwater ecosystems in the region is a diverse group of non-point source pollutants (US EPA, 1994). For example, the Northeast receives among the highest inputs of acid deposition of any region in the United States (Husar *et al.*, 1991; Ollinger *et al.*, 1993). The annual average pH of rain and snow in the Catskill and Adirondack Mountains of New York is about 4·3 (Murdoch, 1991). The anion content of precipitation is dominated by sulfate (roughly two-thirds), with about one-third attributed to nitrate and other minor anions (Driscoll *et al.*, 1991; Murdoch, 1991). There is a gradient of decreasing atmospheric deposition of sulfur and nitrogen from New York to Maine (Ollinger *et al.*, 1993). In western Virginia (Shenandoah River drainage), levels of acidic deposition are as high as those in the north-eastern US, and the mountain streams there are at considerable risk of damage (Cosby *et al.*, 1991).

Additional constituents of non-point pollution (Figure 2) were identified and ranked for the New England/Mid-Atlantic Region using data extracted from the National Water Quality Inventory (1992) (US EPA, 1994). This US inventory is compiled using data submitted by individual states every two years as mandated by the Clean Water Act of 1972. These data, however, must be viewed cautiously

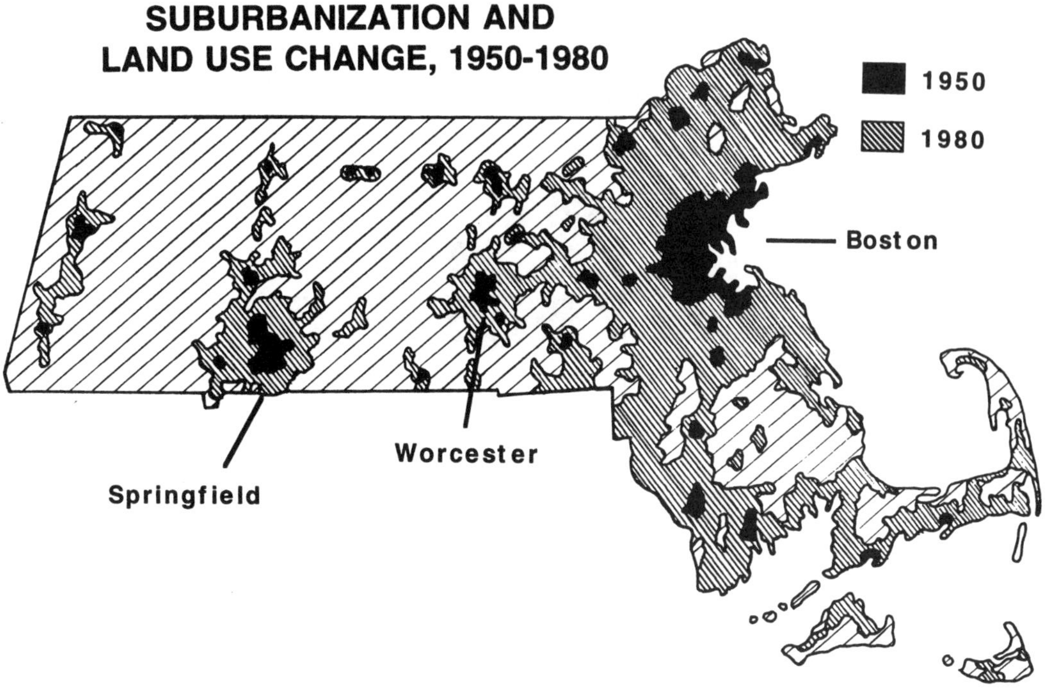

Figure 1. Changes in the spatial extent of suburbanization in the state of Massachusetts during the last half of the twentieth century. The cities of Boston Worcester and Springfield, former agricultural and industrial towns, have converted to residential communities as the human population increased and road transportation improved (Foster, 1995). Modified from Wilkie and Tager (1991) and Foster (1995)

because each state monitors only a small percentage of its aquatic ecosystems, differences in assessment methods exist between and within states and sampling sites are not chosen randomly. Despite the limitations of the data, this inventory is the only information available for anthropogenic stressors across the region.

Nutrients, siltation and pathogens (indicator bacteria) were considered to be among the top four causes of water quality impairment in lakes, streams and estuaries (Figure 2). Nutrients were the top cause of impairment in lakes and estuaries, whereas siltation caused the greatest proportion of impaired river miles throughout the region (Figure 2). Metals caused substantial impairment in rivers and streams (33% of impaired river miles), but not in lakes or estuaries. Eighty-five percent of the metal-impaired streams are located in West Virginia, Pennsylvania and the Ohio River valley where acid mine drainage is a severe problem. The toxicity of heavy metals coupled with the smothering effect of iron precipitates and low pH have rendered many streams sterile (Pennsylvania Department of Environmental Resources, 1994). Organic contaminants caused comparatively little impairment in lakes, streams or estuaries (Figure 2), and metals caused little impairment in lakes and estuaries (2–3·5% of impaired lake acres and km^2 of estuaries, respectively). This, however, reflects a lack of sampling by many states rather than lack of a problem. For example, state-wide fish consumption advisories for mercury were recently adopted for all lakes and streams or all lakes in Massachusetts, Maine, Vermont and New York state (D. Switzer, US EPA, personal communication). A major source of this mercury is fossil fuel combustion (e.g. coal) and subsequent atmospheric deposition on to the landscape (Driscoll *et al.*, 1994).

Clearly, humans continue to influence aquatic ecosystems throughout the region. Consequently, an analysis of the effects of climatic change for this region must acknowledge that climatic effects are likely to interact with direct anthropogenic stressors.

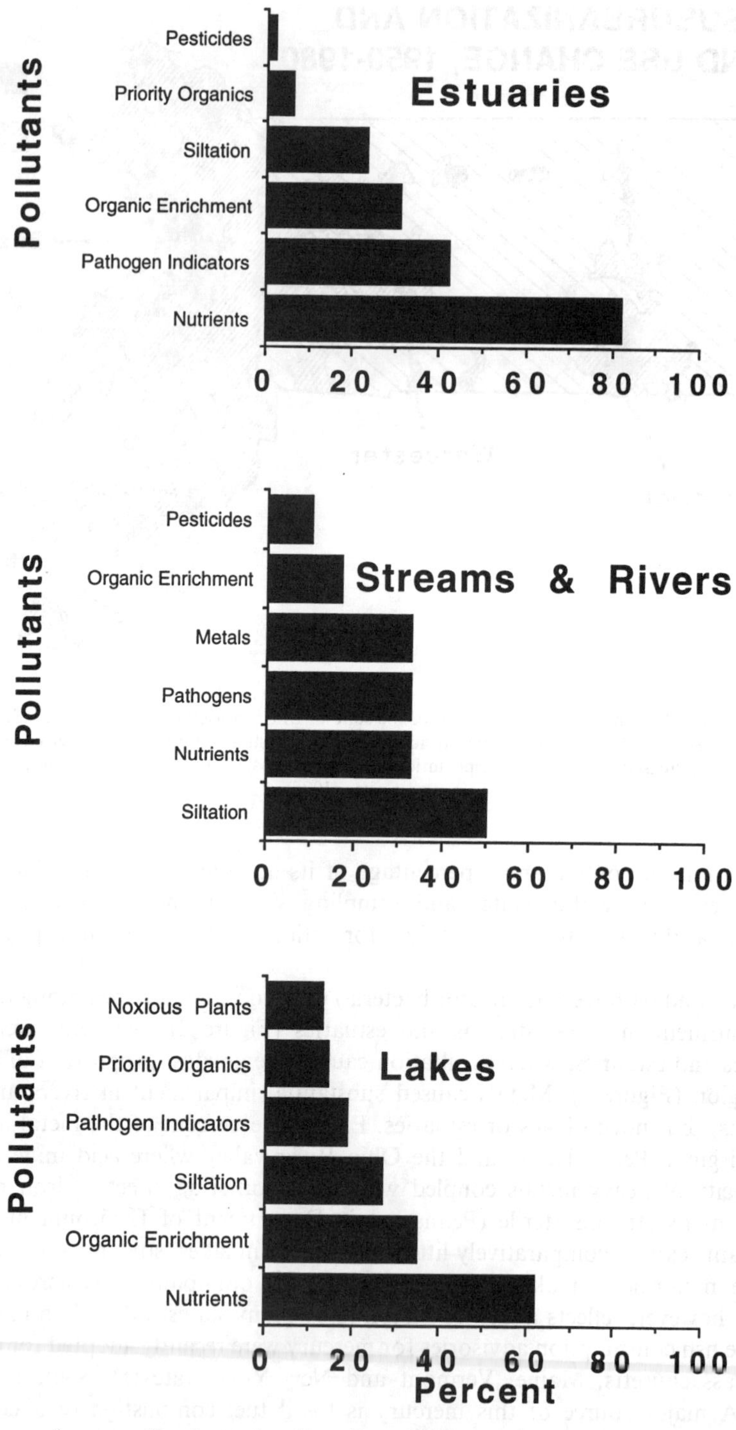

Figure 2. Percentage of assessed square miles of estuaries, river miles, and lake acres impaired (altered enough to affect human use or aquatic life) by pollutants in the New England/Mid-Atlantic Region. Percentages were derived by combining information from the National Water Quality Inventory (1992) (US EPA, 1994) for all states listed in Table I. Only the top six pollutants causing the greatest percentage impairment per habitat are shown. Note that relative percentages may reflect human bias more than reality, because water bodies are non-randomly selected for monitoring

PRESENT AND PROJECTED CLIMATE

The New England/Mid-Atlantic Region was divided into two geographical sections (Figure 3) for climatic analysis. The present climate of the area is generally humid continental with warm to hot summers (mean July temperature = 20–25°C), fairly high humidity and cool to cold winters (mean January temperature = 0–5°C in the southern section and −5 to −10°C in the northern section). The region receives abundant precipitation ranging from 1350 mm/yr in the southern interior states to 1500 mm/yr in the northern coastal areas. Annual variability in precipitation is lower in the Northeast than in any other region of the US, with extremes departing only about 10% from the mean (Miller and Miller, 1992). Snowfall is heavy in the northern section (2000 mm/yr) and in the higher elevation areas of the southern section (1000 mm/yr). The current natural vegetation of the region is deciduous forest grading to coniferous forest in the more northern areas and at higher altitudes.

During the past 150 years, air temperature has increased gradually over the whole region (Jones and Bradley, 1992). For example, in central New England (Hanover, NH), mean annual air temperature has increased (+1·7°C) and mean summer air temperature has increased (+2·2°C) during the last 150 years (Hamburg and Cogbill, 1988). These increases are similar to those observed at Amherst, MA (Bradley *et al.*, 1987). Similarly, water temperatures in the Hudson River have increased 0·12°C per decade between 1920 and 1990 (Ashizawa and Cole, 1994). Normal climatic variability, however, could be responsible for these temperature increases. Alternatively, they might even be related to a general warming from the rather cold climatic conditions existing during the 'Little Ice Age' from the 1600s to the 1800s. If increasing CO_2 is the only forcing factor, the 30 years of decreasing temperatures from 1945 to 1975 become difficult to explain.

Global circulation models (GCMs) can be used to obtain estimates of temperature and precipitation conditions under various future environmental scenarios, and these models provide essentially the only technique available at present to estimate quantitatively the climatic effect of increasing CO_2. GCMs, however, are imperfect, and projections on a regional scale are especially uncertain. Thus, while we are forced to consider the output from GCMs we should always be aware of their limitations (Hayes, 1991).

Climatic projections from four GCMs [GISS, (Goddard Institute of Space Studies), Hansen *et al.*, 1988; OSU (Oregon State University), Schlesinger and Zhao, 1989; UKMO (United Kingdom Meteorological Office), Wilson and Mitchell, 1987; GFDL (Geophysical Fluid Dynamics Laboratory), Manabe and Wetherald, 1987] provide estimates of seasonal changes in air temperature and precipitation under conditions of a future doubling of CO_2. Temperature is projected to increase across the entire region during all seasons (Figure 3). The temperature increase in the southern section of the region ranges from +3·6°C in summer to +4·4°C in autumn. In the northern section, slightly greater temperature increases are expected, ranging from +4·0°C in autumn to +5·0°C in the winter season. Precipitation is generally expected to increase in winter and spring in both sections of the region, but in summer and fall significant decreases in precipitation are predicted in both sections (except for a small increase in summer in the southern portion of the region) (Figure 4). On an annual basis, a small increase in precipitation might be expected in the southern portion of the region and a slight decrease in precipitation in the northern portion.

The temperature and precipitation projections described above were entered into a climatic water budget developed by Thornthwaite and Mather (1955). This water budget was used to project water surplus, water deficit, stream flow and snow cover in the region (Feddema and Mather, 1992; Hodny and Mather, 1995). A water deficit represents the difference between the climatic demand for water and the actual water use by plants, while a water surplus is the amount of precipitation exceeding the climatic water need once the soil is remoistened, and will be ultimately lost as stream flow. According to the water budget, there should be a general shift to drier conditions in the region (Table II), because the projected increase in temperature will result in a greater increase in evapotranspiration than the projected increase in precipitation. The greatest increase in seasonal water deficit is predicted for the summer and autumn, and the present water surplus in the autumn and winter seasons will decrease (Table II). Because there is little deficit in winter and spring

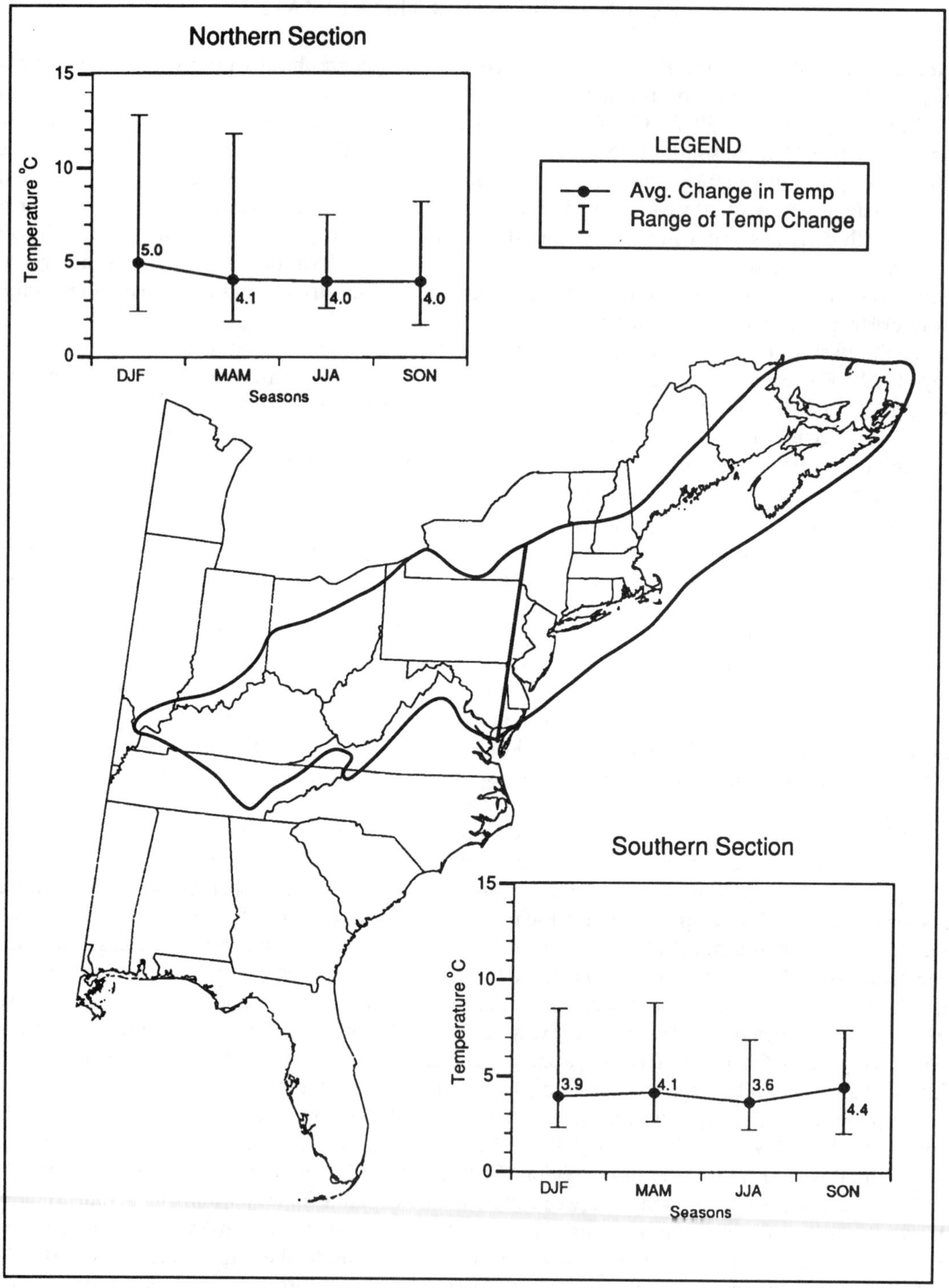

Figure 3. Projected mean change and range of change in air temperature (°C) for $2 \times CO_2$ conditions in the northern and southern sections of the New England/Mid-Atlantic Region. Projections are most likely change generated from four GCMs and based on monthly averages. The vertical line within the region's boundaries separates the northern and southern sections for climatic analysis

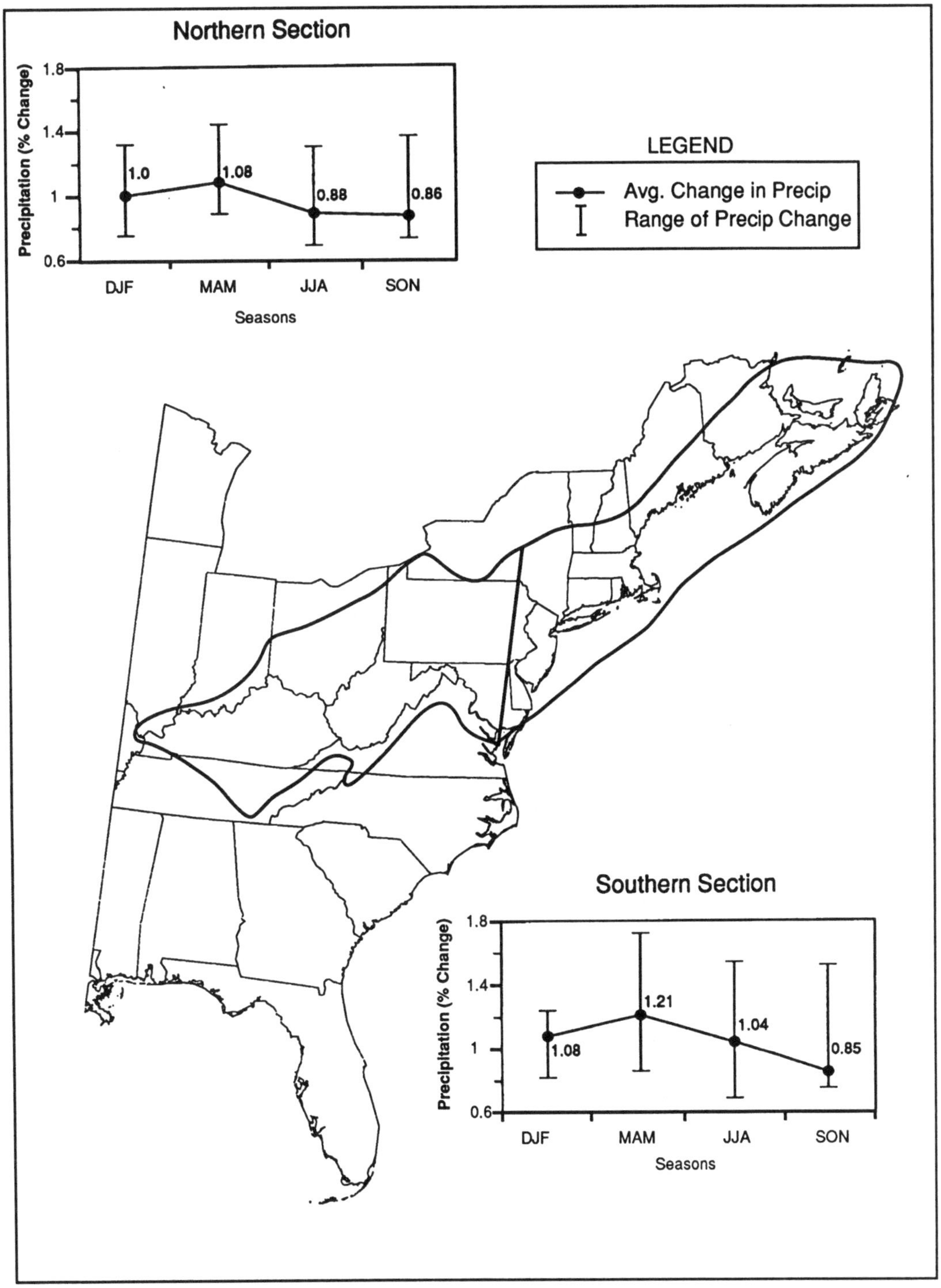

Figure 4. Projected mean change and range of change in precipitation per season for 2 × CO_2 conditions in the northern and southern sections of the New England/Mid-Atlantic Region. Projections are most likely change generated from four GCMs and based on monthly averages. Values less than one indicate a decrease in precipitation and those exceeding one represent an increase. The vertical line within the region's boundaries separates the northern and southern sections for climatic analysis

Table II. Likely seasonal changes in water surplus and deficit in the New England/Mid-Atlantic Region from present with a doubling of CO_2. 'Most likely change' represents an estimated mean using data from all four GCMs (GISS, GFDL, OSU, UK) and the Thornthwaite–Mather water balance model (1955). 'Range of possible change' describes the range of the estimated means described previously. Months of the year are abbreviated and grouped into four seasons, beginning with winter

	Water deficit				Water surplus			
	DJF	MAM	JJA	SON	DJF	MAM	JJA	SON
Present climate (mm)	0	0	9	< 1	310	177	4	94
Most likely change (%)	0	+ < 1	+59	+23	−33	−1	−3	−84
Range of possible change (%)	0	0 to +2	+4 to +106	+1 to +24	+7 to −48	+7 to −41	+1 to −4	+56 to −85

Table III. Projected mean change in total annual stream flow for the southern and northern sections of the New England/Mid-Atlantic Region under conditions of doubled CO_2. Projections obtained as described in Table II

	Southern section	Northern section
Present climate (mm)	372	536
Most likely change (%)	−21	−31
Range of possible change (%)	−32 to +6	−32 to +4

under present conditions, little change in this parameter is predicted under double CO_2 conditions. A decrease in annual stream flow is also projected across all of the region (Table III); however, the greatest seasonal reductions should occur in autumn and winter (Table II).

A significant decrease in snow cover is also expected because of the increase in temperature associated with a doubling of CO_2. Computations using expected temperature data suggest that essentially no snow cover will exist in the southern section of the area, while both the depth and duration of snow cover will be reduced in the northern section (Hodny and Mather, 1995). For example, a computation for the headwaters of the Delaware River in central New York shows a reduction in snow cover amount from 306 mm at present, to 90 mm under conditions of doubled CO_2.

The character of the actual synoptic weather conditions that might exist in the region is more difficult to predict. If we assume, as many climatologists do, that warming in the Arctic will be greater than in the Tropics, this would suggest a decrease in the pole–equator temperature gradient. Normally, this would result in a reduced meridional flow of air masses according to the Rossby wave equation (fewer strong cold, dry air masses flowing southwards over the area as well as fewer warm, moist air masses flowing northwards) (Holton, 1992). The climate of the New England/Mid-Atlantic Region could become less variable in nature although this will probably be characterized by less rapid changes in daily weather rather than in reduced extremes. Summer convective thunderstorms will most likely increase in intensity (similar to those that now occur in the south-eastern US), and there is a suggestion that there may be longer time periods between thunderstorms (Henderson and Robinson, 1994; P. J. Robinson, personal communication). If this occurs, it may lead to greater variability in stream flow, especially in summer, with the smaller streams experiencing more flash flooding as well as longer periods of low flow than they exhibit at present. An increased number of forest fires is also possible owing to increased lightning strikes during periods of dry forest conditions (Price and Rind, 1994).

Table IV. Rate of temperature increase (°C per 100 years) in selected cities, as excess over rural areas of comparable topography. Taken from Changnon (1992) and Landsberg (1981)

City	Winter	Summer
Cleveland	1·0	2·8
Boston	1·6	2·6
Washington, D.C.	0·9	2·4

URBAN CLIMATES

Large metropolitan areas, which are common to this region, generate distinctive climates that, in some ways, may represent a microcosm of global climatic change (Oke, 1993). The magnitude, extremes and rate of change in urban air temperature and precipitation approximate that predicted to occur globally over the next 100 years in response to climatic change (Changnon, 1992). Annual average air temperatures in urban areas (cities with one million or more inhabitants) are 1–3°C warmer than nearby rural areas of comparable topography (Oke, 1993). On some nights, however, urban areas can be as much as 12°C warmer than nearby rural areas (Oke, 1993), and the magnitude of the urban heat island effect increases with population density (Oke, 1973, 1987). Consequently, average urban air temperatures have increased 2–3°C over the last 100 years as cities grew in size (Table IV; Changnon, 1992). Temperatures in urban areas over the last 100 years, however, have increased more during the summer than the winter (Table IV).

Urban climates differ in still other ways from rural or suburban climates. Large cities receive less snow because it is converted to rain, and more total precipitation often falls in the lee of the city (Oke, 1993). Furthermore, in at least some cities, summer rather than winter precipitation is increased (e.g. by 25% in St Louis and Chicago), predominantly because of heavy rainfall events (Changnon, 1992). Thus, the increase in intensity of summer convective storms as predicted for the New England/Mid-Atlantic Region has probably already occurred owing to local reasons in some large metropolitan areas. Greater amounts of runoff flow from urban watersheds owing to the impervious surfaces of asphalt and concrete. Also, in urban areas, solar radiation can be reduced by 1–25% because of air pollutants, and UV radiation can be reduced by 25–90% because of low altitude ozone, aerosols and cloud (Oke, 1993). The latter feature offers the interesting possibility that aquatic systems in urban areas may serve as 'controls' for those located in rural areas and experiencing higher levels of UV-B radiation.

In summary, large metropolitan areas have experienced substantial climatic change over the last 100–150 years, and some of these changes may simulate those projected for a double CO_2 atmosphere (Changnon, 1992; Oke, 1993). Warmer urban climates probably increase the heat content of freshwaters in multiple ways. For example, in urban areas, freshwater ecosystems may be warmed by runoff flowing over hot pavement. Historical records of water temperatures in urban areas could be used to examine hypotheses about warmed aquatic habitats. Aquatic ecosystems located in urban areas may provide 'natural laboratories' or analogues (e.g. Adrian *et al.*, 1995) for measuring how these systems respond to climate change. Of course, using aquatic systems in this way will be challenging because some urban aquatic systems receive increased atmospheric deposition and the water quality of urban runoff is often poor owing to road salt, lead and other contaminants.

INTERACTIVE EFFECTS OF CLIMATE AND DIRECT HUMAN EFFECTS

The direct effect of humans on aquatic ecosystems is likely to increase in this region over the next 50–100 years as the human population of the region increases and climate changes. We present five examples of potential interactive effects between climate change and direct anthropogenic stressors. In these examples, we assume that warmer atmospheric temperatures will increase the temperatures of surface waters. If the frequency of forest fires increases in the northern section of the region (Price and Rind, 1994), the resultant enhanced exposure to wind could transfer the surface heat of lakes to deeper strata via mixing of the water column (Schindler *et al.*, 1990). We also assume reductions in snowpack and stream flow, especially in autumn and winter, as predicted by Feddema and Mather (1992) and Hodny and Mather (1995). Although it is possible that the reductions in flow predicted here may be ameliorated by the anti-transpirent effect of CO_2 enrichment, or the cooling effect of sulfate aerosols, humans may also divert water between basins to meet the water needs of expanding urban areas. Furthermore, the anti-transpirent effect of CO_2 enrichment, observed in laboratory experiments using high rates of air flow, is not expected to occur on a regional scale in a forest (Graham *et al.*, 1990). Thus, the assumption of flow reduction is reasonable.

Estuarine Eutrophication

Global climate change can influence estuaries in a variety of ways, including sea level rise, warming and changes in the salinity of coastal waters (Kennedy, 1990; Peterson *et al.*, 1995). Climate change in the watersheds upstream of estuaries in the New England/Mid-Atlantic Region, however, may also affect primary production and eutrophication of estuarine waters. Eutrophication is a major problem in estuaries of this region, with excess nutrient inputs leading to enhanced rates of primary production and incidences of anoxia and hypoxia in bottom waters (National Research Council, 1993).

Nitrogen is the element that is most limiting to primary production in most estuaries of the northeastern US, and it is therefore the element most responsible for eutrophication (Howarth, 1988; National Research Council, 1993). Most of the nitrogen entering estuaries comes from non-point sources. In the New England/Mid-Atlantic Region, only 25% of nitrogen inputs to the coastal zone comes from sewage, and 75% comes from non-point sources on land, principally from fertilizer and atmospheric deposition (Howarth *et al.*, 1996). Climate change may affect this delivery of nitrogen to estuaries; the projected general decrease in water runoff and greatly decreased snowpack would be expected to result in more nitrogen retention in forests and agricultural lands, and less exported from land to the rivers and estuaries. Exports of organic carbon and phosphorus from land to estuaries would be likely to decline even more than nitrogen in a drier climate because these are largely particle bound, and erosion should decrease (Howarth *et al.*, 1991). Of course, even if the climate becomes generally drier, a possible increase in the intensity of storms may increase erosion and consequent nutrient fluxes, as suggested by the modelling efforts of Howarth *et al.* (1991). None the less, most erosion in the Northeast occurs during the spring runoff period, and generally drier winters and springs are likely to result in less erosion.

A drier climate would also be expected to lead to changes in estuarine functioning owing to changes in hydrological mixing. The circulation patterns in an estuary are controlled by the interaction of tides and freshwater input from rivers, with additional influences from local topography. Salt wedge estuaries have high riverine inputs of water and relatively little tidal influence; these are the most stratified (Postma, 1980). Partially mixed and well-mixed estuaries have proportionately smaller inputs of freshwater from rivers, stronger tidal influences and are less stratified or not stratified at all (Postma, 1980). Tides are fairly strong in much of the New England/Mid-Atlantic Region, and, consequently, most of the estuaries in this region are either partially or well mixed. The projected decreases in freshwater runoff resulting from climatic change would be expected to drive estuaries towards being more mixed and less stratified. This, in turn, might affect primary production and eutrophication in several ways. First, the trapping of materials from both up-river and from coastal waters will tend to increase as estuaries become less stratified (Postma, 1980; Burton, 1988; Howarth *et al.*, 1995); this might lead to increased respiration, nutrient recycling and, therefore, higher primary production. However, a counteracting tendency would be for greater light limitation of production in the more mixed estuaries; the more phytoplankton are mixed to a greater depth in the water column, the less light they experience and the lower their rate of net primary production (Malone *et al.*, 1988; Cole *et al.*, 1991, 1992). Lessened stratification in estuaries also increases the rate of oxygen delivery into waters below the photic zone, decreasing problems from anoxia and hypoxia (Officer *et al.*, 1984; Malone *et al.*, 1988).

The effects of changes in circulation, stratification, primary production and oxygenation would have complex direct and indirect effects on the recruitment and exploitation of estuarine species. Many exploited populations are already under stress from habitat loss related to eutrophication and from excessive harvesting (Houde and Rutherford, 1993). The interactions of climate change, shifting patterns of human effects and eutrophication will most likely make the management of estuarine species even more difficult.

Overall, we expect that a drier climate in the New England/Mid-Atlantic Region would tend to lessen the problem of estuarine eutrophication as non-point nutrient inputs decrease, as light becomes more limiting to phytoplankton production and as more oxygen is advected into bottom waters. Direct support for this prediction comes from long time-series observations in Chesapeake Bay; the intensity of eutrophication and low oxygen levels there are positively related to the intensity of runoff in the spring (Malone *et al.*, 1988). Future changes in land use and in pollutant loading, however, may be even more important (Howarth *et al.*, 1991). Regardless of climate scenario, we note that long-term records of climate and hydrological variability are available for many estuaries in this region. The potential for learning about the interactions between climate variability and estuarine dynamics has not been fully exploited (see Peterson *et al.*, 1995).

Acidification

The magnitude and consequences of acidic deposition across the landscape are influenced by a variety of factors including the distribution and concentration of acids in wet and dry deposition, elevation, precipitation volume, vegetation type and biogeochemical processes that neutralize acidic deposition (Munson and Gherini, 1991; Ollinger *et al.*, 1993). The effects of nitrogen are of particular interest, because this element is a nutrient and its forms and concentrations are sensitive to climatic conditions (Aber *et al.*, 1989; Driscoll *et al.*, 1989; Murdoch and Stoddard, 1992). Therefore, this discussion focuses primarily on the effects of climate change on nitrogen deposition and the cycling of nitrogen within watersheds.

It seems unlikely that the warmer, drier conditions projected for the New England/Mid-Atlantic Region will significantly influence the atmospheric deposition of nitrogen. In the Northeast, rain or snowfall occurs on average only 5% of the time. The remaining 95% of the time the landscape is exposed to inputs of dry deposition (Hicks and Shannon, 1979). Estimates of dry deposition for the region are only 15–25% of the total deposition depending upon the model and geographical location (Ollinger *et al.*, 1993). Thus, total deposition is largely dependent on the distribution, duration and volume of rainfall. If precipitation became more episodic (fewer, larger storms), there would most likely be some decrease in annual wet deposition that would be counterbalanced, only partly, by increases in annual dry deposition (Dr Gary Lovett, Institute of Ecosystem Studies, personal communication). As a result, atmospheric deposition of nitrogen would most likely decrease, but these changes would be small when compared with decreases anticipated from the 1990 Amendments of the Clean Air Act.

Although atmospheric deposition of nitrogen may not change much in response to climate change, nitrogen concentrations in freshwaters could change markedly if the cycling of nitrogen within the watershed is altered. For example, increased concentrations of atmospheric CO_2 and air temperature during the growing season could enhance forest and microbial growth, resulting in increased sequestering of nutrients in terrestrial ecosystems (Peterson and Melillo, 1985; Schindler and Bayley, 1993). Increased assimilation of elements associated with forest growth should decrease solute leaching from forest soils to surface waters. However, 10 years of record in the Catskill Mountains, show that the highest annual average concentrations and yields of stream nitrate generally coincide with highest average annual air temperatures (Murdoch and Burns, 1995). With the exception of the water year 1990 (a year of high nitrogen deposition), stream nitrate concentrations cannot be attributed to changes in atmospheric deposition of nitrogen. The relationship between air temperature and stream nitrate concentrations and yields suggests that microbial mineralization of nitrogen in forest soils may have increased during warm years and exceeded rates of nitrogen assimilation by vegetation. This resulted in higher concentrations of nitrate in soil solutions and surface waters during warm years.

Interestingly, increases in air temperature could also change stream sulfate concentrations during the growing season. Sulfate reduction in ponded or slow-moving waters in the western Adirondacks can be enhanced by increasing air temperature, resulting in a decline in stream sulfate concentrations and an increase in acid-neutralizing capacity (ANC) (Burns, 1996). Likewise, lengthening the growing season would lengthen the period each year when sulfate reduction influences stream chemistry. The effect of increased ANC would be partly offset by attendant increases in dissolved organic carbon, which would, in turn, decrease light penetration and further alter biogeochemical processes (Cirmo and Driscoll, 1993; Burns, 1996).

The hydrological changes associated with a warmer climate will also affect biogeochemical processes and stream acidification. Returning to nitrogen cycling within the watershed, if the growing season is dry, soil moisture deficits could decrease both forest and microbial growth, and lower rates of denitrification. Thus, the effects of climate change on the nitrogen-induced acidity of surface waters will most likely be regulated by changes in processes within the watershed that retain nitrogen (e.g. forest growth) or release it to the atmosphere (e.g. denitrification) relative to processes that supply nitrogen (e.g. microbial mineralization).

Lower snowpack accumulation could also have a significant effect on stream hydrology and acidity in early spring. Snowmelt results in short-term acidification of surface waters, which may be critical to the survival of aquatic organisms. In northern areas of the Northeast as much as 50% of the annual discharge occurs during the months of March, April and May (Likens *et al.*, 1977). Under warmer conditions, the magnitude of the peak flow would diminish and, consequently, so would the episodic acidification caused by spring meltwater.

Changes in stream acidification during spring snowmelt, however, will also depend on rates of soil mineralization during the dormant season, which is influenced by the duration and depth of soil freezing in winter. Shallow soils in the Catskill and Adirondack Mountains of New York often freeze during the late autumn, prior to snow cover, but generally thaw once the insulating barrier of the snowpack has developed (Rascher *et al.*, 1987; Murdoch and Stoddard, 1992). Less accumulation of snowpack could result in more frequent or deeper freezing of the soil, reducing microbial activity during the winter. Deep freezing of soils would lyse root and microbial cells and enhance the release of labile nitrogen (Boutin and Robitaille, 1995). Presumably, this effect would diminish after a few years as the soil pools of available nitrogen become exhausted. However, if temperature increases during winter are high enough to limit soil frost, then microbial activity during the dormant season may exceed that of current conditions resulting in enhanced leaching of nutrients to surface waters.

Finally, less frequent but more severe summer storm events could result in episodes of increased stream acidity. Nitrogen appears to leach from forest soils to streams via a 'wash-out' process in which deposited or mineralized nitrate is flushed from the watershed during periods of increased runoff (Murdoch and Stoddard, 1992). If climate change causes less frequent, but larger rainfall events, the period of dry deposition and microbial nitrification between storms will increase, and this wash-out process may encourage larger pulses of nitrate to surface water. Likewise, in areas of high sulfate deposition, such as western Pennsylvania (Lynch and Corbett, 1989), sulfate pulses during episodic events could become more severe. With increased intensity of storms, dilution of basic cation concentrations during storm events should also increase (Schaefer *et al.*, 1990). Thus, increased acidity during storm events could occur under the projected warmer, drier conditions of the future. However, it is important to stress that these projections are dependent on changes in the seasonal distribution of precipitation and temperature, both of which are difficult to predict from current knowledge. If climate change results in an overall decrease in annual precipitation, but little change in storm frequency, episodic acidification should decrease.

Biomagnification of Toxins

The effects of increased water temperature and elevated concentrations of toxins have been measured separately in aquatic food-webs and are documented in the literature (Spry and Wiener, 1991; Havens and Hanazato, 1993; Moore and Folt, 1993). General predictions of the effect of climate change on the toxicity of aquatic contaminants, however, must be extrapolated from these separate findings, because the combined effects of temperature and toxicity are less well known (Moore and Folt, 1993). At the individual level, the metabolic rates of aquatic ectotherms will increase with increased temperatures. For those toxins taken up via ingestion or respiration (e.g. through the gills of fish), toxicity is likely to be enhanced at warmer temperatures, and organism tolerance to toxins may be diminished (Cairns *et al.*, 1975; Connell and Miller, 1984). In addition, temperature-induced increases in ingestion or uptake of toxins that bioaccumulate (e.g. some metal toxins, chlorinated hydrocarbons) may result in more rapid bioaccumulation.

At the population and community levels, both increased temperature and toxins cause decreases in the body size of aquatic invertebrate communities (Havens and Hanazato, 1993; Moore and Folt, 1993). For example, zooplankton communities containing large cladocerans and copepods are usually more sensitive to a variety of toxins than those comprised of small cladocerans, copepod nauplii and rotifers. This phenomenon has been attributed to the greater sensitivity of larger organisms to toxicity (e.g. owing to enhanced permeability of the animal's cuticle during molting and the greater number of molts to maturity in large *Daphnia*; Gliwicz and Sieniawska, 1986) and to the release of some smaller taxa (e.g. rotifers) from competitive suppression by larger species (Papst and Boyer 1980; Day *et al.*, 1987). At warmer temperatures, food-web structure over the long term will be largely dependent on the relative population growth rates of the most resilient species (Hanazato and Yasuno, 1987). Among algal communities, warmer water temperatures and longer periods of thermal stratification seem to favour large and inedible species (including green and blue-green taxa; Adrian *et al.*, 1995). This, in combination with the loss of larger cladocerans, could have ecosystem-level effects, by decreasing the overall efficiency of carbon and energy transfer through lake food-webs. Furthermore, elevated temperatures and toxins separately cause decreases in species diversity and richness in aquatic systems (Connell and Miller, 1984), and these stressors are likely to ellicit similar effects under the conditions of climate change (Stemberger and Lazorchak, 1994).

The numerous indirect effects of climate change may also alter the concentrations and bioavailability of toxins in aquatic systems. For example, a warmer climate may increase the rates of microbial degradation of toxins thereby reducing their concentrations in the environment. Conversely, predicted decreases of DOC associated with climate change may enhance the toxicity of some metals by decreasing concentrations of meta-binding ligands and increasing metal bioavailability. Lastly, hydrological alterations associated with climate change may also affect *in situ* concentrations of toxins in aquatic systems. Although reduced runoff could result in lower inputs of some contaminants into lakes, atmospherically transported toxins may actually increase in concentration owing to lower flushing rates of lakes and enhanced evaporation.

Mercury is an example of an atmospherically transported contaminant that is present in lakes throughout the developed and pristine areas of the New England/Mid-Atlantic Region (Driscoll *et al.*, 1994). It is so widespread that fish tissue concentrations exceeding 0·2 ppm mercury, a concentration indicative of an unhealthy or impaired ecosystem, have been observed in 30% of the lakes sampled in the Northeast by the EMAP programme (Yeardley and Lazorchak, 1994). The most toxic species, methyl mercury, occurs at such high concentrations in aquatic organisms in many lakes that it constitutes a major public health concern (Bloom, 1992; Driscoll *et al.*, 1994). Bioaccumulation of mercury appears to be related to the age, size and habitat choice of fish (Driscoll *et al.*, 1994; Spry and Wiener, 1991), underscoring the importance of diet to rates of bioaccumulation. Furthermore, greater body burdens of mercury higher in food chains provide evidence of biomagnification (Sorenson *et al.*, 1990). Finally, mercury uptake increases with temperature (Boudou and Ribeyre, 1981).

The potential effects of climate change on mercury toxicity and propagation through aquatic food-webs will probably affect the contamination problem. Increased temperatures and metabolic rates will probably increase bioaccumulation and biomagnification rates, and thereby, increase levels of the contaminant in all organisms. Possible alterations of food-webs (loss of large cladoceran grazers, increases in inedible foods) owing to increased temperatures and toxicity, however, may decrease the efficiency of energy and carbon transfer to higher trophic levels, thereby mitigating the transfer of mercury. Changes in the length of the food chain may also affect the degree of biomagnification. For example, PCB (polychlorinated biphenyls) concentrations in top predators seem to be greater where food chains are long; individuals of the same species in lakes with shorter food chains have lower body burdens (Rasmussen *et al.*, 1990). This predicted scenario may also apply to mercury and other persistent toxins found in aquatic systems.

Wetland desaturation

Most of the wetlands of the region can be placed in one of three categories: bogs, minerotrophic wetlands (e.g. marshes, swamps and fens) and riverine wetlands. Bogs are primarily rain water fed. As such, their water status is controlled by the balance between precipitation and evapotranspiration. Minerotrophic wetlands, by contrast, are fed with surface and subsurface runoff from the surrounding watershed as well as directly from rain. Commonly, water enters such wetlands as groundwater, which has acquired a more mineral-rich chemical composition dictated by the characteristics of the surrounding soils (hence the term, minerotrophic). Riverine wetlands are a type of minerotrophic wetland whose water balance is strongly influenced by the presence of a river or stream, which may periodically flood the wetland. This hydrological categorization is adopted because the most important effects of projected climate change on wetlands in this region are likely to occur via changes in the water balance.

The climate scenario of most concern to wetlands is the scenario of decreased precipitation excess (i.e. a drier climate), exactly what is predicted for this region. Wetlands owe their existence to a high degree of *water saturation* of the soil; if this is eliminated, the wetland ceases to exist as a wetland. However, changes in temperature, the mixing ratio of atmospheric carbon dioxide, atmospheric acids and nutrients and the annual *distribution* of precipitation and evapotranspirational water demand are potentially significant to wetlands.

Intermittent desaturation of a wetland, and especially when limited to upper sediments, is a normal feature of wetland hydrology. However, numerous examples of intentional wetland drainage illustrate the effects of long-term desaturation. Soil oxidation accounts for most of the effects. As oxygen penetrates the sediment, microbes convert reduced chemical species, such as sulfide, ammonium and iron(II), to oxidized forms such as sulfate, nitrate and iron(III). As a result, sulfate and nitrate are leached and drainage waters and pore waters are acidified. Indeed, this release of mineral acids occurs not only as a chemical effect of long-term wetland drainage (Dasberg and Neumann, 1977), but can also be seen in the peaks in concentrations of sulfate and nitrate sometimes observed in late summer in temperate watersheds (e.g. Lazerte, 1993). Other effects include the subsidence of wetlands where the oxidation of organic material in the sediment exceeds the rate of deposition of new detrital organic material, and the loss of aquatic fauna as standing waters disappear from the wetland.

Bogs are probably the most susceptible of this region's wetlands to a drier climate. This is because bogs lack a strong mechanism to maintain water content and water chemistry in the face of increased evapotranspiration. At Thoreau's Bog in eastern Massachusetts, for example, approximately 70% of the annual precipitation is lost to evapotranspiration (Hemond, 1980). If annual evaporation were to increase by an amount equal to 20% of precipitation, an amount that lies within the range of climatic predictions for much of the New England/Mid-Atlantic Region, runoff would be decreased to one-third of its present value. A steady-state analysis of chemical budgets in this bog shows that those solute concentrations governed by hydrology (e.g. chloride) would then triple. Moreover, because annual variability could then result in frequent years when precipitation was *less* than evapotranspiration, the bog would probably experience episodes of extreme drying, with dramatic effects on soil oxidation and on the vegetation. It seems likely that this, and many other bogs in the New England/Mid-Atlantic Region, would not survive as bogs under the larger postulated changes of climate.

Minerotrophic wetlands, which are considerably more common than bogs in the New England/Mid-Atlantic Region, except at the higher elevations and latitudes, should be more resilient than bogs in the face of climate change, mainly because, by definition, they receive water from a larger catchment than do bogs. Many such wetlands are in groundwater discharge areas, receiving inputs from both precipitation and groundwater aquifers. Consequently, their annual water output is usually much larger than that from bogs. Changes in evapotranspiration of such wetlands, therefore, represent a smaller influence on the water budget, and compensation for increased evapotranspiration in such minerotrophic wetlands could often be accomplished by a decrease in surface runoff from the wetland. It should be possible to predict the response of individual minerotrophic wetlands to drier conditions on the basis of present day water balance observations, including the relationship between water storage and the rate of surface water outflow (the stage–discharge curve) and the present water balance. The extent to which desaturation and loss of open water could occur is unknown, because water budgets and stage–discharge relationships have been established for very few wetlands in the region. However, minerotrophic wetlands should be much less affected than bogs.

Riverine wetlands represent perhaps the most resilient of the region's wetlands because of the regulating effect of stream and river stage. Such wetlands are least likely to be extensively desaturated unless river stage is lowered so much that the wetlands are no longer periodically flooded. Less dramatic effects than desaturation and soil oxidation may still result, however. A decrease in open water, caused by a drier climate, can limit wetland use by waterfowl, and possibly favour invasion by introduced plant species. Siltation, habitat fragmentation and eutrophication are common effects of human development. Under the dual impact of climate change and development, even wetlands that remain water saturated may suffer losses of biodiversity and become less valuable as wildlife habitat.

Habitat Fragmentation and Fish Populations

Increased water temperatures and a reduction in stream flow are the effects of climatic change that are most likely to affect stream communities within the region (Poff, 1992; Ward *et al.*, 1990), and each of these variables is strongly influenced by land use patterns. For example, in agricultural areas, lower precipitation may trigger increased irrigation resulting in reduced stream flow (Adams *et al.*, 1990). In forested areas, logging patterns influence instream temperatures through the direct effects of shading and of cover on overland water flow and groundwater temperatures. Reductions in radiative heating by vegetative cover may be particularly important in low-order streams, where the overhanging vegetation significantly reduces the magnitude and variation of the stream temperature compared with that of clear-cut regions (Burton and Likens, 1973; Ringler and Hall, 1975).

For stream-dwelling fish populations, temperature is a critical resource (*sensu* Magnuson *et al.*, 1979), because it affects many important physiological and behavioural pathways that influence growth and survival. Yet, interactions between temperature, food ration, developmental stage, feeding, growth and survival are extremely complex (Elliott, 1982). Thus, predicting the range expansion and contraction of fish species in response to climatic change is difficult. Although the northern distributional limit of many fish species is limited by the fish's ability to store adequate energy for surviving periods of winter starvation (Shuter and Post, 1990), the southern limit of distribution is frequently determined by the fish's tolerance of maximum summer temperatures. Modest increases in stream temperature can increase growth rates of some species, while simultaneously exceeding the thermal maximum of others and driving them from that location. Species distributions may theoretically shift to the north (Shuter and Post, 1990) or to higher elevations (Meisner, 1990); but, shifts in species distributions require connecting streams with suitable habitat, which may not exist.

Stream species that are restricted to a narrow range of cooler temperatures may be particularly vulnerable to climate change. Such species, termed cold water stenotherms, include many stream insects (Vannote and Sweeney, 1980; Sweeney *et al.*, 1992) and a number of fish species that are recreationally important (e.g. salmonids; Elliott, 1982). The potential effects of a warmer climate on cold water populations in the region are perhaps best illustrated by examining predictions for brook trout (*Salvelinus fontinalis*). Meisner (1990) projects that a 3·8°C increase in mean annual temperature would reduce the area suitable for brook trout, primarily in coastal and low elevation areas of their present distribution in the region. Distributions may be further reduced if human activities and vegetational changes result in increased fragmentation of forested areas within watersheds and riparian habitats. Such forest fragmentation could further increase stream temperatures. This same temperature change, however, may result in increased vigour for populations in high elevation and northern portions of the range, because the length of the growing season would increase in these areas (Meisner, 1990).

In the Appalachian Mountains of Virginia and West Virginia, native brook trout are near the southern edge of their distribution in eastern North America (MacCrimmon and Campbell, 1969). Trout distributions in Virginia are already fragmented and essentially restricted to first- to third-order streams (Flebbe, 1994). Here, elevation and latitude interact to provide an acceptable temperature regime for brook, introduced rainbow (*Oncorhynchus mykiss*) and brown trout (*Salmo trutta*) (Meisner, 1990; Flebbe, 1994). Under the GISS warming scenario, the theoretical lower elevation for brook trout occurrence would increase about 700 m for most of Virginia (Meisner, 1990). Because few streams in Virginia are found at these higher elevations, loss of brook trout streams in Virginia would be considerable (Flebbe, 1993). Unreliable summer flows already occur in many high elevation Virginia streams, and the reduced flows predicted by climate change models (Poff, 1992) for the region would further restrict the available trout habitat. Finally, the distribution of remaining brook trout streams would be seriously fragmented as brook trout disappear from lower elevation streams and remain only in isolated patches at the uppermost elevations. Human intervention, in the form of stocking, may be required to maintain fragmented headwater populations in the face of localized extinctions, sometimes exacerbated by anglers.

RECOMMENDATIONS

Clearly, the dual stressors of climate change and direct human impact have the potential to transform aquatic ecosystems dramatically by altering stream flows and nutrient cycles, eliminating habitats (e.g. bogs) and changing community structure. To detect, understand and manage these transformations, both monitoring and research will be necessary because these two approaches complement each other. Well-designed monitoring programmes can help to direct research efforts and provide useful information for interpreting or generalizing from experimental results. Likewise, experiments may reveal new parameters necessary for detecting perturbations via monitoring. Our specific recommendations address both climate change and direct human effects on freshwater ecosystems.

Monitoring

1. *Extend and improve data on the distribution, abundance and effect of anthropogenic stressors (non-point pollution) within the region.* Current data are often substandard for the reasons previously mentioned. Monitoring efforts between state, federal and academic institutions should be better coordinated within the region, meet high standards (e.g. employ statistically valid sampling design) and the data should be disseminated widely.

2. *Improve scientific knowledge regarding the contemporary distribution and abundance of aquatic species in the New England/Mid-Atlantic Region.* Climate change, coupled with either continued or increasing anthropogenic stress, has the potential to alter enormously the distribution and abundance of native species in this region. Yet, knowledge of the present distribution and abundance of aquatic species in this region, as in other areas of North America, is poor (Schindler, 1989). Such baseline information is essential for assessing the biotic effects of climate change and for effective management of aquatic resources. It can be argued that net species diversity may not decline in response to climate change, because species loss may be balanced by the introduction of exotics or by species from the Southeast expanding their ranges northwards. For example, in the Hudson River drainage, total molluscan species richness has not changed substantially over the past 150 years, because species introductions have offset extinctions (Strayer, 1987). Nevertheless, considerable ecological and economic dislocation has resulted from widespread changes in the abundance and distribution of species (US Congress, Office of Technology Assessment, 1993). Intervention by management agencies to restore, protect or replace lost species (e.g. Atlantic salmon), as often demanded by the public, can lead to the expansion of government expense and regulation. Research on the interactions between species changes, human activities and management regulation will be required to understand the

actual ecological and economic costs of altered species distributions and abundances. Finally, conserving the region's present biodiversity is essential to maintain species valuable for restoring and rehabilitating degraded ecosystems.

Future Research

1. *Establish a research centre(s) where field studies designed to understand interactions between freshwater ecosystems and climate change can be conducted.* Although three LTER (Long-Term Ecological Research) sites are located within this region (Hubbard Brook, Harvard Forest and Virginia Coastal Reserve), the potential for major large-scale experiments on freshwater systems is primarily limited to small streams at these locations. This region, however, contains a plethora of lakes and reservoirs, wetlands, and large rivers, some of which provide the entire potable water supply for major metropolitan areas. For example, reservoirs or rivers, rather than wells, supply all of the potable water for the millions of residents of New York City, Boston and Washington D.C. (Schwarz and Dillard, 1990). Managers of such urban water systems warn that 10–20% reductions in runoff and longer hot spells associated with climate change could require costly expansion of supplies as well as treatment facilities to remove contaminants that form more readily in the presence of high temperatures and chlorine (e.g. trichloromethane; Schwarz and Dillard, 1990). Despite the potential effects of climate change on the quantity and quality of surface waters of the region, there is not one scientific institute devoted to freshwater research on a scale commensurate with the significance of the resource.

2. *Project the future distribution, activities and direct effects of humans within the region.* Humans are abundant in this region and their response to climate change will probably greatly influence the structure and functioning of freshwater ecosystems. For example, palaeolimnological studies conducted in Vermont and Connecticut have revealed how major historical shifts in human land use greatly affected some freshwater ecosystems of the region (Brugam, 1978; Engstrom *et al.*, 1985). Because socio-economic factors largely drove these shifts in land use (Whitney, 1994; Foster, 1995), interdisciplinary studies integrating the fields of ecology, sociology and economics will be needed to identify the future distribution, activities and effects of humans as climate changes. Pertinent questions include: what proportion of humans will migrate from the congested coast into upstream forested watersheds in response to changing economics, a deteriorating climate in coastal urban areas or an increase in sea-level? What consequences will such migrations have for freshwater systems?

3. *Develop mathematical analyses, experimental designs and aquatic indicators that distinguish between climatic and anthropogenic effects on aquatic systems.* A mathematical technique called double mass analysis is used by hydrologists to separate the effects of climate from land use activities (e.g. conversion of forest to farmland) on the discharge patterns of rivers (e.g. Verry, 1986). It may be possible to develop similar techniques for separating the effects of climate and land use activities on the processes of aquatic ecosystems. Experimental designs comparing climatically induced changes of community and ecosystem processes among habitats differentially affected by humans should be employed. For example, replicate urban and rural plots over latitudinal or altitudinal gradients might be used. Reductions in flow of headwater streams or lengthened drying of ephemeral ponds may provide indicators or early signals of climatic warming. Such hydrological indicators would be particularly useful for this region because they are less likely to be affected by direct anthropogenic stressors than biotic or chemical indicators. At present, no monitoring of these small aquatic habitats is required of states by the Clean Water Act of 1972 (US EPA, 1994).

4. *Develop and refine projections of climate variability such that the magnitude, frequency and seasonal timing of extreme events can be forecasted.* Inherent in this recommendation is the assumption that freshwater ecosystems will be more greatly affected by changes in the magnitude or frequency of extreme events than changes in the mean. Tests of this assumption, however, are necessary (Grimm, 1993).

5. *Describe, quantitatively, the flux of materials (sediments, nutrients, metals) from watersheds characterized by a mosaic of land uses.* Most of the anthropogenic stressors affecting the freshwater ecosystems of this region are constituents of non-point pollution. Consequently, understanding how the mobility of these

constituents is affected by multiple land uses, as well as altered patterns of rainfall and runoff, is necessary for protecting freshwaters.

Clearly, the direct effect of humans within this region, coupled with the effects of climate change, have the potential to alter aquatic ecosystems and their watersheds to an extent equalling, or surpassing, that which occurred in the late eighteenth and nineteenth centuries. Historical records and palaeoecological studies demonstrate that two major historical transformations of the land and aquatic ecosystems have taken place in this region. The first was deforestation coupled with intensive agriculture. That was followed by farm abandonment and reforestation, which allowed recovery of some aquatic systems. A third major transformation has begun in the coastal areas of this region and is rapidly expanding inland. It is the urbanization of landscapes, involving the interactive effects of urban-induced climate change, pollution and radical physical alteration of hydrological systems. This most radical of the three anthropogenic transformations, coupled with regional climatic change, portends aquatic ecosystems of completely unknown structure and function.

ACKNOWLEDGEMENTS

The authors are indebted to the US Environmental Protection Agency, the US Geological Survey, the American Society of Limnology and Oceanography and the North American Benthological Society for sponsoring and organizing the symposium on the effects of climate change on freshwater ecosystems. The interdisciplinary approach taken promoted the exchange of knowledge and the development of ideas presented here. We also thank Steve Paulsen, Richard Stemberger, Roger Yeardley and Kevin Summers for sharing unpublished data from the EMAP-SW programme funded by the US EPA; Timothy R. Oke for stimulating ideas regarding urban climates; David Foster and Charlie Cogbill for manuscripts and references on historical land use patterns; Laura Conkey who provided maps of human demography; and Ralph Tiner and Barbara Bedford for information regarding the abundance and distribution of wetlands. Finally, the first author is especially grateful for the expert assistance of Ms Lisa Brainard and Irene Laursen, librarians at Wellesley College, who rapidly obtained numerous government documents, maps and publications.

REFERENCES

Aber, J. D., Nadelhoffer, K. J., Steudler, P., and Melillo, J. M. 1989. 'Nitrogen saturation in northern forest ecosystems', *BioScience*, **39**, 378–386.

Adams, R. M., Rosenzweig, C., Peart, R. M., Richie, J. T., McCarl, B. A. Glyer, D., Curry, R. B., Jones, J. W., Boote, K. J., and Allen, L. H. 1990. 'Global climate change and US agriculture', *Nature*, **345**, 219–224.

Adrian, R., Deneke, R., Mischke, U., Stellmacher, R., and Lederer, P. 1995. 'A long term study of the Heiligensee (1975–1992): evidence for effects of climatic change on the dynamics of eutrophied lake ecosystems', *Arch. Hydrobiol.*, **133**, 315–337.

Alampi, G. 1994. *Gale State Rankings Reporter*. Gale Research Inc., Detroit. p. 1657.

Ashizawa, D. and Cole, J. J. 1994. 'Long-term temperature trends of the Hudson River: a study of the historical data', *Estuaries*, **17**, 166–171.

Bloom, N. S. 1992. 'On the chemical form of mercury in edible fish and marine invertebrate tissue', *Can. J. Fish. Aquat. Sci.*, **49**, 1010–1017.

Bosch, J. M. and Hewlett, J. D. 1982. 'A review of catchment experiments to determine the effect of vegetation changes on water yield and evaporation', *J. Hydrol.*, **55**, 3–23.

Boudou, A. and Ribeyre, F. 1981. 'Comparative study of the trophic transfer of two mercury compounds — $HgCl_2$ and CH_3HgCl — between *Chlorella vulgaris* and *Daphnia magna*. Influence of temperature', *Bull. Environ. Contam. Toxicol.*, **27**, 624–629.

Boutin, R. and Robitaille, G. 1995. 'Increased soil nitrate losses under mature sugar maple trees affected by experimentally induced deep frost', *Can. J. For. Res.*, **25**, 588–602.

Bradley, R. S., Eischeid, J. K., and Ives, P. T. 1987. 'The climate of Amherst, Massachusetts', *Contribution No. 50*. Department of Geology & Geography, University of Massachusetts, Amherst. p. 100.

Brugam, R. B. 1978. 'Pollen indicators of land-use change in southern Connecticut', *Quat. Res.*, **9**, 349–362.

Brush, G. S. 1986. 'Geology and paleoecology of Chesapeake Bay: a long-term monitoring tool for management', *J. Washington Acad. Sci.*, **76**, 146–160.

Burns, D. A. 1996. 'The effects of liming an Adirondack lake watershed on downstream water chemistry', *Biogeochemistry*, **32**, 339–362.

Burton, J. D. 1988. 'Riverborne materials and the continent-ocean interface', in Lermand, A. and Meybeck, M. (Eds), *Physical and Chemical Weathering in Geochemical Cycles*. Kluwer, Dordrecht. pp. 299–321.

Burton, T. M. and Likens, G. E. 1973. 'Effects of strip-cutting on stream temperatures in the Hubbard Brook Experimental Forest, New Hampshire', *BioScience*, **23**, 433–435.
Cairns, J., Heath, A.G., and Parker, B. C. 1975. 'The effects of temperature upon the toxicity of chemicals to aquatic organisms', *Hydrobiologia*, **47**, 135–171.
Changnon, S. A. 1992. 'Inadvertent weather modification in urban areas: lessons for global climate change', *Bull. Am. Meteorol. Soc.*, **73**, 619–627.
Cirmo, C. P. and Driscoll, C. T. 1993. 'Beaver pond biogeochemistry: Acid-neutralizing capacity generation in a headwater wetland', *Wetlands*, **13**, 277–291.
Cole, J. J., Caraco, N. F., and Peierls, B. L. 1991. 'Phytoplankton primary production in the tidal, freshwater Hudson River, New York (USA)', *Verh. Int. Ver. Limnol.*, **24**, 1715–1719.
Cole, J. J., Caraco, N. F., and Peierls, B. L. 1992. 'Can phytoplankton maintain a positive carbon balance in a turbid, freshwater, tidal estuary?', *Limnol. Oceanogr.*, **37**, 1608–1617.
Connell, D. W. and Miller, G. J. 1984. *Chemistry and Ecotoxicology of Pollution*. John Wiley and Sons, Inc., New York. pp. 288–332.
Cosby, B. J., Ryan, P. F., Webb, J. R., Hornberger, G. M., and Galloway, J. N. 1991. 'Mountains of Western Virginia', in Charles, D. F. (Ed.), *Acidic Deposition and Aquatic Ecosystems: Regional Case Studies*. Springer-Verlag, New York. pp. 297–318.
Culliton, T. J., Warren, M. A., and Godspeed, T. R. 1990. '50 years of population changes along the nation's coasts 1960–2010', *Technical Report*. US Department of Commerce, NOAA, Strategic Assessment Branch, Rockville, Maryland.
Dahl, T. E. 1990. 'Wetland losses in the United States 1780's to 1980's', *Report*. US Department of the Interior, Fish and Wildlife Service, Washington, D.C.
Dasberg, S. and Neumann, S. P. 1977. 'Peat hydrology in the Hula Basin, Israel: I. Properties of peat', *J. Hydrol.*, **32**, 219–239.
Day, K. E., Kaushik, N. K., and Solomon, K. R. 1987. 'Impact of fenvalerate on enclosed freshwater planktonic communities and on in situ rates of filtration of zooplankton', *Can. J. Fish. Aquat. Sci.*, **44**, 1714–1728.
Driscoll, C. T., Schaefer, D. A., Molot, L. A., and Dillon, P. J. 1989. 'Summary of North American data', in Malanchuk, J. L. and Nilsson, J. (Eds), *The Role of Nitrogen in the Acidification of Soils and Surface Waters*. Nordic Council of Ministers, Copenhagen. pp. 6.1–6.45.
Driscoll, C. T., Newton, R. M., Gubala, C. P., Baker, J. P. and Christensen, S. W. 1991. 'Adirondack Mountains', in Charles, D. F. (Ed.), *Acidic Deposition and Aquatic Ecosystems: Regional Case Studies*. Springer-Verlag, New York. pp. 133–202.
Driscoll, C. T., Yan, C., Schofield, C. L., Munson, R., and Holsapple, J. 1994. 'The mercury cycle and fish in the Adirondack Lakes', *Environ. Sci. Technol.*, **28**, 136–143.
Elliott, J. M. 1982. 'The effects of temperature and ration size on the growth and energetics of salmonids in captivity', *Comp. Biochem. Physiol.*, **73**, 81–91.
Engstrom, D. R., Swain, E. B., and Kingston, J. C. 1985. 'A palaeolimnological record of human disturbance from Harvey's Lake, Vermont: geochemistry, pigments and diatoms', *Freshwat. Biol.*, **15**, 261–288.
Feddema, J. J. and Mather, J. R. 1992. 'Hydrological impacts of global warming over the United States', in Majumdar, S. J., Kalkstein, L. S., Yarnal, B. M., Miller, E. W. and Rosenfeld, L. M. (Eds), *Global Climate Change: Implications, Challenges and Mitigation Measures*. Pennsylvania Academy of Science, Easton, Pennsylvania. pp. 50–62.
Flebbe, P. A. 1993. 'Comment on Meisner (1990): effect of climate warming on the southern margins of the native range of brook trout, *Salvelinus fontinalis*', *Can. J. Fish. Aquat. Sci.*, **50**, 883–884.
Flebbe, P. 1994. 'A regional view of the margin: salmonid abundance and distribution in the southern Apalachian Mountains of North Carolina and Virginia', *Trans. Am. Fish. Soc.*, **123**, 657–667.
Foster, D. 1992. 'Land-use history (1730–1990) and vegetation dynamics in central New England, USA', *J. Ecol.*, **80**, 753–772.
Foster, D. 1995. 'Land-use history and four hundred years of vegetation change in New England', in Turner, B. L., II, Sal, A. G., Bernaldez, F. G., and di Castri, F. (Eds), *Global Land Use Change: A Perspective from the Columbian Encounter*. Scientific Research Council of Spain, Madrid, Spain. pp. 253–339.
Gliwicz, W. Z. and Sieniawska, A. 1986. 'Filtering activity of *Daphnia* in low concentrations of a pesticide', *Limnol. Oceanogr.*, **31**, 1132–1137.
Graham, R. L., Turner, M. G., and Dale, V. H. 1990. 'How increasing CO_2 and climate change affect forests', *Bioscience*, **40**, 575–587.
Greeley, W. B. 1925. 'The relation of geography to timber supply', *Econ. Geogr.*, **1**, 1–4.
Grimm, N. B. 1993. 'Implications of climate change for stream communities', in Kareiva, P. M., Kingsolver, J. G., and Huey, R. B. (Eds), *Biotic Interactions and Global Change*. Sinauer Associates Inc., Sunderland, Massachusetts. pp. 293–314.
Hamburg, S. P. and Cogbill, C. V. 1988. 'Historical decline of red spruce populations and climatic warming', *Nature*, **331**, 428–431.
Hanazato, T. and Yasuno. M. 1987. 'Effects of carbamate insecticide, carbaryl, on the summer phyto- and zooplankton communities in ponds', *Environ. Pollut.*, **48**, 145–159.
Hansen, J., Fung, I., Lacis, A., Lebedeff, S., Rind, S., Ruedy, R., Russell, G., and Stone, P. 1988. 'Global climate changes as forecast by the GISS 3-D model', *J. Geophys. Res.*, **93**, 9342–9364.
Havens, K. E. and Hanazato, T. 1993. 'Zooplankton community responses to chemical stressors: a comparison of results from acidification and pesticide contamination research', *Environ. Pollut.*, **82**, 277–288.
Hayes, J. T. 1991. 'Global climate change and water resources', in Wyman, R. L. (Ed.), *Global Climate Change and Life on Earth*. Chapman and Hall, Inc., New York. pp. 18–42.
Hemond, H. F. 1980. 'Biogeochemistry of Thoreau's Bog', *Ecol. Monogr.*, **50**, 507–526.
Henderson, K. G. and Robinson, P. J. 1994. 'Relationships between the Pacific/North American teleconnection patterns and precipitation events in the south-eastern USA', *Int. J. Clim.*, **14**, 307-323.
Hick, B. B. and Shannon, J. D. 1979. 'A method for modelling the deposition of sulfur by precipitation over regional scales', *J. Appl. Meteorol.*, **18**, 1415–1420.

Hodny, J. W. and Mather, J. R. 1995. 'Climate change and the water resources of Delmarva and the Delaware River basin', *Publ. Climatol., Lab. Climatol.*, **48**, 1–87.

Holton, J. R. 1992. *An Introduction to Dynamic Meterology*. Academic Press, San Diego. p. 511.

Houde, E. D. and Rutherford, E. S. 1993. 'Recent trends in estuarine fisheries: predictions of fish production and yield', *Estuaries*, **16**, 161–176.

Howarth, R. W. 1988. 'Nutrient limitation of net primary production in marine ecosystems', *Annu. Rev. Ecol. Syst.*, **19**, 89–110.

Howarth, R. W., Fruci, J. R., and Sherman, D. 1991. 'Inputs of sediment and carbon to an estuarine ecosystem: influence of land use', *Ecol. Appl.*, **1**, 27–39.

Howarth, R. W., Jensen, H., Marino, R., and Postma. H. M. 1995. 'Transport and processing of phosphorus in near-shore and oceanic waters', in Tiessen, H. (Ed.), *Phosphorus in the Global Environment*, SCOPE #54. Wiley & Sons, Chichester. pp. 323–345.

Howarth, R. W., Billen, G., Swaney, D., Townsend, A., Jaworski, N, Lajtha, K., Downing, J. A., Elmgren, R., Caraco, N., Jordan, T., Berendese, F., Freney, J., Kudeyarov, V., Murdoch, P., and Zhao-liang, Zhu, 1996. 'Regional nitrogen budgets and riverine N and P fluxes for the drainages to the North Atlantic Ocean: natural and human influences', in Howarth, R. W. (Ed.), *Nitrogen Cycling in the North Atlantic Oceans and its Watersheds*. Kluwer, Boston. pp. 75–140.

Husar, R. B., Sullivan, T. J., and Charles, D. F. 1991. 'Historical trends in atmospheric sulfur deposition and methods for assessing long-term trends in surface-water chemistry', in Charles, D. F. (Ed.), *Acidic Deposition and Aquatic Ecosystems: Regional Case Studies*. Springer-Verlag, New York. pp. 65–82.

Jones, P. D. and Bradley, R. S. 1992. 'Climatic variations in the longest instrumental record', in Bradley, R. S. and Jones, P. D. (Eds), *Climate Since A.D. 1500*. Routledge, London. pp. 246–268.

Kennedy, V. S. 1990. 'Anticipated effects of climate change on estuarine and coastal fisheries', *Fisheries*, **15**, 16–24.

Landsberg, H. E. 1981. 'The urban climate', *Int. Geophys. Ser.*, **28**, 1–275.

Lazerte, B. D. 1993. 'The impact of drought and acidification on the chemical exports from a minerotrophic conifer swamp', *Biogeochemistry*, **13**, 153–175.

Leavesley, G. H., Turner, K., D'Agnese, F. A., and McKnight, D. 1997. 'Regional delineation of North America for the assessment of freshwater ecosystems and climate change', *Hydrol. Process.*, **11**, 819–824.

Lee, R. 1980. *Forest Hydrology*. Columbia University Press, New York.

Likens, G. E., Bormann, F. H., Pierce, R. S., Eaton, J. S., and Johnson, N. M. 1977. *Biogeochemistry of a Forested Ecosystem*. Springer-Verlag, New York. p. 146.

Lynch, J. A. and Corbett, E. S. 1989. 'Hydrologic control of sulfate mobility in a forested watershed', *Wat. Resour. Res.*, **25**, 1695–1703.

MacCrimmon, H. R. and Campbell, J. S. 1969. 'World distribution of brook trout, *Salvelinus fontinalis*', *J. Fish. Res. Bd Can.*, **26**, 1699–1725.

Magnuson, J. J., Crowder, L. B., and Medvick, P. A. 1979. 'Temperature as an ecological resource', *Am. Zool.*, **19**, 331–343.

Malone, T. C., Crocker, L. H., Pike, S. E., and Wendler, B. W. 1988. 'Influence of river flow on the dynamics of phytoplankton production in a partially stratified estuary', *Mar. Ecol. Prog. Ser.*, **48**, 235–249.

Manabe, S. and Wetherald, R. T. 1987. 'Large-scale changes of soil wetness induced by an increase in atmospheric carbon dioxide', *J. Atmos. Sci.*, **44**, 1211–1235.

Meisner, J. D. 1990. 'Effect of climatic warming on the southern margin of the native range of brook trout, *Salvelinus fontinalis*', *Can. J. Fish. Aquat. Sci.*, **47**, 1065–1070.

Miller, E. W. and Miller, R. M. 1992. *Water Quality and Availability: a Reference Handbook*. ABC-CLIO, Inc. Santa Barbara, California. p. 430.

Moore, M. V. and Folt, C. L. 1993. 'Zooplankton body size and community structure: effects of thermal and toxicant stress', *Trends Ecol. Evol.*, **8**, 178–183.

Morgan, K. O., Morgan, S., and Quitno, N. 1993. *State Rankings 1993*. Morgan Quitno Corporation, Lawrence, Kansas. p. 511.

Munson, R. K. and Gherini, S. A. 1991. 'Processes influencing surface water acidification', in Charles, D. F. (Ed.), *Acid Deposition and Aquatic Ecosystems: Regional Case Studies*. Springer-Verlag, New York. pp. 9–34.

Murdoch, P. S. 1991. 'Chemical budgets and stream chemistry dynamics of a headwater stream in the Catskill Mountains of New York, October 1, 1983 through September 30, 1985', US Geological Survey Water Resources Investigation Report 88-4035. USGS, Albany, New York. p. 66.

Murdoch, P. S. and Burns, D. A. 1995. 'Effect of climate on nitrate concentrations in watersheds of the Catskill Mountains, New York', in Gvennfelt, P., Rodhe, H., Thörnelöf, E., and Wisniewski, J. (Eds), *Acid Reign '95, Proceedings, Conference on Acid Deposition, Gotenborg, Sweden, 26–30 June, 1995*. Kluwer Academic Publishers, Amsterdam. p. 357.

Murdoch, P. S. and Stoddard, J. L. 1992. 'The role of nitrate in the acidification of streams in the Catskill Mountains of New York', *Wat. Resour. Res.*, **28**, 2707–2720.

National Research Council, 1993. *Managing Wastewater in Coastal Urban Areas*. National Academy Press, Washington, D.C. p. 477.

Officer, C. B., Biggs, R. B., Taft, J. L., Cronin, L. E., Tyler, M. A., and Boynton, W. R. 1984. 'Chesapeake Bay anoxia: origins, development, and significance', *Science*, **223**, 22–27.

Oke, T. R. 1973. 'City size and the urban heat island', *Atmos. Environ.*, **7**, 769–779.

Oke, T. R. 1987. *Boundary Layer Climates*, 2nd edn. Methuen, London, p. 435.

Oke, T. R. 1993. 'Global change and urban climates', *Proc. 13th Internat. Congress Biometeorol.*, **1**, 123–131.

Ollinger, S. V., Aber, J. D., Lovett, G. M., Millham, S. E., Lathrop, R. G., and Ellis, J. M. 1993. 'A spatial model of atmospheric deposition for the forested northeastern United States', *Ecol. Appl.*, **3**, 459–472.

Papst, M. H. and Boyer, M. G. 1980. 'Effects of two organophosphorus insecticides on the chlorophyll *a* and pheopigment concentrations of standing ponds', Hydrobiologia, **69**, 245–250.

Paulsen, S. G., Larsen, D. P., Kaufman, P., Whittier, T. R., Baker, J. R., Peck, D. V., McGue, J., Hughes, R. M., McMullen, D., Stevens, D., Stoddard, J. L., Lazorchak, J. M., Kinney, W., Selle, A. R., and Hjort, R. 1991. 'EMAP-surface waters monitoring and research strategy fiscal year 1991', *EPA/600/3-91/022*. US EPA, Corvallis, Oregon. p. 184.

Pennsylvania Environmental Resources Department, 1994. *Water Quality Assessment*, 305 (b) Report. Water Quality Management Bureau, Harrisburg, Pennsylvania. p. 186.

Peterson, B. J. and Melillo, J. M. 1985. 'The potential storage of carbon caused by eutrophication of the biosphere', *Tellus*, **37B**, 117–127.

Peterson, D., Cayan, D., DiLeo, J., Noble, M., and Dettinger, M. 1995. 'The role of climate in estuarine variability', *Am. Sci.*, **83**, 58–67.

Poff, N. L. 1992. 'Regional hydrologic response to climate change: an ecological perspective', in Firth, P. and Fisher, S. G. (Eds), *Global Climate Change and Freshwater Ecosystems*. Springer-Verlag, New York. pp. 88–115.

Postma, H. 1980. 'Sediment transport and sedimentation,' in Olausson, E. and Cato, I. (Eds), *Chemistry and Biogeochemisty of Estuaries*. Wiley, Chichester. pp. 153–187.

Price, C. and Rind, D. 1994. 'The impact of a 2 × CO_2 climate on lightning-caused fires', *J. Clim.*, **7**, 1484–1494.

Rascher, C. M., Driscoll, C. T., and Peters, N. E. 1987. 'Concentration and flux of solutes from snow and forest floor during snowmelt in the west-central Adirondack region of New York', *Biogeochemistry*, **3**, 209–224.

Rasmussen, J. B., Rowan, D. J., Lean, D. R. S., and Carey, J. H. 1990. 'Food chain structure in Ontario Lakes determines PCB levels in lake trout (*Salvelinus namaycush*) and other pelagic fish', *Can. J. Fish. Aquat. Sci.*, **47**, 2030–2038.

Ringler, N. H. and Hall, J. D. 1975. 'Effects of logging on water temperature and dissolved oxygen in spawning beds', *Trans. Am. Fish. Soc.*, **1**, 111–121.

Robinson, W. F. 1988. *Mountain New England — Life Past and Present*. Little, Brown and Co., Boston. p. 206.

Schaefer, D. A., Driscoll, C. T., van Dreason, R., and C. P. Yatsko. 1990. 'The episodic acidification of Adirondack lakes during snowmelt', *Wat. Resour. Res.*, **26**, 1639–1647.

Schindler, D. W. 1989. 'Biotic impoverishment at home and abroad', *BioScience*, **39**, 426.

Schindler, D. W. and Bayley, S. E. 1993. 'The biosphere as an increasing sink for atmospheric carbon: estimates from increased nitrogen deposition', *Global Biogeochem. Cycles*, **7**, 717–733.

Schindler, D. W., Beaty, K. G., Fee, E. J., Cruikshank, D. R., DeBruyn, E. R., Findlay, D. L., Linsey, G. A., Shearer, J. A., Stainton, M. P., and Turner, M. A. 1990. 'Effects of climatic warming on lakes of the central boreal forest', *Science*, **250**, 967–970.

Schlesinger, M. E. and Zhao, Z. 1989. 'Seasonal climatic changes induced by doubled CO_2 as simulated by the OSU atmospheric GCM/Mixed-layer ocean model', *J. Clim.*, **2**, 459–495.

Schwarz, H. E. and Dillard, L. A. 1990. 'Urban water', in Waggoner, P. E. (Ed.), *Climate Change and U.S. Water Resources*. Wiley, New York. pp. 341–366.

Shuter, B. J. and Post, J. R. 1990. 'Climate, population viability, and the zoogeography of temperate fishes', *Trans. Am. Fish. Soc.*, **119**, 314–336.

Smith, D. S. 1990. 'The Adirondack Park in the 21st Century', *Technical Report No. 20*. Forest Resources of the Adirondacks.

Sorensen, J. A., Glass, G. E., Schmidt, K. W., Huber, J. K., and Rapp, G. R., Jr. 1990. 'Airborne mercury deposition and watershed characteristics in relation to mercury concentrations in water sediments plankton and fish of eighty northern Minnesota lakes, USA', *Environ. Sci. Technol.*, **24**, 1716–1727.

Spry, D. J. and Wiener, J. G. 1991. 'Metal bioavailability and toxicity to fish in low alkalinity lakes: a critical review', *Environ. Pollut.*, **71**, 243–304.

Stemberger, R. S. and Lazorchak, J. M. 1994. 'Zooplankton assemblage responses to disturbance gradients', *Can. J. Fish. Aquat. Sci.*, **51**, 2435–2447.

Strayer, D. 1987. 'Ecology and zoogeography of the freshwater mollusks of the Hudson River basin', *Malacol. Rev.*, **20**, 1–68.

Sweeney, B. W., Jackson, J. K., Newbold, J. D., and Funk, D. H. 1992. Climate change and the life histories and biogeography of aquatic insects in eastern North America', in Firth, P. and Fisher, S. G. (Eds), *Global Climate Change and Freshwater Ecosystems*. Springer-Verlag, New York. pp. 143–176.

Thornthwaite, C. W. and Mather, J. R. 1955. 'The water balance', *Publ. Climatol.*, **8**, 1–86.

Tiner, R .W. 1984. *Wetlands of the United States: Current Status and Recent Trends*. US Fish and Wildlife Service, National Wetlands Inventory, Washington, D.C. GPO.

US Congress, Office of Technology Assessment, 1993. 'Harmful non-indigenous species in the United States', *OTA-F-565*. US Governent Printing Office, Washington, D.C.

US DOA (Department of Agriculture), 1994. *Summary Report, 1992 National Resources Inventory*. U.S. Department of Agriculture, Soil Conservation Service, Washington, D.C.

US EPA (Environmental Protection Agency), 1994. 'National water quality inventory: 1992 report to Congress', *EPA 841-R-94-001* US Environmental Protection Agency, Washington, D.C.

Vannote, R. L. and Sweeney, B. W. 1980. 'Geographic analysis of thermal equilibria', *Am. Nat.*, **5**, 667–695.

Verry, E. S. 1986. 'Forest harvesting and water: the lake states experience', *Wat. Resour. Bull.*, **22**, 1039–1047.

Ward, A. K., Ward, G. M., Harlin, J. and Donahoe, R. 1990. 'Geological mediation of stream flow and sediment and solute loading to stream ecosystems due to climate change, in Firth, P. and Fisher, S. G. (Eds), *Global Climate Change and Freshwater Ecosystems*. Springer-Verlag, New York. pp. 116–142.

Whitney, G. G. 1994. *From Coastal Wilderness to Fruited Plain: A History of Environmental Change in Temperate North America, 1500 to the Present*. Cambridge University Press, Cambridge. p. 451.

Wilkie, R. W. and Tager, J. 1991. *Historical Atlas of Massachusetts*. University of Massachusetts Press, Amherst, MA.

Wilson, C. A. and Mitchell, J. F. B. 1987. 'A doubled CO_2 climate sensitivity experiment with a global climate model including a simple ocean', *J. Geophys. Res.*, **91**, 13315–13343.

Wofsy, S., Goulden, M. L., Munger, J. W., Fan, S. M., Bakwin, P. S., Daube, B. C., Bassow, S. L. and Bazzaz, F. A. 1993. 'Net exchange of CO_2 in a midlatitude forest', *Science*, **260**, 1314–1317.

Yeardley, R. B. and Lazorchak, J. M. 1994. 'Results of the 1993 EMAP-Surface waters Northeast Lakes Pilot: fish tissue contaminants', Abstract and poster, 1994. *Annual Meeting, Society of Environmental Techology and Chemistry (SETAC)*.

6

EFFECTS OF CLIMATE CHANGE ON FRESHWATER ECOSYSTEMS OF THE SOUTH-EASTERN UNITED STATES AND THE GULF COAST OF MEXICO

PATRICK J. MULHOLLAND,[1] G. RONNIE BEST,[2] CHARLES C. COUTANT,[1] GEORGE M. HORNBERGER,[3] JUDY L. MEYER,[4] PETER J. ROBINSON,[5] JOHN R. STENBERG,[6] R. EUGENE TURNER,[7] FRANCISCO VERA-HERRERA[8] AND ROBERT G. WETZEL[9]

[1]*Environmental Sciences Division, Oak Ridge National Laboratory, PO Box 2008, Oak Ridge, TN 37831-6036, USA*

[2]*Wetlands Ecology Branch, Natural Resources Division, US Geological Survey, 700 Cajundome Blvd, Lafayette, La 70506-3152, USA*

[3]*Department of Environmental Sciences, Clark Hall, University of Virginia, Charlottesville, VA 22903, USA*

[4]*Institute of Ecology, 711 Biological Science Building, University of Georgia, Athens, GA 30602, USA*

[5]*Department of Geography, CB#3220, 203 Saunders Hall, University of North Carolina, Chapel Hill, NC 25599-3220, USA*

[6]*Center for Wetlands, Phelps Lab, University of Florida, Gainesville, FL 32611-2061, USA*

[7]*Department of Marine Sciences, Louisiana State University, Baton Rouge, LA 70803, USA*

[8]*Estacion el Carmen, Ap Postal 30, Cd. del Carmen C.P. 24140, Campeche, Mexico*

[9]*Department of Biological Sciences, University of Alabama, Tuscaloosa, AL 35487-0206, USA*

ABSTRACT

The south-eastern United States and Gulf Coast of Mexico is physiographically diverse, although dominated by a broad coastal plain. Much of the region has a humid, warm temperate climate with little seasonality in precipitation but strong seasonality in runoff owing to high rates of summer evapotranspiration. The climate of southern Florida and eastern Mexico is subtropical with a distinct summer wet season and winter dry season. Regional climate models suggest that climate change resulting from a doubling of the pre-industrial levels of atmospheric CO_2 may increase annual air temperatures by 3–4°C. Changes in precipitation are highly uncertain, but the most probable scenario shows higher levels over all but the northern, interior portions of the region, with increases primarily occurring in summer and occurring as more intense or clustered storms. Despite the increases in precipitation, runoff is likely to decline over much of the region owing to increases in evapotranspiration exceeding increases in precipitation. Only in Florida and the Gulf Coast areas of the US and Mexico are precipitation increases likely to exceed evapotranspiration increases, producing an increase in runoff. However, increases in storm intensity and clustering are likely to result in more extreme hydrographs, with larger peaks in flow but lower baseflows and longer periods of drought.

The ecological effects of climate change on freshwaters of the region include: (1) a general increase in rates of primary production, organic matter decomposition and nutrient cycling as a result of higher temperatures and longer growing seasons: (2) reduction in habitat for cool water species, particularly fish and macroinvertebrates in Appalachian streams; (3) reduction in water quality and in suitable habitat in summer owing to lower baseflows and intensification of the temperature–dissolved oxygen squeeze in many rivers and reservoirs; (4) reduction in organic matter storage and loss of organisms during more intense flushing events in some streams and wetlands; (5) shorter periods of inundation of riparian wetlands and greater drying of wetland soils, particularly in northern and inland areas; (6) expansion of subtropical species northwards, including several non-native nuisance species currently confined to southern Florida; (7) expansion of wetlands in Florida and coastal Mexico, but increase in eutrophication of Florida lakes as a result of greater runoff from urban and agricultural areas; and (8) changes in the flushing rate of estuaries that would alter their salinity regimes, stratification and water quality as well as influence productivity in the Gulf of Mexico.

Many of the expected climate change effects will exacerbate current anthropogenic stresses on the region's freshwater systems, including increasing demands for water, increasing waste heat loadings and land use changes that alter the quantity and quality of runoff to streams and reservoirs. Research is needed especially in several critical areas: long-term monitoring of key hydrological, chemical and biological properties (particularly water balances in small, forested catchments and temperature-sensitive species); experimental studies of the effects of warming on organisms and ecosystem processes under realistic conditions (e.g. *in situ* heating experiments); studies of the effects of natural hydrological variation on biological communities; and assessment of the effects of water management activities on organisms and ecosystem processes, including development and testing of management and restoration strategies designed to counteract changes in climate.

INTRODUCTION

The region consisting of the south-eastern United States and the Gulf Coast of Mexico is physiographically diverse. A dominant feature is the broad coastal plain bordering the Atlantic Ocean and Gulf of Mexico from Virginia to eastern Mexico. The region also includes the south-eastern Piedmont and the southern Appalachian Mountains, the latter including areas of relatively high relief. This range in physiography provides a morphologically diverse array of freshwater ecosystems from the low gradient, black water streams and wetlands of the Coastal Plain to high gradient streams in the mountains (see Hackney *et al.*, 1992). The region was not glaciated during the Pleistocene and consequently has fewer natural lakes than regions to the north. The exception is Florida, which has numerous, generally shallow lakes, many of which are karst features. Land use in the region includes urban/residential, agriculture and forested areas. Forested lands are extensive in many portions of the region, more so than in many other regions of North America.

The climate of the region is generally humid and warm temperate, although the southern part of Florida and eastern Mexico are subtropical and areas of higher elevation in the southern Appalachians are similar in climate to more northern locales (see Leavesley *et al.*, 1997 for maps of the present climate regime of the region). Average January temperatures are generally $> 5°C$ and average July temperatures are $25°-30°C$ over all but the extreme northern and southern portions of the region. Precipitation is generally abundant across the region year round, with average annual values within the range 100–180 cm across most of the region. However, annual precipitation is considerably lower (45–75 cm) over a band along the Rio Grande River in southern Texas and north-eastern Mexico. Also, the subtropical areas of Florida and Mexico experience a distinctive seasonality in precipitation with wet summers and dry winters. Snow contributes very little to the annual precipitation in the region, although the southern Appalachians can receive considerable snowfall during some years. While seasonality in precipitation is minimal over most of the region north of Florida and east of Texas, the dominant type of precipitation event switches from large frontal systems generally from the west during winter, to convective storm activity with increasing dominance of southern and south-easterly air flow in the summer.

The region's hydrology reflects the large annual precipitation, almost all as rain, and the high rates of evapotranspiration. Annual runoff is generally <50% of annual precipitation, although in the southern Appalachians it is somewhat higher (48–75%; Swift *et al.*, 1988) than in the other areas. Although precipitation is distributed relatively evenly throughout the year over much of the region, runoff and stream flow are highly seasonal because of high rates of evapotranspiration during the growing season. While high runoff and stream flows can occur during all seasons, baseflows and storm flow peaks are higher and storm flow events more numerous during winter and spring than during summer and autumn. Consequently, wetland inundation is more extensive during the winter and spring months, particularly in the floodplains of Coastal Plain rivers and streams. However, in the subtropical region of Florida, wetland hydrology is primarily dependent on runoff produced by summer rains.

Freshwater ecosystems are abundant in the region, and include both natural and man-made systems. Streams and rivers, impoundments and reservoirs, and wetlands are the dominant freshwater ecosystems, although in Florida natural lakes are also numerous. Among the most distinctive regional features are the extensive land–water interfaces associated with streams, particularly low gradient streams, wetlands and impoundments and reservoirs, most of which are shallow with highly irregular shorelines. Thus, the region's freshwater ecosystems are generally characterized by large edge : volume ratios and the extensive influence of inputs from and exchanges with the adjoining terrestrial landscape. Food-webs are often dominated by allochthonous organic inputs from terrestrial systems or primary production of emergent and benthic communities. Changes in terrestrial ecosystems that affect the transport of organic matter or sediments into freshwaters are particularly important issues in considerations of aquatic biological communities in the region.

Estuaries are also a dominant feature, because of the region's long coastline and broad coastal plain. We consider estuaries in our regional assessment because of their regional significance and their extensive interactions with, and dependence on, up-gradient freshwater ecosystems.

The region's freshwater ecosystems are also distinctive in their high biological diversity. The high physiographical diversity, and consequently high diversity of habitats, large range in climate from subtropical to temperate and the lack of Pleistocene glaciation have resulted in a rich assemblage of freshwater species, many of which are endemic. Despite a generally warm climate, cool water species (relicts from past periods of cooler climate) are relatively common in the southern Appalachians. Introduction of exotics and displacement of native species is a serious problem in some areas, particularly in the subtropical portions of Florida.

RECENT TRENDS AND FUTURE SCENARIOS

Climate

Recent trends. A cursory examination of temperature anomalies for the south-eastern United States indicates that the warming anticipated as a result of increased CO_2 is not reflected in the record (see Karl *et al.*, 1994a). In fact, temperatures over the past 50 years have shown a decreasing trend for the south-eastern US (Plantico *et al.*, 1990; Lettenmaier *et al.*, 1994). This temperature decline is probably linked to an increase in cloudiness, which may also be related to an observed decreasing trend in the diel *range* of temperatures in the region (Karl *et al.*, 1993).

Regional precipitation records appear to indicate increasing trends, at least at some sites. Although Karl *et al.* (1994b), using records for the period 1900–1992, found no significant trends in precipitation for much of the south-eastern US, in their 'South Coastal Plain Region' there was evidence of a generally increasing trend in precipitation. Also, although precipitation records for their 'Gulf Coast Region' showed no long-term trend, the data do suggest an increasing trend for the period after about 1950. Lettenmaier *et al.* (1994) also reported a significant upwards trend in precipitation for the Gulf Coast over the past 50 years.

Some of the trends of most importance to freshwater ecosystems, which can be examined using currently available data, are those for stream flow. There is little evidence of significant trends in stream flow, except for increases in runoff for portions of the Gulf Coast (Lettenmaier *et al.*, 1994). Although the complexities of the full hydrological cycle may preclude any direct interpretation of this finding, it does appear to be consistent with the upward trend in precipitation for the area. Karl and Riebsane (1989) showed that the observed decadal and long-term changes in runoff were not correlated with corresponding trends in temperature, but were correlated with trends in precipitation.

Future scenarios. Estimates of future climate cannot be made by simple extension of recent trends, or even by analysis of the more distant past, because the changes in atmospheric composition resulting from human activities are creating unique conditions. Instead, the basic information will come from physically based mathematical simulations of climate processes, i.e. the general circulation models (GCMs). At present their resolution restricts the usable information to monthly temperature averages and precipitation totals for areas of approximately 4° latitude by 4° longitude. Thus five or six cells cover the whole of region 5. Finer spatial and temporal detail comes from combining model results with analyses of past conditions (Robinson and Finkelstein, 1991).

Several GCMs have been developed, using a variety of approaches, to predict climate changes resulting from a doubling of pre-industrial levels of atmospheric CO_2 (Gates, 1992). The models used to develop the climate scenario presented here were those used for the Intergovernmental Panel on Climate Change (IPCC) scientific assessment of 1990 (Houghton *et al.*, 1990). These models provided a range of possible future conditions from which a single generalization for our region was developed. For the southern United States and eastern Mexico all models indicate a future warming. For an effective doubling of atmospheric CO_2, temperatures throughout the region should increase by about 3–4°C, with slight seasonal differences within the various parts of the region (Figure 1a). Very recent modelling efforts, performed as an update of the IPCC assessment, suggest that the global temperature change predictions of previous assessments may be high by about a factor of two (Houghton *et al.*, 1996). Therefore, our $2 \times CO_2$ temperature scenario may be somewhat high, but probably represents a plausible scenario at some future time with somewhat larger increases in atmospheric CO_2.

Climate model results are much less clear for precipitation, but general indications are that there may be small changes in winter, but substantial increases in summer, everywhere except along the western coast of the Gulf of Mexico (Figure 1b). These models suggest that future atmospheric circulation patterns can be related to those of the present climate (Henderson and Robinson, 1994). For the extra-tropical portion of region 5 (the south-eastern US, except for peninsular Florida), the most important circulation features are the jet stream position and the location of depression tracks (mid-latitude, low pressure, storm systems). Future tracks may be further north. In summer this is likely to create a lower frequency of extended periods of gentle, widespread rain from passage of depressions, but will increase the frequency of shorter, sharper and more localized, showers from passage of fronts. The hotter summers are likely to increase thunderstorm frequency and intensity as well as to stimulate evapotranspiration. Overall, in summer, the extra-tropical regions, may experience a change towards a higher frequency of relatively short and strong rainfall events with longer dry periods between them, leading to a runoff regime with more extreme conditions than at present. This can be visualized as an increase in the temporal clustering of events. In winter, there is likely to be a similar northwards movement of the jet stream, decreasing the number of depressions and their associated precipitation. Warmer winter conditions may stimulate increased precipitation from thunderstorms, but only along the Atlantic coast is this likely to compensate for the decrease from depressions. Therefore, changes in winter precipitation may range from a slight increase along the Atlantic coast to a decline in the northern and western portions of the region (Figure 1b).

In the tropical areas of region 5, where there is commonly a summer precipitation maximum, GCMs predict only small changes in the atmospheric circulation. For much of Florida and the Yucatan Peninsula, the major change appears to be a more vigorous hydrological cycle that is likely to stimulate precipitation and evaporation in all seasons. The increased precipitation will be from more intense events rather than from a change in their number, suggesting a more peaked and extreme hydrological regime. In the somewhat drier areas inland of the western Gulf coast the general lack of atmospheric moisture will minimize this effect. Indeed, in the winter dry season the increased evaporation and drier surface feeds back into a decrease in rainfall, indicating increased periods of water stress at this time.

Neither the models nor the analyses of current climate are sufficiently sophisticated to estimate future hurricane conditions. However, there is some relationship between Atlantic Ocean water temperature and hurricane frequency which suggests that hurricanes may become more frequent in the future warmer conditions (Hess and Eisner, 1994).

Scenarios of future changes in runoff are particularly difficult to make at this time owing to uncertainty in changes in evapotranspiration (ET), as well as uncertainty in changes in precipitation. Changes in runoff are potentially very large in this region as a result of the relatively high current rates of ET and potentially large changes in ET with climate change. Increases in air temperatures and lengthening of the growing season may increase ET substantially, reducing soil moisture levels and runoff, particulary in summer. For example, at

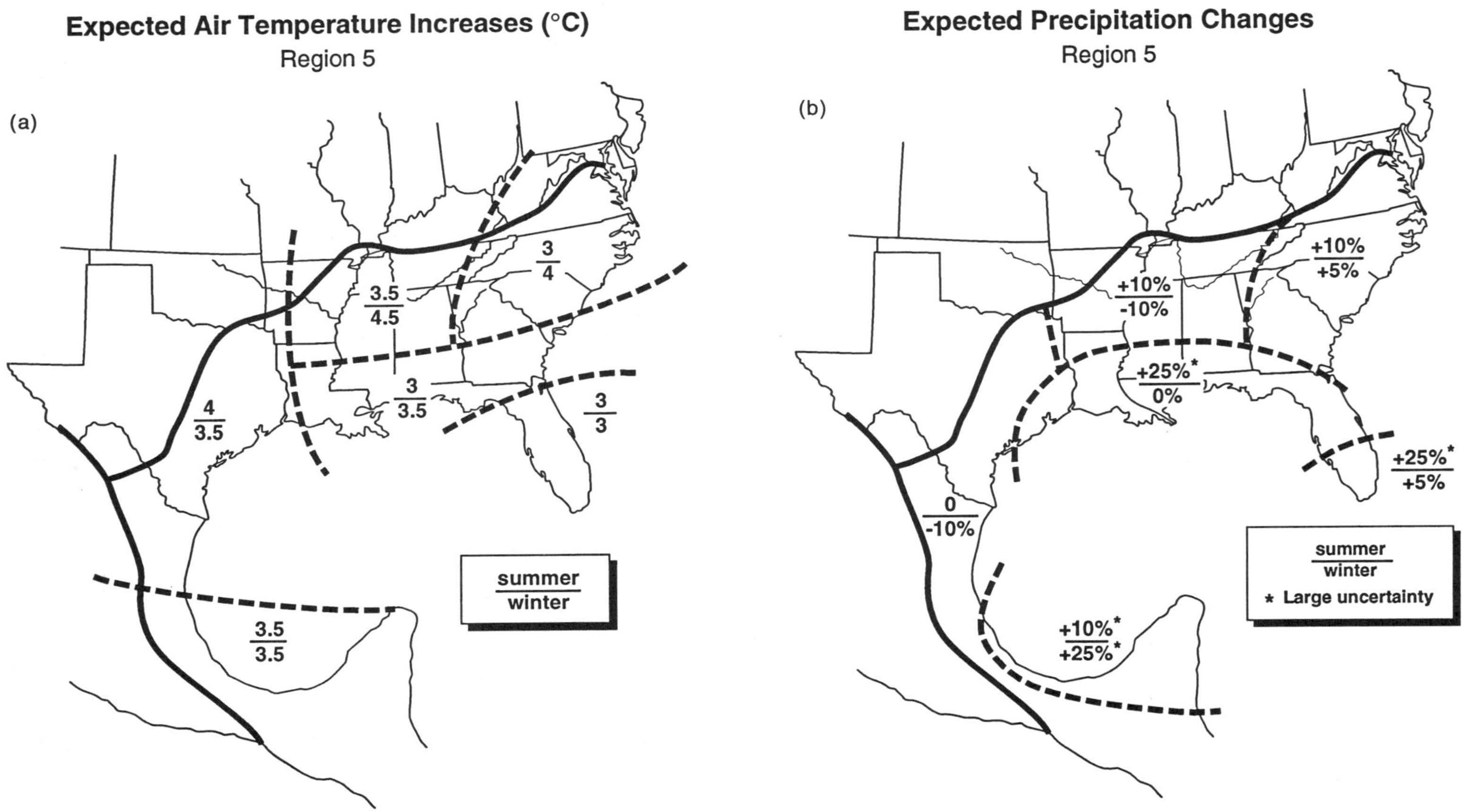

Figure 1. Current estimates of potential change in region 5 climate with a doubling of atmospheric CO_2: (a) seasonal air temperature increases ($\pm 1°C$), and (b) seasonal precipitation changes ($\pm 10\%$, unless noted by *)

the Experimental Lakes Area in north-western Ontario, a 1.5°C increase in air temperatures during the period 1970–1990 was accompanied by a 50% increase in ET (Schindler *et al.*, 1996). In a regional analysis of climate change effects, Neilson and Marks (1994) predicted that much of the south-eastern US is likely to experience increases in ET and declines in runoff under a future warmer climate, regardless of changes in precipitation. However, vegetation shifts that result in lower leaf area index or increases in plant water use efficiency owing to the direct effects of increasing CO_2 concentrations on stomatal resistance would tend to reduce ET, thus counteracting the effect of temperature. Nonetheless, in our analysis we assume that regional ET will most likely increase, resulting in a general drying of soils and generally lower runoff during the growing season, even with small to modest increases in summer precipitation. Only in areas experiencing relatively large increases in summer precipitation (25% or more) is it possible that runoff will increase. However, although average runoff rates across much of the region are likely to decline with climatic warming, the incidence of large runoff events may actually increase owing to the increase in storm intensity or clustering, particularly during the summer.

Anthropogenic Stresses

Some of the more significant trends in freshwater ecosystems in the region are not linked directly to climate, but reflect anthropogenic stresses. Between 1970 and 1980 the population of the south-eastern US increased by 21%, more than double the national rate (Healy, 1985). Growth has continued in the region. An extreme example is Florida, where the 1990 population of 12.92 million is 370% higher than the 1950 population of 2.77 million, and fully 33% higher than the 1980 population of 9.75 million (Marella, 1992)! The population growth within the region affects water and land use directly. In addition, population growth in the rest of the US and in the world, results in increased demands for agricultural products, timber and minerals from the south-east, all of which are important to the region's economy. Consequently, increasing demands for crops, wood and minerals will place increasing stresses on land and water resources in the region. Addressing these problems, many of which may be exacerbated by changing climate, will be a formidable task for policy makers in the future.

The south-eastern US has abundant water resources. Only in the far south-western parts of the region (e.g. Texas) does the consumptive use of water exceed 10% of the renewable supply (US Congress, 1993). Nevertheless, the seasonality of precipitation and evapotranspiration is not ideal for many of the water uses, and leads to considerable stresses in terms of water quantity. For example, irrigation is becoming increasingly important in the region. Of the total water withdrawals of 18 million gallons per day in Florida in 1990, the largest use was for agriculture (Marella, 1992). Water use has already led to moderate, if not critical, overdrafts of groundwater in much of the south-east (US. Congress, 1993).

The effects of anthropogenic stresses on water quality are well documented and can be expected to become more severe in the future. Erosion, which is inextricably linked to crop agriculture and to harvesting of forests, is of considerable importance throughout the south-east (e.g. see Healy, 1985). Atmospheric deposition of strong acids is very high over much of the south-east (Irving, 1991). Increases in nitrate concentrations in streams and rivers reflect increasing fertilizer use as well as increasing atmospheric deposition of nitrogen compounds resulting from increasing automobile emissions (Smith *et al.*, 1987). Mining activities have lead to numerous water quality problems, among them the acidification of surface waters (e.g. Cushman *et al.*, 1990). Intrusion of saltwater into coastal aquifers is of significant concern throughout coastal areas of the south-eastern US as pressures on the use of ground waters increase (US Congress, 1993).

Freshwaters of the south-eastern US are highly managed. Almost all major river basins have been dammed, usually with deep storage reservoirs in headwaters and shallower main stem reservoirs serving the needs of navigation downstream. Most flows are managed for hydropower and flood control, with recreation also often influencing the control of water levels. The biota of major rivers is determined largely by artificial habitats, whether in reservoirs or in tailwaters, where flows fluctuate markedly. Fisheries are highly managed to conform to available habitats. The high degree of water management means that seasonal precipitation and runoff variations are not primary factors in determining the flows in many rivers, except as they affect the total availability of water during extremes (droughts, floods). Water management introduces social considerations in evaluations of climate change, such as additional storage capacity needed to maintain present flow regimes, alterations in the prescribed seasonal flows and reservoir levels and trade-offs between competing water interests.

EFFECTS OF CLIMATE ON FRESHWATERS AND FUTURE CHANGE

Climatological Determinants of Freshwater Ecosystems

Climate variables are strong determinants of the structure and function of freshwater ecosystems. Regional and local climate, together with catchment characteristics (e.g. topography, vegetation, soils), establish the water temperature and hydrological regimes of freshwaters to which biological communities and processes are adapted. Climate also influences freshwater ecosystems indirectly through its effect on human activities that alter catchment characteristics (e.g. agriculture) or characteristics of freshwater systems themselves (e.g. channelization, water withdrawals, heat discharges). Therefore, changes in climate (e.g. air temperature, precipitation) will result in changes in the temperature and hydrological regimes, and in turn in the biology, of freshwaters that are both direct effects of climate change, as well as indirect effects resulting from changes in human activities in response to climate change (Figure 2). Feedbacks from climate-induced changes in freshwaters may also be important in further modifying local climate or human activities.

We have identified six climate characteristics that are particularly important in determining key features of freshwater ecosystems in the region, either directly or indirectly (Table I). Air temperature and radiative heating strongly influence available habitat and species ranges, rates of organic matter decomposition and nutrient cycling, and flow regimes, by means of effects on evapotranspiration. Air temperatures also influence human activities, such as energy usage (highest demands in much of the region are for space cooling in summer) and agriculture, which in turn have strong effects on the thermal regimes, hydrology and chemistry of freshwater ecosystems. The precipitation regime strongly affects the quality and quantity of runoff, wetland distribution and hydroperiod, flushing rates and incidence of anoxia, and land–water interactions in riparian environments. The precipitation regime also influences human water demands and flow management. A distinctive feature of the region is the abundance of reservoirs and the importance of

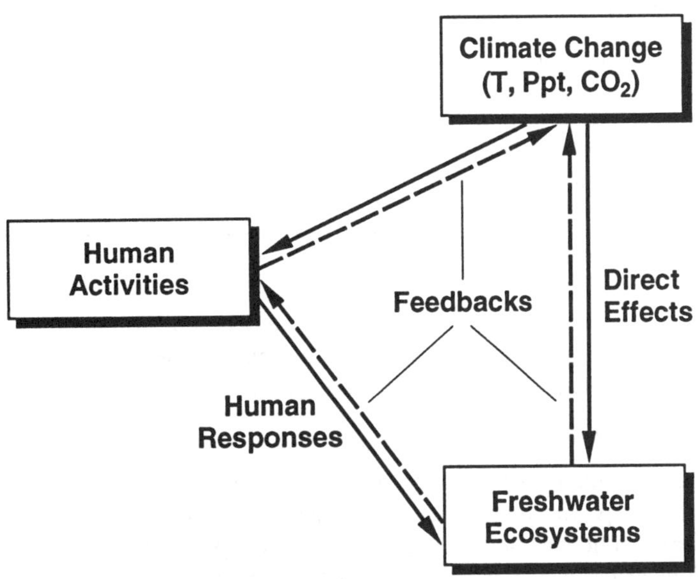

Figure 2. Effects and feedbacks in the interacting systems of climate, freshwater ecosystems and human societies

Table I. Climate characteristics that strongly influence aquatic ecosystem characteristics in region 5

Climate characteristic	Aquatic ecosystem characteristic: Direct effects	Indirect effects
1. Average annual air temperature	Habitat for cool water species (via groundwater temperature)	
	Growing season (particularly for wetland vegetation)	
	Organic matter decomposition (CO_2, CH_4 emissions)	
	Nutrient cycling (via respiration/growth balance of consumers)	
2. Summer air temp., radiative heating	Runoff (via evapotranspiration)	Electricity demand waste heat discharges increase in water temperatures
	Habitat in open water bodies (via temperature/ dissolved oxygen squeeze)	
	Increase in fish pathogens and parasites	
3. Minimum winter air temperature	Range of subtropical species (including nuisance species)	Agriculture response (e.g. expansion of citrus) eutrophication hydrological changes
4. Summer drought	Habitat (via effects on dissolved oxygen)	Water demands (flow management) lower flows
	Water quality (increase in toxics and nutrients concerntrations)	
	Florida wetlands (size, flow, fire frequency)	
5. Intensity/duration of rain events	Erosion and sedimentation stress (particularly fish, molluscs, macroinvertebrates)	Channelization, levées exacerbate hydrograph extremes
	Eutrophication (via surface runoff from urban/ agricultural lands)	
	Stream acidification episodes (southern Applachians)	
6. Winter/spring rainfall	Wetland hydroperiod and size	
	Riverine system productivity (via effect on river/ floodplain interactions)	
	Reservoir and estuary flushing (stratification, bottom anoxia)	

reservoir management in freshwater ecology. Reservoirs have been built on almost all of the larger rivers, including multiple reservoirs in series on some rivers. Flow management is an important determinant of habitat and water quality in these reservoirs as well as in the riverine environments downstream.

Scenarios of Future Change in Freshwater Ecosystems

Given the uncertainties in climate change expected for the region, particularly in the precipitation regime and hydrological responses, specific predictions of changes in freshwater ecosystems are very uncertain. However, we do have some confidence that annual average air temperatures will increase and that significant changes in the hydrological cycle will occur (discussed in section on recent trends above). Using our knowledge of the relationships between climate variables and freshwater characteristics, and assuming that temperatures will increase and precipitation regimes will change, we can develop scenarios of future change in freshwater ecosystems that are plausible, if not probable. We consider the major types of freshwaters in the region separately and limit our discussion to the changes that are thought to be most significant or most likely to alter the structure or functioning of these ecosystems.

Streams and rivers. Under a warmer climate, the habitat available to cool water species is expected to decrease and the distributions of these species will become more spatially fragmented. The projected changes in the distributions of native brook trout (*Salvelinus fontinalis*) exemplify this trend. By relating the southern limits of brook trout distribution to the 15°C groundwater temperature isotherm, Meisner (1990) projected that a 3.8°C increase in mean annual air temperature would result in elimination of this species in Georgia, South Carolina and parts of Virginia, and increased fragmentation of its distribution in Tennessee and North Carolina. In addition to pushing some cool water species beyond their upper lethal temperature limits, increases in water temperature may also increase susceptibility to pathogens and parasites.

The life history characteristics, biogeography, and productivity of aquatic insects are strongly influenced by temperature (Sweeney *et al.*, 1992; Benke, 1993). Hence, changes in stream water temperatures are also likely to alter the distribution and production of some of the aquatic insect species. With a warming of 4°C, the thermal regime of a stream will become similar to that presently characteristic of streams 640 km to the south (Sweeney *et al.*, 1992). For example, the thermal regimes of streams in Virginia, North Carolina and Tennessee would become similar to those now in southern Alabama, southern Georgia and northern Florida. Changes in voltinism, temporal and spatial patterns of abundance and productivity of aquatic insects are likely. However, shifts in species distributions might result in minimal changes in the overall abundance of aquatic insects if warm water species rapidly replaced cool water species. The ratio of production to biomass (P : B) of aquatic insects is positively related to temperature, especially in the following groups: Simuliidae, Chironomidae, Odonata, Ephemeroptera, Trichoptera and Plecoptera (Benke, 1993). These groups dominate the benthos in streams, and hence increases in P : B, as well as in total productivity, are expected under a warmer climate. Although increases in insect productivity would be expected to result in increases in fish productivity, shorter life cycles and lower average biomass may prevent fish from effectively exploiting the increased productivity.

Changes in organic matter decomposition and the detritus-based food-web in streams are expected as a result of temperature increases and perhaps the direct effects of higher atmospheric CO_2 concentrations on riparian vegetation (Meyer and Pulliam, 1992; Oswood *et al.*, 1992). Increases in temperature will accelerate microbial decomposition rates. However, the net effect of a more intense but shorter period of microbial decomposition on the productivity of detritivores is unknown. Changes in riparian vegetation resulting from drier soils in summer may have significant effects on the amount and chemistry of litter inputs to streams. In general, leaf fall mass and the nitrogen content of leaves are lower and the lignin content of leaves is higher under drier conditions than under wetter conditions (Pastor and Post, 1986; Garten *et al.*, 1994). Higher atmospheric CO_2 concentrations may also reduce leaf nitrogen content, particularly in bottomland hardwood species (Williams *et al.*, 1986; Bazzaz, 1990). These changes, which may occur under drier soil conditions, are likely to reduce the rate of litter decomposition in streams (Webster and Benfield, 1986) and, consequently, decrease food quality for invertebrate detritivores. Thus, the net effect of increases in temperature and atmospheric CO_2 on stream organic matter dynamics and detritus-based food-webs is uncertain, but potentially is substantial.

Significant changes in the biota of streams and rivers of the region are also expected if there are substantial changes in hydrological regimes. In areas near the Gulf Coast and in Florida, where annual precipitation may increase substantially, increased flooding of riparian wetlands should increase macroinvertebrate and fish production in the low gradient streams and rivers (Meyer, 1990). However, much of the region may experience lower runoff rates in summer if increases in evapotranspiration exceed increases in precipitation. In addition, the increased clustering of storms may produce longer periods of low flow conditions in most streams and rivers during the summer. This condition would be exacerbated by increased water withdrawal to meet human demands for irrigation water and public water supply in a warmer climate. Many streams in the region have very low summer flows owing to high summer evapotranspiration rates and geological characteristics that limit the size of shallow groundwater aquifers. Analyses of climate change effects have suggested that south-eastern US streams may be highly vulnerable to extended periods with no rainfall in summer, shifting from perennial to intermittent flow regimes (Poff, 1992; Ward *et al.*, 1992). Flow intermittency presents severe physiological and life history problems to stream flora and fauna, resulting in more simplified biological communities favouring species with short, multivoltine life histories (e.g. chironomids, mayflies) rather than longer life cycles (e.g., caddis-flies, molluscs). Less habitat area is available to stream fauna at low flow, and biotic interactions (e.g. predation) intensify in the remaining habitat, which can result in local elimination of some species.

Changes in climate that produce longer periods of low flow in summer also reduce stream water quality. Summer low flows are currently the periods of poorest water quality in many south-eastern rivers and streams (Meyer, 1992). Not only is there less dilution of anthropogenic additions of nutrients, organics and toxic substances, but also the warm temperatures and biomass accumulations (resulting from longer periods between episodes of scouring) increase community respiration rates resulting in lower dissolved oxygen concentrations. For example, during a summer drought, oxygen concentrations in the main channel of the Ogeechee River, a relatively unpolluted black water river in Georgia, fell to 3.4 mg/l (Meyer, 1992). Many of the black water rivers of Florida fall below the dissolved oxygen water quality criteria for recreation of 5 mg/l during low flow (Belanger *et al.*, 1983). In a recent study conducted in the Ozark region of Missouri, Smale and Rabeni (1995) reported that stream fish assemblages were more strongly affected by minimum dissolved oxygen levels than by maximum water temperatures. Low dissolved oxygen conditions would become more extensive in a warmer climate with longer periods of low flow, and the consequences for regional stream biota might be severe.

Streams and rivers of the south-eastern US have very high biotic diversity as a result of the lack of Pleistocene glaciation and a long history of drainage development going back as far as 500 million years (Isphording and Fitzpatrick, 1992). The region has the highest diversity of stream fish species in the US (McAllister *et al.*, 1986) as well as a high diversity of molluscs and aquatic insects. For example, of the 1653 species of caddis-flies in North America, 63 are endemic to the southern Appalachians (Morse, 1993). This diversity is currently at risk from exotic species invasions, excess loading of nutrients and/or toxics and sedimentation, which in the past stemmed largely from agriculture and forestry, but is now also coming from urbanization (including second home development), road construction and gravel mining. The direct and indirect effects of climate change will exacerbate these existing threats to biodiversity. A greater clustering of storms with the potential for higher flood peaks will increase erosion and sediment loading to stream channels, particularly in catchments disturbed by human activities, thus compounding an already serious problem. Increased storm clustering may also stimulate human responses, such as construction of additional reservoirs and levées, that will have important indirect effects on the biota of rivers and streams. In a region where most drainages have already been altered by reservoir construction and levées that have damaged or eliminated riparian wetlands, this could result in further loss of biodiversity.

Episodic stream acidification may also be exacerbated by climate change, particularly in the southern Appalachians and Florida where many streams are very low in acid-neutralizing capacity (ANC). In contrast to the Mid-Atlantic and north-eastern regions of the US, chronic (baseflow) acidification is uncommon in the south-east, with the exception of some sections of Florida, where natural sources of acidity may be more important (Kaufmann *et al.*, 1988). However, episodic acidification during high flow events has been documented in some streams in the Southern Blue Ridge Province (Elwood, 1991). A regional analysis based on an empirical model predicts that 2–9% of streams in the southern Appalachians, and up to 55% of the streams in the north-western and north-central areas of Florida, experience episodic acidification (Eshleman, 1988). Increased clustering of storms will most likely lead to longer and/or more severe acidic episodes in these streams. This could lead to elimination of key, acid-sensitive species, particularly trout, and many of the common macroinvertebrates in the southern Appalachians, where historical acidification has been minimal (Mulholland *et al.*, 1992). Increasing rates of atmospheric nitrogen deposition and successional maturity of vegetation may lead to increasing nitrogen saturation of forests in the region and may further contribute to increases in stream acidification (Stoddard, 1994).

Reservoirs and impoundments. The human propensity for developing water resources for hydropower, navigation, flood control and power station cooling has turned the nearly lake-free south-eastern US into a mosaic of artificial reservoirs along natural river channels. These water bodies are now dominant features in the regional hydrosystem. They support characteristic aquatic ecosystems and often intense recreational uses of the fish resources. As man-made and highly managed systems, reservoirs will experience not only the direct effects of altered climate on their physicochemical and biological processes, but also the effects of changes in human management as we attempt to use these hydrosystems to adjust to climate effects on society. To these larger water bodies are added a plethora of tiny impoundments (farm ponds and small recreational lakes) that pepper the rural (and sometimes urban) terrain. Although not natural features of the landscape, these impoundments have become an integral part of the present human environment and their alteration through climate change would also be of concern.

Reservoir morphology and placement of reservoirs on a river system play a major part in determining the vulnerability to changes in temperature and precipitation associated with climate change. Reservoirs on large river systems such as the Tennessee, Arkansas and Savannah are of two general types: (1) storage (in the headwaters, where flow rate is low relative to the water volume stored); and (2) main stem (in the main large river channels, where flow is large relative to the volume stored). Storage reservoirs most resemble natural lakes, except that they often have discharges from deep layers and widely fluctuating water elevations owing to water releases to maintain river flows downstream. Main stem reservoirs are often just slowed down rivers for much of their length, and often exist in chains in which upstream water bodies greatly affect those downstream (Kimmel *et al.*, 1988; Thornton *et al.*, 1990).

In general, we expect that storage reservoirs will be more vulnerable to climate change than main stem reservoirs because of their longer water residence times and less variable flow and temperature regimes. Storage reservoirs generally have lake-like summer thermal stratification that establishes layers of warm surface, intermediate (in temperature and depth) and cold deep water that provide distinct thermal habitats for fish and other aquatic life. Because of deep discharges (usually for hydropower), the cold water usually dominates the riverine reaches immediately downstream. Summer warming of surface waters and dissolved oxygen depletion in deeper waters forces cool water species such as trout, striped bass, sauger and walleye into a habitat 'squeeze' as areas with suitable temperatures and adequate dissolved oxygen contract (Figure 3). Increases in summer temperatures and changes in both natural and managed hydrological regimes that increase water residence times (reduce flow) will intensify the summer habitat squeeze as cool water refugia become more restricted and microbial metabolism depletes oxygen levels faster in the deeper layers. Increased habitat constriction will increase the vulnerability of cool water fishes to food depletion, overharvest and disease. Higher water temperatures and lower dissolved oxygen concentrations in dam tailwaters may reduce the riverine habitat for cool water fishes downstream from many storage reservoirs.

A secondary effect of increased summer electricity demands may be important in altering reservoir ecosystems. Much of the river–reservoir hydrodynamics in the region is now dictated by hydropower demand schedules. In addition, some reservoirs are heavily affected by power plants that use their waters for condenser cooling. For example, some reservoirs in Texas have little flow through and have temperatures and circulation patterns that are dominated by power plant cooling circuits. As climate warms in summer, more electricity will be required to supply air conditioning in the region. This demand will require more thermal discharges and release of more impounded water through turbines and will exacerbate the seasonal reservoir drawdowns that cause barren shorelines. A small increment in summer air temperature may thus translate into large changes in aquatic habitats affected by electricity generation.

Altered runoff will most likely be the main climate change effect on small impoundments, such as farm ponds. Farm ponds are an important feature in the south-eastern US, occupying approximately 0.3% of the

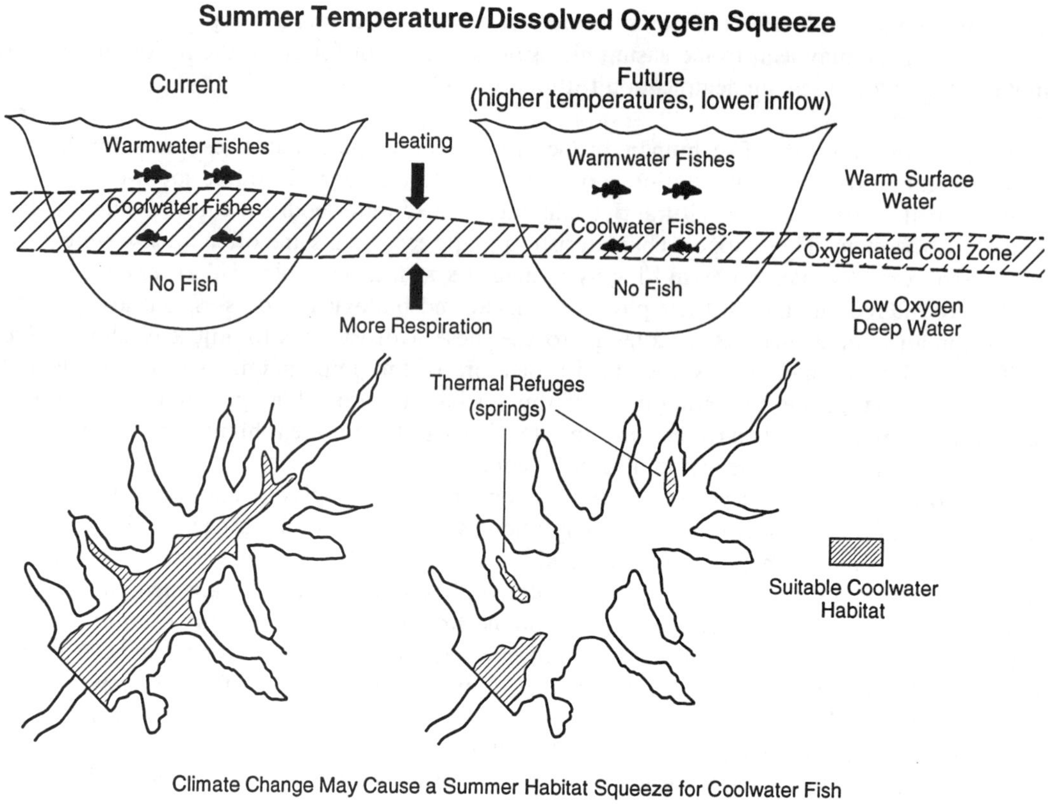

Figure 3. Potential increases in the severity of the summer temperature–dissolved oxygen squeeze with future climate change

total land area (Menzel and Cooper, 1992). These water bodies are fed primarily by surface runoff. Reductions in summer runoff resulting from higher evapotranspiration rates and longer periods with no rainfall could cause drying or severe water quality degradation, as discussed for rivers and streams.

Wetlands. Wetlands are a common feature of the landscape in the broad Coastal Plain of the region. Wetlands are important interfaces between the terrestrial and aquatic ecosystems of the region, functioning as metabolic sieves that modify nutrient and organic carbon inputs to many streams, rivers and estuaries (Wetzel, 1990). The region's wetlands are also important sources of reduced gases, such as CH_4 and H_2S, to the atmosphere. Wetlands have a large storage potential for organic carbon because of their high rates of plant photosynthesis and relatively low rates of decomposition under inundation. While the importance of hydrology in regulating the functions of wetland ecosystems has long been recognized (e.g. Mitsch and Gosselink, 1993), recent interest in the roles of wetlands in global atmospheric changes has stimulated evaluations of hydroperiod in regulating relative rates of CO_2 and CH_4 fluxes under dry versus wet conditions (e.g. Nyman and DeLaune, 1991; Happell and Chanton, 1993; Funk *et al.*, 1994). The extent and duration of wetland inundation is crucial to primary and secondary productivity, decomposition, quantity and timing of gas effluxes and the biogeochemical cycling of major nutrients and pollutants, particularly heavy metals.

Regional climate changes resulting in 3–4°C increases in mean annual temperature and generally reduced summer runoff but more intense and clustered storms, would affect wetlands in several ways. In much of the region, lower summer runoff would most likely reduce water levels and the spatial extent of inundation in riparian wetlands, although water levels might fluctuate more widely than under current conditions of more evenly distributed rainfall. However, in wetlands along the Gulf Coast, where summer precipitation increases may be large, widely fluctuating water levels could exceed the water retentive capacities of wetland organic matter and result in a greater frequency of flushing events. Under such conditions nutrients and organic matter accumulations may not be retained in the wetland long enough for efficient decomposition.

Lower baseflows and increased flow variabilities will alter plant species and competitive interactions. Among truly hydrophytic species, increased dominance of deep-rooted emergent species with well-developed aeration systems, replacement of submersed species with heterophyllous species and improved competition of deciduous species that are semi-flood tolerant would be expected. Riparian species (e.g. mesic alders) and certain conifers and hardwoods would be lost, possibly reducing shading in riparian wetlands.

Higher temperatures, particularly in winter, would extend the growing season. If water levels were maintained at reasonable levels, net plant productivity should increase markedly as a result of the extended growing season (except for those species for which photoperiod is the dominant factor controlling growing season). The productivity of some wetland species may also be increased by the direct effects of increased atmospheric CO_2 concentration. Higher plant productivity should result in increased deposition of organic matter. However, the higher temperatures and fluctuating water levels will also lead to: (1) higher rates of decomposition; (2) intermittently drained sediments with a greater percentage of aerobic decomposition; and (3) less organic accumulation and higher evasion of CO_2 to the atmosphere. The combined effect of climatic warming on plant productivity, organic matter decomposition and evapotranspiration rates will produce intensified carbon and nutrient cycles in wetlands with greater cohort turnovers (higher P : B ratios) and lower capacity for long-term nutrient and carbon storage.

Increased methane emission to the atmosphere is commonly found in wetlands with increasing temperatures, provided water levels are relatively constant and hydrosoils are saturated. Decreasing water levels, such as during summer droughts, would decrease wetland production of CH_4 but increase production of CO_2. In addition, under conditions of occasional aerobic decomposition within formerly anoxic sediments, biogeochemical processes of denitrification and sulfate reduction would decline, with concomitant reductions in N_2, N_2O and H_2S production.

Although higher trophic levels of wetland ecosystems constitute only a very small portion of energy flows and organic carbon budgets, many of the animals of wetlands are exceedingly valuable in economic and aesthetic terms. Decreased baseflows and increased spates would alter species composition, with a shift from species with long life cycles (e.g. molluscs) to those with short, multivoltine cycles. Such volatile, ephemeral hydrological conditions of wetlands would be less suitable for many economically important species (e.g. many waterfowl). The more variable conditions would lead to a greater propensity of r-selected species, and a tendency towards more rapid, unstable and short-lived community responses.

Florida lakes and wetlands. The climate of Florida is somewhat distinct from most other areas of the south-eastern US in its more seasonal rainfall regime, high proportion of wetlands (about 21% of total area), large number of natural lakes (the majority being solution features) and subtropical biota in the southern portion of the state (Myers and Ewel, 1991). Rainfall patterns exhibit a marked latitudinal effect with well-defined wet (80% of annual rainfall during May–October) and dry (20% of annual rainfall during November–April) seasons in south Florida grading into a bimodal pattern with wet summer and winter seasons and dry spring and autumn seasons (Chen and Gerber, 1991). Climate appears to be the most important factor controlling plant species distribution (Box *et al.*, 1993), with winter minimum temperatures defining the northern distribution of subtropical species and summer maximum temperatures limiting southern ranges of temperate species.

Unlike other regions of the US, Florida contains many subtropical lakes (Crisman, 1992). The transition from warm temperate lakes in the northern portions of the state to subtropical conditions in the central and southern areas is marked by a reduction in the length in the mixing period from about 4–5 months to 1–2 months, higher and less seasonally variable biomass and productivity of phytoplankton and macrophytes, and a shift in the size structure of plankton communities towards small zooplankton, protozoa and bacteria, particularly in eutrophic lakes (Beaver and Crisman, 1991; Crisman, 1992). While a majority of the 119 native fish species of Florida are temperate taxa near the southern limit of their distribution, almost all of the 28 exotic species established in recent years are subtropical or tropical (Courtenay, 1994).

Increasing winter minimum temperatures (or more probably reduction in the frequency and severity of freezing conditions) will most likely produce a northward shift in the range of subtropical species. Range shifts would be expected for several recently introduced invasive species, such as the tree *Melaleuca quinquenervia* (Cajeput) and the shrub *Schinus terebinthifolius* (Brazilian pepper), that can quickly suppress native species. In central and northern portions of the state, freshwater marshes may become dominated by *Melaleuca* and hardwood swamps by *Schinus*, as has occurred in the south. General climatic warming would also favour the spread northwards of the subtropical/tropical exotic fishes, such as *Clarias batrachus* (Walking catfish), as native species become stressed by higher water temperatures and reduced dissolved oxygen concentrations, the factors generally controlling their distributions (Loftus and Kushlan, 1987). The high connectivity of lakes, streams and wetlands in Florida (both natural and man-made surface–water connections) further enhances the potential for the rapid spread of invasive subtropical aquatic species northwards as the climate becomes warmer.

The relatively large increases in summer precipitation, proposed in our climate change scenario for Florida (Figure 1b), would most likely result in increases in wetland surface area, increased flow in streams and rivers and shorter water turnover times in lakes (Ewel and Smith, 1992). Wetlands formerly covered about 50% of Florida prior to extensive drainage and it is likely that some of that wetland area would be restored under a wetter climate. Expansion of cypress fringes and swamps adjacent to lakes is likely to increase DOC inputs and expand the number of acidic, highly coloured lakes (Crisman *et al.*, 1995b). However, if land managers respond by increasing drainage structures, then receiving waterways and downstream wetlands may become deeper with vegetation shifts away from rooted macrophytes towards floating vegetation and phytoplankton. Invasive floating species, such as *Eichhornia crassipes*, may become a greater problem (Reddy and DeBusk, 1991). Increased runoff from expanded agricultural and urban areas may lead to higher nutrient loadings and more severe eutrophication of some lakes, such as Apopka and Okeechobee, and wetlands. In coastal zones, and in the southern Everglades region, increased rainfall, coupled with sea level rise, may also increase water depth and lengthen the hydroperiod of wetlands. The extensive pine flatwoods/cypress swamp areas may become increasingly dominated by *Taxodium distichum* var. *nutans* (pond cypress) and extensive areas of *Cladium jamaicense* (sawgrass) marsh may be converted to open water (Herndon *et al.*, 1991).

Increases in primary productivity and subsequent organic matter accumulation rates resulting from a warmer climate and increases in atmospheric CO_2 might offset, at least in part, the effect of increased rainfall on wetland water depths and hydroperiods. If future precipitation increases in Florida are limited primarily to the current summer wet season, increased organic matter accumulation rates might even promote succession from systems with relatively long hydroperiods (e.g. marshes and cypress swamps) to systems with shorter hydroperiods (e.g. bayheads and wet prairies). This succession may lead to ecosystems that are more susceptible to short-term drought and fire.

Florida lakes, particularly those in the northern areas, are likely to undergo significant changes with climatic warming. Warming will shift the warm temperate lakes of north Florida to subtropical conditions, resulting in a reduction in length of mixing, an increase in productivity and nutrient cycling rates, simplification and reduction of the macrozooplankton community, and an increase in protozoa and bacteria. Management problems caused by blooms of the blue-green alga *Microcystis*, and by growths of the exotic macrophyte *Hydrilla verticillata*, primarily confined to the subtropical lakes, would expand northwards with a similar shift in subtropical conditions. Many of the native, temperate fish species will be lost and replaced with exotic, subtropical species, as has occurred in southern Florida.

Estuaries. Estuaries along the Atlantic and Gulf of Mexico coastlines of the south-eastern US and Mexico are generally broad and shallow. As a result, freshwater turnover times (estuary volume/freshwater inflow) tend to be somewhat lower than those along the north and middle Atlantic coasts and the Pacific coast of the US (Turner, 1992). Freshwater flushing rate, through its negative effect on salinity, is a critical parameter governing biological processes in estuaries. For example, the annual yield of penaeid shrimp in the Gulf of Mexico is inversely related to the annual discharge of the Mississippi River (Figure 4a), perhaps because of reduced salinities in estuarine nursery areas at higher river flows (Turner, 1992).

If the runoff along the Gulf Coast increases, then estuarine flushing rates should increase, possibly leading to reduced yields of shrimp and other species favouring higher estuarine salinities. Increased river flow may also lead to increased density stratification (freshwater at the surface and saline water near the bottom), decreasing vertical mixing and the transport of oxygen to bottom waters of estuaries. Currently, a large mass of hypoxic bottom water (up to 9500 km^2) forms in the northern Gulf of Mexico during April–October as a result of the increased density stratification and nutrient loading from high, late winter and early spring discharge of the Mississippi River (Rabalais *et al.*, 1991; Justic *et al.*, 1993). Mississippi River discharge is primarily a function of the hydrological balance in inland areas of the US, particularly the Ohio River Basin, thus climate change effects in these areas are most important in making projections of future response of

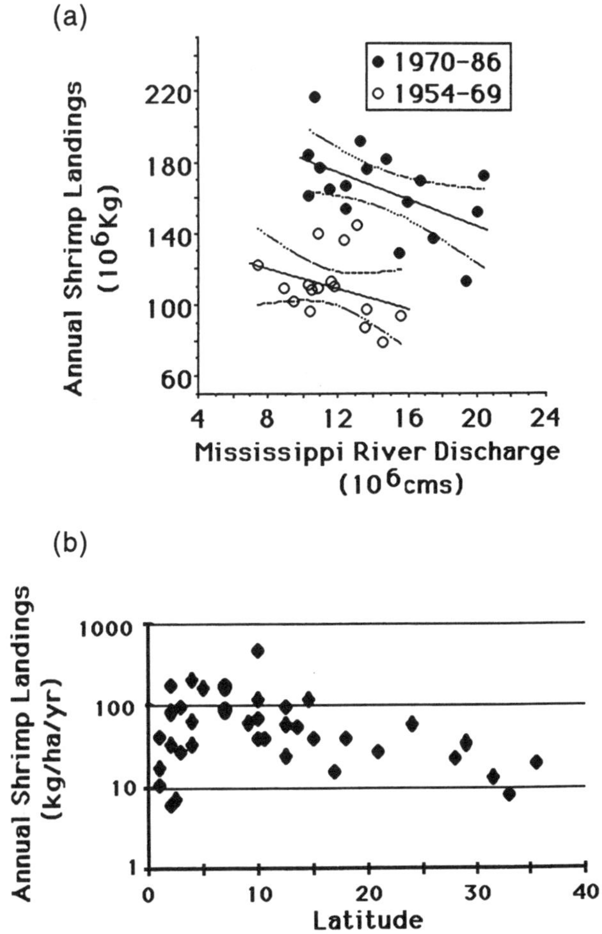

Figure 4. (a) Relationship between annual yield of shrimp in the Gulf of Mexico and annual discharge of the of the Mississippi River (95% confidence intervals are also shown). Different relationships for the two different periods are probably the result of equipment changes that enhanced yield (more nets, larger boats, electronic equipment). (b) Relationship between annual shrimp yield and latitude. Adapted from Turner (1992)

estuaries at the mouth of the Mississippi River. In one analysis, Mississippi River discharge is projected to increase with climate change, particularly during the late spring and summer (Miller and Russell, 1992). Increasing discharge would most likely increase nutrient loads and water column stratification in the northern Gulf of Mexico, thus exacerbating the already serious problems of eutrophication and hypoxia (Turner and Rabalais, 1994; Justic *et al.*, 1996).

Climate change could have positive effects on estuarine productivity as a result of a general warming. Several commercially valuable estuarine-dependent species have higher growth rates and annual harvests at lower latitudes where temperatures are higher (Copeland and Bechtel, 1974). For example, shrimp harvest is generally highest within the latitudes of 3°N to 15°N, and declines northwards (Figure 4b). Global warming may expand this region of high shrimp yield northwards, increasing shrimp harvests throughout the region, assuming that salt marsh nursery areas are not negatively affected by other factors, such as water level changes.

A key uncertainty in projections of future climate change effects on estuaries is change in sea level. Sea levels are expected to rise with global warming, but the rate of increase is highly uncertain. Rapid increases in sea level may result in coastal wetland loss, increasing open water areas and increasing estuarine depths. These morphometric changes will tend to reduce freshwater flushing rates and increase salinities, and thus may counteract the effects of increased river discharge on these characteristics. However, the loss of coastal wetlands and marshes with rapid sea level rise will most likely reduce estuarine productivity because many estuarine species are dependent on wetlands as nursery areas or sources of organic matter.

Freshwaters and estuaries of eastern Mexico. The dominant aquatic ecosystems of eastern Mexico include several major river systems (Bravo, Panuco, Papaloapan, Grijalva and Usumacinta) and a number of large reservoirs (e.g., Vicente Guerrero in the Panuco River Basin, Miguel Aleman and Cerro de Oro in the Papaloapan Basin, and the Angostura, Chicoasen, Malpaso and Penitas in the Guijalva Basin). Amongst the largest rivers of eastern Mexico, only the Usumacinta River is unimpounded. Extensive floodplain wetlands border the lower portions of several of these rivers, particularly the Grijalva and Usumacinta. The precipitation regime is highly seasonal, with the summer being the period of highest precipitation and river flow; the area is similar in this regard to southern Florida but distinct from other areas of the region.

Coastal lagoons and wetlands, particularly mangrove wetlands, are also important aquatic ecosystems of eastern Mexico. In particular, the Terminos Lagoon and its surrounding mangrove forest, located in the Usumacinta River Delta in Cempeche State, are an important reserve for many species of subtropical flora and fauna that have been eliminated elsewhere (e.g. crocodiles, manatees, jaguars, etc.). The Taiahua Lagoon, located near the mouth of the Panuco River, supports an important commercial fishery. The extensive seagrass communities in the coastal lagoons and the mangrove wetlands of eastern Mexico are extremely important for many coastal species and the local fishery, as they are in other subtropical and tropical areas (Yanez-Arancibia *et al.*, 1993).

Natural lakes in the region are generally absent, with the exception of Catemaco Lake in southern Veracruz State. However, the northern part of the Yucatan Peninsula contains a number of limestone sinks, termed cenotes, which are surface expressions of the groundwater table in areas where the ground surface has collapsed. Cenotes can be cave-like or steep-walled, range in depth from <1 m to over 50 m, have cool water temperatures, and, in many cases, are interconnected by groundwater flow (Cole, 1963). The cenotes contain a variety of aquatic biota, including a common catfish (*Rhamdia* spp.) and an eyeless fish (*Tiphliasina piercei*, Brotulidae).

The freshwaters of eastern Mexico are likely to experience different effects of climate change, depending on whether they are located inland or near the coast. Increases in air temperatures (Figure 1a) and decreases in precipitation in northern and inland areas (Figure 1b) will most likely result in reduced runoff and water quality stresses (dissolved oxygen declines, nutrient concentration increases, habitat reduction from the temperature-dissolved oxygen squeeze) in rivers and reservoirs that should be similar to those described for rivers and reservoirs of the western areas of the south-eastern US described above.

Coastal areas of eastern Mexico, including the Yucatan Peninsula, are likely to be most affected by changes in precipitation (Figure 1b), and by a rise in sea level. In the northern portion of eastern Mexico, declines in river discharge with declines in precipitation are likely to result in salinity increases in coastal lagoons fed by rivers. However, substantial increases in precipitation in coastal regions in the south may result in expansion of floodplain wetlands and a rise in the regional groundwater tables and flow rates, perhaps influencing the cenotes of the Yucatan. Effects on the freshwater wetlands of the river deltas, and the brackish coastal lagoons and mangrove wetlands, are likely to be extensive, and will depend on the interactive effects of changes in precipitation and sea level rise. This is particularly important for the wetlands of the Usumacinta Delta, the largest of the coastal wetlands and lagoons of eastern Mexico, and the least affected by human activities. The area of freshwater wetlands could be reduced by rising sea levels, probably resulting in elimination of freshwater flora (particularly the submersed and emergent freshwater macrophyte communities) and fauna (e.g. many cichlid fishes) from these areas, as well as a decline in the biomass of species requiring a freshwater habitat for a portion of their life cycles. Also, many important freshwater species (e.g. the manatee, *Trichechus manatus*) will be unable to shift their range further inland because of the more extensive human effects there. However, if precipitation and runoff increases compensate for sea level rise, then coastal freshwater wetlands may expand, benefitting freshwater species.

The seasonal variation in salinity may increase dramatically in some areas because of a combination of increases in precipitation during the wet season and sea level rise. Areas that are now freshwater or brackish zones may remain so during the wet season if increases in freshwater inputs offset sea level rise, but may change to saline zones during the dry season as high salinity water moves inland. These changes could have profound consequences for coastal species requiring a relatively narrow range of salinity for part or all of their life cycle and unable to find a suitable habitat (e.g. the shrimp, *Penaeus setiferus*, *P. duorarum* and *P. aztecus*). In contrast to the south-eastern US, the coastal plain of eastern Mexico is relatively narrow in many areas, preventing the freshwater and brackish tidal zone from simply shifting inland with sea level rise. Also, greater water depths owing to increased freshwater input during the rainy season and sea level rise may result in conversion of vegetated wetlands, including mangrove swamps, to open water areas, which would have a negative effect on species dependent on these wetlands for physical habitat or food resources provided by the vegetation. Finally, increases in freshwater inflows to coastal lagoons may increase water column stratification and lead to increased hypoxia and reduced habitat, as described for estuaries in the northern Gulf of Mexico above.

RECOMMENDATIONS FOR FUTURE RESEARCH AND MANAGEMENT

It has become clear in this exercise that most, if not all, of our considerations of changes in freshwater ecosystems resulting from expected future climate change require rather specific predictions of future climate and the rate of climate change. While regional analyses of climate change are becoming more common and improving in accuracy, there still remains considerable uncertainty in these analyses. Indeed, while the magnitude of the temperature change is unclear, even the sign of the precipitation change is uncertain. Therefore, foremost among the recommendations for future research is the development of more precise regional climate change analyses, particularly with regard to precipitation changes. The regional analyses must focus on within-year variability (e.g. changes in summer vs winter) and on changes in the intensity and temporal pattern of precipitation, rather than on annual averages alone. Significant ecological changes are at least as likely to be a result of changes in subannual climate variability as they are a result of changes in annual averages. However, even without precise scenarios of future climate, considerable improvements in our ability to understand, predict and adapt to future climate-driven changes in freshwater ecosystems of the region can be made.

Long-term Records and Monitoring Studies

Detecting changes in freshwater ecosystems as a result of climate change is likely to be difficult. Part of the difficulty stems from the natural stochastic variability in climate. In a study by the Office of Technology Assessment, it is reported that if average rainfall over the south-east increases by 10%, there is only a 7% chance of detecting that trend in 25 years (US Congress 1993, p. 75). A similar conclusion can be drawn from the work by Wolock and Hornberger (1991). The other difficulty stems from the fact that changes in freshwater ecosystems will occur as a result of very significant changes in land use and other human activities. Because the detection of trends, and the analysis of trends to determine causes is so difficult, the careful design and implementation of long-term scientific studies and the acquisition of appropriate data are critically important.

We believe that well-integrated, multidisciplinary studies on relatively undisturbed, forested catchments are critically important for developing the understanding and the data required for analysis of change in freshwater ecosystems. Among the variables that should be included in the data collection programme at selected sites are precipitation, runoff, water temperature, concentrations of a variety of chemical constituents and biological variables. The analysis of rainfall–runoff data for catchments in which the data collection programme has been done carefully will allow for a number of important analyses, among them the consideration of changes in evapotranspiration. Water temperatures in headwater streams are strongly controlled by groundwater temperatures, which in turn reflect air temperature regimes at multi-year scales. A regional network of stream temperature monitors in headwaters will provide a regional assessment of changes in air temperatures less affected by short-term variation. Regier and Meisner (1990) propose a wild salmonid watch to monitor the effects of climate change and other human disturbances on economically important species. Brook trout distributions should be monitored in the southern Appalachians as an indicator of climate change effects. Community-level indices of macroinvertebrate populations, such as the PET (Plecoptera, Ephemeroptera, Trichoptera) index or the North Carolina biotic index (Lenat, 1993), are important integrative measures of ecological status and should be monitored across a regional network of sites in order to identify climate change effects.

Correlations of the occurrence of species with certain physicochemical features of aquatic habitats provide the baseline for making predictions about future changes. Long-term records and monitoring studies should be developed explicitly so that, after several years of records, the conditions necessary to support a species can be estimated. For example, simultaneous records of fish and stream temperatures at several locations over several years would allow conclusions such as: 'species x is not found in the field at temperatures above y°C', as has been attempted by Eaton *et al.* (1995).

Although several existing programmes have as their aim the evaluation of trends in the quality of ecosystems (Leahy and Wilber, 1991; US EPA, 1992), the linked data collection that we envision for selected catchments would not be redundant. Collection of data on temperature, water quality, and biological indicators at intensively studied sites will provide the high quality database that will be essential for a critical scientific evaluation of broad regional assessments.

Finally, an expanded effort must be made to synthesize existing records of ecological and hydrological data, with particular emphasis on identifying relationships between ecological and hydrological patterns. Considerable amounts of data have been collected by state and federal agencies, as well as individual researchers and research groups. However, many of these data may exist in a form, or formats, that make comparative analyses difficult. Support should be provided to search out, compile and reformat important data sets that may provide evidence of long term trends. Support for new, innovative trend analyses of long-term data must also be provided.

Experimental Studies

Experimental studies of the effects of water temperature increases on organisms and ecosystem processes are needed. These studies should include laboratory studies to determine temperature limits of key species, as well as field experiments to determine physiological, behavioural and population responses to water temperature increases. Relationships between ecological processes (e.g. primary production, organic matter decomposition, nutrient cycling) and temperature must be determined under field conditions, or relationships determined in the laboratory must be field tested. Experimental warming studies may be feasible in some ecosystems or ecosystem enclosures (e.g. streams, small impoundments, reservoir coves, small wetlands) and such large-scale, *in situ* studies are critically needed. For example, headwater stream ecosystems in the southern Appalachians may be particularly vulnerable to climatic warming owing to their relatively low summer temperatures, and whole-stream warming experiments are feasible and should be undertaken.

Experiments to determine the effects of changes in hydrology on populations or ecosystem processes are also needed but are likely to be difficult to conduct. However, studies of responses to natural hydrological events and comparative studies of ecosystems with contrasting hydrology will be useful for identifying the effects of hydrological changes. The direct effect of atmospheric CO_2 on evapotranspiration is an important uncertainty preventing more accurate predictions of runoff response to climate change, and chamber experiments with whole trees or free air circulation experiments (FACE) are needed to determine CO_2–transpiration relationships. Development of population or ecosystem models that include hydrological variables more explicitly is needed for more accurate predictions of the effects of hydrological changes.

Management of the hydrology of freshwaters in the region will become more intensive as human demands on water resources increase (e.g. drinking water supplies, waste heat discharges, waste water loadings, catchment land use changes). Significant changes in the amount or variability of precipitation will further intensify water management, and will require new strategies and control structures. Studies that identify and quantify the responses of populations or ecosystem processes to specific water management activities or strategies, particularly in reservoirs and the freshwater ecosystems downstream, must be initiated.

Management Strategies to Ameliorate Climate Change Effects

Reduction of current and future anthropogenic stresses that will exacerbate the effects of expected climate change can be of major benefit in reducing climate change effects on freshwaters. For example, the loss of riparian shading has had a large and well-defined effect on stream temperatures. Restoration of riparian vegetation in key areas (e.g. salmonid streams in the southern Appalachians) can be an important strategy for ameliorating climate-induced stream warming. Similarly, thermal discharges that now raise the temperature of some waters many more times than is expected under climate change could be reduced through beneficial uses of waste heat or redirection of waste heat to the atmosphere. Finally, restoration of the morphology of stream and river channels and wetlands to conditions more closely approximating their natural state is gaining momentum in some areas of the region (e.g. Kissimmee River project in Florida). Restoration of hydrological features that dampen runoff response to precipitation events (e.g. headwater and riparian wetlands) should be a priority. Much of the historical deterioration of habitat and water quality in streams, rivers and wetlands in the region is the result of large rates of erosion accompanying forest clearing and agriculture, and more recently urbanization. The potential for erosion will increase, both directly as a result of climate change (e.g. more intense storms), and indirectly as a result of continued urbanization and expansion of warm weather agriculture. More sophisticated, ecologically based erosion control strategies, including both source and near-channel controls, must be developed.

ACKNOWLEDGEMENTS

The authors wish to thank the US Environmental Protection Agency, the US Geological Survey, the American Society of Limnology and Oceanography and the North American Benthological Society for sponsoring and organizing the symposium on effects of climate change on freshwater ecosystems in North America from which this paper originated. We also thank Tom Crisman for his contributions on the ecology of Florida lakes. Support during manuscript preparation for P. J. Mulholland and C. C. Coutant was provided by the Environmental Sciences Division, US Department of Energy, under contract DE-AC05-96OR22464 with Lockheed Marietta Energy Research Corp.

REFERENCES

Bazzaz, F. A. 1990. 'The response of natural ecosystems to the rising global CO_2 levels', *Annu. Rev. Ecol. Syst.*, **21**, 167–197.

Beaver, J. R. and Crisman, T. L. 1991. 'Temporal variability in algal biomass and primary productivity in Florida lakes relative to latitudinal gradients, organic color and trophic state', *Hydrobiologia*, **224**, 89–97.

Belanger, T. V., Dierberg, F. E. and Roberts, J. 1983. 'Dissolved oxygen concentrations in Florida's humic-colored waters and water quality standard implications', *Flor. Sci.*, **48**, 107–119.

Benke, A. C. 1993. 'Concepts and patterns of invertebrate production in running waters', *Verh. Int. Verein. Limnol.*, **25**, 15–38.

Box, E. O., Crumpacker, D. W., and Hardin, E. D. 1993. 'A climatic model for location of plant species in Florida, U.S.A', *Biogeogr.*, **20**, 629–644.

Chen, E. and Gerber, J. F. 1991. 'Climate', in Meyers, R. L. and Ewel, J. J. (Eds), *Ecosystems of Florida*. University of Central Florida Press, Orlando.

Cole, G. A. 1963. 'The american southwest and middle America,' in Frey, D. G. (Ed.), *Limnology in North America*. The University of Wiconsin Press, Madison. pp. 393–434.

Copeland, B. J. and Bechtel, T. J. 1974. 'Some environmental limits of six gulf-coast estuarine organisms', *Contr. Mar. Sci.*, **18**, 169–204.

Courtenay, W. R., Jr. 1994. 'Non-indigenous fishes in Florida', in Schmitz, D. C. and Brown, T. C. (Eds), *An Assessment of Invasive Non-indigenous Species in Florida's Public Lands*, Tech. Rept. TSS-94-100. Florida Department of Environmental Protection, Tallahassee. pp. 57–63.

Crisman, T. L. 1992. 'Natural lakes of the southeastern United States', in Hackney, C. T., Adams, S. M., and Martin, W. A. (Eds), *Biodiversity of the Southeastern United States: Aquatic Communities*. John Wiley & Sons, New York. pp. 475–538.

Cushman, R. M., Gough, S. B., Moran, M. S., and Craig, R. B. 1990. *Sourcebook of Hydrologic and Ecological Features: Water Resources Regions of the Conterminous United States*. Ann Arbor Science, Ann Arbor.

Eaton, J. G., McCormick, J. H., Goodno, B. E., O'Brien, D. G., Stephan, H. G., Hondzo, M., and Scheller, R. M. 1995. 'A field information-based system for estimating fish temperature tolerances', *Fisheries*, **20**, 10–18.

Elwood, J. W. 1991. 'The Southern Blue Ridge Province', in Charles, D. F. (Ed.), *Acidic Deposition and Aquatic Ecosystems*. Springer-Verlag, New York. pp. 319–364.

Eshleman, K. N. 1988. 'Predicting regional episodic acidification of surface waters using empirical models', *Wat. Resour. Res.*, **24**, 1118–1126.

Ewel, K. C. and Smith, J. E. 1992. 'Evapotranspiration from Florida pond cypress swamps', *Wat. Resour. Bull.*, **28**, 299–304.

Funk, D. W., Pullman, E. R., Peterson, R. M., Crill, P. M., and Billings, W. D. 1994. 'Influence of water table on carbon dioxide, carbon monoxide, and methane fluxes from taiga bog microcosms', *Glob. Biogeochem. Cycles*, **8**, 271–278.

Garten, C. T., Huston, M. A., and Thoms, C. A. 1994. 'Topographic variation of soil nitrogen dynamics at Walker Branch Watershed, Tennessee', *For. Sci.*, **40**, 497–512.

Gates, W. L. 1992. 'AMIP: the atmospheric model intercomparison project', *Bull. Am. Meteorol. Soc.*, **73**, 1962–1970.

Hackney, C. T., Adams, S. M., and Martin, W. H. (Eds), 1992. *Biodiversity of the Southeastern United States, Aquatic Communities*. John Wiley & Sons, New York. 779 pp.

Happell, J D. and Chanton, J. P. 1993. 'Carbon remineralization in a north Florida swamp forest: effects of water level on the pathways and rates of soil organic matter decomposition', *Glob. Biogeochem. Cycles*, **7**, 475–490.

Healy, R. G. 1985. *Competition for Land in the American South*. The Conservation Foundation, Washington, D.C.

Henderson, K. G. and Robinson, P. J. 1994. 'Relationships between the Pacific/North American teleconnection patterns and precipitation events in the south-eastern USA', *Int. J. Climatol.*, **14**, 307–323.

Herndon, A., Gunderson, L., and Stenberg, J. 1991. 'Sawgrass (*Cladium jamaicense*) survival in a regime of fire and flooding', *Wetlands*, **11**, 17–27.

Hess, J. C. and Eisner, J. B. 1994. 'Historical developments leading to current forecast models of annual Atlantic hurricane activity', *Bull. Am. Meteorol. Soc.*, **75**, 1611–1621.

Houghton, J. T., Jenkins, G. J., and Ephraums, J. J. 1990. *Climate Change: the IPCC Scientific Assessment*. Cambridge University Press, Cambridge. 365 pp.

Houghton, J. T., Callander, B. A., and Varney, S. K. 1995. *Climate Change 1995: The Science of Climate Change*. Contribution of Working Group I to the Second Assessment Report of the Intergovernmental Panel on Climate Change. Cambridge University Press, Cambridge.

Irving, P. M. (Ed.), 1991. *Acidic Deposition: State of Science and Technology*, Vol. II, *Aquatic Processes and Effects*, US National Acid Precipitation Assessment Program. US Government Printing Office, Washington D.C.

Isphording, W. C. and Fitzpatrick, J. F. 1992. 'History of drainage systems in the southeastern United States', in Hackney, C. T., Adams, S. M., and Martin, W. H.(Eds), *Biodiversity of the Southeastern United States: Aquatic Communities*. John Wiley & Sons, New York. pp. 19–56.

Justic, D., Rabalais, N. N., Turner, R. E., and Wiseman, W. J. 1993. 'Seasonal coupling between riverborne nutrients, net productivity and hypoxia', *Mar. Pollut. Bull.*, **26**, 184–189.

Justic, D., Rabalais, N. N., and Turner, R. E. 1996. 'Effects of climate change on hypoxia in coastal waters: a doubled CO_2 scenario for the Gulf of Mexico', *Limnol. Oceanogr.*, **41**, 992–1003.

Karl, T. R. and Riebsame, W. E. 1989. 'The impact of decadal fluctuations in mean precipitation and temperature on runoff: a sensitivity study over the United States', *Climatic Change*, **15**, 423–448.

Karl, T. R., Jones, P. D., Knight, R. W., Kukla, G., Plummer, N., Razuvayev, V., Gallo, K. P., Lindseay, J., Charlson, R. J., and Peterson, T. C. 1993. 'Asymmetric trends of daily maximum and minimum temperatures', *Bull. Am. Meteorol. Soc.*, **74**, 1007–1023.

Karl, T. R., Easterling, E. R., Knight, R. W., and Hughes, P. Y. 1994a. 'U.S. national and regional temperature anomalies', in Boden, T. A., Kaiser, D. P., Sepanski, R. J., and Stoss, F. W. (Eds), *Trends '93: A Compendium of Data on Global Change*, ORNL/CDIAC-65, Carbon Dioxide Information and Analysis Center. Oak Ridge National Laboratory, Oak Ridge. pp. 686–736.

Karl, T. R., Easterling, D. R., and Groisman, P. Ya. 1994b. 'United States historical climatology network — national and regional estimates of monthly and annual precipitation', in Boden, T. A., Kaiser, D. P., Sepanski, R. J., and Stoss, F. W. (Eds), *Trends '93: A Compendium of Data on Global Change*, ORNL/CDIAC-65, Carbon Dioxide Information and Analysis Center. Oak Ridge National Laboratory, Oak Ridge. pp. 830–905.

Kaufman, P. R., Herilhy, A. T., and Elwood, J. W. 1988. 'Chemical characteristics of streams in the Mid-Atlantic and Southeastern United States. Vol I: population descriptions and physico-chemical relationships', EPA/600/3-88/021a. US Environmental Protection Agency, Washington, D.C.

Kimmel, B. L., Soballe, D. M., Adams, S. M., Palumbo, A. V., Ford, C. J., and Bevelhimer, M. S. 1988. 'Inter-reservoir interactions: effects of a new reservoir on organic matter production and processing in a multiple-impoundment series', *Verh. Int. Verein. Limnol.*, **23**, 985–994.

Leahy, P. P. and Wilber, W. G. 1991. 'National water-quality assessment program', *Open-file Report 91-54*. US Geological Survey. U.S. Government Printing Office, Washington, DC.

Leavesley, G. H., Turner, K., D'Agnese, F. A., and McKnight, D. 1997. 'Regional delineation of North America for the assessment of freshwater ecosystems and climate change', *Hydrol. Process.*, **11**, this issue.

Lenat, D. R. 1993. 'A biotic index for the southeastern United States: derivation and list of tolerant values, with criteria for assigning water-quality ratings', *J. North Am. Benthol. Soc.*, **12**, 279–290.

Lettenmaier, D. P., Wood, E. F., and Wallis, J. R. 1994. 'Hydro-climatological trends in the continental United States, 1948–88', *J. Clim.*, **7**, 586–607.

Loftus, W. F. and Kushlan, J. A. 1987. 'The fresh water fishes of southern Florida', *Bull. Flor. State Mus.*, **31**, 147–344.

Marella, R. L. 1992. 'Water withdrawals, use, and trends in Florida, 1990', *US Geological Survey Water Resources Investigations Rept. 92-4140*. U.S. Government Printing Office, Washington, DC.

McAllister, D. E., Platania, S. P., Schuler, F. W., Baldwin, M. E., and Lee, D. S. 1986. 'Ichthyofaunal patterns on a geographic grid', in Hocutt, C. H. and Wiley, E. O. (Eds), *The Zoogeography of North American Freshwater Fishes*. John Wiley & Sons, New York. pp. 17–51.

Meisner, J. D. 1990. 'Effect of climatic warming on the southern margins of the native range of brook trout, *Salvelinus fontinalis*', *Can. J. Fish. Aquat. Sci.*, **47**, 1065–1070.

Menzel, R. G. and Cooper, C. M. 1992. 'Small impoundments and ponds', in Hackney, C. T., Adams, S. M., and Martin, W. H. (Eds), *Biodiversity of the Southeastern United States, Aquatic Communities*. John Wiley & Sons, New York. pp. 389–420.

Meyer, J. L. 1990. 'A blackwater perspective on riverine ecosystems', *BioScience*, **4**, 643–651.

Meyer, J. L. 1992. 'Seasonal patterns of water quality in blackwater rivers of the Coastal Plain, Southeastern United States', in Becker, C. D. and Neitzel, D. A. (Eds), *Water Quality in North American River Systems*. Batelle Press, Columbus. pp. 249–276.

Meyer, J. L. and Pulliam, W. M. 1992. 'Modifications of terrestrial-aquatic interactions by a changing climate', in Firth, P. and Fisher, S. G. (Eds), *Global Climate Change and Freshwater Ecosystems*. Springer-Verlag, New York. pp. 177–191.

Miller, J. R. and Russell, G. L. 1992. 'The impact of global warming on river runoff', *J. Geophys. Res.*, **97**, 2757–2764.

Mitsch, W. J. and Gosselink, J. G. 1993. *Wetlands*, 2nd edn. Van Nostrand Reinhold, New York. 722 pp.

Morse, J. C. 1993. 'A checklist of the Trichoptera of North America, including Greenland and Mexico', *Trans. Am. Entomol. Soc.*, **119**, 47–93.

Mulholland, P. J., Driscoll, C. T., Elwood, J. W., Osgood, M. P., Palumbo, A. V., Rosemond, A. D., Smith, M. E., and Schofield, 1992. 'Relationships between stream acidity and bacteria, macroinvertebrates, and fish: a comparison of north temperate and south temperate mountain streams, USA', *Hydrobiologia*, **239**, 7–24.

Myers, R. L. and Ewel, J. J. (Eds), 1991. *Ecosystems of Florida*. University of Central Florida Press, Orlando.

Neilson, R. P. and Marks, D. 1994. 'A global perspective of regional vegetation and hydrologic sensitivities from climatic change', *J. Vegetation Sci.*, **5**, 715–730.

Nyman, J. A. and DeLaune, R. D. 1991. 'CO_2 emission and soil Eh responses to different hydrological conditions in fresh, brackish, and saline marsh soils. *Limnol. Oceanogr.*, **36**, 1406–1414.

Oswood, M. W., Milner, A. M., and Irons, J. G., III 1992. 'Climate change and Alaskan rivers and streams', in Firth, P. and Fisher, S. G. (Eds), *Global Climate Change and Freshwater Ecosystem*. Springer-Verlag, New York. pp. 192–210.

Pastor, J. and Post, W. M. 1986. 'Influence of climate, soil moisture, and succession on forest carbon and nitrogen cycles', *Biogeochemistry*, **2**, 3–17.

Plantico, M. S., Karl, T. R., Kukla, G., and Gavin, J. 1990. 'Is recent climate change across the United States related to rising levels of anthropogenic greenhouse gases?', *Geophys. Res.*, **95**, 16617–16637.

Poff, N. L. 1992. 'Regional hydrologic response to climate change: an ecological perspective', in Firth, P. and Fisher, S. G. (Eds), *Global Climate Change and Freshwater Ecosystems*. Springer-Verlag, New York. pp. 88–115.

Rabalais, N. N., Turner, R. E., Wiseman, W. J., Jr, and Boesch, D. F. 1991. 'A brief summary of hypoxia on the northern Gulf of Mexico continental shelf: 1985–1988', *J. Geol. Soc. (London) Spec. Publ.*, **58**, 35–47.

Reddy, K. R. and DeBusk, W. F. 1991. 'Decomposition of water hyacinth detritus in eutrophic lake water', *Hydrobiologia*, **211**, 101–109.

Regier, H. A. and Meisner, J. D. 1990. 'Anticipated effects of climate change on freshwater fishes and their habitat', *Fisheries*, **15**, 10–15.

Robinson, P. J. and Finkelstein, P. L. 1991. 'The development of impact-oriented climate scenarios. *Bull. Am. Meteorol. Soc.*, **72**, 481–490.

Schindler, D. W., Bayley, S. E., and Parker, B. R. 1996. 'The effects of climatic warming on the properties of boreal lakes and streams at the Experimental Lakes Area, Northwestern Ontario', *Limnol. Oceanogr.*, **41**, 1004–1017.

Smale, M. A. and Rabeni, C. F. 1995. 'The influence of hypoxia and hyperthermia on fish species composition in headwater streams', *Trans. Am. Fish. Soc.*, **124**, 711–725.

Smith, R. A., Alexander, R. B., and Wolman, M. G. 1987. 'Water-quality trends in the nation's rivers', *Science*, **235**, 1607–1615.

Stoddard, J. L. 1994. 'Long-term changes in watershed retention of nitrogen, its causes and aquatic consequences', in Baker, L. A. (Ed.), *Environmental Chemistry of Lakes and Reservoirs*, Advances in Chemistry Series No. 237. American Chemical Society, Washington, D.C. pp. 223–284.

Sweeney, B. W., Jackson, J. K., Newbold, J. D., and Funk, D. H. 1992. 'Climate change and the life histories and biogeography of aquatic insects in eastern North America', in Firth, P. and Fisher, S. G. (Eds), *Global Climate Change and Freshwater Ecosystems*. Springer-Verlag, New York. pp. 143–176.

Swift, L. W., Jr, Cunningham, G. B., and Douglass, J. E. 1988. 'Climatology and hydrology', in Swank, W. T. and Crossley, D. A., Jr (Eds), *Forest Hydrology and Ecology at Coweeta*. Springer-Verlag, New York. pp. 35–56.

Thornton, K. W., Kimmel, B. L., and Payne, F. E. (Eds), 1990. *Reservoir Limnology: Ecological Perspectives*. John Wiley & Sons, New York.

Turner, R. E. 1992. 'Coastal wetlands and penaeid shrimp habitat', in Stroud, R. H. (ed.), *Stemming the Tide of Coastal Fish Habitat Loss. Proc., 14th Annual Marine Recreational Fisheries Symposium, Baltimore, Maryland.* National Coalition for Marine Conservation, Savannah. pp. 97–104.

Turner, R. E. and Rabalais, N. N. 1994. 'Evidence for coastal eutrophication near the Mississippi River plume', *Nature*, **368**, 619–621.

US Congress, 1993. *Preparing for an Uncertain Climate*, Volume I. Office of Technology Assessment, OTA-O-567, US Government Printing Office, Washington, D.C.

US EPA (Environmental Protection Agency), 1992. *EMAP Monitor. EPA 600/M-91-051.* U.S. Environmental Protection Agency, Washington, DC.

Ward, A. K., Ward, G. M., Harlin, J., and Donahoe, R. 1992. 'Geological mediation of stream flow and sediment and solute loading to stream ecosystems due to climate change', in Firth, P. and Fisher, S. G. (Eds), *Global Climate Change and Freshwater Ecosystems*. Springer-Verlag, New York. pp. 116–142.

Webster, J. R. and Benfield, E. F. 1986. 'Vascular plant breakdown in freshwater ecosystems', *Annu. Rev. Ecol. Syst.*, **17**, 567–594.

Wetzel, R. G. 1990. 'Land-water interfaces: metabolic and limnological regulators', *Verh. Int. Verein. Limnol.*, **24**, 6–24.

Williams, W. E., Garbutt, K., Bazzaz, F. A., and Vitousek, P. M. 1986. 'The response of plants to elevated CO_2. IV. Two deciduous-forest tree communities', *Oecologia*, **69**, 454–459.

Wolock, D. M. and Hornberger, G. M. 1991. 'Direct and indirect effects of atmospheric CO_2 levels on catchment hydrological response', *J. Forecast.*, **10**, 105–116.

Yanez-Arancibia, A., Lara-Dominguez, A. L., and Day, J. W. 1993. 'Interactions between mangrove and seagrass habitats mediated by estuarine nekton assemblages: coupling of primary and secondary production. *Hydrobiologia*, **264**, 1–12.

7

EFFECTS OF CLIMATE CHANGE ON INLAND WATERS OF THE PACIFIC COASTAL MOUNTAINS AND WESTERN GREAT BASIN OF NORTH AMERICA

JOHN M. MELACK,[1] JEFF DOZIER,[2] CHARLES R. GOLDMAN,[3] DAVID GREENLAND,[4] ALEXANDER M. MILNER[5] AND ROBERT J. NAIMAN[6]

[1]*Department of Ecology, Evolution and Marine Biology, and Institute for Computational Earth System Science, University of California, Santa Barbara, CA 93106, USA*
[2]*School of Environmental Science and Management, University of California, Santa Barbara, CA 93106, USA*
[3]*Division of Environmental Studies, University of California, Davis, CA 95616, USA*
[4]*Department of Geography, University of Oregon, Eugene, OR 97403-1251, USA*
[5]*School of Geography, University of Birmingham, Birmingham B15 2TT, UK*
[6]*Center for Streamside Studies, University of Washington, Seattle, WA 98195, USA*

ABSTRACT

The region designated as the Pacific Coastal Mountains and Western Great Basin extends from southern Alaska (64°N) to southern California (34°N) and ranges in altitude from sea level to 6200 m. Orographic effects combine with moisture-laden frontal systems originating in the Pacific Ocean to produce areas of very high precipitation on western slopes and dry basins of internal drainage on eastern flanks of the mountains. In the southern half of the region most of the runoff occurs during winter or spring, while in the northern part most occurs in summer, especially in glaciated basins. Analyses of long-term climatic and hydrological records, combined with palaeoclimatic reconstructions and simulations of future climates, are used as the basis for likely scenarios of climatic variations. The predicted hydrological response in northern California to a climate with doubled CO_2 and higher temperatures is a decrease in the amount of precipitation falling as snow, and substantially increased runoff during winter and less in late spring and summer. One consequence of the predicted earlier runoff is higher salinity in summer and autumn in San Francisco Bay. In saline lakes, the incidence of meromixis and the associated reduction in nutrient supply and algal abundance is expected to vary significantly as runoff fluctuates. In subalpine lakes, global warming will probably will lead to increased productivity. Lacustrine productivity can also be altered by changes in wind regimes, drought-enhanced forest fires and maximal or minimal snowpacks associated with atmospheric anomalies such as El Niño–Southern Oscillation (ENSO) events. Reduced stream temperature from increased contributions of glacial meltwater and decreased channel stability from changed runoff patterns and altered sediment loads has the potential to reduce the diversity of zoobenthic communities in predominately glacier-fed rivers. Climatic warming is likely to result in reduced growth and survival of sockeye salmon in freshwater, which would, in turn, increase marine mortality. Further research activities should include expanded studies at high elevations and of glacier mass balances and glacial runoff, applications of remote sensing to monitor changes, further refinement of regional climatic models to improve forecasts of future conditions and continued analyses of long-term physical, chemical and biological data to help understand responses to future climates.

INTRODUCTION

Climatic conditions affecting inland waters of the Pacific Coastal Mountains and Western Great Basin encompass an extraordinary range. Orographic effects combine with moisture-laden frontal systems originating in the Pacific Ocean to produce areas of very high precipitation on western slopes and dry basins of internal drainage on eastern flanks of the coastal and interior mountains. While seasonal temperature variation along the coast is moderated by the proximity of the ocean, low temperatures occur at high latitudes and altitudes within the region. In the southern half of the region most of the runoff occurs during the winter or early spring, while in the northern part most of the runoff occurs in the summer, especially in glaciated basins. Climatic anomalies, such as the El Niño–Southern Oscillation (ENSO), can produce large interannual differences in runoff. These widely varying climatic conditions make regional forecasts of likely climate change, induced by increased CO_2 concentrations, difficult.

Numerous small to medium-sized rivers and small, high elevation lakes occur throughout the region. Several large freshwater and saline lakes are also prominent, as are San Francisco Bay and Puget Sound. To the north, glaciated basins increase and are especially common in Alaska. Anthropogenic effects on the aquatic habitats of the region are considerable in the more populated southern portion but are smaller in much of the northern portion and at high elevations. Therefore, the potential effects of climatic changes on the inland waters are expected to differ between the aquatic habitats and to be modulated by human activities in parts of the region.

Our contribution is divided into four sections. First, we review the regional characteristics of current climatic and hydrological conditions. Secondly, we discuss the climatic variability documented in historical records and derived from palaeoclimatic evidence as well as scenarios of future climates based on general circulation models (GCMs). Thirdly, we present the possible responses of aquatic environments to the potential consequences of climatic change. Finally, we make recommendations for research and monitoring activities.

REGIONAL CHARACTERISTICS

The region designated as the Pacific Coastal Mountains and Western Great Basin spans 30 degrees of latitude from south-central Alaska (64°N) to southern California (34°N). Distances from the Pacific coast to the eastern boundary vary with latitude: 400 km in California, 200 km in Oregon and Washington, 300–400 km in British Columbia and up to 600 km in Alaska where the boundary becomes northern. Tectonic, volcanic and glacial activity has created a region of great relief characterized by mountain ranges and inland valleys. While the ranges nearest the coast are typically 500–2000 m, the mountains forming the eastern and northern boundary are 3000–4000 m with peaks reaching 5000–6000 m in northern British Columbia and Alaska, where Denali attains 6200 m (Hunt, 1974).

Within California, the Central Valley is bordered on the west by the Coast Ranges and on the east by the Sierra Nevada, which also forms the western edge of the Great Basin. The Klamath Mountains link the Cascades and the Coast Range in southern Oregon. Further north, the Cascades and the Coast Range are separated by the Willamette Valley and the Puget Sound Trough. The Olympic Mountains are a major feature at the northern end of the Coast Range in Washington. Insular and coastal Mountains extend through British Columbia to south-eastern Alaska where elevations exceed 4000 m in the heavily glaciated St Elias Mountains. The Chugach Mountains lie in south-central Alaska, where further inland are the Wrangell and Talkeetna Mountains and the Alaska Range. To the west lies the Aleutian Range (Hunt, 1974; Muhs *et al.*, 1987).

Major urban areas, including Vancouver, Seattle, Portland, San Francisco, Sacramento, Fresno and the northern edge of Los Angeles, contain the majority of people in the region. Irrigated agriculture has changed the landscape in the Central Valley of California, and a mixture of grazing and agriculture continues on the coastal plains and inland valleys northwards to southern British Columbia. The mountainous terrain throughout the region supports extractive forestry and some silviculture; large tracts are designated as wilderness areas or parks.

After a general overview of the climatic and hydrological features of the whole region, we will describe several subregions in more detail. Regional temperature and rainfall patterns for the Pacific coastal region described below, are evident in Plates 2–4 in Leavesley *et al.*, (1997). Mean July temperatures are 15–20°C along the California coast and throughout most of Oregon, Washington and southern British Columbia. Within the southern interior of California, July temperatures range from 20–30°C, while inland in British Columbia and throughout southern Alaska mean July temperatures are 10–15°C. Mean January temperatures in California and coastal Oregon are 5–10°C, and those in Washington and southern British Columbia and inland in northern California and Oregon are 0–5°C. Along the coast of central and northern British Columbia and southern Alaska mean January temperatures are 0 to −5°C, while interior temperatures in northern British Columbia and south-central Alaska range from −5 to −20°C.

Average annual precipitation in southern California and on the western edge of the Great Basin ranges from 150 to 600 mm, while in the Sierra Nevada and northern California it is 600–1800 mm. With the exception of the Olympic Mountains, which receive 2400–4000 mm, annual precipitation in western Oregon and western Washington is 1000–3000 mm. Annual precipitation in British Columbia and southern and south-central Alaska ranges up to 5000 mm.

The quantity, variability and timing of runoff, which may be altered if climate changes, influence many ecological processes (Poff and Ward, 1989; Naiman *et al.*, 1992; Statzner *et al.*, 1992). When compared with streams of the rest of the coterminous United States, three patterns of stream flow variability dominate in California, Oregon and Washington: intermittent and flashy, winter rain generated and snowmelt generated (Poff and Ward, 1989). Strong latitudinal patterns in stream flow and water temperatures are evident along the Pacific coast (Figures 1 and 2). Mean annual runoff and its variability increase with latitude. The proportion of annual discharge occurring in winter decreases from California to Alaska, and, conversely, the proportion in summer increases with latitude. Mean annual water temperature and the range of temperatures decreases with latitude. The temperature extremes reached in the southern and northern portions of the region are ecologically significant. Water temperatures near or above 25°C in Californian streams are physiologically limiting to many aquatic organisms adapted to cool conditions (Bjornn and Reiser, 1991). With winter temperatures at 0°C above a latitude of 48°N, surface, anchor and frazil ice may form and influence the survival of aquatic organisms (Oswood *et al.*, 1991).

The *Sierra Nevada*, 700 km in length, is the longest continuous mountain range in the coterminus United States. The climate is characterized by warm, dry summers and cool, wet winters with strong altitudinal variation (Miller, 1955). Eleven major rivers flow west from the Sierra Nevada; the eastern side of the range is drained by many streams, all of which terminate in the Great Basin, except those diverted to the city of Los Angeles.

The Sierra Nevada generates about 25 km^3 of runoff annually, and supplies most of the water used by California's cities, agriculture, industry and hydroelectric facilities (Kahrl, 1978). Stream flow in the Sierra Nevada is generated by seasonal rainfall and snowmelt, and is modified by impoundments and diversion schemes. Basin-wide means of 1 April water equivalence for snow courses above 2500 m indicate that peak snowpack water equivalence for the high Sierra Nevada averages 750–850 mm, decreases from north to south and is lower on the east side of the crest than on the west side (California Cooperative Snow Survey, 1986). Snow courses between 1800 and 2500 m have an average peak water equivalence of about 600 mm.

Snowmelt runoff becomes more important, and midwinter rainfall-generated runoff becomes less important with increasing elevation (Cayan and Riddle, 1993). For example, in the American River basin, less than half of the annual runoff occurs from April to July in the lower two-thirds of the basin, while in catchments adjoining the Sierra Nevada crest, more than two-thirds of the annual runoff occurs during this period (Elliott *et al.*, 1978). Overall, about half of the precipitation in the major river basins of the western slope of the Sierra Nevada becomes stream flow (Kattelmann *et al.*, 1983).

Stream flow, both in absolute magnitude and as a proportion of precipitation, increases with elevation. In the American River Basin, stream flow data from 25 sub-basins (Armstrong and Stidd, 1967) indicate an increase in stream flow of about 30 mm per 100 m gain in elevation. Runoff efficiency increases from about 30% in the foothills to more than 80% near the crest (Elliott *et al.*, 1978).

The flow in Sierra Nevada rivers is highly variable in time, both within and between years. Annual volumes can be 20 times greater in very wet years than in very dry years. Peak flows in the Sierra Nevada

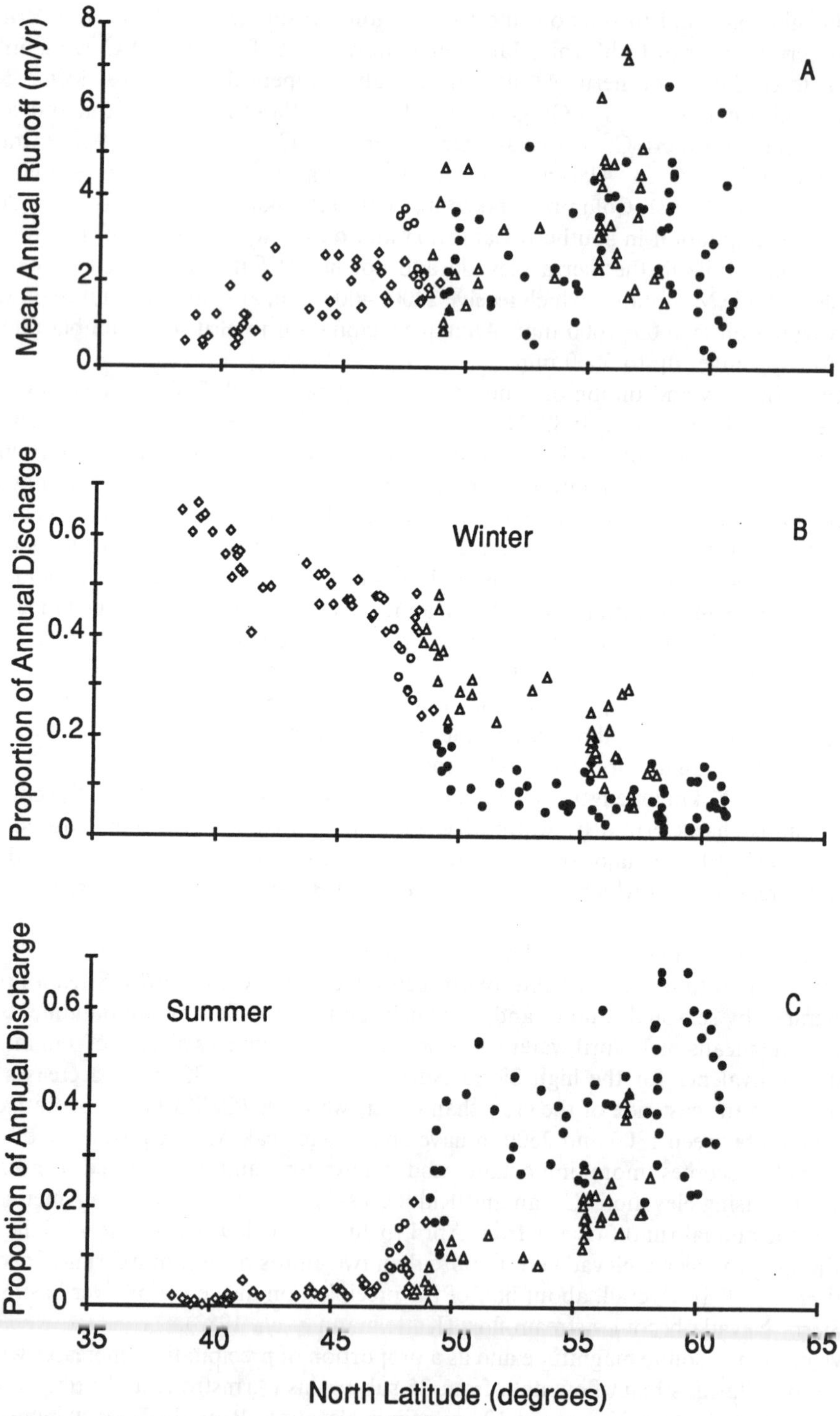

Figure 1. (A) Mean annual runoff as a function of latitude for 151 rivers of the Pacific coast from central California to southern Alaska divided into four subregions: (◇) southern coastal mountains, (○) Olympic mountains, (△) northern lowlands and islands, and (●) northern mainland mountains. (B) Proportion of mean annual discharge occurring during three winter months. (C) Proportion of mean annual discharge occurring during three summer months. (From Naiman and Anderson, 1997)

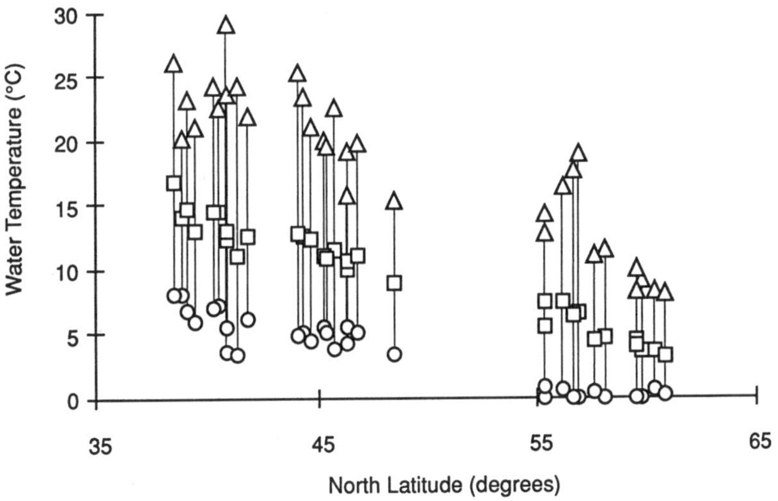

Figure 2. Stream water temperature as a function of latitude. (From Naiman and Anderson, 1997)

result from snowmelt, warm winter storms and summer and early autumn convective storms (Kattelmann, 1990). In rivers with headwaters at high elevation, snowmelt floods occur each spring as periods of sustained high flow, long duration and large volume. However, they rarely produce the highest instantaneous peaks. Large snowmelt floods occurred in 1906, 1938, 1952, 1969 and 1983. Midwinter rainfall on snow cover has produced all the highest flows in major Sierra Nevada rivers during this century (Kattelmann *et al.*, 1991). When subtropical air masses move into the Sierra Nevada in summer and early autumn, sufficient moisture is available to generate extreme rainfall. These storms may generate the greatest floods in some alpine basins that are high enough to avoid midwinter rain on snow events.

Stream flow in Sierra Nevada rivers may be low during intense or extended droughts. For example, during. 1977 when average snow water equivalence in early April was only 25% of the long-term mean, stream flow as a proportion of average annual flow ranged from 0.08 to 0.26. Dry periods may last for several years. From 1928 to 1937, runoff was below average in each year. The past two decades have included record droughts for one year (1977), two years (1976–1977), three years (1990–1992) and six years (1987–1992).

The *Cascade Mountains* extend 1100 km south to north and link the Sierra Nevada in northern California to the Coast Mountains in southern British Columbia (Nelson, 1991). The Cascades separate a mild, maritime climate with high precipitation and dense forests to the west from the colder, high elevation deserts of the Columbia–Snake River Plateau and Great Basin to the east. Mean annual temperatures in the Cascades range from 4 to 10°C, and average annual precipitation along the crest ranges from 2500 mm/yr in the north to about 1200 mm/yr in the south (Hunt, 1974). About 80% of annual precipitation (above 600 m) falls as snow from October to April, and runoff peaks during snowmelt in May and June (Rasmussen and Tangborn, 1976).

Along the *Pacific coast of Canada* two similar climatic regions occur in insular and coast mountains (Slaymaker, 1990). The mountains of Vancouver and Queen Charlotte islands experience mean annual precipitation above 4000 mm, almost all as rain, and regional runoff of about 3500 mm, 64% of which occurs in winter. At sea level, the mean annual temperature is 8.5°C. In the coast mountains, mean annual precipitation (3500 mm), runoff (3150 mm, 30% of which occurs in winter) and temperature (5–8°C) are only slightly less than on the islands. Permafrost is sporadic and only at the highest elevations.

In *south-central Alaska*, average annual rainfall varies from 480 mm in the Kenai area to 310 mm in more northerly areas away from coastal influences. Further south along the coast, annual rainfall increases, reaching annual peaks of 5000 mm in south-east Alaska. Permafrost is restricted to south-central Alaska in areas away from coastal influence and is discontinuous, being mainly confined to north-facing slopes. In south-central Alaska, winter flows in non-glacial streams are low (<5% of total runoff), and the principal peak is during spring snowmelt, often with a second lower peak in the autumn. Further from the maritime influence, the autumn peak is typically smaller and streams do not return to baseflow in the summer because of the prolonged snowmelt season. Non-glacial rivers in south-east Alaska have significant winter flow (10–15% of total runoff) and two peaks, one during spring snowmelt and a second larger peak during the autumn rainy season (Milner *et al.*, 1997).

Glaciated basins are important along the northern Pacific coast of North America. In Alaska, estimates of glacial area vary from 65 800 km^2 (Brown, 1989) to 74 700 km^2 (Post and Meier, 1980). The majority of glaciers in the coterminous US are located in the Olympic Mountains (266 glaciers) and the North Cascade Mountains (742 glaciers) of Washington (total area 420 km^2) and the Sierra Nevada (497 glaciers) of California (total area 55 km^2).

Runoff from glaciated basins arises from both glaciers and ice-free areas (Collins, 1985), and glacial runoff continues throughout the summer months, usually reaching a peak in July (Milner and Petts, 1994). Glacial runoff in Alaska is considerable with an estimated runoff rate of 220 km^3 annually, equivalent to 35% of the total runoff in Alaska, 14% of the total runoff from the coterminous states and 10% of the total runoff in the US (Mayo, 1986).

CLIMATIC VARIABILITY AND CHANGE

To evaluate plausible scenarios for changes in climate and its variability in the next century, as concentrations of greenhouse gases increase, several complementary approaches, each with strengths and weaknesses, are useful. Examination of historical records permits statistical analysis of relationships between atmospheric circulation, temperature, precipitation and runoff (Cayan and Peterson, 1989; Redmond and Koch, 1991). Moreover, trends and anomalies observed over prior decadal periods are similar to those expected as climate changes in the next century (Karl and Riebsame, 1989). However, the factors causing the observed climatic variations could differ from those related to the radiative effects of greenhouse gases. Analyses of climatic changes during the geological past indicate climatic sensitivities to radiative forcing and oceanic circulation (Broecker and Denton, 1989), but are not exact analogies to the rate and nature of current increases in greenhouse gases (Schneider *et al.*, 1990). Large-scale models of climate, i.e. GCMs, attempt to represent mathematically the physical interactions between the atmosphere, oceans and land. Because of computational limitations and imprecise parameterizations of hydrological and meteorological processes, however, GCMs do not include important complexities. Their ability to simulate regional climates and their estimates of climatic variability at the regional level, especially, are in need of improvement.

Historical records and palaeoclimatic evidence

Instrumental climatological records for the Pacific coast of North America date back to the 1820s (Roden, 1989). The longest quasi-continuous data are from Sitka, Alaska, where Russian measurements began in 1828. Many meteorological records are available for the Pacific coast since the mid-1800s, and these data have received considerable statistical analysis (e.g. Roden, 1966; Granger, 1977a; McGuirk, 1982; Roden, 1989; Goodridge, 1991a, b).

Climatic variability at a site results from a combination of local and distant processes, the relative importance of which can vary in space and time. Hence, large-scale physical forcing need not always produce spatially coherent effects. For example, while temperature at exposed coastal sites is usually coherent over long distances, precipitation and winds in mountainous terrain are not (Roden, 1989). Therefore, when evaluating climatological trends, variations or patterns statistically or mechanistically, the spatial and temporal coherence of the data must be carefully considered.

Climatic variability along the Pacific coast of North America on an interannual scale is often associated with mid and high latitude variations in the polar jet stream, which are greatest during the winter

(Roden, 1989). One to three times a decade, low latitude disturbances, i.e. ENSO events, link to mid-latitude circulation and cause climatic anomalies such as especially warm and rainy periods. In contrast, when the Aluetian low pressure is well developed with low central pressure values, below average precipitation often occurs in the Pacific Northwest because storm systems are diverted northwards.

Variations in runoff can result from changes in temperature and precipitation. Since about 1950 in the Sierra Nevada, the proportion of total annual stream flow occurring during April–July has decreased, while the portion during the autumn and winter has increased (Aguado *et al.*, 1992). Although climatic warming may be implicated, concomitant increases in autumn precipitation and decreases in precipitation in May–July were also recorded. Moreover, based on 50 years of historical records for two streams in the northern Sierra Nevada, Pupacko (1993) reported increased winter and early spring runoff during the period 1965–1990 compared with 1939–1964, which he attributed to small increases in temperature that enhanced the rain to snow ratio.

In rivers of the eastern slope of the Sierra Nevada, events at both extremes have been evident in recent years (Kattelmann, 1992). Five of the largest snowmelt floods since the 1920s (in terms of volume) occurred from 1978 to 1986. Five of the smallest snowmelt floods, occurred from 1987 to 1991. These events support the theories of some climatologists that extreme events are becoming more common in the western United States (Granger, 1977b; Michaelsen *et al.*, 1987).

Furthermore, studies of past climate based on widths of tree rings have detected multiple wet and dry periods and warm and cool intervals in the Coastal Mountains and Sierra Nevada of California and in the Pacific Northwest that have persisted from years to decades to more than two centuries (Graumlich and Brubaker, 1986; Graumlich, 1987, 1993; Michaelsen *et al.*, 1987; Haston and Michaelsen, 1994; Stine, 1994). The presence of tree stumps well below modern lake levels provides strong evidence for very arid conditions in the past (Lindström, 1990; Stine, 1994). Conversely, the period 1937–1986 was an anomalously wet period in a 1000-year reconstruction of precipitation from dendrochronological evidence (Graumlich, 1993). Furthermore, a warm period from *ca.* 1100 to 1375 AD in California, corresponding to widespread warmth in the north temperate latitudes when pre-industrial, low CO_2 concentrations existed, indicates the need to consider climatic forcing other than increased concentrations of greenhouse gases. Tree ring studies undertaken on the central coast of California also indicate major fluctuations in the variability of precipitation over the last 600 years (Haston and Michaelsen, 1994).

Another rich source of information about climate during the geological past are the sediments and exposed shorelines of lakes (Street and Grove, 1979; Benson and Thompson, 1987). An exceptionally long palaeolimnological record of Tule Lake in northern California documents the last three million years of the lake's history (Bradbury, 1992). Besides indicating warmer and wetter conditions beginning about 15 000 years ago, the inference of a warm, monomictic lake experiencing a climate with low seasonality during the Pliocene may be analogous to the projected period of climate warming in the next century. A detailed analysis of deltaic deposits in combination with other sedimentary, geomorphological, tephrostratigraphic and historical evidence at Mono Lake, California, documented lake level fluctuations of 40 m over the past 3800 years (Stine, 1990). Throughout most of the last 2000 years the level of Mono Lake has been above its natural historical levels, and, perhaps more importantly, on at least two occasions, has experienced climatic conditions as dry as those during the severe drought between 1928 and 1934.

Large-scale atmospheric patterns have been linked statistically to variability in precipitation and stream flow along the Pacific coast of North America (Cayan and Roads, 1984, Cayan and Peterson, 1989; Schonher and Nicholson, 1989; Leathers *et al.*, 1991; Redmond and Koch, 1991; Cayan and Riddle, 1993; Dettinger and Cayan, 1995). Oceanic and atmospheric interconnections relevant to the synoptic meteorology of the Pacific coast have been examined with principal component analysis and teleconnective studies that include the Southern Oscillation (SOI), Pacific–North America (PNA) and Central North Pacific (CNP) indices (Cayan and Peterson, 1989; Leathers and Palecki, 1992; Figure 3). Strongly positive and negative PNA indices are associated with warm event (low SOI values) and cold event (high SOI values) ENSOs, and with precipitation and temperature anomalies on the west coast of North American

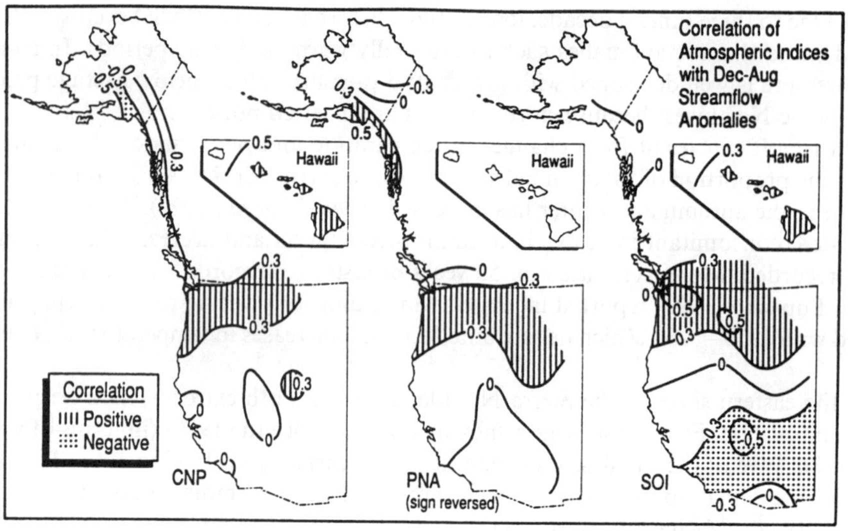

Figure 3. Correlation between December–August stream flow anomalies and winter CNP (left), PNA middle) and SOI (right). The sign on PNA is reversed to aid comparison with other indices. (From Cayan and Peterson, 1989)

(Yarnal and Diaz, 1986). During periods with a weak CNP index, stream flows are high in Washington and Oregon, and, conversely, stream flows are often low during times with a strong CNP index (Cayan and Peterson, 1989). The strength of the teleconnective patterns can vary. For example, the Pacific Northwest may be inside or on the southern edge of a region with lower rainfall during and after an ENSO event (Ropelewski and Halpert, 1986).

Persistent climatic anomalies during the Holocene, such as the 'Little Ice Age ' and 'Medieval Warm Period' may be explained by changes in the frequency of anomalous circulation patterns. For example, the, existence of shallow lakes for several decades *ca.* 3600 and 400 BP in the Mojave Desert may have resulted from a weakening of the subtropical high in the eastern Pacific, which led to an anomalously low pressure system and heavy precipitation along the Pacific coast of North America (Enzel *et al.*, 1989). Potentially, analogous changes in large-scale atmospheric circulation could occur in the future, but are difficult to predict with extant models.

GCMs

Results from GCMs suggest that global warming, induced by a doubling of atmospheric CO_2 concentrations, would range from 0 to 10°C along the Pacific coast of North America, depending on the version of GCM used and the season or latitude considered (Schlesinger and Mitchell, 1987). Projections of changes in precipitation along the Pacific coast range from, slight decreases during the summer in southern portions to no change, or to increases elsewhere (Schlesinger and Mitchell, 1987).

Outputs from five GCM simulations of climatic conditions in coastal Canada (50–60°N) under a doubled atmospheric concentration of CO_2 by 2050 all indicate warming, and most suggest increased precipitation (Slaymaker, 1990). The warmest and driest scenario predicted winter temperatures of +6°C and summer temperatures of +4.2°C with no change in precipitation. The coolest and wettest scenario calculated winter temperatures of +2.4°C and summer temperatures of +0.6°C with winter precipitation of +200 mm and summer precipitation of +100 mm.

Recent developments, which have nested limited-area models (LAMs) within GCMs, are especially useful for the Pacific coast because of the importance of complex topography and oceanic influences (Giorgi and Mearns, 1991). In a promising effort, Giorgi *et al.* (1993) have tested a LAM-GCM with 60 km resolution over the western United States for periods that included the 1982–1983 El Niño and a drought. Seasonally averaged biases for daily average temperatures ranged from +0.81 to −0.61°C for California and the Pacific Northwest. The seasonal precipitation biases were less than 20% of observations for all seasons in the Pacific Northwest, and, over California, precipitation was underestimated by 10–30% except in summer, when it was overestimated by 50%. Snow depth was most accurately predicted for the Pacific Northwest. Further effort with a LAM nested in a GCM has projected climatic changes for the US after CO_2 doubling as another step in testing the methodology (Giorgi *et al.*, 1994). Although the simulations forecast scenarios such as an increase in winter precipitation along the Pacific coast, Giorgi *et al.* (1994) caution against using such scenarios in assessments of the effects of potential climate change.

EFFECTS OF CLIMATE CHANGE ON AQUATIC ECOSYSTEMS

Potential responses of aquatic ecosystems to climatic change have been considered in general terms in several recent reviews (e.g. Carpenter *et al.*, 1992; Firth and Fisher, 1992). Our approach is to emphasize results specifically pertinent to the Pacific Coastal Mountains and Western Great Basin. We have included results from mechanistic and empirical models, analyses of long-term data sets and discussion of qualitative hypotheses. We consider rivers and estuaries, saline and montane lakes, glaciated basins and riparian zones and salmon ecology.

Rivers and estuaries

In principle, hydrological models of runoff could be coupled with climatic information generated from GCMs to evaluate the potential effects of climate change on river discharge. In practice, incompatibilities of space and, to some extent, time between the hydrological and atmospheric models complicate such a coupling (Miller and Russell, 1992; Hostetler, 1994). Moreover, some parameterizations and step jumps of CO_2 levels used in the models are simplifications. Hence, hydrological results derived from the output of GCMs should not be judged as predictions. With these caveats in mind, Lettenmaier and Gan (1990) examined the hydrological sensitivities of four medium-sized montane catchments in the Sacramento and San Joaquin river basins to scenarios for long-term warming derived from three GCMs.

The four catchments [McCloud, Thomes, American (North Fork) and Merced, Figure 4] considered by Lettenmaier and Gan (1990) currently experience a mediterranean climate with almost all the precipitation occurring between November and April and with snow predominating at the higher elevations. Runoff is derived mainly from snowmelt in spring and summer. Hydrological conditions were simulated by coupling the snowmelt and soil moisture models of the US National Weather Service River Forecasting System; calibration and verification of the coupled models were performed for different periods using historical data. Steady-state climate scenarios after a CO_2 doubling were obtained from three GCMs [Geophysical Fluid Dynamics Laboratory (GFDL), Goddard Institute of Space Science (GISS) and Oregon State University (OSU)], and simulated changes in temperature and precipitation were used to drive the hydrological models. Other climatic variables such as relative humidity and wind were held constant because of the difficulty in obtaining consistent and reliable values from the GCMs. Moreover, interactions between hydrological responses and potential changes in vegetation of the catchments caused by climatic changes were not considered.

All three GCMs simulated increases in temperatures that averaged between 2 and 5°C. Although increases and decreases in precipitation were forecast by the GCMs, both the GFDL and GISS models generally simulated augmented precipitation in the autumn and winter. The primary hydrological effect of the climate having higher temperatures after doubled CO_2 was a decrease in the amount of precipitation falling as snow and less snow available to generate runoff in spring and summer. In fact, the temperature increases simulated by the GFDL and GISS models were of sufficient magnitude to eliminate almost completely water storage as snow in all the catchments except the Merced, which has the highest average elevation. As a consequence, runoff increased substantially during winter and decreased in late spring and summer (Figure 5). Site-specific

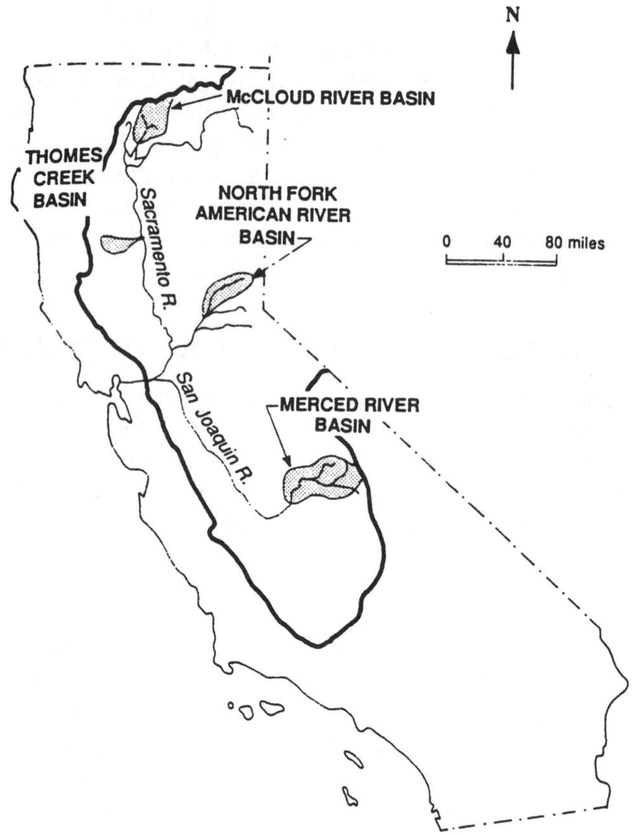

Figure 4. Catchments included in study by Lettenmaier and Gan (1990)

differences stem from differences in elevation and the resulting proportions of rain and snow. Further consequences of simulated climatic warming were large increases in flood maxima and the occurrence of many large floods in winter instead of spring. These shifts in the seasonal distribution and variability of runoff have major implications for the management of water resources (Gleick, 1987, 1989; Lettenmaier and Sheer, 1989; Vankatwijk *et al.*, 1993), for transport of solutes and particles and for the ecology of aquatic biota (Regier and Meisner, 1990).

At the mouth of the Sacramento–San Joaquin River system lies San Francisco Bay, one of the largest estuaries in western North America. The spatial and temporal variability of salinity in the Bay is fundamental to the physics, chemistry and ecology of the estuarine ecosystem (Nichols *et al.*, 1986) and affects all levels of the food-web, from phytoplankton to predators (Jassby *et al.*, 1995). Because the main control on salinity fluctuations is the volume of freshwater flowing into the Bay (Peterson *et al.*, 1989), the earlier spring runoff expected by the climate warming scenarios examined by Lettenmaier and Gan (1990) are likely to produce higher summer and autumn salinity in the Bay.

An alternative approach to assess how changes in climate might effect the Bay's salinity was applied by Cayan and Peterson (1993). They used about 60 years of historical records to investigate how extreme categories of spring climates (i.e. cool and wet, warm and wet, cool and dry, and warm and dry), produced by distinct atmospheric circulation patterns, influence the Bay's salinity. Anomalously warm and/or dry springs tended to cause earlier snowmelt with less runoff in summer and higher salinities in the autumn. Furthermore, diversions for human uses tended to reduce spring inflows to the Bay and to do so disproportionately during drier periods.

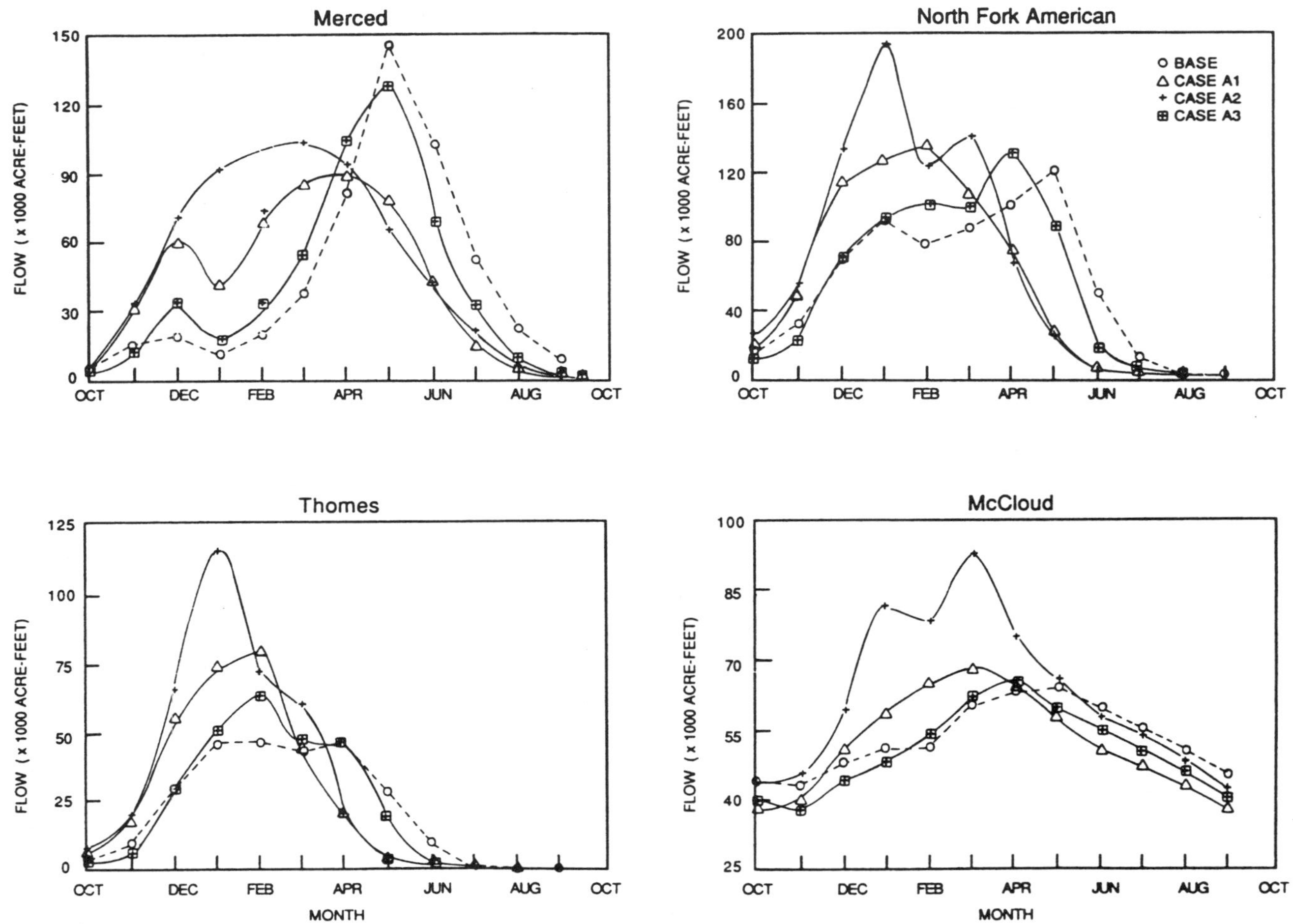

Figure 5. Monthly mean stream flow for baseline condition and with three GCM forecasts (A1, GFDL; A2, GISS; A3, OSU) of climate change after doubling of CO_2 concentration. (From Lettenmaier and Gan, 1990)

Saline lakes

Within the endorheic catchments of the Western Great Basin occur several large saline lakes (e.g. Mono, Walker, Pyramid and Abert) which have undergone major changes in size and salinity as the regional climate has varied (Benson and Thompson, 1987; Stine, 1990). During the last few decades, diversion of inflows for human uses have caused lake levels to decline and salinities to increase in several of these lakes. The observed and anticipated ecological effects of the increasing salinities in Mono and Pyramid lakes have resulted in many long-term, experimental and modelling studies that can be used to infer potential responses to climatically induced changes in size and salinity (Patten *et al.*, 1987; Galat *et al.*, 1988).

Aquatic organisms living in saline waters expend a greater portion of their energy on osmotic regulation and less on growth and reproduction as salinity increases. For example, salinity bioassays on algal (Melack *et al.*, 1985; Herbst and Castenholz, 1994) and invertebrate species (Herbst *et al.*, 1993) from Mono Lake showed declining productivity with increasing salinity. However, results from physiological experiments do not translate necessarily to fully natural conditions. In fact, a model of plankton dynamics in Mono Lake, which incorporated nitrogen cycling, predicted lower reductions in productivity at higher salinities than the laboratory experiments (Jellison *et al.*, 1993a).

Abrupt increases in lake level have also occurred in Mono and Pyramid lakes because of exceptional runoff associated with ENSO events such as those observed in the 1980s (Galat *et al.*, 1990; Jellison and Melack, 1993a). One consequence of the large volume of freshwater inflow was the lakes becoming meromictic. Meromixis is persistent chemical stratification with incomplete vertical mixing over the course of a year. In Mono Lake, meromixis persisted from 1983 to 1988 and was characterized by a marked decrease in vertical mixing (Jellison and Melack, 1993a) and nutrient transport across the pycnocline (Jellison *et al.*, 1993b) which resulted in diminished algal productivity (Jellison and Melack, 1993b).

To assess the incidence of meromixis during a period of declining lake levels punctuated with occasional ENSO events, Romero and Melack (1996) applied a one-dimensional vertical mixing model to Mono Lake using 50 years of historical stream flow and precipitation data. They assumed that the historical records, which included strong and weak ENSO events and 50 years of stream diversions, would simulate the expected long-term reduction in regional precipitation. A greater incidence of meromixis was predicted to occur at low lake levels because of greater dilution of the mixolimnion per unit freshwater input and higher initial salinities of the lake. Furthermore, a more variable interannual hydrology increased the likelihood of meromixis.

The reduction in nutrient supply to the euphotic zone during meromixis will lead to changes in algal abundance, composition and productivity. For example, meromixis and even temporary winter stratification in Pyramid Lake led to nitrogen limitation and to nitrogen-fixing cyanobacterial blooms (Galat *et al.*, 1990). Following the onset of meromixis in Mono Lake, algal biomass in the spring and autumn and annual productivity were less than during periods with complete winter mixing (Jellison and Melack, 1993b). In contrast, annual production was very high the year meromixis ended. Hence, predictions of the regional climate having a more strongly seasonal stream flow with increased peak runoff, earlier snowmelt and lower baseflows (Cooley, 1990; Lettenmaier and Gan, 1990; Pupacko, 1993) will probably lead to greater variance in planktonic dynamics in saline lakes.

Montane lakes

Changes in snowmelt in response to altered precipitation and energy fluxes are likely to modify the timing and quantity of hydrological and chemical fluxes in montane catchments (Williams *et al.*, 1996). The potential consequences to the montane lakes and streams may be altered flushing rates and residence times, stratification and heat content and inputs of nutrients and other solutes, including those causing acidification. Time-series spanning two decades and results from simulations available for montane lakes in California offer the possibility of assessing the magnitude of such responses to changes in snowmelt.

Emerald Lake, a representative high elevation lake in the southern Sierra Nevada (Melack and Stoddard, 1991), has experienced extreme variations in snowfall during the past decade (Kattelmann and Elder, 1991; Melack *et al.*, 1993). The El Niño of 1982–1983 was the strongest of the twentieth century and produced extraordinary precipitation in California and elsewhere (Quiroz, 1983). The subsequent period included a prolonged, severe drought and two additional years with very large snowfall.

To assess the responses of stream chemical composition to an increased rate of snowmelt, as expected from global warming, Wolford and Bales (1996) have applied a mechanistic hydrological and geochemical model to the Emerald Lake catchment. Inputs from two contrasting years, one with high and one with low snowfall, were used to drive the model under the observed conditions and with snowmelt rates increased by 30%. During the period of rapid snowmelt, modelled concentrations of calcium and acid-neutralizing capacity (ANC) were lower at the higher snowmelt rate. Conversely, the higher rate of snowmelt caused the snow to disappear sooner, which led to very low discharge in late summer and much higher concentrations of solutes. When a 30% higher rate of snowmelt was combined with a doubled wet atmospheric deposition of solutes, ANC was depressed to near or below zero. In the exceptionally dilute waters of the Sierra Nevada, these changes in solute concentrations may influence the aquatic biota, as was demonstrated experimentally in streams and lakes of the region (Barmuta *et al.*, 1987; Kratz *et al.*, 1993).

Temperature profiles measured regularly for two decades in Castle Lake, California, a small subalpine lake, have revealed large interannual differences in heat content associated with the amount of winter snowfall (Strub *et al.*, 1985). In all years with anomalously large heat content, the snowfall was lower than average; while in all years (except 1973) with anomalously small heat contents, snowfall was higher than average. ENSO events produced both anomalies, but some anomalous years were without ENSOs. Hence, simple interpretations of effects from ENSO events would not be appropriate.

Ecological processes in montane lakes such as Castle Lake may be strongly influenced by climatic conditions, and, therefore, be sensitive to changes in climate caused by global warming. For example, during the extraordinary El Niño of 1983, Castle Lake remained frozen until early July and summer primary productivity was only 25% of the long-term average (Strub *et al.*, 1985). In years with incomplete vertical mixing, productivity is often less than in years with spring holomixis (Goldman and de Amezaga, 1984). A 27-year record of summer primary productivity had no long-term trend, but considerable interannual variability, with extreme years often occurring when ENSOs occurred (Jassby and Goldman, 1992). Further statistical analysis of interactions between the early summer productivity and climate revealed negative associations with the time of ice-out and total precipitation preceding the spring bloom (Goldman *et al.*, 1989; Jassby *et al.*, 1990). The time of ice-out sets the length of the growing season, and total precipitation affects flushing rates.

Additional evaluation of the potential effects of global warming on primary productivity in montane lakes is possible by linking GCM projections of climate change with a model of lacustrine productivity. Byron and Goldman (1990) attempted such an approach by using almost three decades of limnological data from Castle Lake to develop a regression model that related algal productivity to regional temperature and precipitation. When used in combination with GCM calculations of temperature and precipitation under doubled CO_2 concentrations, their model predicted increased algal productivity. These increases were driven by the GCM-projected higher temperatures, which were linked to earlier melt and a longer growing season. Because the GCMs calculated January–May precipitation as being similar to historical records, no effect of doubled CO_2 on runoff was observed; this result deserves additional study using newer, regional predictions of climate change.

In larger lakes, longer residence times and larger volumes buffer some short-term climatic variability that affects smaller lakes such as Castle and Emerald lakes. However, even in Lake Tahoe, climate-related factors are apparent in the three decades of limnological data available. Through a careful statistical analysis, Goldman *et al.* (1989) demonstrated that the May peak in primary productivity was associated with the maximum depth of spring mixing. The depth of mixing reflected the intensity of storms in March. Hence, projections of climate change would need to face the significant challenge of predicting wind regimes to allow forecasts of effects on some lakes.

Atmospheric deposition of nutrients can influence algal abundance and productivity in oligotrophic lakes (Jassby *et al.*, 1994), and climatic conditions can be linked to atmospheric deposition in several ways (Lesack and Melack, 1991). Large forest fires can result occasionally in considerable nutrient inputs to lakes with a stimulating effect on productivity (e.g. Goldman *et al.*, 1990). Hence, the potential influence of climate change on the incidence of fires should be evaluated.

Glaciated basins

A common hypothesis is that global warming will cause mid-latitude glaciers to recede as a result of increased melting and, consequently, sea levels will rise (Roots, 1989). However, this may be a simplified hypothesis, and the effect may vary according to the geographical area and whether the glacier's terminus is on land or at tide water (Oswood *et al.*, 1992). Evidence from the European Alps indicates that the twentieth century has been characterized by a general retreat of alpine valley glaciers (Groves, 1988) with a consequent overall reduction in glacial runoff (Collins, 1985). However, this widespread retreat of glaciers in the European Alps appeared to be generally arrested in the mid 1960s, and, subsequently, the termini of many glaciers have slowly re-advanced, although overall increases in area have been small (Kasser, 1983). Loss of mass has been shown for glaciers in the Cascade Mountains of Washington (Krimmel, 1989).

In Alaska, many tide water glaciers along the south-east and south-central coast have shown a rapid retreat within the last century, but this phenomenon is considered to be related more directly to local fjord topography than to climate-induced effects (Powell, 1990). For Alaskan glaciers with their termini on land, records for the period of the mid-1970s to the late 1980s would appear to indicate an increase in mass balance for the Wolverine Glacier (maritime) in south-central Alaska and the Gulkana Glacier (continental) in the Alaska Range (Mayo and Trabant, 1986; Mayo and March, 1990). However, when examining the 30-year record for these glaciers, it is apparent that there is an overall net negative balance for the period 1966–1994 (Hodge, personal communication).

The relationship between glacier mass balance and glacial runoff is complex. Glacier recession will initially lead to runoff additional to that produced if the glacier was in a steady state, but sustained deglaciation will result in diminished flows (Collins, 1987). Unfortunately, runoff data for the Wolverine and Gulkana glaciers are not complete during the years of record for the glacier mass balances. During positive mass balances in the late 1970s, total runoff appeared to increase, but runoff patterns in relation to negative mass balances are not readily apparent (Hodge, personal communication). Clearly, there is a lag between alterations in runoff and changes in glacier mass balance.

Other factors, mediated by climatic variations, may influence the relationship between runoff and glacier mass. Increased snow accumulation from winter precipitation may increase the time before the underlying ice is exposed to radiation, and thus may retard the onset of ice melt and, since snow has a higher albedo than ice, the amount of melting under direct radiation will be reduced (Collins, 1985, 1987). A matter of some debate is the likely role of cloud cover in climate change. Increased cloud cover would reduce the amount of radiation reaching the ice surface, potentially reducing ice melt. Associated with increased cloud cover could be increased precipitation, which would elevate the contribution of non-glacial sources to runoff.

Glacier-fed rivers

The ecology of glacier-fed rivers has been relatively unstudied compared with rivers originating from other sources. Milner and Petts (1994) proposed a qualitative model to predict the gradient of zoobenthic communities in rivers downstream of glacier margins as determined by two principal variables, water temperature and channel stability. Water temperature conveys deterministic trends to the typical pattern of invertebrate communities in glacial rivers, and determines the point at which certain taxa may colonize. However, even if temperatures are suitable, channel instability may delay colonization and allow some of the cold-tolerant taxa to remain dominant in the community. Chironomid larvae of the genus *Diamesa* are typically the sole taxa present when $T_{max} < 2°C$, and are joined by other Diamesinae, Orthocladiinae and Simuliidae when T_{max} is 2–4°C. *Diamesa* larvae typically possess long posterior pro-legs with strong claws that enable them to grip the substrata in unstable channels. Increased glacial runoff may have a significant effect on flow, temperature and sediment regimes in glaciated basins. Reduced stream temperature from increased contributions of glacial meltwater and decreased channel stability from changed runoff patterns and altered sediment loads, will, potentially, reduce the diversity of zoobenthic communities in glacier-fed rivers and cause an increase in the relative abundance of a number of key taxa, most notably *Diamesa* spp.

Oswood *et al.* (1992) considered the biogeographical implications for freshwater invertebrates of a less rigorous thermal environment at higher latitudes. Increased water temperature and decreased permafrost in areas south of the Alaska Range allow for potential redistributions of benthic invertebrate communities with the possible arrival of new predators (e.g. Megaloptera) and the loss of some cold water stenotherms at their southerly ranges. The altitudinal range of invertebrates in mountain streams may also change with the extension of some species to higher elevations and the reduction in the amount of habitat for cold water stenotherm taxa.

Riparian zones

Biological communities at the boundary between terrestrial and freshwater ecosystems are particularly sensitive to climate change. Examples include riparian forests, marginal wetlands, littoral zones in lakes, floodplain lakes and forests, and areas with significant groundwater–surface water exchange (Naiman *et al.*, 1988). For example, in response to climate change, it is expected that there will be changes in the ratios of the number of vegetative patch types and their areas.

Riparian zones associated with rivers in the Pacific Coastal Mountains and Western Great Basin are likely candidates for alteration as climate changes. For example, in the Queets River, Washington, and other coastal rivers, variability in sediment delivery, discharge and benthic shear stress determine the structure, composition and spatial distribution of riparian vegetation (Fetherston *et al.*, 1995). Climatically induced changes in the hydrological regime are expected to have substantial consequences for riparian forests, in terms of their structure, biogeochemical processes and resiliency (Naiman *et al.*, 1993). Much of the possible variability is apparent in the species, and their life history strategies in the temperate coastal rainforest (Naiman and Anderson, 1997). Rivers with high discharge favour willow, sexually reproducing black cottonwood and many exotic species (DeFerrari and Naiman, 1994). In contrast, rivers with little variability favour more upland species such as western hemlock, sitka spruce and Douglas fir, asexually reproducing cottonwood and less exotic species.

Salmon ecology

Salmon, a major natural resource in the Pacific Northwest respond to climate-induced changes in temperature and runoff (Neitzel *et al.*, 1991; Levy, 1992). However, the potential responses of each salmonid species and subspecies may vary owing to different life cycle adaptations to environmental conditions in its natal river system. Furthermore, changes in both inland waters and the Pacific Ocean must be considered for these anadromous fish.

Neitzel *et al.* (1991) evaluated the potential effects of changes in hydrographs, total annual discharge, water temperature and texture of the stream bed on salmonoid reproductive cycles resulting from a hypsithermal-type warmer, drier climate in the Columbia River Basin. Spring and summer chinook responded negatively to the hypsithermal-type climate because of changes in the timing and size of spring runoff. Autumn chinook and coho are affected by temperature, discharge and the timing of snowmelt; hence, positive responses were indicated in rivers on the west side of the Cascades while negative effects appeared likely on the east side. In contrast to other species, sockeye salmon live in lakes during part of their life cycle; the warmer summer temperatures in lakes and lower spring flows expected after climatic warming would probably cause negative effects on the sockeye salmon the Columbia River Basin. Alterations in the regional allocation of water resources to hydroelectric generating facilities and irrigation may well compound the direct effects of climate change on the salmon.

Changes to the climates of British Columbia and the north-east Pacific may influence the production of Fraser River sockeye salmon (*Oncorhynchus nerka*), the most valuable salmon resource in British Columbia. These salmon are especially susceptible to climatic warming because they are at the southern range of sockeye in North America and because they make long migrations in near-surface waters of the Pacific Ocean. Bioenergetic models (Hinch *et al.*, 1994) and qualitative analyses (Henderson *et al.*, 1992; Levy, 1992) have been applied to assess the probable consequences of climate warming on production of sockeye salmon.

Henderson *et al.* (1992) examined the probable affects of atmospheric warming on the Adams River stock, a large and commercially important population in the Fraser River Basin, using a 2.5°C summer and 4.0°C winter increase in temperature derived from GCMs run for western Canada. Anadromous adult salmon return to the Adams River to spawn in late summer and early autumn, the fertilized eggs incubate in river gravels and alevins hatch and emerge in spring. The young sockeye enter Shuswap Lake, where they feed until the following spring. Then, they move into the Pacific Ocean, where they remain for two years before returning to their natal spawning ground as four-year-olds.

The warmer atmospheric temperatures predicted to occur with doubling of atmospheric CO_2 concentrations were surmised to influence the ecological conditions in Shuswap Lake and the river in several ways. Greater input of nutrients during the winter isothermal period instead of during the spring, and a longer period of thermal stratification, were deemed likely to lead to lower planktonic productivity, and smaller juvenile sockeye. Based on the work of Brett *et al.* (1969), which determined the optimal temperatures for juvenile sockeye growth as a function of food supply, and an examination of the vertical distribution of the fish and their zooplankton prey, Henderson *et al.* (1992) concluded that climatic warming would result in reduced growth and survival of sockeye salmon in freshwater, which would, in turn, lead to a decrease in smolt to adult survival (Henderson and Cass, 1991). Moreover, climatic warming may change aspects other than temperature in the river system utilized by the salmon. For example, if flow in the Fraser River is diminished, as is anticipated, adult fish would be hindered in their spawning runs and some spawning gravels could desiccate.

During the time the Fraser River sockeye spend in the ocean almost all their growth occurs and the surplus energy reserves needed for their migration and spawning are acquired. GCMs forecast increased sea surface temperatures (SST) and weaker north–south pressure gradients over the north-east Pacific Ocean, which could weaken oceanic upwelling and reduce secondary productivity (Hsieh and Boer, 1992). In turn, warmer, SSTs and less zooplankton would probably result in smaller adult sockeye with fewer and smaller eggs and less energy reserves (Hinch *et al.*, 1994, 1995). Hence, during both the freshwater and marine stages of their life history, sockeye salmon are likely to be adversely affected by climatic warming.

A recent study of lakes on the Alaskan Peninsula examined sockeye salmon returns from spawning adults to Lake Becharof, via the Egegik River, and to Lake Ugashik (Mathiesen *et al.*, 1997). An earlier study reported that the primary productivity of Lake Becharof tended to be nutrient and, probably, light limited (Goldman, 1960). In attempting to account for the increased escapement beginning in the 1970s, Mathiesen *et al.* (1997) found that the surface temperatures at the outflows of the lakes had increased from 7.6 to 8.8°C for Lake Becharof from 1962 to 1989, and from 8.8 to 9.2°C in the smaller Ugashik Lake. Both lakes, probably because of their proximity to the sea, had shorter periods of winter ice cover than the much larger Lake Iliamna or Naknek Lake, which are situated north and inland. The decaying carcasses from the larger return of adults from the sea also provided fertilizing nutrients to the nursery lakes, helping to sustain the food-web during the longer growing season.

In south-central Alaska, several major systems of lakes support large runs of sockeye salmon and include the Kenai, Tustamena and Skilak lakes, all glacially influenced. Climate-induced increases in glacial runoff into these lakes in summer could cause concomitant increases in turbidity and reductions in primary productivity. The offsetting effects of a shorter period of ice cover and warmer water temperatures could enhance zooplankton and salmon fry growth.

With increasing latitude, overwintering mortality becomes a significant factor in the production of salmonids. Consequently, a reduction in the length of ice cover in streams may enhance overwintering survival. Conversely, increased winter flows and spring peaks may reduce salmonid egg to fry survival, particularly if the cumulative effects of altered land use are considered, e.g. increased erosion and runoff. Higher spring peaks in flow and warmer water temperatures may cause earlier emergence of fry and migration of pink and chum salmon fry to estuaries at a time when their food sources have not developed adequately. Lower summer flow may reduce the amount of suitable spawning and rearing habitat.

RECOMMENDATIONS FOR RESEARCH

Further research activities should include expanded studies at high elevations and of mass balances of glaciers, applications of remote sensing to monitor changes, further refinement of regional climatic models to improve forecasts of future conditions and continued analyses of long-term physical, chemical and biological data to help understand responses to future climates.

The selection of appropriate scales of measurement can be difficult because of discordant scales amongst atmospheric, hydrological and ecological systems (O'Neill, 1988; Hostetler, 1994). Furthermore, detection of gradual changes requires consistent monitoring for long periods at a range of spatial scales. One alternative to comprehensive, and hence expensive, monitoring programme is to focus on unusual circumstances that can indicate how a system responds to perturbations. Actual experimental treatments can do just that, but, in many situations, the large scale of the system precludes replicated, experimental designs.

Along the Pacific coast of North America, high elevation environments, whilst a conspicuous feature and one susceptible to climatic changes, have received insufficient attention and, in general, lack long-term data. Similarly, while glaciers are abundant in the northern portion of the region, more studies of mass balance and analyses of glacial runoff are needed to assess glacial responses to varying climate. Moreover, maintenance of existing stations gauging the discharge of rivers throughout the region is crucial.

Remote sensing has great potential for augmenting ground-based studies in the analysis of regional aspects of hydrological and ecological conditions (e.g. Sellers, 1995). Relevant variables amenable to remote sensing include snow covered area, water vapour and fluxes, cloud cover, radiation and land cover. A combination of optical and microwave sensors carried on several satellites, improved algorithms for calculating relevant variables and electronic communication networks offer many opportunities for merging of measurements and modelling.

Increased coupling of regional climate models with hydrological models operated at the scale of catchments or lakes (e.g. Hostetler and Giorgi, 1993) would provide valuable information for subsequent ecological studies. One challenge to such efforts is the inclusion of wind regimes, which are very difficult to simulate, especially in mountainous terrain. Furthermore, because of the importance of ENSO events along the Pacific coast, special attention should be given to modelling (e.g. Meehl and Branstator, 1992) and palaeoclimatic research (e.g. Enfield, 1992) that considers the robustness of ENSOs to major changes in atmospheric boundary conditions. A related issue, in need of improved GCM modelling, is the extent to which climatic variability will change as concentrations of greenhouse gases increase (e.g. Rind *et al.*, 1989). Hydrological models also require improvements to increase their utility for assessing the effects of climatic changes (Leavesley, 1994). Besides conceptual advances and better parameter estimation techniques, innovative approaches to generate climate scenarios (e.g. Dettinger and Cayan, 1992) are needed. Furthermore, because of interactions between vegetation and runoff (e.g. Kattelmann, 1991), hydrological models should be linked with ecological models (e.g. Neilson and Marks, 1994).

While many usual environmental measurements are appropriate and necessary to evaluate ecological responses to climatic changes, several are worthy of special mention based on their pertinence to the Pacific coastal region. Fire is a common occurrence in the region, and optical and nutritional conditions in lakes and streams can be influenced by burning of vegetation in their catchments. Moreover, because the frequency of fire is highly dependent on climatic conditions, both measurements and modelling should include the role of fire. Snowmelt and associated ice cover play dominant roles in hydrological and hydrochemical conditions in the high elevation streams and lakes of the region, and are particularly sensitive to climatic perturbations. Hence, experimental and monitoring programmes should target the annual extent of ice cover on lakes and the snowmelt period. To evaluate the probable consequences of climatic warming for salmon requires examination of the responses of all life history stages to the cumulative effects of likely environmental changes in the lakes, rivers and oceans inhabited by the fish. Overall, these activities will require careful coupling of reliable limnological and climatic data and should provide valuable insights into the complex results of possible future climatic change.

ACKNOWLEDGEMENTS

We thank R. Kattelmann for valuable material on stream flow in the Sierra Nevada. Preparation of the manuscript benefited from comments by two anonymous reviewers and was supported in part by NASA grant NTW-2602. Permission to use figures 1 and 2 was granted by Island Press, Washington, DC.

REFERENCES

Aguado, E., Cayan, D., Riddle, L., and Roos, M. 1992. 'Climatic fluctuations and the timing of west coast streamflow', *J. Clim.*, **5**, 1468–1483.

Armstrong, C. F. and Stidd, C. K. 1967. 'A moisture balance profile on the Sierra Nevada', *J. Hydrol.*, **5**, 258–268.

Barmuta, L. A., Cooper, S. D., Hamilton, S. K., Kratz, K. W., and Melack, J. M. 1990. 'Responses of zooplankton and zoobenthos to experimental acidification in a high-elevation lake (Sierra Nevada, California, U.S.A.)', *Freshwat. Biol.*, **23**, 571–586.

Benson, L. C. and Thompson, R. S. 1987. 'Lake-level variation in the Lahontan Basin for the past 50,000 years', *Quat. Res.*, **28**, 69–85.

Bjornn, T. C. and Reiser, D. W. 1991. 'Habitat requirements of salmonids in streams', in Meehan, W. R. (Ed.), *Influences of Forest and Rangeland Management on Salmonid Fishes and their Habitats.* American Fisheries Society Spec. Publ. 19. pp. 83–138.

Bradbury, J. P. 1992. 'Late Cenozoic lacustrine and climatic environments at Tule Lake, northern Great Basin, USA', *Climate Dynam.*, **6**, 275–285.

Brett, J. R., Shelbourn, J. E., and Shoop, C. T. 1969. 'Growth rate and body composition of fingerling sockeye salmon, *Oncorhynchus nerka*, in relation to temperature and ration size', *J. Fish. Res. Bd Can.*, **26**, 2363–2394.

Broecker, W. S. and Denton, G. H. 1989. 'The role of ocean-atmosphere reorganization in glacial cycles', *Geochim. Cosmochim. Acta*, **53**, 2465–2501.

Brown, C. S. 1989. 'A description of the United States' contribution to the World Glacier Inventory', in Oerlemans, J. (Ed.), *Glacier Fluctuations and Climate Change.* Kluwer Academic Publishers, Dordrecht. pp. 103–108.

Byron, E. R. and Goldman, C. R. 1990. 'The potential effects of global warming on the primary productivity of a subalpine lake', *Wat. Resour. Bull.*, **26**, 983–989.

California Cooperative Snow Survey, 1986. 'Water conditions in California', *Bulletin 120-86*. California Department of Water Resources, Sacramento.

Carpenter, S. R., Fisher, S. G., Grimm N. B., and Kitchell, J. F. 1992. 'Global change and freshwater ecosystems', *Annu. Rev. Ecol. Syst.*, **23**, 119–139.

Cayan, D. R. and Peterson, D. H. 1989. 'The influence of north Pacific atmospheric circulation on streamflow in the west', in Peterson, D. H. (Ed.), *Aspects of Climatic Variability in the Pacific and the Western Americas.* Geophys. Monogr. Ser. 55. pp. 375–397.

Cayan, D. R. and Peterson, D. H. 1993. 'Spring climate and salinity in the San Francisco Bay Estuary', *Wat. Resour. Res.*, **29**, 293–303.

Cayan, D. R. and Riddle, L. G. 1992. 'A multi-basin seasonal streamflow model for the Sierra Nevada', in Redmond, K. T. and Tharp, V. L. (Eds), *Proceedings, Ninth Pacific Climate Workshop.* Interagency Ecological Studies Program Technical Report 34. California Department of Water Resources, Sacramento. pp. 141–152.

Cayan, D. R. and Riddle, L. G. 1993. 'The influence of precipitation and temperature on seasonal streamflow in California', *Wat. Resour. Res.*, **29**, 1127–1140.

Cayan, D. R. and Roads, J. O. 1984. 'Local relationships between United States west coast precipitation and monthly mean circulation parameters', *Mon. Weath. Rev.*, **112**, 1276–1282.

Collins, D. N. 1985. 'Climatic variation and runoff from Alpine glaciers', *Z. Gletsch. Glazialgeol.*, **20**, 127–145.

Collins, D. N. 1987. 'Climatic variations and runoff from glacierized Alpine basins', *IAHS Publ.*, **168**, 77–89.

Cooley, K. R. 1990. 'Effects of CO_2-induced climatic changes on snowpack and streamflow', *Hydrol. Sci. J.*, **35**, 511–522.

Dana, G. L., Jellison, R., Melack, J. M., and Starrett, G. L. 1993. 'Relationships between *Artemia monica* life history characteristics and salinity', *Hydrobiologia*, **263**, 129–143.

DeFerrari, C. M. and Naiman, R. J. 1994. 'A multiscale assessment of the occurrence of exotic plants on the Olympic Peninsula, Washington', *J. Veg. Sci.*, **5**, 247–258.

Dettinger, M. D. and Cayan, D. R. 1992. 'Climate-change scenarios for the Sierra Nevada, based on winter circulation patterns', in *Managing Water Resources during Global Change, American Water Resources Association Tech. Publ. Ser.* TPS-92-4. pp. 681–690.

Dettinger, M. D. and Cayan, D. R. 1995. 'Large-scale atmospheric forcing of recent trends toward early snowmelt runoff in California'. *J. Clim.*, **8**, 606–623.

Elliott, R. D., Griffith, D. A., Hannaford, J. F., and Flueck, J. A. 1978. 'Special report on background and supporting material for the Sierra cooperative pilot project', North American Weather Consultants, Report 78-22, vol. 2.

Enfield, D. B. 1992. 'Historical and prehistorical overview of El Niño/Southern Oscillation', in Diaz, H. F. and Markgraf, V. (Eds), *El Niño — Historical and Paleoclimatic Aspects of the Southern Oscillation*. Cambridge University Press, Cambridge. pp. 95–117.

Enzel, Y., Cayan, D. R., Anderson, R. Y., and Wells, S. G. 1989. 'Atmospheric circulation during Holocene lake stands in the Mojave Desert: evidence of regional climate change', *Nature*, **341**, 44–47.

Fetherston, K. L., Naiman, R. J., and Bilby, R. E. 1995. 'Large woody debris, physical process, and riparian forest development in montane river networks of the Pacific Northwest', *Geomorphology*, **13**, 133–144.

Firth, P. and Fisher, S. G. (Eds). 1992. *Global Climatic Change and Freshwater Ecosystems*. Springer-Verlag, New York.

Galat, D. L., Coleman, M., and Robinson, R. 1988. 'Experimental effects of elevated salinity on three benthic invertebrates in Pyramid Lake, Nevada', *Hydrobiologia*, **158**, 133–144.

Galat, D. L., Verdin, J. P., and Simis, L. L. 1990. 'Large-scale patterns of *Nodularia spumigena* blooms in Pyramid lake, Nevada, determined from Landsat imagery: 1972–1986', *Hydrobiologia*, **197**, 147–164.

Giorgi, F. and Mearns, L. O. 1991. 'Approaches to the simulation of regional climate change: a review', *Rev. Geophys.*, **29**, 191–216.

Giorgi, F., Bates, G. T., and Nieman, S. J. 1993. 'The multiyear surface climatology of a regional atmospheric model over the western United States', *J. Clim.*, **6**, 75–95.

Giorgi, F., Brodeur, C. S., and Bates, G. T. 1994. 'Regional climate change scenarios over the United States produced with a nested regional climate model', *J. Clim.*, **7**, 375–399.

Gleick, P. H. 1987. 'The development and testing of a water balance model for climate impact assessment: modeling the Sacramento basin', *Wat. Resour. Res.*, **23**, 1049–1061.

Gleick, P. H. 1989. 'Climate change, hydrology, and water resources', *Rev. Geophys.*, **27**, 329–344.

Goldman, C. R. 1960. 'Primary productivity and limiting factors in three lakes of the Alaska Peninsula', *Ecol. Monogr.*, **30**, 207–230.

Goldman, C. R. and de Amezaga, E. 1984. 'Primary productivity and precipitation at Castle Lake and Lake Tahoe during twenty-four years, 1959–1982', *Verh. Int. Verein. Limnol.*, **22**, 591–599.

Goldman, C. R., Jassby, A., and Powell, T. 1989. 'Interannual fluctuations in primary production: meteorological forcing of two subalpine lakes', *Limnol. Oceanogr.*, **34**, 310–323.

Goldman, C. R., Jassby, A. D., and deAmezaga, E. 1990. 'Forest fires, atmospheric deposition and primary productivity at Lake Tahoe, California-Nevada', *Verh. Int. Verein. Limnol.*, **74**, 499–503.

Goodridge, J. D. 1991a. 'Air temperature trends in California', in Betancourt, J. L. and Tharp, V. L. (Eds), *Proc., Seventh Annual Pacific Climate Workshop*, Interagency Ecological Studies Program Tech. Rep. 26. California Department of Water Resources, Sacramento. pp. 101–108.

Goodridge, J. D. 1991b. 'One hundred years of rainfall trends in California'. in Betancourt, J. L. and Tharp, V. L. (Eds), *Proc., Seventh Annual Pacific Climate Workshop*, Interagency Ecological Studies Program Tech. Rep. 26. California Department of Water Resources, Sacramento. pp. 109–114.

Granger, O. E. 1977a. 'Secular fluctuations of seasonal precipitation in lowland California', *Mon. Weath. Rev.*, **105**, 386–397.

Granger, O. 1977b. 'Increasing variability in California precipitation', *Ann. Assoc. Am. Geogr.*, **69**, 533–543.

Graumlich, L. J. 1987. 'Precipitation variation in the Pacific Northwest (1675–1975) as reconstructed from tree rings', *Ann. Assoc. Am. Geogr.*, **77**, 19–29.

Graumlich, L. J. 1993. 'A 1000-year record of temperature and precipitation in the Sierra Nevada', *Quat. Res.*, **39**, 249–255.

Graumlich, L. J. and Brubaker, L. 1986. 'Reconstruction of annual temperature (1590–1979) for Longmire, Washington, derived from tree rings', *Quat. Res.*, **25**, 223–234.

Groves, J. M. 1988. *The Little Ice Age*. Methuen, London.

Haston, L. and Michaelsen, J. 1994. 'Long-term central coastal California precipitation variability and relationships to El Niño-Southern Oscillation', *Water Resour. Res.*, **7**, 1373–1387.

Henderson, M. A. and Cass, A. J. 1991. 'Effect of smolt size on smolt-to-adult survival for Chilko lake sockeye salmon (*Oncorhynchus nerka*)', *Can. J. Fish. Aquat. Sci.*, **48**, 988–994.

Henderson, M. A., Levy, D. A., and Stockner, J. S. 1992. 'Probable consequences of climate change on freshwater production of Adams River sockeye salmon (*Oncorhynchus nerka*)', *GeoJournal*, **28**, 51–59.

Herbst, D. B. and Castenholz, R. W. 1994. 'Growth of the filamentous green alga *Ctenocladus circinnatus* (Chaetophorales, Chlorophyceae) in relation to environmental salinity', *J. Phycol.*, **30**, 588–593.

Herbst, D. B., Conte, F. P., and Brookes, V. J. 1988. 'Osmoregulation in an alkaline salt lake insect, *Ephydra* (*Hydropyrus*) *hains* Say (Diptera: Ephydridae) in relation to water chemistry', *J. Insect Physiol.*, **34**, 903–909.

Hinch, S. G., Healey, M. C., Diewert, R. E., and Henderson, M. A. 1994. 'Climatic change and ocean energetics of Fraser River sockeye (*Oncorhynchus nerka*)', *Can. J. Fish. Aquat. Sci., Spec. Publ.*, **121**, 439–445.

Hinch, S. G., Healey, M. C., Diewert R. E., Thomson, K. A., Hourston, R., Henderson, M. A., and Juanes, F. 1995. 'Potential effects of climate change on marine growth and survival of Fraser River sockeye', *Can. J. Fish. Aquat. Sci.*, **52**, 2651–2659.

Hostetler, S. W. 1994. 'Hydrologic and atmospheric models: the (continuing) problem of discordant scales', *Climatic Change*, **27**, 345–350.

Hostetler, S. W. and Giorgi, F. 1993. 'Use of output from high-resolution atmospheric models in landscape-scale hydrologic models: an assessment', *Water Resour. Res.*, **29**, 1685–1695.

Hsieh, W. H. and Boer, G. J. 1992. 'Global climate change and ocean upwelling', *Fish. Oceanogr.*, **1**, 333–338.

Hunt, C. B. 1974. *Natural Regions of the United States and Canada*. W. H. Freeman and Co, San Francisco, CA.

Jassby, A. D. and Goldman, C. R. 1992. 'Meteorological signals in primary productivity at two mountain lakes', in Redmond, K. T. (Ed.), *Proc., Eighth Annual Pacific Climate Workshop*, Interagency Ecological Studies Program Tech. Rep. 31. California Department of Water Resources, Sacramento. pp. 71–80.

Jassby, A. D., Powell, T. M., and Goldman, C. R. 1990. 'Interannual fluctuations in primary production: direct physical effects and the trophic cascade at Castle Lake, California', *Limnol. Oceanogr.*, **35**, 1021–1038.

Jassby, A. D., Reuter, J. E., Axler, R. P., Golman, C. R., and Hackley, S. H. 1994. 'Atmospheric deposition of nitrogen and phosphorus in the annual nutrient load of Lake Tahoe (California-Nevada)', *Wat. Resour. Res.*, **30**, 2207–2216.

Jassby, A. D., Kimmerer, W. J., Monismith, S. G., Armor, C., Cloern, J. E., Powell, T. M., Schubel, J. R., and Vendlinski, T. J. 1995. 'Isohaline position as a habitat indicator for estuarine populations', *Ecol. Appl.*, **5**, 272–289.

Jellison, R. and Melack, J. M. 1993a. 'Meromixis in hypersaline Mono Lake, California. 1. Stratification and vertical mixing during the onset, persistence, and breakdown of meromixis', *Limnol. Oceanogr.*, **38**, 1008–1019.

Jellison, R. and Melack, J. M. 1993b. 'Algal photosynthetic activity and its response to meromixis hypersaline Mono Lake, California', *Limnol. Oceanogr.*, **38**, 818–837.

Jellison, R., Melack, J. M., and Dana, G. L. 1993a. 'A modeling analysis of *Artemia* dynamics in Mono Lake', *Mono Basin EIR Auxiliary Report No. 13*. Report to the California State Water Resources Control Board, Sacramento, CA.

Jellison, R., Miller, L. G., Melack, J. M., and Dana, G. L. 1993b. 'Meromixis in hypersaline Mono Lake, California. 2: nitrogen fluxes', *Limnol. Oceanogr.*, **38**, 1020–1039.

Kahrl, W. (Ed.) 1978. *The California Water Atlas*. Governor's Office of Planning and Research, Sacramento.

Karl, T. R. and Riebsame, W. E. 1989. 'The impact of decadal fluctuations in mean precipitation and temperature on runoff: a sensitivity study over the United States', *Climatic Change*, **15**, 423–447.

Kasser, P. 1983. 'Influences of changes in the glacierized area on summer runoff in the Porte du Scex drainage of the Rhone', *IAHS Publ.*, **95**, 221–225.

Kattelmann, R. 1990. 'Floods in the high Sierra Nevada, California, USA', in Sinniger, R. O. and Monbaron, M. (Eds), *Hydrology in Mountainous Regions, II — Artificial Reservoirs; Water and Slopes. IAHS Publ.*, **194**, 311–318.

Kattelmann, R. 1991. 'Peak flows from snowmelt runoff in the Sierra Nevada, USA', *IAHS Publ.*, **205**, 203–211.

Kattelmann, R. 1992. 'Historical floods in the eastern Sierra Nevada', in Hall, C. A., Doyle-Jones, V., and Widawski, B. (Eds), *The History of Water in the Eastern Sierra Nevada, Owens Valley and White Mountains*. University of California Press, Berkeley. pp. 74–86.

Kattelmann, R. and Elder, K. 1991. 'Hydrologic characteristics and water balance of an alpine basin in the Sierra Nevada', *Wat. Resour. Res.*, **27**, 1553–1562.

Kattelmann, R., Berg, N. H., and Rector, J. 1983. 'The potential for increasing streamflow from Sierra Nevada watersheds', *Wat. Resour. Bull.*, **19**, 395–402.

Kattelmann, R., Berg, N., and McGurk, B. 1991. 'A history of rain-on-snow floods in the Sierra Nevada', *Proc. Western Snow Conference*, **59**, 138–141.

Kratz, K., Cooper, S. D., and Melack, J. M. 1993. 'Effects of single and repeated experimental acid pulses on invertebrates in a high altitude Sierra Nevada stream', *Freshwat. Biol.*, **32**, 161–183.

Krimmel, R. M. 1989. 'Mass balance and volume of South Cascade Glacier, Washington 1958–1985', in Oerlemans, J. (Ed.), *Glacier Fluctuations and Climate Change*. Kluwer Academic Publishers, Dordrecht. pp. 193–206.

Leathers, D. J., and Palecki, M. A. 1992. 'The Pacific/North American teleconnection pattern and the United States climate. Part II: temporal characteristics and index specification', *J. Clim.*, **5**, 707–716.

Leathers, D. J., Yarnal, B., and Palecki, M. A. 1991. 'The Pacific/North American teleconnection pattern and United States climate. Part I: regional temperature and precipitation associations', *J. Clim.*, **4**, 517–528.

Leavesley, G. H. 1994. 'Modeling the effects of climate change on water resources — a review', *Climatic Change*, **28**, 159–177.

Leavesley, G. H., Turner, K., D'Agnese, F. A., McKnight, D. 1997. 'Regional delineation of North America for the assessment of freshwater ecosystems and climate change', *Hydrol. Process.*, **11**, 819–824.

Lesack, L. F. W. and Melack, J. M. 1991. 'The deposition, composition, and potential sources of major ionic solutes in rain of Central Amazon basin', *Wat. Resour. Res.*, **27**, 2953–2977.

Lettenmaier, D. P. and Gan, T. Y. 1990. 'Hydrologic sensitivities of the Sacramento–San Joaquin River basin, California, to global warming', *Wat. Resour. Res.*, **26**, 69–86.

Lettenmaier, D. P. and Sheer, D. P. 1989. 'Assessing the effects of global warming on water resources systems: an application to the Sacramento–San Joaquin River Basin, California', *J. Wat. Resour. Plann. Manage. Div. Am. Soc. Civ. Eng.*, **117**, 108–125.

Levy, D. A. 1992. 'Potential impacts of global warming on salmon production in the Fraser River watershed', *Can. Tech. Rep. Fish. Aquat. Sci., 1889*. Department of Fisheries and Oceans, Vancouver, B.C.

Lindström, S. 1990. 'Submerged tree stumps as indicators of mid-Holocene aridity in the Lake Tahoe basin', *J. Calif. Great Basin Anthropol.*, **12**, 146–157.

Mathiesen, O. A., Goering, J. J., and Russell, R. 1997. 'Sockeye salmon production in Egegik and Ugashik watersheds', *Proc. Int. Assoc. Appl. Theor. Limnol.*, **25**, in press.

Mayo, L. R. 1986. 'Annual runoff rate from glaciers in Alaska: a model using the altitude of glacier mass balance equilibrium', *Cold Regions Hydrology Symposium*, American Water Resources Association, Tech. Publ. Ser. TPS-86-1. pp. 509–517.

Mayo, L. R. and March, R. S. 1990. 'Air temperature and precipitation at Wolverine Glacier, Alaska; Glacier growth in a warmer, wetter climate', *Ann. Glaciol.*, **14**, 191–194.

Mayo, L. R. and Trabant, D. C. 1986. 'Recent growth of the Gulkana Glacier, Alaska Range, and its relation to glacier-fed river runoff', in Subitzky, S. (Ed.), *Selected Papers in the Hydrological Sciences*, US Geological Survey Water Supply Paper 2290. pp. 91–99.

McGuirk, J. P. 1982. 'A century of precipitation variability along the Pacific coast of North America and its impact', *Climatic Change*, **4**, 41–56.

Meehl, G. A. and Branstator, G. W. 1992. 'Coupled climate model simulation of El Niño/Southern Oscillation: implications for paleoclimate', in Diaz, H. F. and Markgraf, V. (Eds), *El Niño — Historical and Paleoclimatic Aspects of the Southern Oscillation*. Cambridge University Press, Cambridge, pp. 69–91.

Melack, J. M. and Stoddard, J. L. 1991. 'Sierra Nevada', in Charles, D. F (Ed.), *Acidic Deposition and Aquatic Ecosystems: Regional Case Studies*. Springer-Verlag, New York. pp. 503–530.

Melack, J. M., Cooper, S. D., and Lenz, P. H. 1985. 'The ecology of Mono Lake, California', *Nat. Geogr. Soc. Res. Rep. 1979 Projects*, 461–470.

Melack, J. M., Hamilton, S. K., and Sickman, J. O. 1993. 'Interannual solute variations in a high elevation lake of the Sierra Nevada, California', *Verh. Int. Verein. Limnol.*, **25**, 374–377.

Michaelsen, J., Haston, L., and Davis, F. W. 1987. '400 years of central California precipitation variability reconstructed from tree rings', *Wat. Resour. Bull.*, **23**, 809–818.

Miller, D. H. 1955. 'Snow cover and climate in the Sierra Nevada, California', *Univ. Calif. Public. Geogr.*, **11**.

Miller, J. R. and Russell, G. L. 1992. 'The impact of global warming on river runoff', *J. Geophys. Res.*, **97**, 2757–2764.

Milner, A. M. and Petts, G. E. 1994. 'Glacial rivers: physical habitat and ecology', *Freshwat. Biol.*, **32**, 295–307.

Milner, A. M., Irons, J. G. III, and Oswood, M. W. 1997. 'The Alaskan landscape: An introduction for limnologists', in Milner, A. M. and Oswood, M. W. (Eds), *Freshwater of Alaska: Ecological Syntheses*. Springer-Verlag, New York. in press.

Muhs, D. R., Thorson, R. M., Clague, J. J., Mathews, P. W., McDowell, H. F., and Kelsey, H. M. 1987. 'Pacific coast and mountain system', in Graf, W. L. (Ed.), *Geomorphic Systems of North America*. Geological Society of America. pp. 517–581.

Naiman, R. J. and Anderson, E. C. 1997. 'Ecological implications of latitudinal variations in watershed size, streamflow, and water temperatures in the coastal temperate rain forest of North America', in Schoonmaker, P., von Hagen, B. and Wolf, E. C. (Eds), *The Rain Forests of Home: Profile of a North American Bioregion*. Island Press, in press.

Naiman, R. J., Decamps, H., Pastor, J., and Johnston, C. A. 1988. The potential importance of boundaries to fluvial systems', *J. North Am. Benthol. Soc.*, **7**, 289–306.

Naiman, R. J., Beechie, T. J., Benda, L. E., Berg, D. R., Bisson, P. A., MacDonald, L. H., O'Connor, M. D., Olson P. L., and Steel, E. A. 1992. 'Fundamental elements of ecologically healthy watersheds in the Pacific northwest coastal ecoregion', in Naiman, R. J. (Ed.), *Watershed Management: Balancing Sustainability with Environmental Change*. Springer-Verlag, New York. pp. 127–188.

Naiman, R. J., Decamps, H., and Pollock, M. M. 1993. 'The role of riparian corridors in maintaining regional biodiversity', *Ecol. Appl.*, **3**, 209–212.

Neilson, R. P. and Marks. D. 1994. 'A global perspective of regional vegetation and hydrologic sensitivities from climatic change', *J. Veg. Sci.*, **5**, 715–730.

Neitzel, D. A., Scott, M. J., Shankle, S. A., and Chatters, J. C. 1991. 'The effect of climate change on stream environments: the salmonid resource of the Columbia River basin', *Northwest Environ. J.*, **7**, 271–293.

Nelson, P. O. 1991. 'Cascade mountains', in Charles, D. F. (Ed.), *Acidic Deposition and Aquatic Ecosystems: Regional Case Studies*. Springer-Verlag, New York. pp. 531–563.

Nichols, F. H., Cloern, J. F., Luoma, S. N., and Peterson, D. H. 1986. 'The modification of an estuary', *Science*, **31**, 367–573.

O'Neill, R. V. 1988. 'Hierarchy theory and global change', in Rosswall, T., Woodmansee, R. G., and Risser, P. G. (Eds), *Scales and Global Change*. John Wiley Sons Ltd., New York, pp. 29–45.

Oswood, M. W., Miller, L. K., and Irons, J. G. III 1991. 'Overwintering of freshwater benthic invertebrates', in Lee R. E. Jr. and Deninger, D. L. (Eds), *Insects at Low Temperatures*. Chapman and Hall, New York. pp. 360–375.

Oswood, M. W., Milner, A. M., and Irons, J. G. III 1992. 'Climate change and Alaskan rivers and streams', in Firth, P. and Fisher, S. G. (Eds), *Global Climate Change and Freshwater Ecosystems*. Springer-Verlag, New York. pp. 192–210.

Patten, D. T., Conte, F. P., Cooper, W. F., Dracup, J., Dreiss, S. S., Harper, K., Hunt, G. L., Kilham, P., Klieforth, H. E., Melack, J. M., and Temple, S. A. 1987. *The Mono Basin Ecosystem, Effects of Changing Lake Level*. National Academy Press, Washington, D.C.

Peterson, D. H., Cayan, D. R., Festa, J. F., Nichols, F. H., Walters, R. A., Slock, J. V., Hager, S. E., and Schemel, L. E. 1989. 'Climate variability in an estuary: effects of river flow on San Francisco Bay', in Peterson, D. H. (Ed.), *Aspects of Climate Variability in the Pacific and Western Americas*, Geophys. Monogr. Ser. 55. pp. 419–442.

Poff, N. L. and Ward, J. V. 1989. 'Implications of streamflow variability and predictability for lotic community structure: a regional analysis of streamflow patterns', *Can. J. Fish. Aquat. Sci.*, **46**, 1805–1818.

Post, A. and Meier, M. F. 1980. 'World glacier inventory', *IAHS Publ.*, **126**, 45–47.

Powell, R. D. 1990. 'Advance of glacial tidewater fronts in Glacier Bay, Alaska', in Milner, A. M. and Wood, J. D., Jr. (Eds), *Proceedings, of the Second Glacier Bay Science Symposium*. US Department of the Interior, Washington, D.C. pp. 67–73.

Pupacko, A. 1993. 'Variations in northern Sierra Nevada streamflow: implications of climate change', *Wat. Resour. Bull.*, **29**, 283–290.

Quiroz, R. S. 1983. 'Seasonal climate summary. The climate of the "El Niño" winter of 1982–83 — a season of extraordinary climatic anomalies', *Mon. Weath. Rev.*, **111**, 1685–1705.

Rasmussen, L. A. and Tangborn, W. V. 1976. 'Hydrology of the north Cascades region, Washington: 1. Runoff, precipitation, and storage characteristics', *Wat. Resour. Res.*, **12**, 187–202.

Redmond, K. T. and Koch, R. W. 1991. 'Surface climate and streamflow variability in the western United States and their relationship to large-scale circulation indices', *Wat. Resour. Res.*, **27**, 2381–2399.

Regier, H. A. and Meisner, J. D. 1990. 'Anticipated effects of climate change on freshwater fishes and their habitat', *Fisheries*, **15**, 10–15.

Rind, D., Goldberg, R., and Ruedy, R. 1989. 'Change in climate variability in the 21st century', *Climatic Change*, **14**, 5–37.

Roden, G. I. 1966. 'A modern statistical analysis and documentation of historical temperature records in California, Oregon and Washington, 1821–1964', *J. Appl. Meteorol.*, **5**, 3–24.

Roden, G. I. 1989. 'Analysis and interpretation of long-term climatic variability along the west coast of North America', in Peterson, D. H. (Ed.), *Aspects of Climatic Variability in the Pacific and the Western Americas*, Geophys. Monogr. Ser. 55. pp. 93–111.

Romero, J. R. and Melack, J. M. 1996. 'Sensitivity of long-term and seasonal vertical mixing in a large saline lake to variations in runoff', *Limnol. Oceanogr.*, **41**, 955–965.

Roots, E. F. 1989. 'Climate change: high latitude regions', *Climate Change*, **15**, 223–253.
Ropelewski, C. F. and Halpert, M. S. 1986. 'North American precipitation and temperature patterns associated with the El Niño/ Southern Oscillation (ENSO)', *Mon. Weath. Rev.*, **114**, 2352–2362.
Schlesinger, M. E. and Mitchell, J. F. B. 1987. 'Climate model simulation of the equilibrium climatic response to increased carbon dioxide', *Rev. Geophys.*, **25**, 760–790.
Schneider, S. H., Gleick P. H., and Mearns, L. O. 1990. 'Prospects for climate change', in Waggoner, P. E. (Ed.), *Climate Change and U.S. Water Resources.* John Wiley & Sons, New York. pp. 41–73.
Schonher, T. and Nicholson, S. E. 1989. 'The relationship between California rainfall and ENSO events', *J. Clim.*, **2**, 1258–1269.
Sellers, P. 1995. 'Forward: Remote sensing of land surface for studies of global change', *Remote Sens. Environ.*, **51**, 1–2.
Slaymaker, O. 1990. 'Climate change and erosion processes in mountain regions of western Canada', *Mount. Res. Develop.*, **10**, 171–182.
Statzner, B., Gore, J. A., and Resh, V. H. 1992. 'Hydraulic stream ecology: observed patterns and potential applications', *J. North Am. Benthol. Soc.*, **7**, 307–360.
Stine, S. 1990. 'Late Holocene fluctuations of Mono Lake, eastern California', *Palaeogeogr. Palaeoclimatol. Palaeoecol.*, **78**, 333–381.
Stine, S. 1994. 'Extreme and persistent drought in California and Patagonia during mediaeval time', *Nature*, **369**, 546–549.
Street, F. A. and Grove, A. T. 1979. 'Global maps of lake-level fluctuations since 30000 yr B.P.', *Quad. Res.*, **12**, 83–118.
Strub, P. T., Powell, T., and Goldman, C. R. 1985. 'Climatic forcing: effects of El Niño on a small, temperate lake', *Science*, **227**, 55–57.
Vankatwijk, V. F., Rango, A., and Childress, A. E. 1993. 'Effect of simulated climate change on snowmelt runoff modeling in selected basins', *Wat. Resour. Bull.*, **29**, 755–766.
Williams, M. W., Losleben, M., Caine, N., and Greenland, D. 1996. 'Changes in climate and hydrochemical responses in a high-elevation catchment in the Rocky Mountains, USA', *Limnol. Oceanogr.*, **41**, 939–946.
Wolford, R. A. and Bales, R. C. 1996. 'Hydrochemical modeling of Emerald Lake Watershed, Sierra Nevada, California: sensitivity of stream chemistry to changes in fluxes and model parameters', *Limnol. Oceanogr.*, **41**, 947–954.
Yarnal, B. and Diaz, H. F. 1986. 'Relationships between extremes of the Southern Oscillation and the winter climate of the Anglo-American Pacific Coast', *J. Climatol.*, **6**, 197–219.

8

POTENTIAL EFFECTS OF CLIMATE CHANGE ON AQUATIC ECOSYSTEMS OF THE GREAT PLAINS OF NORTH AMERICA

A. P. COVICH,[1] S. C. FRITZ,[2] P. J. LAMB,[3] R. D. MARZOLF,[4] W. J. MATTHEWS,[5] K. A. POIANI,[6] E. E. PREPAS,[7] M. B. RICHMAN[8] AND T. C. WINTER[9]

[1] *Department of Fishery and Wildlife Biology, Colorado State University, Ft. Collins, CO 80523, USA*
[2] *Department of Earth and Environmental Sciences, Lehigh University, Bethlehem, PA 18015, USA*
[3] *Department of Meteorology, University of Oklahoma, Norman, OK 73019, USA*
[4] *US Geological Survey, 3215 Marine Street, Boulder, CO 80303, USA*
[5] *Department of Zoology, University of Oklahoma, Norman, OK 73019, USA*
[6] *The Nature Conservancy, 1313 5th Street SE, Minneapolis, MN 55414, USA*
[7] *Department of Biological Sciences, University of Alberta, Edmonton, AB T6G 2E9, Canada*
[8] *Department of Meteorology, University of Oklahoma, Norman, OK 73019, USA*
[9] *US Geological Survey, PO Box 25046 MS 413, Denver, CO 80225, USA*

ABSTRACT

The Great Plains landscape is less topographically complex than most other regions within North America, but diverse aquatic ecosystems, such as playas, pothole lakes, ox-bow lakes, springs, groundwater aquifers, intermittent and ephemeral streams, as well as large rivers and wetlands, are highly dynamic and responsive to extreme climatic fluctuations. We review the evidence for climatic change that demonstrates the historical importance of extremes in north–south differences in summer temperatures and east–west differences in aridity across four large subregions. These physical driving forces alter density stratification, deoxygenation, decomposition and salinity. Biotic community composition and associated ecosystem processes of productivity and nutrient cycling respond rapidly to these climatically driven dynamics. Ecosystem processes also respond to cultural effects such as dams and diversions of water for irrigation, waste dilution and urban demands for drinking water and industrial uses. Distinguishing climatic from cultural effects in future models of aquatic ecosystem functioning will require more refinement in both climatic and economic forecasting. There is a need, for example, to predict how long-term climatic forecasts (based on both ENSO and global warming simulations) relate to the permanence and productivity of shallow water ecosystems. Aquatic ecologists, hydrologists, climatologists and geographers have much to discuss regarding the synthesis of available data and the design of future interdisciplinary research.

INTRODUCTION

Purpose and Scope

Despite the relatively low topographical complexity of the region, the Great Plains straddles major climatic and vegetational ecotones and experiences great extremes in temperature and moisture. These extremes affect surface and groundwater hydrology and influence the distribution and quality of the region's many types of surface waters and associated biota. Here, we describe this region's characteristic physiography, geological features and aquatic ecosystems. The goal of this review is to provide a basis for predicting accurately the responses of aquatic ecosystems to directional climate change that is predicted to occur in response to global warming. We review past changes in climate that shifted the locations of the vegetational ecotones and altered the salinities and water levels in many aquatic habitats.

Our objectives are to summarize the available information on patterns of climatic spatial variability and to identify subregions of importance to ecological dynamics. We review examples of recent applications of meso-scale climatic models that relate to aquatic ecosystems and their biotic communities. We provide some examples of specific biotic communities in wetlands, springs, lakes and rivers that may be altered by directional climatic change. We emphasize that recent studies provide evidence for other sources of predictable climatic variability in this region. The global patterns of El Niño-induced shifts in precipitation are evident in the centre of this mid-continental region, especially during March (see section on 'ENSO connections'). Finally, we define some areas of research that require further study to relate the unique characteristics of this region to the likely effects of climatic change.

Our discussion is intended to complement previous reviews (Firth and Fisher, 1991; Turner and Rabalais, 1991; Carpenter *et al.*, 1992; Grimm, 1993) that provided global and regional approaches (Gleick, 1989; Schmandt and Clarkson, 1992; Packard, 1995) to synthesizing related information on the large-scale effects of climatic change. Interpreting extremes in precipitation and temperature in the Great Plains is on-going (Karl *et al.*, 1991), and in this review we do not attempt to evaluate the assumptions or bases for global circulation models or the complexities of matching their large-scale predictions to the smaller scaled studies on specific aquatic ecosystems.

Cultural Effects on Water Balances

We emphasize that any analysis of potential effects of climatic change on freshwater ecosystems in the Great Plains region, as in many others, must also consider the simultaneous effects of changing demands for groundwater and surface water resources by an expanding human population and increased demand for agricultural production, waste removal and water supplies (Covich, 1993; Naiman *et al.*, 1995; Changnon and Demisse, 1996; Changnon and Glantz, 1996; Postel *et al.*, 1996).

Identifying recent hydrological trends caused solely by climatic change is difficult because of past and on-going changes in water resources resulting from impoundment for various purposes and withdrawal for irrigation (Haragan, 1990; Hirsch *et al.*, 1990; Cohen, 1991; Glantz, 1994). Increased growth of human populations and anthropogenic diversions of water for urban and industrial uses place into perspective the added importance of future agricultural demands for irrigation (e.g. Waggoner and Schefter, 1990). These withdrawals have already greatly altered stream ecosystems (Benke, 1990; Covich, 1993; Ferrington, 1993).

If major shifts in precipitation occur in the Great Plains as a result of global climate change (Mearns *et al.*, 1990; Giorgi and Mearns, 1991; Giorgi *et al.*, 1994), the additional competition for water resources could result in further degradation or elimination of freshwater habitats. Recent droughts and subsequent changes in land use have greatly altered spatial and temporal distributions of wetlands and their biota (Swanson *et al.*, 1988). Moreover, when shallow water habitats dry up they are exposed to wind erosion. Transfer of dust and salts to agricultural areas results in economic losses (e.g. Ballantyne, 1978; Glantz, 1994).

Prolonged drought may also lead to increased shifts from dryland farming to irrigated farming methods that further increase demands for surface and groundwater resources. The hot, dry summer of 1988 reduced the US corn yield by nearly 40% and simulation models of drought effects on crop production predict major economic losses (Rosenzweig, 1990; Gates, 1993; Rosenzweig and Hillel, 1993; Rosenzweig and Parry, 1993). Despite the region's climatic variability, many river basins are among the continent's most agriculturally productive zones. Future climate changes are likely to be coupled with the additional cumulative effects of intensive land use and may accelerate the effects on biotic communities and associated ecosystem processes.

REGIONAL OVERVIEW

Physiography, Geological, and Land Use Characteristics

The overall landscape of the region is characterized by relatively low topographic relief. Elevation increases from approximately 200 m in the south-east to over 1200 m in the north-west. Runoff ranges from about 100 mm/yr in central Minnesota to about 500 mm/yr east of the Mississippi River and to <10 mm/yr along the Rocky Mountains (Brown and Matthews, 1995). Soils and runoff waters range from neutral to acid in the eastern plains and from neutral to alkaline in the western plains. Grasslands comprise the dominant terrestrial vegetation throughout the mid-continent as a result of low rainfall, high evaporation, frequent natural fires and grazing (historically by migratory bison herds). An east to west transition zone from forest to prairie spans the northern subregion. As discussed below, this transition zone is known to have moved eastwards and then westwards over the last *ca.* 800 years in response to climatic change.

The largest number of lakes in the region is in the north-eastern portions of the Great Plains where Cretaceous and Tertiary sedimentary bedrock is overlain by unconsolidated glacial deposits up to several hundred metres thick. In these glaciated regions the majority of lakes occupy morainal depressions or dammed valleys carved by glacial meltwater. They range from less than a few hectares to several hundred square kilometres in size and from 0·1 to >30 m in depth. Some of the most saline lakes are seasonal playas in the southern, and pot-holes in the northern, subregions; the majority of basins contain water throughout most years. During the ice-free season most of the shallow lakes do not stratify thermally as a result of frequent, intense winds. Several of the deepest lakes are meromictic (i.e. the waters do not circulate fully and remain permanently stratified).

A large number of lakes also occurs in the Nebraska Sandhills region, an area characterized by stabilized dunes that are believed to be the result of aeolian activity in the late Tertiary to the Holocene (Loope *et al.*, 1995). In valleys the water table reaches the surface and gives rise to permanent lakes, which are uniformly quite shallow (<4 m) and for the most part highly eutrophic (McCarraher, 1977).

River basins in the Great Plains are nearly all underlain by thick deposits of sedimentary rock that contain extensive groundwater aquifers. Groundwater from these regional aquifer systems eventually discharges into major rivers. Groundwater inflows have an effect on the magnitude and duration of stream flow as well as on the chemical quality of stream flow. The interaction of surface water and shallow groundwater with the atmosphere is of major regional significance. The net interchange of water between surface water and the atmosphere is negative for most of the region; i.e. evaporation exceeds precipitation except for the area east of the Mississippi River and the Ozark Plateau (Winter and Woo, 1990). The excess of evaporation over precipitation is as great as 1·5 m in the southern high plains.

The Great Plains can be divided into four subregions that have distinctive hydrological landscapes, physiography and climate (Winter, 1995). These four subregions are: glaciated prairie, high plains, eastern plains and the Ozark Plateau (Figure 1).

Glaciated Prairie

The northernmost subregion is the glaciated prairie. Glacial deposits overlie sedimentary rocks, and landforms are largely flat to irregular plains. The deficit of precipitation relative to evaporation ranges from 0 to 60 cm. The northern limit of this subregion is defined by the limit of discontinuous permafrost. The most common surface water features are small prairie wetlands (pot-holes or sloughs) that are present in the numerous depressions characteristic of glacial terrain (Winter and Woo, 1990; van der Valk, 1989). Most of these depressional wetlands are isolated from one another and have no interaction with streams. Furthermore, the underlying glacial deposits have very low permeability (Winter, 1995) and groundwater exchange is relatively small. Therefore, these wetlands are highly dependent on precipitation for their water supply, and they are very vulnerable to changes in climate. Although groundwater fluxes to and from these wetlands are small, their chemical and biological characteristics are determined to a large extent by their interaction with groundwater (LaBaugh, *et al.*, 1987).

Along the eastern edge of the glaciated prairie subregion there are, commonly, deeper and much more permanent lakes. Many of the small but deep lakes contain a palaeoecological record that is continuous from the time of glacial retreat (about 10 000 years ago). Elk Lake (Minnesota), for example, contains more than 10 000 annual laminations (varves) that have been studied extensively for their palaeoecological information (Bradbury, *et al.*, 1993).

A prominent surface water feature farther to the west, Devils Lake (North Dakota), also contains a long palaeoecological record in its sediments (Fritz, 1990; Fritz *et al.*, 1991). This lake is particularly significant because it is a large natural lake that receives surface inflow from a large closed drainage basin, and therefore integrates complex hydrological dynamics occurring over a large prairie landscape subject to climatic shifts (see section on 'Aquatic ecosystems').

In addition to the large area of depressional wetlands, and transient, local groundwater flow systems in the glaciated prairie, large rivers are the other major surface water features. Several large rivers traverse the region in Canada. The subregion is bounded on the south by the Missouri River. All of the large rivers have their source in the Rocky Mountains, but they also have major sources from relatively stable regional groundwater flow systems.

Many dams and water diversions have been constructed on the large rivers. Much of their natural flow regime has been altered by the presence of large storage reservoirs. The other very significant human effect on the freshwater resources of this subregion has been extensive drainage and tillage of the small depressional wetlands.

High Plains

The landscape of the northern high plains, which includes eastern Montana and Wyoming and the western Dakotas, is characterized by tablelands and high hills composed of thin residuum overlying sedimentary rock. South of the northern high plains, the central high plains subregion extends from the Missouri River to central Nebraska, and sandhills cover most of central and western Nebraska. The landscape of the central high plains is composed of plains and low hills. Sedimentary bedrock is overlain by loess in most of the subregion, and by dune sand in the sandhills. The landscape of the southern high plains, which includes most of Kansas and Oklahoma, north-west Texas and eastern Colorado and New Mexico, consists of flat to irregular plains. Sedimentary bedrock is overlain by alluvial deposits.

The climate of the high plains is cool to cold in the north, and warm to cool in the south. The deficit of precipitation relative to evaporation ranges from 40–60 cm in the north and from 20–160 cm in the south. The uplands in the high plains have many ephemeral streams that serve an important hydrological function because they are commonly focal points for groundwater recharge.

The most common surface water features in the high plains are streams of small to moderate size. This subregion is also transected by large rivers fed by runoff from the Rocky Mountains. Most of the rivers in this western landscape have small to large floodplains, and most receive groundwater inflow. However, because of the low quantities of precipitation, recharge to groundwater is also low. As discussed below, riparian wetlands exist along most streams in the high plains, and water loss to evapotranspiration can be large. Because of the minimal to moderate groundwater inflow and the large loss of both groundwater and surface water to evapotranspiration, streams in the high plains had little to no flow under pre-settlement, natural conditions, except during spring floods.

The greatest cultural impact on these western riverine systems has been the development of both groundwater and surface water for irrigation. Ranching has also accelerated stream bank erosion where riparian corridors were damaged by watering and overgrazing by cattle. In addition, the construction of large dams in the Rocky Mountain headwaters areas, as well as a few on the plains, has altered much of the natural flow regime of large rivers. Changes in climate, which could affect source areas as well as agricultural water use, could considerably alter these streams and their contiguous riparian wetlands.

Eastern Plains

The landscape of the eastern plains includes much of Iowa, northern Missouri and Illinois. It is characterized by flat to irregular plains that contain deeply incised stream valleys. A mantle of loess covers most of the interfluvial uplands. The climate of the eastern plains is warm to cold. The deficit of precipitation relative to evaporation ranges from zero along the Mississippi River to 20 cm in western Iowa. However, precipitation exceeds evaporation by as much as 20 cm in part of the subregion east of the Mississippi River.

Figure 1. Physiographical subregions of the Great Plains: (1) glaciated prairie; (2) high plains; (3) eastern plains; (4) Ozark Plateau

The most common surface water features of the eastern plains are small to moderate sized streams. The subregion is transected by the Missouri, Mississippi, Arkansas, and Illinois rivers, and it is bounded in the south-east by the Ohio River. Streams in the eastern plains have larger and more sustained flows than streams to the west, because the eastern plains have a relatively wet climate and more permeable aquifers compared with the prairies to the west.

The pre-settlement floodplains of the large rivers had extensive wetlands that were sustained by regional groundwater discharge and by flooding. Regulation of flow by dams and levées and conversion of wetlands to agricultural land by construction of dikes and levées greatly altered riverine wetlands. Furthermore, intensive agriculture in the uplands of this subregion has resulted in extensive runoff of agricultural chemicals into surface water bodies.

Ozark Plateau

The landscape of the Ozark Plateau in Missouri, Oklahoma and Arkansas is characterized by medium to high hills, and the geological framework is largely carbonate bedrock over sandstone. The climate of this subregion is warm to cool, and the ratio of precipitation to evaporation ranges from -20 cm in the west to $+20$ cm in the east.

The most common surface water features in this karst landscape are small to medium sized rivers that emerge as springs from solution cavities in the limestone. Rivers also commonly re-enter the subsurface by way of solution cavities. This region has perhaps the largest magnitude and most concentrated interchange of groundwater and surface water compared with any other landscape. This unique hydrological environment has led to ecosystems that are unique to karst terrain.

The greatest cultural impact on water resources in this region has been the development of both groundwater and surface water; development of one usually has an immediate and direct effect on the other. Furthermore, water resources in karst regions are easily contaminated by land use practices because of the direct connection of the land surface and the groundwater system through sink-holes.

Aquatic Ecosystems

The climate-sensitive aquatic habitats in the Great Plains range from springs, and spring-fed streams with uniform, relatively cool waters and permanent flows, to temporary, intermittent streams, ponds, pot-hole lakes and playas with warm waters throughout the summer months. The major natural aquatic ecosystems are large river drainage networks with numerous ponds and wetlands, which are distributed extensively across the region. The Athabasca, Saskatchewan, and Qu'Appelle Rivers flow into Hudson Bay. The Mississippi River drainage covers most of the interior of this region. The James and Kansas rivers arise in the plains, while major drainages, i.e. the Missouri, Platte and Arkansas rivers, arise in the Rocky Mountains and cross the plains as part of the Mississippi River drainage system. Historically, these large rivers had wide and braided channels. However, as previously discussed, most large rivers have been greatly modified by dams, dikes, channelization and groundwater development.

Deep lakes, floodplain lakes and shallow pot-holes of glacial origin are typical of the northern subregion (Eddy, 1963; van der Valk, 1989). Shallow ox-bow lakes, deflation basins (Reeves, 1966), playas (Sublette and Sublette, 1967; Guthery and Bryant, 1982; Osterkamp and Wood, 1987; Bolen *et al*., 1989; Haukos and Smith, 1993) and buffalo wallows are also typical aquatic habitats in the southern subregion (Carlander *et al*., 1963). It is estimated that there are approximately 25 000 playa lakes in the High Plains (Steiert, 1995).

These numerous shallow water habitats tend to be eutrophic, although many are highly turbid and have their aquatic productivity limited by light availability. Many rivers in this region receive runoff from fertilized drainages and localized nutrient additions from feed lots and urban centres, as well as point source additions from industries and waste water treatment facilities. Floodplain dynamics in riparian wetlands are also a major controlling factor over nutrient cycling in this region (e.g. Johnsgard, 1984; Galatowitsch and van der Valk, 1994).

Water balances and heat budgets are available for several well-studied lakes and wetlands, although flood control storage reservoirs and water diversions complicate the estimates of runoff and discharge. A high interannual variability of precipitation characterizes the region (Akin, 1991; Opie, 1993). Frequent droughts and floods provide striking evidence of this extreme variability. As discussed below, historical, extreme dry periods and/or high temperatures have significantly lowered lake and stream levels, reduced some wetland areas and increased salinities.

On a broad scale, salinity and ionic composition reflect both geology and gradients of precipitation minus evaporation. However, water chemistry can vary greatly within a small geographical area and is dependent on groundwater sources and topographical position within the drainage system (Almendinger, 1990). Water chemistry surveys of lakes and prairie pot-holes have been made in the north-central US (Winter, 1977; Gorham *et al*., 1983; LaBaugh *et al*., 1987, 1996), in the Nebraska Sandhills region (McCarraher, 1977; LaBaugh, 1986) and in Saskatchewan (Hammer, 1978; Hammer and Haynes, 1978; Last, 1992). Lakes within the northern Great Plains show a great diversity in ionic concentration and composition. These lakes range from dilute freshwater to hypersaline, with a salinity gradient from $<0{\cdot}1$ to >300 g/l and a variety of brine types (Last, 1992; Fritz *et al*., 1993; Hammer, 1990). Freshwater lakes (salinity <3 g/l) have sulfate, or, less commonly, carbonate, as the dominant anion, whereas sulfate is dominant in the majority of saline lakes in the Dakotas and Saskatchewan. Carbonate is abundant in many of the saline systems in the northern Plains and is dominant in a small number of saline lakes, whereas relatively few lakes show high proportions of chloride. In general, the relative proportions of $HCO_3 + CO_3$ decrease with increasing salinity, and SO_4 typically shows the opposite trend. Salts of calcium, magnesium and sodium may characterize freshwater lakes. In general, calcium proportions decrease with increasing salinity, whereas proportions of sodium and magnesium show no consistent relationship with salinity. Carbonate-dominated saline lakes, typically, have sodium as the dominant cation, whereas either magnesium or sodium may be dominant in sulfate-dominated systems.

In contrast to the northern Great Plains, the lakes of the Nebraska Sandhill region are uniformly dominated by sodium or sodium + potassium carbonate–bicarbonate brines, and in this respect are more similar to some of the alkaline Alberta lakes. Despite this relatively similar brine chemistry, the lakes span a large salinity gradient from <0·1 to *ca.* 100 g/l (McCarraher, 1977).

CLIMATE

The climate of the Great Plains is generally characterized by its continental nature and by steep gradients in temperature and precipitation. Temperature increases from north to south, with a northern regional boundary delimited by a discontinuous permafrost line. The region is bounded to the west by the north–south oriented Rocky Mountains, which reduce the amount of Pacific moisture that enters the Great Plains, creating a rain shadow and steep precipitation gradients. Mean annual precipitation decreases westwards from >1000 mm east of the Mississippi River to about 200 mm in the west-central part of the region, with a sharp increase to about 600 mm in the Pacific-influenced strip just east of the Rocky Mountains. Precipitation in the Canadian Great Plains also decreases from north (approximately 450 mm) to south (<250 mm in some areas). These factors create a relatively arid climate in the west-central portion of the region and to the north (near the Canada–US border). This pattern was well known in the nineteenth century and the 100th meridian was viewed as the westward limit of non-irrigated agriculture (Powell, 1878; Akin, 1991).

Most moisture and warmth in the south-east and central areas of the Great Plains originates from the Gulf of Mexico, with air flow that extends farther northwards during the spring and summer. In autumn and winter, cold Arctic air from the polar region flows southwards. The contrasting air masses meet frequently, producing precipitation and severe weather. Shifts between these air masses can change meteorological conditions rapidly. The northern areas, to the lee of the Rocky Mountains, are also noted for their extensive high winds (Richman *et al.*, 1991a), particularly during winter. In this area, the major source of moisture and winter warmth comes from Pacific systems that have passed over the mountains. The adiabatic compression of these air masses as they descend from the mountains produces warm, often severe winds (chinooks) and can give rise to rapid temperature increases in winter. In addition, collisions between warm Pacific air and cold Arctic air are responsible for much of the precipitation in the north-western areas, often in the form of severe storms.

In Canada and the western border states, the influence of Pacific moisture is seen along the western edge of the region (Fisheries and Environment Canada, 1978). The south-central area of the Canadian Great Plains has both the lowest annual precipitation and the lowest mean annual snowfall, with increasing precipitation to the west, east and north (reaching a maximum in the forested region that lies between the permafrost and the prairie).

In general, weather patterns are dominated by the interactions of several air masses: (1) westerly air streams with moist air that move eastwards from the Pacific and lose much (in the northern subregions) or most (in the southern subregions) of their moisture as snowfall over western mountains; (2) warm, moist tropical air streams from the Gulf of Mexico, which are the major source of moisture for the south-western and central subregions; and (3) cold, dry Arctic air streams from the north. All of these factors contribute to the variability of both temperature and precipitation over the entire region.

Much of the region has exceptionally fine-scaled, spatially explicit, long-term climatic records. To document the spatial distribution and variability of mean temperature and precipitation, the Lamb–Richman daily precipitation and daily maximum and minimum temperature data sets for 1949–1992 were analysed. These data sets are derived from heavily quality controlled observations from a cooperative network operated by the National Weather Service (Richman and Lamb, 1985, 1987; Lamb and Richman, 1990). These data demonstrate steep, north-south gradients in mean daily maximum temperature and a north-west–south-east gradient of mean daily minimum temperature, particularly in winter. Extreme mean daily temperature values range from −23°C (winter minimum, near Lake Winnipeg) to +37°C (summer maximum, near the Texas–Mexico border).

Precipitation over the US part of the region exhibits a predominant strong gradient towards the west and north-west from maxima along or near the Gulf Coast. Extremes in winter precipitation range from a high of

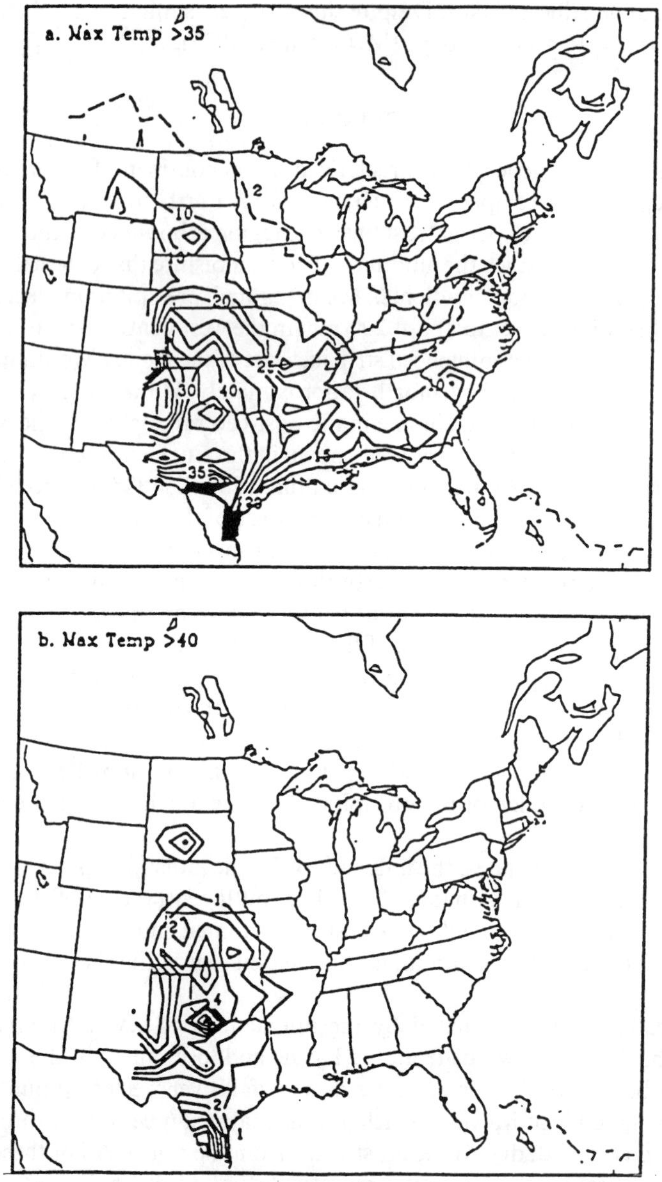

Figure 2. Percentage of summer days with maximum temperature exceeding (a) 35°C and (b) 40°C, 1949–1992, for the eastern two-thirds of North America

450 mm (near the Alabama–Mississippi border) to <50 mm in the central and northern plains. The summer precipitation pattern is characterized by a surge of moisture up into the interior regions, which begins to reverse in the autumn with a re-emergence of the gradient into winter. The Gulf of Mexico is an important source of moisture for much of the eastern half of the United States.

Because much of the southern subregion has high temperatures in summer, frequencies of extreme high temperatures exceeding 35 and 40°C are presented (Figure 2). The 35°C pattern (Figure 2a) includes much of the southern Great Plains at a value of 10% of days with this high temperature. Maximum values occur on the Texas–Oklahoma border near the Red River where >50% of days during the summers of 1949–1992

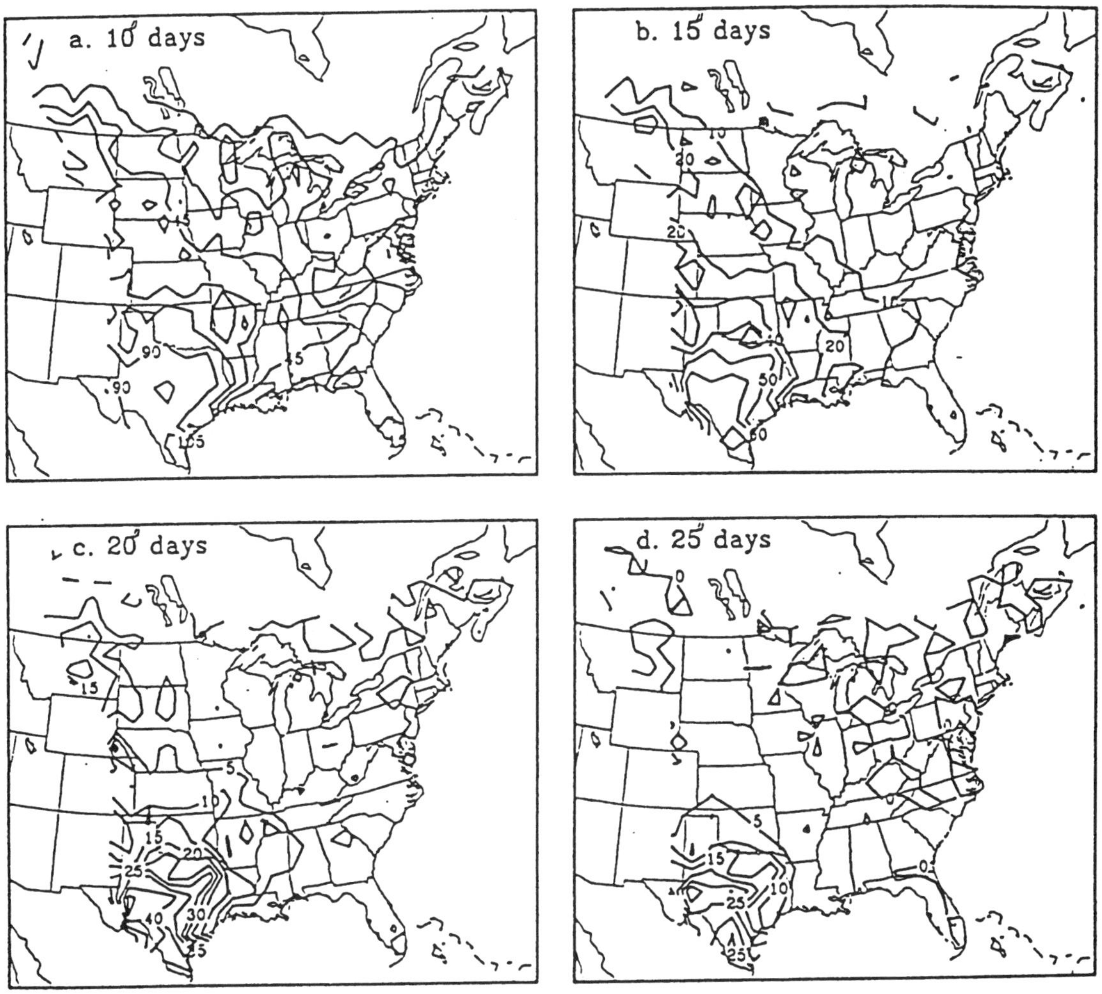

Figure 3. Frequency of occurrence of summer dry spells (no precipitation) greater than or equal to (a) 10, (b) 15, (c) 20, (d) 25 days, 1949–1992, for the eastern two-thirds of North America

exceeded the 35°C limit. These extreme thermal conditions intensely stress shallow water ecosystems in the southern half of the region in terms of water temperatures and lowering of water levels by evapotranspiration. When the number of occurrences of high temperatures in excess of 40°C was calculated (Figure 2b), the Kansas–Oklahoma–Texas region emerged as the most affected. Approximately 8% of summer days in southern and northern (but not central) Texas have maximum temperatures that exceed 40°C.

The predisposition of the southern subregion to drought can be assessed by examining the frequency of occurrence of consecutive days without precipitation. These frequencies were calculated for summer periods of 10, 15, 20, and 25 days (Figure 3). A complex pattern emerges, but with a predominant gradient oriented from south-west to north-east. The highest probabilities of dry spells are always concentrated in the southern and western portions of the region. These range up to one period per year for 20-day dry spells. Texas and Oklahoma are at the greatest risk for this type of stress as was evident in widespread grassland fires in 1996. When combined with the temperature extremes, a picture of extreme ecosystem stress emerges from the southern Great Plains.

Drought Frequency

The Great Plains had the most severe drought in the continental United States during the 1930s (Skaggs, 1975) and, in general, has more persistent droughts and wet intervals in comparison with the rest of the United States (Diaz, 1983; Karl, 1983; Stockton and Meko, 1983; Riebsame *et al.*, 1990). During the last century severe droughts in the Great Plains occurred in the late 1880s, early 1890s, 1910s, 1930s, 1950s, late 1970s, late 1980s, and most recently, in 1996. The 1930s drought was long lived and widespread (Borchert, 1950; Case and MacDonald, 1995). Yet, within the southern and central Great Plains, the 1950s drought was more severe and affected a much larger area (McGregor, 1985). In contrast, the 1988 drought was most severe in the northern and eastern Plains (Trenberth *et al.*, 1988).

ENSO Connections

Relationships between regional precipitation and the El Niño–Southern Oscillation (ENSO) phenomenon have received considerable attention over the past decade and have the potential to provide the basis for seasonal prediction (Ropelewski and Halpert, 1996). Early studies (e.g. Ropelewski and Halpert, 1986, 1987, 1989) indicated that precipitation in the south-eastern US, along the Gulf Coast, in Southern Texas and in the Rocky Mountains has a robust signal associated with ENSO. Several weaker ENSO–precipitation relationships were also suggested.

Richman *et al.* (1991b) sought to extend earlier results by applying multivariate statistical techniques to the Lamb–Richman data set and the Southern Oscillation index. They discovered an ENSO signal for the Great Plains: spring wetness/dryness over the southern and central Great Plains was found to accompany the warm/cold tropical Pacific El Niño/La Niña events. Furthermore, summer wetness/dryness in the upper Midwest was found to be associated with El Niño/La Niña.

Montroy (1993, 1997) extended the study of Richman *et al.* (1991b) by examining actual sea surface temperature (SST) anomaly patterns in the tropical Pacific and their relationship to precipitation anomalies over North America, east of the Rocky Mountains. His findings confirmed the pattern of springtime wetness (March–June) and identified those areas of the tropical Pacific with the most important anomalous SST values. An example of such a precipitation–SST relationship for March is depicted in Figure 4. Separate ENSO–SST coherency modes were first isolated statistically (using principal component analysis) for the central and eastern tropical Pacific (Figures 4a and b), and then these were regressed against the Lamb-Richman precipitation data set to resolve the respective precipitation anomaly patterns associated with each SST mode (Figure 4c and d). As a further check of the physical consistency of these findings, the station rainfall anomalies for the Great Plains regions (delineated in Figure 4c and d) were related back to the ocean SST data for individual grid points. The SST coherencies (Figures 4e and f) were highly consistent with the original principal component-based modes, indicating that large coherent regions of anomalously warm/cold water in the central and eastern tropical Pacific were related to the observed wetness/dryness in the southern and central Great Plains in March. A full description of these relationships is presented in Montroy (1997).

Because tropical Pacific SST anomalies tend to be long lived, this newly recognized set of relationships provides an opportunity to derive potential predictability of lead times of several months, to seasons. Furthermore, there may be substantial predictability in this spring ENSO–precipitation relationship for the central Great Plains, as the incidence of above normal precipitation in the central Great Plains has a good correspondence to El Niño events (Figure 5). All precipitation stations north of the Red River on the Texas–Oklahoma border, for which the correlation with tropical Pacific SST pattern in Figure 4a exceeded 0·3, are isolated in Figure 5a. Note that six of the ten El Niño events during 1950–1992 were associated with positive precipitation anomalies, whereas all six of the La Niña events coincided with a drier than normal March over the region (Figure 5b).

REPRESENTATIVE AQUATIC HABITATS

Springs

In the southern portion of the western and northern Great Plains, springs are widespread. Throughout Missouri, Oklahoma, Arkansas and Texas large numbers of springs exist in the uplands (Bowles and Arsuffi,

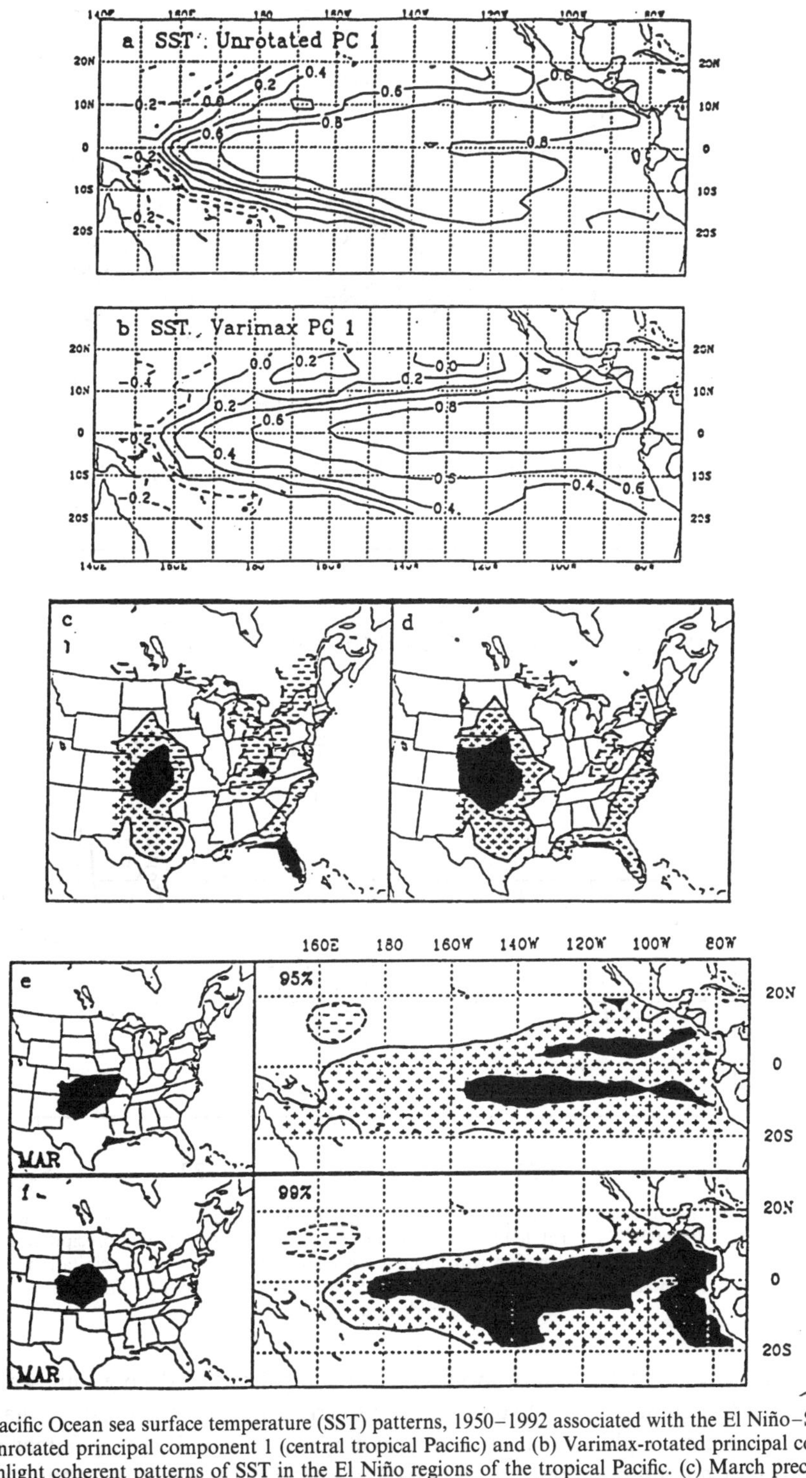

Figure 4. Tropical Pacific Ocean sea surface temperature (SST) patterns, 1950–1992 associated with the El Niño–Southern Oscillation phenomenon. (a) Unrotated principal component 1 (central tropical Pacific) and (b) Varimax-rotated principal component 1 (eastern tropical Pacific) highlight coherent patterns of SST in the El Niño regions of the tropical Pacific. (c) March precipitation correlation patterns associated with SST unrotated PC1 and (d) Varimax SST PC1 indicate precipitation signal over the Great Plains. The +/− symbols indicate positive/negative correlations > +0·2, solid shading represents correlations > +0·4. Regions shaded in the left-hand side of (e) and (f) were correlated with tropical Pacific grid cell SST anomalies (shown to right-hand side) of (e) and (f)

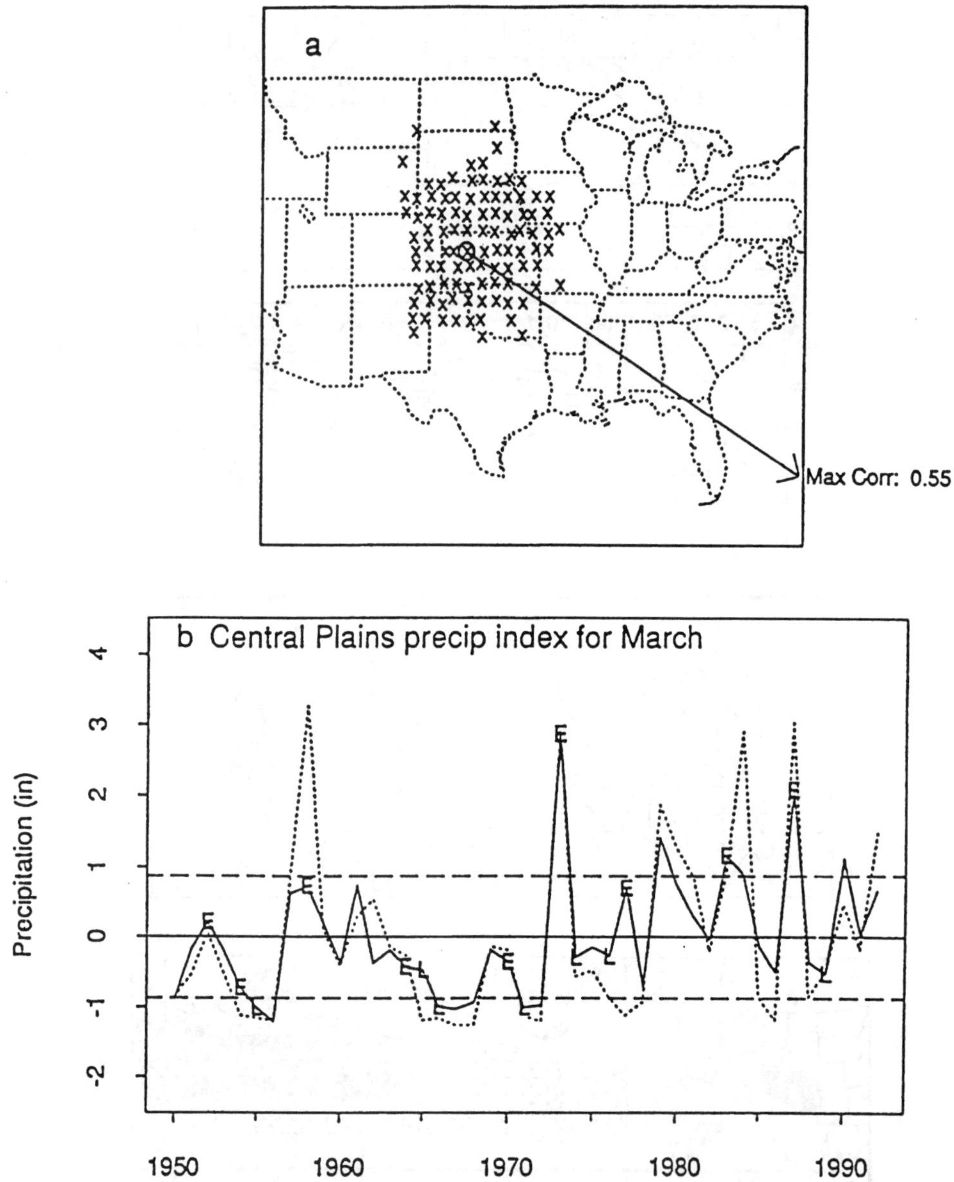

Figure 5. Space–time relationships of central Plains precipitation region for March. (a) Stations with greater than 0·2 correlation between March precipitation and March SST unrotated PC1 are identified as 'X'. (b) Average March precipitation anomaly for stations in (a) (solid line) and for single station in (a) identified with highest correlation (0·55) between precipitation and SST (dotted line). Dashed lines in (b) represent +1 standard deviation about the 1950–1992 area average mean. El Niño years 'E' and La Niña years 'L' are identified

1993; Brown and Matthews, 1995). Because springs are often regulated by deep groundwater sources, these ecosystems can integrate a combination of surface and subsurface phenomena (such as changes in temperatures or flow rates) into a detectable signal of long-term climatic change. If spring faunas remain relatively unchanged when uniform conditions prevail, any marked changes in species distributions or abundances could serve as early warning sensors of climatic change in regional aquatic ecosystems. Most of these spring faunas occur in relatively cool waters and are adapted to uniform environmental conditions (e.g. Danks and

Williams, 1991; Glazier, 1991; Varza and Covich, 1995). Only a few species are adapted for living in warm, thermal springs that are restricted to isolated locations (e.g. Waring, 1965). Some extremely warm water springs range from 30 to 55°C and provide unique habitats for specialized crustaceans (reviewed by Covich and Thorp, 1991).

Springs whose flow is strongly regulated by anthropogenic activities (e.g. water withdrawal from aquifers, groundwater pumping and elimination of recharge potential) may vary widely in discharge within or between years. Low flows during natural drought or warming periods might be exacerbated by the increased cultural withdrawal of additional groundwater to make up for loss of surface sources of water supply. Even without climatic change, there are already serious threats to some spring-fed streams and their restricted faunas (e.g. Comal and San Marcos springs that flow into the Guadalupe River, Texas) as a consequence of anthropogenic withdrawal of water (Longley, 1981).

In the Edwards Plateau region of Texas there exist numerous large springs that support a unique and largely irreplaceable fish (Hubbs, 1994) and invertebrate fauna (Edwards *et al.*, 1989; Kroschewsky, 1990). There are 90 animal species that are apparently endemic to the permanent waters in this region (Bowles and Arsuffi, 1993). The unique spring head fauna includes fishes (such as *Gambusia* and *Ethesotoma*) and numerous locally endemic crustaceans (Holsinger and Longley, 1980; Longley, 1981) and gastropods (Hershler and Longley, 1986). Some unique spring fishes, such as *Gambusia amistedensis* and *Gambusia georgei*, are already known to be extinct. Hubbs (1995) emphasizes the overall stability or physico-chemical constancy of streams fed by these large springs, in contrast to the marked instability of physical and chemical conditions in small streams of the region that are fed only by runoff or groundwater seeps.

Matthews *et al.* (1985) evaluated between-year variation in the invertebrate and fish fauna in 50 springs throughout Oklahoma. Fishes existed in approximately one-third of these springs or spring-run habitats, most often *Gambusia affinis* or darters of the genus *Ethesotoma*, with numerous minnows (Cyprinidae) as well. Rather than affecting fishes by virtue of absolute limits of thermal tolerance of individuals (e.g. Matthews and Zimmerman, 1990), altered thermal conditions could 'uncouple' adaptive complexes of reproductive traits, by shifting reproduction times of fish to periods when food availability (aquatic invertebrates) might be less than optimal (e.g. Marsh, 1980)..

Devils Lake

One of the best examples of the effects of changing climate on lacustrine systems is from Devils Lake, North Dakota. Continuous measurements of lake level have been made over the last century, and salinity measurements are occasional until 1960 and generally monthly thereafter. These data (Figure 6), compiled by the US Army Corps of Engineers, clearly show the decline in lake level and coincident rise in salinity associated with drought intervals and the inverse response during periods of high precipitation (Fritz, 1990). Lake level appears to be controlled primarily by climate, although the complex hydrology of the lake, which includes a number of upstream storage basins, complicates the relationship. Analysis of climate data indicates that periods of very low lake level are correlated with both decreased summer precipitation and increased summer temperature for the first half of the twentieth century. Water level fluctuations from 1931 to 1983 are also correlated with changes in the cumulative departure from average winter precipitation until 1969; but the relationship breaks down after this date, perhaps because of changes in basin storage and drainage area (Wiche *et al.*, 1986). Based on the sensitivity of the modern lake to climate the stratigraphic record of palaeosalinity fluctuations has been used to reconstruct the climate history of the Devils Lake region, and this record clearly demonstrates that severe drought is a common climatic feature of the northern Great Plains (Fritz *et al.*, 1991, 1994; Haskell *et al.*, 1996).

Prairie Potholes

The prairies of the American Midwest and western Canada have several hundred thousand small (< 100 ha), shallow (maximum depth <5 m) pothole lakes (Krapu and Duebbert, 1989). These lakes lie in shallow depressions in fertile glacial till or fluvial deposits that extend over approximately 776 900 km^2. Such

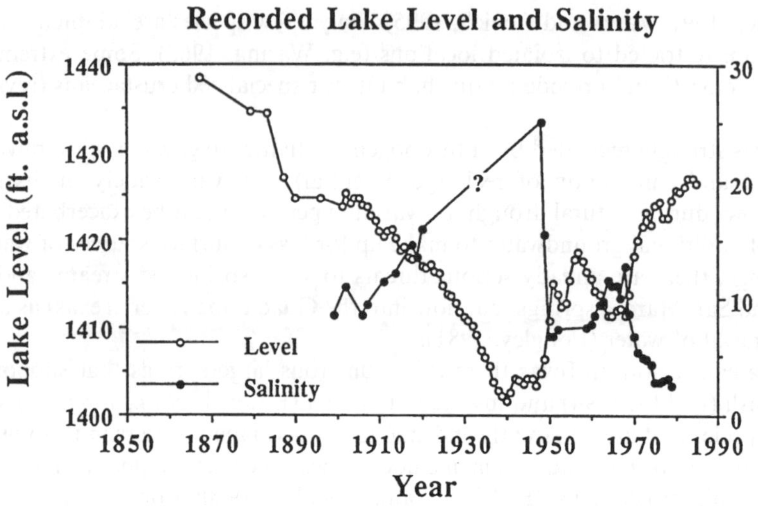

Figure 6. Relationship between water level and salinity for Devils Lake, North Dakota. Data compiled by US Army Corps of Engineers

closed-basin wetlands are found throughout the glaciated portion of the northern Great Plains and comprise some of the most ecologically valuable freshwater resources of the region (Hubbard, 1988; Swanson, 1988; van der Valk, 1989).

Water levels in pothole lakes fluctuate widely in response to long-term climatic wet/dry cycles, and even annual water level fluctuations can be significant (Barica, 1980). Thus, prairie potholes vary significantly in water permanence and water quality. They range from small, fresh, temporary and seasonal ponds that dry up yearly, to larger, brackish semipermanent and permanent ponds that hold water during most years (Stewart and Kantrud, 1971). These lakes have small effective watersheds (Lawler *et al.*, 1974) owing to the low relief of the region and their land-locked nature. Unlike more rapidly flushing water bodies lying along active river systems, prairie potholes typically have no permanent inflow or outflow streams. Rather, they receive water input from precipitation, diffuse surface runoff and groundwater seepage (Barica, 1980). Water quality can vary between pot-holes on a relatively fine spatial scale.

Prairie wetland water budgets are dominated by atmospheric processes, in particular, direct precipitation and evapotranspiration (Winter, 1989; Winter and Rosenberry, 1995). Although atmospheric exchanges and groundwater are dominant components of the annual water budget in prairie potholes, spring runoff becomes increasingly important at higher latitudes (Schwartz and Gallup, 1978; Burn, 1994). Most runoff at higher latitudes occurs over a relatively short time period, when the soils are still frozen and impermeable to infiltration. However, inflow is intense enough during this brief interlude that it contributes the largest pool of organic matter relative to other precipitation events (S. Cooke, University of Alberta, personal communication). Thus, runoff in spring from rain and melting snow on frozen or saturated soils supplies wetlands with an important source of water early in the season (Woo and Winter, 1993).

Of the 100 lakes in the *Atlas of Alberta Lakes* (Mitchell and Prepas, 1990), 76 are in the northern portion of the Great Plains (south-central Alberta). Evaporation exceeds precipitation in all 76 lakes; the difference between evaporation and precipitation (Figure 7) is almost threefold greater in southern than in northern lakes (Swanson, 1988). Precipitation has declined overall from 1905 to the present in central Alberta as a consequence of decreased snowfall rather than rainfall (Evans and Prepas, 1996). The relative importance of spring runoff in the annual nutrient budget of pothole lakes may be enhanced by an increase in the importance of snowfall.

There is a trend for decreasing water levels evident in about 50% of lakes in the *Atlas of Alberta Lakes* that do not have water level control structures (i.e. weirs, dams). For example, the water level in Cooking Lake

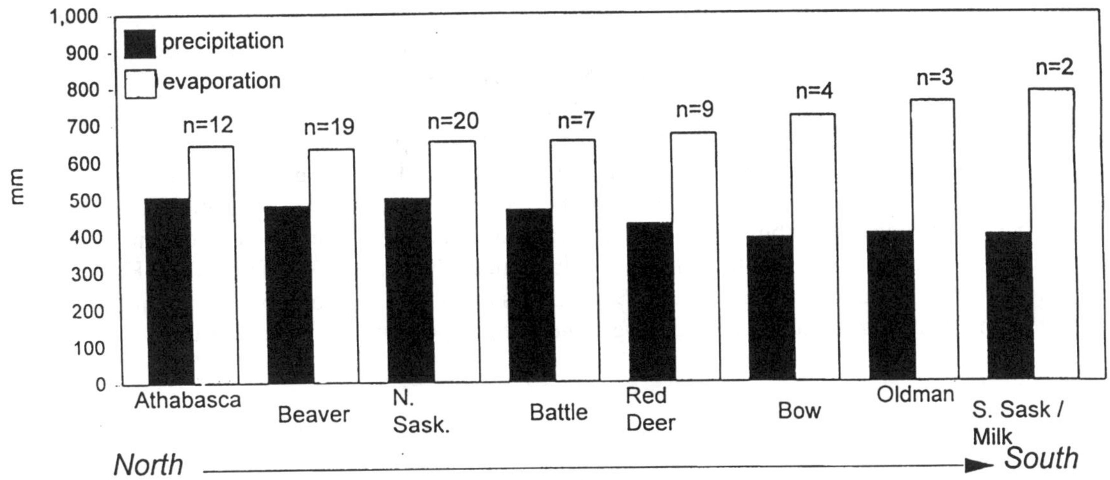

Figure 7. Mean annual evaporation and precipitation for 76 lakes by river basin within Alberta. (From *Atlas of Alberta Lakes*)

has declined by 2 m since 1919. This decline was attributed to decreased precipitation and spring runoff, and to high rates of lake evaporation (Alberta Environment, 1977).

An increase in salinity is apparent in some lakes in central and southern Alberta and Saskatchewan. Redberry Lake, in central Saskatchewan, has increased in salinity from 12 to 21 g/l in the last 70 years. As salinity increases, species diversity in aquatic communities tends to decrease (Swanson *et al.*, 1988).

In central Alberta, mean daily temperature during summer has increased overall from 1905 to the present, a trend apparently driven by an increase in the daily minimum temperature (i.e. warmer nights; Evans and Prepas, 1996). In shallow lakes, nighttime water column mixing owing to onshore breezes could be reduced if the land surrounding the lakes remains relatively warm during the night. Mean annual temperatures for May–August are extremely variable, but the increasing trend is stronger in May and June than in July and August (Evans and Prepas, 1996).

In summer, high insolation and shallow depths result in warm water (>20°C) that is relatively uniform, within 1–3°C (Barica, 1975). Deeper lakes, however, may be thermally stratified for extended periods (>four months). Thus, shallow lakes tend to be polymictic, while deeper lakes may be partially meromictic or dimictic.

Most prairie potholes and lakes are eutrophic (Barica, 1980), owing to relatively rapid recycling of phosphorus from the bottom sediments (Mitchell and Prepas, 1990) and allochthonous nutrient loading from agriculture. Frequent water level fluctuations may enhance the release of nutrients from shoreline soils, and thus promote eutrophication, especially in shallow systems (Trew, 1983). Internal nutrient cycling generally determines the trophic status of most of these systems on an annual basis; external nutrient loading is less important.

Vegetation changes in prairie wetlands are also highly dynamic and respond to changing water levels (Kantrud *et al.*, 1989). Wetland cover cycles have been described in detail (e.g. van der Valk, 1989; Kantrud *et al.*, 1989), and, in semipermanent ponds, range from lake-like conditions during high water years to completely vegetated basins during drought.

There are more than 200 species of phytoplankton in prairie pot-hole lakes (Barica, 1980). Diatoms, which require mixed conditions and available N and P, can dominate in spring or autumn when water temperatures are relatively low. However, in summer, high P concentrations and water temperatures (>15°C) promote the growth of cyanobacteria, which rapidly out-compete diatoms as a result of the ability of cyanobacteria to fix N and alter their depth within the water column by controlling intracellular buoyancy. These cyanobacterial blooms can form surface scums that diminish light penetration and may contain toxin-producing species (Kotak *et al.*, 1993). Under optimal conditions, cyanobacterial biomass can double each day and peak biomass can be extremely high (Hammer, 1964 1986; Barica, 1980).

The complexity of aquatic animal communities between pot-hole lakes has a broad range that depends on local conditions (Barica, 1980). Although productive enough to support dense invertebrate and vertebrate communities, pot-hole lakes have highly variable nutrient, dissolved oxygen, water level, salinity and temperature regimes, which result in less diverse and productive animal communities relative to other regions. The geographical distributions of invertebrate communities reflect differences in water depth, permanence, chloride and sulfate content of pot-hole lakes (e.g. Cvancara, 1983; Timms and Hammer, 1988; Hammer *et al.*, 1990; Bataille and Baldassarre, 1993). Shifts in community composition are likely to be most sensitive to climatic changes that alter water levels and habitat permanence or seasonality.

EVIDENCE FOR PAST CLIMATIC CHANGE

Water level data are available from streams with relatively long records (> 50 years) of river discharge and from historical documentation of lake level changes in pot-holes and lakes. Detailed estimates of annual precipitation over the past 300 years have been reconstructed from tree ring studies in some basins.

Deep lakes with long sedimentary records and tree ring chronologies provide high resolution information on rapidly changing regional climates. For example, annual laminae (varves) have been deposited in the sediments of Elk Lake, Minnesota, for more than 10 000 years. This fine time-scale of deposits has provided important evidence for rates of climatic change since the last ice age. As an example, thick varves and the abundance of quartz in the sediments are indicators of aeolian transport during dry periods. These data indicate that the middle Holocene had two extended periods of dry and windy conditions. A significant feature of this record is the sharp change in climatic conditions, spanning only decades, from dry to wet about 5400 years ago, from wet to dry about 4800 years ago, and from dry to wet about 4000 years ago (Figure 8). Another, less severe, dry period occurred less than 1000 years ago. These data are particularly significant because they indicate that some changes were very rapid (Bradbury *et al.*, 1993).

The broad-scale climatic history of the northern Great Plains is known from pollen studies that document vegetation change over the last 12 000 years following the retreat of the Laurentide Ice Sheet. Palaeolimnological studies also reconstruct the climatic effects on limnological variables, such as lake level changes. The majority of these studies document century-scale climatic variability over the Holocene and provide a general framework for assessing the extent to which recent or future climate changes are unique or unusual, in comparison with the relatively recent geological past. Only a few recent studies are sufficiently detailed to examine patterns and rates of climate change on a decadal or subdecadal time-scale.

In general, pollen studies from the northern Great Plains show climatic warming following the retreat of the Laurentide Ice Sheet, culminating in a mid-Holocene warm/dry interval (Bartlein *et al.*, 1984). At the southern edge of the northern Plains, spruce forest was replaced directly by prairie *ca.* 12 000 BP (Watts and Wright, 1966), and around 10 000 BP in Canada (Ritchie, 1976; Teller and Last, 1981). In intermediate locations, spruce forest gave way to transitional communities of spruce–aspen parkland or deciduous woodland before the expansion of prairie between 10 000 and 7000 BP (Watts and Bright, 1968; Barnosky *et al.*, 1987; Radle *et al.*, 1989; Baker *et al.*, 1992). The variability in timing of prairie expansion suggests regional variability in the early Holocene climate, particularly with respect to moisture balance, which controls the dominance of forest versus grassland.

Palaeolimnological studies from lakes in the region (Radle *et al.*, 1989; Last, 1990; Fritz *et al.*, 1991; Vance and Mathewes, 1994; Laird *et al.*, 1996) show a transition from hydrologically open freshwater to closed saline conditions, approximately coincident with the transition to prairie. A subsequent shift to moister/cooler conditions occurs as early as 6000 BP in north-central Montana and western Canada (Wells, 1970; Sauchyn, 1990; Vance *et al.*, 1992; Last and Sauchyn, 1993) but not until after 4500 BP in some parts farther

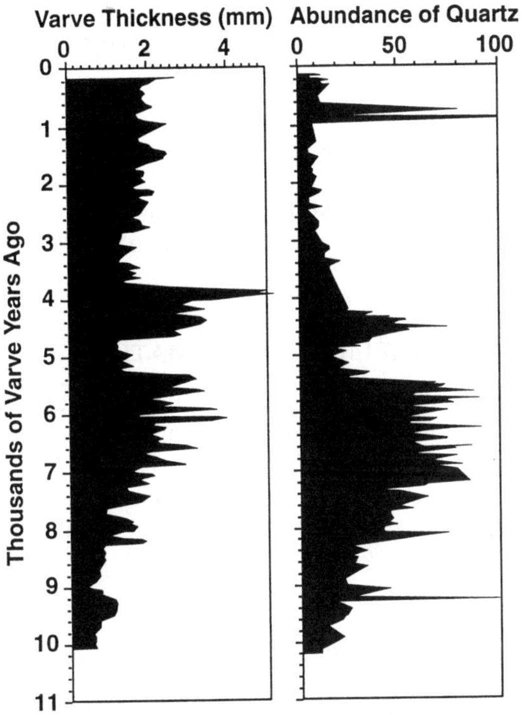

Figure 8. Thickness of laminated lake sediments (varves) and abundance of quartz grains demonstrating rapid changes in rates of deposition resulting from climatic changes, for Elk Lake, Minnesota. (From Bradbury *et al.*, 1993)

to the east (Watts and Bright, 1968; Last and Schweyen, 1985; Last and Slezak, 1988; Vance *et al.*, 1992; Laird *et al.*, 1996).

Most pollen-based studies in the northern Great Plains show relatively stable conditions over the last few thousand years. Lake records, however, suggest that oscillations between high and low salinity occurred during the last 2000 years (Last and Schweyen, 1985; Last and Slezak, 1986; Fritz *et al.*, 1991; Vance *et al.*, 1992; Van Stempvoort *et al.*, 1993). A number of droughts occurred during this period that were at least as severe as the historical droughts of the twentieth century (Fritz *et al.*, 1994). Similarly, tree ring studies, which extend over the last few centuries, suggest that the twentieth century droughts were not unusual in magnitude, and may in fact have been shorter in duration than those prior to written records (Meko, 1992; Case and MacDonald, 1995).

Palaeoecological data from the central Plains are relatively sparse (Fritz, 1996). Pollen studies from Kansas (Gruger, 1973) and the Nebraska Sandhills region (Watts and Wright, 1966) show a late glacial environment dominated by spruce forest and a rapid transition from spruce forest to grassland in the late Pleistocene as the climate warmed. Alluvial data suggest that the middle Holocene was a period of major stream instability and more intense erosion in comparison with earlier and subsequent intervals (May, 1989; Martin, 1992; Arbogast and Johnson, 1994). At the same time, aeolian evidence indicates increased aridity and landscape instability (Forman and Maat, 1990). In the Sandhills (Wright, 1985) a subsequent transition from marsh sediments to lake gyttja *ca.* 3700 BP suggests a rise in water table associated with increased moisture, similar to the pattern seen to the north. Tree ring data suggest an interval of drought in the central Great Plains *ca.* 800 BP, and this is approximately coincident with evidence of increased aeolian activity (Muhs, 1985; Madole, 1994) and the onset of stream entrenchment (May, 1989; Martin, 1992; Arbogast and Johnson, 1994).

The history of the southern Great Plains is known from a small number of studies that include century-scale pollen (Holliday, 1989) and palaeohydrological (Haynes, 1991) studies of bog and lake sediments, and, more recently, isotopic studies of alluvial sediments and soils (Humphrey and Ferring, 1994; Nordt *et al.*, 1994). Pollen studies in north and central Texas show a transition from deciduous forest, during the glacial period, to oak savannah and grassland, which persisted from the late Pleistocene to the present day. A cool and dry late Pleistocene, followed by an increase in moisture about 10 800 BP, is suggested by both pollen and isotope records. A mid-Holocene drought characterizes the southern Plains, although here too the data suggest variability in the timing and duration, and it is unclear whether temperature during this interval was actually higher (Humphrey and Ferring, 1994). More mesic conditions followed after *ca.* 4000 BP and have continued to the present day, although there is some indication of a brief warm/dry episode *ca.* 2000 BP (Nordt *et al.*, 1994).

POTENTIAL EFFECTS OF CLIMATIC CHANGE

By 2030, several GCMs (general circulation models) project that the Great Plains could be on average 2–4°C warmer in winter and 2–3°C warmer in summer. GCMs also project a less productive growing season as a result of less precipitation (Mearns *et al.*, 1990; Gates, 1993; Halpin, 1995; Hogg and Williams, 1996). Although projections remain uncertain and regional biotic responses to predicted climatic shifts are even more uncertain, ecologists have begun to explore several alternative simulations regarding regional climatic change (e.g. Hogg *et al.*, 1995).

Prairie Lakes and Wetlands

Studies that have addressed the potential effects of a changing climate on prairie wetlands are few, particularly those focused on the shallow, closed-basin wetlands typical of northern glaciated region. Most studies of prairie wetland hydrology and vegetation dynamics have characterized wetlands under current climatic conditions (e.g. Shjeflo, 1968; LaBaugh *et al.*, 1987; Winter, 1989; Kantrud *et al.*, 1989; Woo and Rowsell, 1993; Rosen *et al.*, 1995). Crowe (1993) used a coupled water balance–salinity model to examine the sensitivity of a large, groundwater-dominated lake in Alberta to climatic variability. He found that even small changes in temperature (1–2°C increase) or precipitation (5–10% decline) had a significant effect on lake water quality and the availability of groundwater resources. The major effect of these climate perturbations was to increase significantly the salinity of the lake, as a result of: (1) decreased inputs of surface runoff and direct precipitation for dilution; (2) increased evapotranspiration, which concentrates salts; and (3) reduced surface discharge as lake levels fall below the elevation of the outlet, which removes dissolved mass from the lake (Crowe, 1993). Although this model was developed for a large, permanent lake with surface outflow, similar changes will probably occur in the brackish and slightly brackish intermediate sized pot-holes.

In a second hydrological study in the region, Hands and Stefan (1993) used a lake water quality model linked to a daily weather database to simulate water temperature profiles under changing climate conditions in Minnesota lakes. Lakes ranged from $<0{\cdot}4$ to >5 km^2 and <6 to >20 m deep. Simulations using output from a general circulation model (GISS model) showed that epilimnetic temperatures were higher but increased less than air temperature, hypolimnetic temperatures in seasonally stratified dimictic lakes were unchanged or even lower, evaporative water loss increased significantly, onset of stratification occurred earlier and overturn later in the season, and overall lake stability was greater in spring and summer (Hands and Stefan, 1993).

Studies on the effects of a changing climate on prairie wetland vegetation are even fewer. Several field experiments in northern Alberta have examined the effects of fire and drought on wetland vegetation with implications for potential climate change (Hogenbirk and Wein, 1991, 1992). Although outside the Prairie pot-hole region, this study demonstrated that increased drought and fire, which would be expected with climatic warming, had a significant effect on wetland vegetation. Eurasian dicots increased 5- to 13-fold after treatments and the authors hypothesized that Eurasian species might dominate early successional communities in these mid-boreal wetlands (Hogenbirk and Wein, 1991). In a second study, the authors

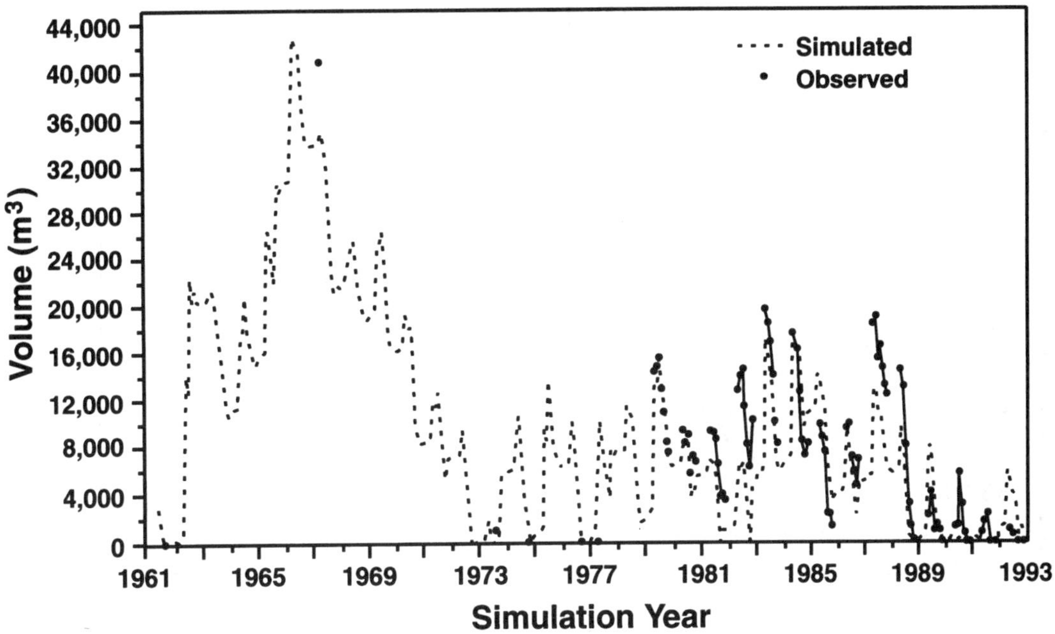

Figure 9. Observed water volume (m^3) in wetland P1 versus WETSIM model calculations, 1961–1992. Wetland P1 is a semi permanent prairie wetland located at the Cottonwood Lake Site, North Dakota. (From Poiani *et al.*, 1996)

showed that the effects of increased temperatures on seedling emergence from wetland seed banks again favoured introduced weedy species (Hogenbirk and Wein, 1992). Other relevant issues related to climate change and prairie wetland vegetation, bird habitat and bird populations (e.g. direct effects of CO_2, UV radiation) are reviewed thoroughly in Larson (1994).

A mathematical simulation model of wetland vegetation and hydrology was developed by Poiani and Johnson (1993a) and updated recently (Poiani *et al.*, 1996). The wetland simulation model (WETSIM) calculated changes in water level and vegetation cover for semipermanent prairie wetlands. WETSIM was developed, calibrated and tested using data from semipermanent wetlands on the Cottonwood Lake site, central North Dakota (Poiani and Johnson, 1993a; Poiani *et al.*, 1996), where extensive, long-term climate, hydrology and vegetation data exist (Winter and Carr, 1980; LaBaugh *et al.*, 1987; Winter and Rosenberry, 1995). Results for an historical 30-year simulation of wetland P1 at Cottonwood Lake using WETSIM (Poiani *et al.*, 1996) showed good agreement between simulated and actual values. The wetland model depicted known wet/dry periods over the three decades and simulated water levels, and cover ratios compared favourably with actual conditions for the period in which these data were recorded (Figure 9).

WETSIM was also used to explore the sensitivity of prairie wetland dynamics to potential changes in climate (Poiani and Johnson, 1991, 1993b; Poiani *et al.*, 1995). Climate change simulations were performed for semipermanent wetland P1 using climate data for one decade adjusted for doubled CO_2. Climate adjustments included: (1) projections from the GISS climate model ('GISS simulation'); (2) a range of hypothetical temperature (+2 and +4°C) and precipitation changes (−20, −10, 0, +10, +20%) ('sensitivity simulations'); and (3) a +2°C increase in temperature combined with a 10% increase or decrease in total growing season precipitation that varied by season (all months, spring, summer, autumn) ('seasonal simulations'). Simulations of different climatic changes were compared with water level and cover changes in wetland P1 using data with no adjustments ('current climate').

All climate change simulations showed significant changes in wetland hydrology and cover dynamics compared with current climate conditions. For example, under GISS projections, peak water levels were

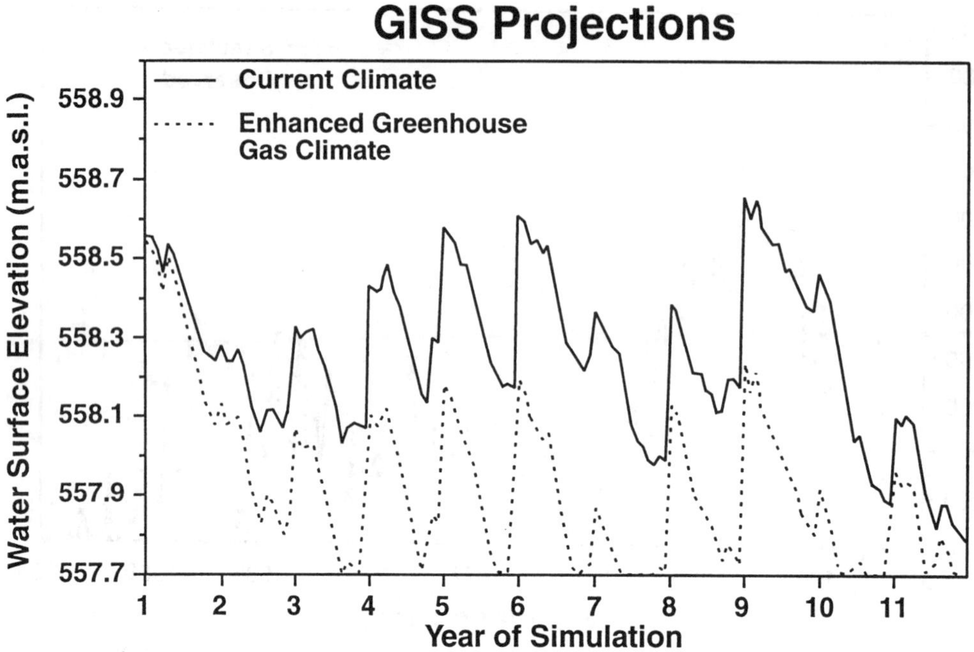

Figure 10. WETSIM model output of changes in water surface elevation (metres above sea level) in wetland P1 for an 11-year simulation under current climate and enhanced greenhouse gas climate conditions. Temperature and precipitation adjustments for the greenhouse gas scenario were from the GISS general circulation model. The entire wetland was dry at approximately 557·7 m above sea level. (From Poiani and Johnson, 1993b)

almost 0·5 m lower than under current climate conditions and the wetland dried nearly every year (Figure 10). Open water declined slightly under the current climate owing to drought late in the decade. In contrast, open water area decreased quickly in the GISS simulations from the initial 51% to 0% by year 4 (Poiani and Johnson, 1993b). The implications of these potential changes in hydrology and cover to the quality of waterfowl breeding habitat are severe and are discussed in Poiani and Johnson (1991).

Sensitivity simulations elucidated several important aspects of wetland behaviour. Hydrological processes were, in general, more sensitive than vegetation processes to changes in temperature and precipitation. Sensitivity simulations also showed that there may be a threshold temperature beyond which changes in precipitation become less important to wetland hydrology. Precipitation change was much less influential on wetland water levels with an increase of +4°C compared with +2°C (Poiani and Johnson, 1993b). Wetland hydrology was more sensitive, however, to increases versus decreases in precipitation at the more moderate temperature increase (+2°C). These preliminary results stress the need for accurate projections of climate change, particularly at the regional scale (Sulzman *et al.*, 1995).

Seasonal simulations also identified several important aspects of wetland sensitivity to climate (Figure 11). First, the response of P1 was similar to precipitation changes imposed over all months, summer and autumn. Even a 10% increase in precipitation in these seasons produced lower water levels compared with the current climate, as a result of increased evapotranspiration (Poiani *et al.*, 1995). Secondly, wetland hydrology was most affected by changes in spring precipitation. Thus, potential changes in climate that affect spring refill, such as changes in spring precipitation and snowmelt may have the greatest impact on prairie wetland hydrology and vegetation. Finally, results showed that relatively small changes in water level during dry years can significantly affect the proportion of emergent vegetation in the basin. Emergent vegetation is difficult to eliminate under dry conditions once it is established owing to its ability to persist in shallow, fluctuating water.

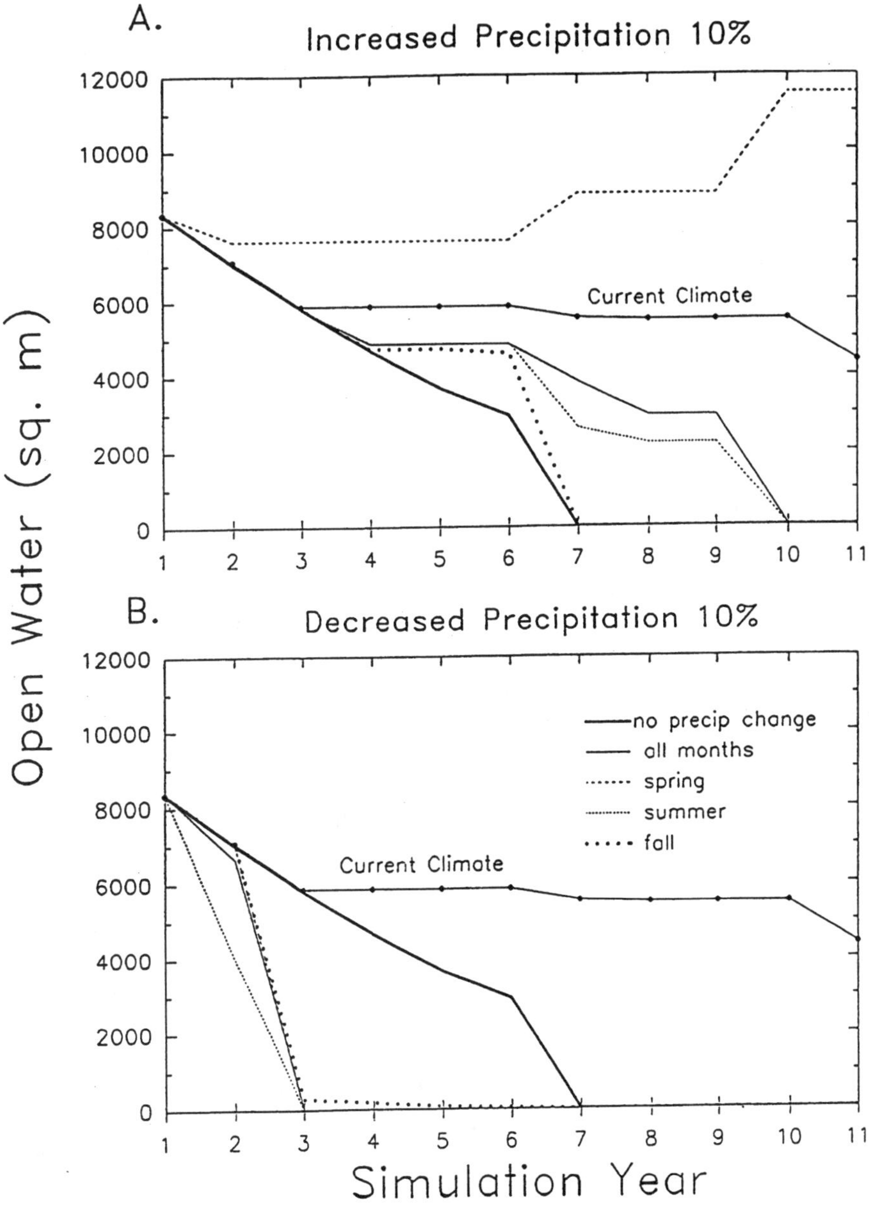

Figure 11. WETSIM model output of open water areas (m^2) in wetland P1 under nine hypothetical climate simulations and the current climate. Each simulation was run for 11 years. Size of open water is on 1 August of each year. In the climate change simulations, temperatures were elevated +2°C above current climate conditions and growing season precipitation was: unchanged, (A) increased or (B) decreased by 10%, with those changes distributed between different sets of months in the growing season. (From Poiani *et al.*, 1995)

A second study of climate change and northern prairie wetlands supports the general simulation results described above. Larson (1995) found that climate variables such as spring and autumn temperatures, yearly precipitation, the previous year's count of wet basins, and in grassland regions, the previous fall precipitation, accounted for 63 to 65% of the variation in number of wet basins under the current climate. When changes in temperature and precipitation consistent with increased greenhouse gases were imposed on models, the number of wet basins declined between 12 and 74% depending on the climate change simulation and region (Larson, 1995). Parkland wetlands were most vulnerable to increased temperatures, while sensitivity to increased precipitation did not vary geographically.

Wetland Waterfowl

The wetlands in the pothole region provide the continent's most important habitat for breeding waterfowl (Swanson, 1988; Kantrud *et al.*, 1989; Swanson and Duebbert, 1989). Prairie wetlands produce 50–80% of the total duck populations of North America, including 20 of 34 species of ducks that breed in North America (Batt *et al.*, 1989; Greenwood *et al.*, 1995). The number of wet basins and their cover conditions influence breeding bird habitat, e.g. nesting cover, brood rearing areas, food supply (Kantrud *et al.*, 1989; Swanson and Duebbert, 1989). For example, waterfowl breeding strategies are altered in response to drought, including migration to different geographical areas, or remaining on permanent lakes in a non-breeding status (Swanson and Duebbert, 1989).

Currently, wetland habitat loss in the prairie pothole region is high owing to competing agricultural land use (Hubbard and Linder, 1986; Hubbard, 1988; Rosen *et al.*, 1995). Thousands of wetlands have been drained, ditched or tilled, and many others receive sediments, chemicals and nutrients from surrounding cropland (Kantrud *et al.*, 1989; Neely and Baker, 1989). Potential climatic changes, in combination with existing stresses from agriculture, may further increase the loss of wetland value and function. Thus, assessment of the potential effects of climate change on this valuable freshwater resource remains imperative.

Long-term climatic data suggest that variable precipitation will greatly alter the population dynamics of waterfowl, especially in the northern prairie pothole region. As discussed above, simulations of projected water level changes as a result of global warming have suggested that semipermanent prairie pothole wetlands will be sensitive to changes in evaporation and precipitation. These simulations provide a basis for estimating future declines in waterfowl production resulting from lower water levels. Spatially specific models of population dynamics will be needed to manage duck hunting and to analyse the responsiveness of migratory routes to variable habitats (Nichols *et al.*, 1995; Johnson, 1996).

Stream Fishes

During late summer in the Great Plains, many shallow water habitats are already at extremely high temperatures and low flows, yet native fishes are adapted to these conditions (Matthews and Zimmerman, 1990; Smale and Rabeni, 1995; Fausch and Bestgen, 1996; Rabeni, 1996). Any additional increases in water temperatures, lowering of water levels or increases in salinity are likely to be highly detrimental to most native fish species. For example, streams in the southern Great Plains currently have some of the hottest free-flowing water on earth, with summer maxima of 38–40°C. Native fishes that survive in these warm waters are already near their lethal thermal limits. Although, some species persist and even reproduce under such harsh conditions, it seems likely that they are already functioning at or near their limits of tolerance, and it is unlikely that these species could adapt, genetically, in a sufficiently rapid manner to permit their continued existence if directional warming occurs (Matthews and Zimmerman, 1990).

Increased extinction rates for endemic species and isolated local populations are likely to occur if water temperatures exceed current maxima. Such 'local' extinctions will be significant if future climatic extremes are caused by large-scale, synchronous, warmer and drier climates throughout the entire region. Previous extreme events have had relatively local effects, especially in headwater streams, and recolonization of habitats by fishes and invertebrates from downstream refugia were relatively rapid (e.g. Larimore *et al.*, 1959).

The southern Great Plains represents a biogeographical transition between the very diverse fish faunas of eastern uplands and the very depauperate faunas of the extreme western plains (McAllister *et al.*, 1986). With the exception of the Edwards Plateau in Texas, there are few endemic fishes that occur in the Great Plains (McAllister *et al.*, 1986). However, there are a few endemic species that are characteristic of Great Plains streams: e.g. *Notropis bairdi* and *Notropis girardi*, located in the western Red and Arkansas river drainages, respectively; *Fundulus zebrinus*, which occurs broadly throughout the plains; and *Cyprinodon rubrofluviatilis*, a pupfish endemic to the Red and Colorado river drainages.

In the eastern portion of the region (approximately the eastern half of Nebraska, Kansas, Oklahoma and Texas), local fish assemblages may contain as many as 20–25 species within a few hundred metres of stream (W. J. Matthews, personal records). In contrast, as one passes westwards to streams in regions with lower rainfall and higher evapotranspiration, stream flows become much more erratic, and streams are intermittent or ephemeral (intermittent reaches of streams cease flow, while ephemeral reaches dry completely). Mean temperature fluctuates in excess of 10°C daily, and oxygen concentrations are commonly less than 2 ppm in isolated pools. In these habitats, fishes may be limited most by tolerance for low oxygen conditions (Matthews, 1987; Zale *et al*., 1989). The effects of extreme temperatures and oxygen depletion on game fishes and other associated species is also of considerable importance to the recreational uses of reservoirs, rivers and ponds (Davis, 1975; Armour, 1993; Smale and Rabeni, 1995).

In extreme south-western Oklahoma and the Texas panhandle, mainstream and tributaries of the Red River carry a very high salt load, with salinities exceeding that of sea water. At such sites it is common to find only two fish species, the Red River pupfish (*Cyprinodon rubrofluviatilis*) and plains killifish (*Fundulus zebrinus*), which are highly tolerant of extremes of heat and salinity (Echelle *et al*., 1973).

Many of these native fish species are in confined habitats and cannot migrate to cooler, less saline waters. Given the east–west orientation of the many tributaries of the main drainages (Matthews, 1988), there is limited access to migratory corridors for fish (and associated fauna such as riverine mussels) to reach cooler, northern habitats. Dams and water diversions further limit any likelihood of movement to northern habitats.

Warming of groundwater and spring-fed streams is also a distinct possibility that could affect fish species diversity. In a warmer, drier climate the increased cultural demand for diverting surface and ground water resources could eliminate some shallow water habitats. In the event of global warming and increased desiccation, many kilometres of what are now small prairie streams will cease to exist, i.e. they will lack water habitable by fishes, year-round. Monitoring of temperatures, dissolved oxygen and flow rates in these habitats is needed, along with dynamic modelling in order to evaluate whether directional changes are significant.

CONCLUSIONS AND RECOMMENDATIONS

Studies of past climatic changes in the Great Plains indicate that the spatial and temporal distributions of rainfall and temperature have changed relatively rapidly several times in the last 10 000 years. These climatic changes not only influenced the distribution of terrestrial plant communities, but also altered the distribution of aquatic plants and animals, water levels and salinities. Regional, high resolution palaeolimnological data provide a solid base for 'backcasting' simulations based on current and future global circulation models.

We conclude that sufficient data are available for several subregions to generate some likely consequences of increased warming and decreased precipitation on shallow water habitats. The effects of decreased water levels range from changes in biotic diversity in springs, playas and ox-bow lakes in the south, to shifts in salinity and productivity in prairie pothole lakes and wetlands in the north. Despite the availability of considerable data on aquatic ecosystems, these studies have generally focused on relatively small spatial scales. Their relationships to large-scale predictions based on global circulation models have yet to be adequately explored. Initial simulations of water levels in some prairie wetlands provide evidence that this approach is useful as a basis for generalization. However, there are complex characteristics of geomorphology, soil distributions, groundwater flows, biota and cultural demands on water budgets that create unique differences between morphologically similar types of habitats.

Trends in climatic variables are apparently influenced by ENSO teleconnections that may provide a basis for examining short-term extremes, as well as a perspective on long-term variability. Whenever these climate change effects are compounded by changes in land use and policy shifts over the next several decades, we expect that some significant and clearly observable changes can be documented.

To identify the specific effects of anthropogenic climatic and land use change we recommend a coordinated research programme focused on broad-scale comparisons of matched catchments of similar size and regional location but with different land use. Correctly interpreting relationships between the various sources of ecosystem change will require maintaining current long-term data collection sites. Moreover, it may become necessary to establish new sites that are selected to accomplish comparative studies of specific land use changes related to agriculture (such as shifts from irrigated to dryland farming) and to contrast cultural and natural sources of variation in water supplies. An early assessment of current data is needed to determine which subregions are lacking adequate data for larger scale regional comparisons, and which ecosystem mechanisms are currently well understood. We further recommend that comparative analyses of well-studied ecosystems be planned with participation from several disciplines, so that gaps in our understanding of the complex interactions can be rapidly filled.

REFERENCES

Akin, W. E. 1991. *Global Patterns.* University of Oklahoma Press, Norman.

Alberta Environment, 1977. *Cooking Lake Area Study*, Vol. 1, *Planning Report*. Plan. Div. Edmonton, Alberta.

Almendinger, J. E. 1990. 'Groundwater control of closed-basin lake levels under steady-state conditions', *J. Hydrol.*, **112**, 293–318.

Arbogast, A. F. and Johnson, W. C. 1994. 'Climatic implications of the late Quaternary alluvial record of a small drainage basin in the central Great Plains', *Quat. Res.*, **41**, 298–305.

Armour, C. L. 1993. 'Evaluating temperature regimes for protection of smallmouth bass', *Resource Publication 191*. U.S. Fish and Wildlife Service, Washington, D.C.

Baker, R. G., Maher, L. J., Chumbley, C. A., and Van Zandt, K. L. 1992. 'Patterns of Holocene environmental change in the midwestern United States', *Quat. Res.*, **37**, 379–389.

Ballantyne, A. K. 1978. 'Saline soils in Saskatchewan due to wind deposition', *Can. J. Soil Sci.*, **58**, 107–108.

Barica, J. 1975. 'Summerkill risk in prairie ponds and possibilities of its prediction,' *J. Fish. Res. Bd Can.*, **32**, 1283–1288.

Barica, J. 1980. 'Some biological characteristics of plains aquatic ecosystems and their effect on water quality', in Waite, D. T. (Ed.) *Prairie Surface Waters: Problems and Solutions, Proceedings, 1979 PARC Symposium*, Canadian Plains Proceedings No. 7. Canadian Plains Center, Regina.

Barnosky, C. W., Grimm, E. C., and Wright, H. E. 1987. 'Towards a postglacial history of the northern Great Plains: a review of the paleoecological problems', *Ann. Carnegie Mus.*, **56**, 259–273.

Bartlein, P. J., Webb, T., and Fleri, E. 1984. 'Holocene climatic change in the northern midwest: pollen-derived estimates', *Quat. Res.*, **222**, 361–374.

Bataille, K. J. and Baldassarre, G. A. 1993. 'Distribution and abundance of aquatic macroinvertebrates following drought in three prairie pothole wetlands', *Wetlands*, **13**, 260–269.

Batt, B. D., Anderson, M. G. Anderson, C. D., and Caswell, F. D. 1989. 'The use of prairie potholes by North American ducks', in van der Valk, A. G. (Ed.), *Northern Prairie Wetlands*. Iowa State University Press, Ames. pp. 204–227.

Benke, A. C. 1990. 'A perspective on America's vanishing streams', *J. North Am. Benthol. Soc.*, **9**, 77–88.

Bolen, E. G., Smith, L. M., and Schramm, H. L., Jr. 1989. 'Playa lakes: prairie wetlands of the southern high plains', *BioScience*, **39**, 615–623.

Borchert, J. R. 1950. 'The climate of the central North American grassland', *Ann. Assoc. Am. Geogr.*, **40**, 1–39.

Bowles, D. E. and Arsuffi, T. L. 1993. 'Karst aquatic ecosystems of the Edwards Plateau region of central Texas, U.S.A.: a consideration of their importance, threats to their existence, and efforts for their conservation', *Aquat. Conserv. Mar. Freshwat. Ecosyst.* **3**, 317–330.

Bradbury, J. P., Dean, W. E., and Anderson, R. Y. 1993. 'Holocene climatic and limnologic history of the north-central United States as recorded in the varved sediments of Elk Lake, Minnesota: a synthesis', in Bradbury, J. P. and Dean, W. E. (Eds), *Elk Lake, Minnesota: Evidence for Rapid Climate Change in the North-Central United States, Special Paper 276. Geological Society of America, Boulder, Colorado. pp. 309–328.*

Brown, A. V. and Matthews, W. J. 1995. 'Stream ecosystems of the central United States', in Cushing, C. E., Cummins, K. W., and Minshall, G. W. (Eds), *River and Stream Ecosystems, Ecosystems of the World*, Vol. 22. Elsevier Press, New York. pp. 89–116.

Burn, D. H. 1994. 'Hydrologic effects of climatic change in west-central Canada', *J. Hydrol.*, **160**, 53–70.

Carlander, K. D., Campbell, R. S., and Irwin, W. H. 1963. 'Mid-continent states', in D. G. Frey (Ed.), *Limnology in North America.* University of Wisconsin Press, Madison. pp. 317–348.

Carpenter, S. R., Fisher, S. G., Grimm, N. B., and Kitchell, J. K. 1992. 'Global change and freshwater ecosystems', *Annu. Rev. Ecol. Syst.*, **23**, 119–139.

Case, R. A. and MacDonald, G. M. 1995. 'A dendroclimatic reconstruction of annual precipitation on the western Canadian prairie since A.D. 1505 from *Pinus flexilis* James', *Quat. Res.*, **44**, 267–275.

Changnon, S. A. and Demissie, M. 1996. 'Detection of changes in streamflow and floods resulting from climate fluctuations and land use-drainage changes', *Climatic Change*, **32**, 411–421.

Changnon, S. A. and Glantz, M. H. 1996. 'The Great Lakes diversion at Chicago and its implications for climatic change', *Climatic Change*, **32**, 199–214.

Cohen, S. J. 1991. 'Possible impacts of climatic warming scenarios on water resources in the Saskatchewan River sub-basin, Canada', Climatic Change, **19**, 291–317.

Covich, A. P. 1993. 'Water and ecosystem', in Gleick, P. H. (Ed.), *Water in Crisis.* Oxford University Press, Oxford. pp. 40–55.

Covich, A. P. and Thorp, J. H. 1991. 'Crustacea: introduction and Peracarida', in Thorp, J. H. and Covich, A. P. (Eds), *Ecology and Classification of North American Freshwater Invertebrates*. Academic Press, San Diego. pp. 665–689.

Crowe, A. S. 1993. 'The application of a coupled water-balance–salinity model to evaluate the sensitivity of a lake dominated by groundwater to climatic variability', *J. Hydrol.*, **141**, 33–73.

Cvancara, A. M. 1983. 'Aquatic mollusks of North Dakota', *Report of Investigation No. 78*. North Dakota Geological Survey, Fargo.

Danks, H. V. and Williams, D. D. 1991. 'Arthropods of springs, with particular reference to Canada: synthesis and needs for research', *Mem. Entomol. Soc. Can.*, **155**, 203–217.

Davis, J. C. 1975. 'Minimal dissolved oxygen requirements of aquatic life with emphasis on Canadian species: a review', *J. Fish. Res. Bd Can.*, **32**, 2295–2332.

Diaz, H. F. 1983. 'Some aspects of major dry and wet periods in the contiguous United States', 1895–1981', *J. Climate Appl. Meteorol.*, **22**, 3–16.

Echelle, A. A., Echelle, A. F., and Hill, L. G. 1973. 'Interspecific interactions and limiting factors of abundance and distribution in the Red River pupfish, *Cyrinodon rubrofluviatilis*', *Am. Midland Nat.* **88**, 109–130.

Eddy, S. 1963. 'Minnesota and the Dakotas', in Frey, D. G. (Ed.), *Limnology in North America*. University of Wisconsin Press, Madison. pp. 301–315.

Edwards, R. J., Longley, G., Moss, G., Ward, R., Matthews, J., and Stewart, B. 1989. 'A classification of Texas aquatic communities with special consideration toward the conservation of endangered and threatened taxa', *Texas, J. Sci.*, **41**, 231–240.

Evans, J. C. and Prepas, E. E. 1996. 'Potential effects of climate change on ion chemistry and phytoplankton communities in prairie saline lakes', *Limnol. Oceanogr.* **41**, 1063–1076.

Fausch, K. D. and Bestgen, K. R. 1996. 'Ecology of fishes indigenous to the central and southwestern Great Plains', in Knopf, F. L. and Samson. F. B. (Eds), *Ecology of Great Plains Vertebrates and Their Habitats*. Springer-Verlag, New York. pp. 131–166.

Ferrington, L. C. Jr. 1993. 'Endangered rivers: a case history of the Arkansas River in the Central Plains', *Aquat. Conserv. Mar. Freshwat. Ecosyst.*, **3**, 305–316.

Firth, P. and Fisher, S. G. (Eds), 1991. *Climate Change and Freshwater Ecosystems*. Springer-Verlag, New York.

Fisheries and Environment Canada, 1978. *Hydrological Atlas of Canada*. Supply and Services Canada, Ottawa, Ontario. 34 maps.

Forman, S. L. and Maat. P. 1990. 'Stratigraphic evidence for late Quaternary dune activity near Hudson on the Piedmont of northern Colorado,' *Geology*, **75**, 745–748.

Fritz, S. C. 1990. 'Twentieth-century salinity and water-level fluctuations in Devils Lake, North Dakota: a test of a diatom-based transfer function', *Limnol. Oceanogr.*, **35**, 1771–1781.

Fritz, S. C. 1996. Paleolimnological records of climatic change in North America', *Limnol. Oceanogr.*, **41**, 882–889.

Fritz, S. C., Juggins, S., Battarbee, R. W., and Engstrom, D. R. 1991. 'Reconstruction of past changes in salinity and climate using a diatom-based transfer function', *Nature*, **352**, 706–708.

Fritz, S. C., Juggins, S., and Battarbee, R. W. 1993. 'Diatom assemblages and ionic characterization of lakes of the Northern Great Plains, North America: a tool for reconstructing past salinity and climate fluctuations', *Can. J. Fish. Aquat. Sci.*, **50**, 1844–1856.

Fritz, S. C., Engstrom, D. R., and Haskell, B. J. 1994. 'Little Ice Age aridity in the North American Great Plains: a high-resolution reconstruction of salinity fluctuations from Devils Lake, North Dakota, USA', *Holocene*, **4**, 69–73.

Galatowitsch, S. M. and van der Valk, A. G. 1994. *Restoring Wetlands. An Ecological Approach*. Iowa State University Press, Ames.

Gates. D. M. 1993. *Climate Change and its Biological Consequences*. Sinauer Associates, Sunderland.

Giorgi, F. and Mearns, L. O. 1991. 'Approaches to the simulation of regional climate change: a review', *J. Geophys. Res.*, **29**, 191–216.

Giorgi, F., Brodeur, C. S., and Bates, G. T. 1994. 'Regional climate change scenarios over the United States produced with a nested regional climate model', *J. Climate*, **7**, 375–399.

Glantz, M. H. 1994. 'Drought, desertification and food production', in Glantz, M. H. (Ed.), *Drought Follows the Plow: Cultivating Marginal Areas*. Cambridge University Press, Cambridge. pp. 9-30.

Glazier, D. S. 1991. 'The fauna of North American temperate cold springs: patterns and hypotheses', *Freshwat. Biol.*, **26**, 527–542.

Gleick, P. H. 1989. 'Climate change, hydrology, and water resources', *Rev. Geophys.*, **27**, 329–344.

Gorham, E., Dean, W. E., and Sanger, J. E. 1983. 'The chemical composition of lakes in the north-central United States', *Limnol. Oceanogr.*, **28**, 287–301.

Greenwood, R. J., Sargeant, A. B., Johnson, D. H., Cowardin, L. W., and Shaffer, T. L. 1995. 'Factors associated with duck nest success in the prairie pothole region of Canada', Wildlife Monogr. **128**, 1–57.

Grimm, N. B. 1993. 'Implications of climate change for stream communities,' in Kareiva, P. M., Kingsolver, J. G., and Huey, R. B. (Eds), *Biotic Interactions and Global Change*. Sinauer, Sunderland. pp. 293–314.

Gruger, J. 1973. 'Studies on the late Quaternary vegetation history of northeastern Kansas', *Geol. Soc. Am. Bull.*, **84**, 239–250.

Guthery, F. S. and Bryant, F. C. 1982. 'Status of playas in the Southern Great Plains', *Wildlife Soc. Bull.*, **10**, 309–317.

Halpin, P. N. 1995. 'Modeling potential impacts of climate change on northern landscapes', in Peterson, D. L. and Johnson, D. R. (Eds), *Human Ecology and Climate Change*. Taylor and Francis, Washington, D.C. pp. 57–74.

Hammer, U. T. 1964. 'The succession of 'bloom' species of blue-green algae and some causal factors', *Verh. Int. Verein. Limnol.*, **15**, 829–836.

Hammer, U. T. 1978. 'The saline lakes of Saskatchewan. III. Chemical characterization', *Int. Rev. Ges. Hydrobiol.*, **63**, 311–335.

Hammer, U. T. 1986. *Saline Lake Ecosystems of the World*. Dr. W. Junk, B. V. Publishers, The Hague.

Hammer, U. T. 1990. 'The effects of climate change on the salinity, water levels, and biota of Canadian prairie lakes', *Verh. Int. Verein. Theoret. Angew. Limnol.* **24**, 321–326.

Hammer, U. T. and Haynes, R. C. 1978. 'The saline lakes of Saskatchewan. II. Locale, hydrography, and other physical aspects', *Int. Rev. Ges. Hydrobiol.*, **63**, 179–203.

Hammer, U. T., Sheard, J. S., and Kranabetter, J. 1990. 'Distribution and abundance of littoral benthic fauna in Canadian prairie saline lakes', *Hydrobiologia*, **197**, 173–192.

Hands, M. and Stefan, H. G. 1993. 'Regional water temperature characteristics of lakes subjected to climate change', *Climatic Change*, **24**, 187–211.

Haragan, D. R. 1990. *Human Intervention in the Climatology of Arid Lands.* University of New Mexico Press, Albuquerque.

Haskell, B., Engstrom, D. R., and Fritz, S. C. 1996. 'Late Quaternary paleohydrology in the North American Great Plains inferred from the geochemistry of endogenic carbonates and fossil ostracodes from Devils Lake, North Dakota', *J. Paleolimnol.*, **124**, 179–193.

Haukos, D. A. and Smith, L. M. 1993. 'Moist soil management of playa lakes for migrating and wintering ducks', *Wildlife Soc. Bull.*, **21**, 228–298.

Haynes, C. V. 1991. 'Geoarchaeological and paleohydrological evidence for a Clovis-age drought in North America and its bearing on extinction', *Quat. Res.*, **35**, 438–450.

Hershler, R. and Longley, G. 1986. 'Phreatic hydrobiids (Gastropoda: Prosobranchia) from the Edwards (Balcones Fault Zone) aquifer region, south-central Texas', *Malacologia*, **27**, 127–172.

Hirsch, R. M., Walker, J. F., Day, J. C., and Hallio, R. 1990. 'The influence of man on hydrologic systems', in Wolman, M. G. (Ed.), *Surface Water Hydrology*. Geological Society of America, Boulder, pp. 329–359.

Hogenbirk, J. C. and Wein, R. W. 1991. 'Fire and drought experiments in northern wetlands: a climate change analogue', *Can. J. Bot.*, **69**, 1991–1997.

Hogenbirk, J. C. and Wein, R. W. 1992. 'Temperature effects of seedling emergence from boreal wetland soils: implications for climate change', *Aquat. Bot.* **42**, 361–373.

Hogg, I. D. and Williams, D. D. 1996. 'Response of stream invertebrates to a global-warming thermal regime: an ecosystem-level manipulation', *Ecology*, **77**, 395–407.

Hogg, I. D., Williams, D. D., Eadie, J. M., and But, S. A. 1995. 'The consequences of global warming for stream invertebrates: a field simulation', *J. Therm. Biol.*, **20**, 199–206.

Holliday, V. T. 1989. 'Middle Holocene drought on the southern Great Plains', *Quat. Res.*, **31**, 74–82.

Holsinger, J. R. and Longley, G. 1980. 'The subterranean amphipod crustacean fauna of an artesian well in Texas', *Smithson. Contrib. Zool.*, **308**, 1–62.

Hubbard, D. E. 1988. 'Glaciated prairie wetland functions and values: a synthesis of the literature', *US Fish Wildl. Serv. Biol. Rep.*, **88**(43)., Ft. Collins, Colorado. 50 pp.

Hubbard, D. E. and Linder, R. L. 1986. 'Spring runoff retention, in prairie pothole wetlands', *Soil Wat. Conserv.* **41**, 122–125.

Hubbs, C. 1995. 'Springs and spring runs as unique aquatic systems', *Copeia*, 1995: 989–991.

Humphrey, J. D. and Ferring, C. R. 1994. 'Stable isotopic evidence for latest Pleistocene and Holocene climatic change in north-central Texas,' *Quat. Res.*, **41**, 200–213.

Johnsgard, P. A. 1984. *The Platte. Channels in time.* University, of Nebraska Press, Lincoln.

Kantrud, A., Millar, J. B., and van der Valk, A. G. 1989. 'Vegetation of wetlands of the prairie pothole region', in van der Valk, A. G. (Ed.), *Norhern Prairie Wetlands.* Iowa State University Press, Ames. pp. 132–187.

Karl, T. R. 1983. 'Some spatial characteristics of drought duration in the United States', *J. Climate Appl. Meteorol.*, **22**, 1356–1366.

Karl, T. R., Heim, R. R., Jr., and Quayle, R. G. 1991. 'The greenhouse effect in central North America: if not now when?', *Science*, **251**, 1058–1061.

Kotak, B. G., Kenefick, S. L., Fritz, D. L., Rousseaux, C. G., Prepas, E. E., and Hrudey, S. E. 1993. 'Occurrence and toxicological evaluation of cyanobacterial toxins in Alberta lakes and farm dugouts', *Wat. Res.*, **27**, 495–506.

Krapu, G. L. and Duebbert, H. F. 1989. 'Prairie wetlands: characteristics, importance to waterfowl, and status', in Sharitz, R. R. and Gibbons, J. W. (Eds), *Freshwater Wetlands and Wildlife*, US Department of Energy Symposium, Series 61. US Department of Energy, Washington D.C. pp. 811–828.

Kroschewsky, J. R. 1990. 'The Edwards Aquifer Research and Data Center; objectives and accomplishments', *Stygologia*, **5**, 213–220.

LaBaugh, J. W. 1986. 'Limnological character of selected lakes in the Nebraska Sandhills', *J. Hydrol.*, **86**, 279–298.

LaBaugh, J. W., Winter, T. C., Adomaitis, V. A., and Swanson, G. A. 1987. 'Hydrology and chemistry of selected prairie wetlands in the Cottonwood Lake Area, Stutsman County, USGS, North Dakota, 1979–82', *US Geol. Surv. Prof. Pap. 1431*.

LaBaugh, J. W., Winter, T. C., Swanson, G. A., Rosenberry, D. O., Nelson, R. D., and Euliss, N. H. Jr. 1996. 'Changes in atmospheric circulation patterns affect midcontinent wetlands sensitive to climate', *Limnol. Oceanogr.*, **41**, 871–881.

Laird, K., Fritz, S. C., and Grimm, E. C. 1996. 'The paleoclimatic record of a closed-basin lake in the northern Great Plains: Moon Lake, Barnes County, North Dakota', *Limnol. Oceanogr.*, **41**, 890–902.

Laird, K., Fritz, S. C., Grimm, E. C., and Mueller, P. G. 1996. Century-scale paleoclimatic reconstruction from Moon Lake, a closed-basin lake in the northern Great Plains.

Lamb, P. J. and Richman, M. B. 1990. 'Use of cooperative weather station data in contemporary climate research', *Trans. Ill. State Acad. Sci.*, **83**, 70–81.

Larimore, R. W., Childers, W. F., and Heckrotte, C. 1959. 'Destruction and re-establishment of fish and invertebrates affected by drought', *Trans. Am. Fish. Soc.*, **88**, 261–285.

Larson, D. L. 1994. 'Potential effects of anthropogenic green house gases on avian habitats and populations in the northern Great Plains', *Am. Midland Nat.* **131**, 330–316.

Larson, D. L. 1995. 'Effects of climate on numbers of prairie wetland', *Climatic Change*, **30**, 169–180.

Last, W. M. 1990. 'Paleochemistry and paleohydrology of Ceylon Lake, a salt-dominated playa basin in the northern Great Plains, Canada', *J. Paleolimnol.*, **4**, 219–238.

Last, W. M. 1992. 'Chemical composition of saline and subsaline lakes of the northern Great Plains, western Canada', *Int. J. Salt Lake Res.*, **1**, 47–76.

Last, W. M. and Sauchyn, D. J. 1993. 'Mineralogy and litho-stratigraphy of Harris Lake, southwestern Saskatchewan, Canada', *J. Paleolimnol.*, **9**, 23–39.

Last, W. M. and Schweyen, T. H. 1985. 'Late Holocene history of Waldsea Lake, Saskatchewan, Canada', *Quat. Res.*, **24**, 219–234.

Last, W. M. and Slezak. L. A. 1986. 'Paleohydrology, sedimentology, and geochemistry of two meromictic saline lakes in southern Saskatchewan', *Geog. Phys. Quat.*, **40**, 5–15.

Last, W. M. and Slezak, L. A. 1988. 'The salt lakes of western Canada: a paleolimnological overview', *Hydrobiol.*, **158**, 301–316.

Lawler, G. H., Sunde, L. A., and Whitaker, J. 1974. 'Trout production in prairie ponds', *J. Fish. Res. Bd Can.*, **31**, 929–936.

Longley, G. 1981. 'The Edwards Aquifer: Earth's most diverse groundwater ecosystem?' *Int. J. Speleol.*, **11**, 123–128.

Loope, D. B., Swinehart, J. B. and Mason, J. P. 1995. 'Dune-dammed paleovalleys of the Nebraska Sand Hills: intrinsic versus climatic controls on the accumulation of lake and marsh sediments', *Geol. Soc. Am. Bull.* **107**, 396–406.

Madole, R. F. 1994. 'Stratigraphic evidence of desertification in the west-central Great Plains within the past 1000 years', *Geology*, **22**, 483–486.

Marsh, E. 1980. 'The effects of temperature and photoperiod on the termination of spawning in the orangethroat darter (*Ethesotoma spectabile*) in Central Texas', *Texas J. Sci.*, **32**, 129–142.

Martin, C. W. 1992. 'Late Quaternary sedimentation and paleo-environmental change in the Republican River basin, southcentral Nebraska', *Quat. Res.*, **37**, 315–322.

Matthews, W. J. 1987. 'Physicochemical tolerance and selectivity of stream fishes as related to their geographic ranges and local distributions', in Matthews, W. J. and Heins, D. C. (Eds), *Community and Evolutionary Ecology of North American Stream Fishes.* University of Oklahoma Press, Norman. pp. 111–120.

Matthews, W. J. 1988. 'North American prairie streams as systems for ecological study', *J. North Am. Benthol. Soc.*, **7**, 387–409.

Matthews, W. J. and Zimmerman, E. G. 1990. 'Potential effects of global warming on native fishes of the southern Great Plains and the Southwest', *Fisheries*, **15**, 26–32.

Matthews, W. J., Hoover, J. J., and Milstead, W. B. 1985. 'Fishes of Oklahoma springs', *Southwest. Nat.* **30**, 23–32.

May, D. W. 1989. 'Holocene alluvial fills in the South Loup valley, Nebraska', *Quat. Res.*, **32**, 117–120.

McAllister, D. E., Platania, S. P., Schueler, F. W., Baldwin, M. E., and Lee, D. S. 1986. 'Ichthyofaunal patterns on a geographic grid', in Hocutt, C. H. and Wiley, E. O. (Eds), *The Zoogeography of North American Freshwater Fishes.* John Wiley, New York. Chapter 2.

McCarraher, D. B. 1977. *Nebraska's Sandhills Lakes.* Nebraska Game and Parks Commission., Lincoln.

McGregor, K. M. 1985. 'Drought during the 1930s and 1950s in the central United State', *Phys. Geogr.*, **6**, 288–301.

Mearns, L. O., Gleick, P. H., and Schneider, S. H. 1990. 'Climate forecasting', in Waggoner, P. E. (Ed.), *Climate Change and U.S. Water Resources.* Wiley Interscience, New York. pp. 87–137.

Meko, D. M. 1992. 'Dendroclimatic evidence from the Great Plains of the United States', in Bradley, R. S. and Jones, P. D. (Eds), *Climate Since A.D. 1500.* Routledge, New York. pp. 312–330.

Mitchell, P. and Prepas, E. E. (Eds). 1990. *Atlas of Alberta Lakes.* University of Alberta Press, Edmonton. 675 pp.

Montroy, D. L. 1993. 'On the relation of North American precipitation to the El Nino-Southern Oscillation', *BSc Honors Thesis*, School of Meteorology, University of Oklahoma, 71 pp.

Montroy. D. L. 1997. 'Linear-relation of central and eastern North American precipitation to tropical Pacific sea surface temperature anomalies', *J. Climate*, **10**, 541–558.

Muhs, D. R. 1985. 'Age and paleoclimatic significance of Holocene sand dunes in northeastern Colorado', *Ann. Assoc. Am. Geogr.*, **75**, 566–582.

Naiman, R. J., Magnuson, J. J., McKnight, D. M., and Stanford, J. A. 1995. *The Freshwater Imperative. A Research Agenda.* Island Press, Washington, D.C.

Neely, R. K. and Baker, J. L. 1989. 'Nitrogen and phosphorus dynamics and the fate of agricultural runoff', in van der Valk. A. G. (Ed.), *Northern Prairie Wetlands.* Iowa State University Press, Ames. pp. 92–131.

Nichols, J. D., Johnson, F. A., and Williams, B. K. 1995. 'Managing North American waterfowl in the face of uncertainty', *Ann. Rev. Ecol. System* **26**, 177–199.

Nordt, L. C., Boutton, T. W., Hallmark, C. T., and Waters, M. R. 1994. 'Late Quaternary vegetation and climate changes in central Texas based on the isotopic composition of organic carbon', *Quat. Res.*, **41**, 109–120.

Opie, J. 1993. *Ogallala. Water for a Dry Land.* University of Nebraska Press, Lincoln.

Osterkamp, W. R. and Wood, W. W. 1987. 'Playa-lake basins on the southern High Plains of Texas and New Mexico: I. Hydrologic, geomorphic, and geologic evidence for their development', *Geol. Soc. Am. Bull.* **99**, 215–223.

Packard, J. 1995. 'Biodiversity', in North, G. R., Schmandt, J., and Clarkson, J. (Eds), *The Impact of Global Warming on Texas.* University of Texas Press, Austin. pp. 119–149.

Poiani, K. A. and Johnson, W. C. 1991. 'Global warming and prairie wetlands: potential consequences for waterfowl habitat', *BioScience*, **41**, 611–618.

Poiani, K. A. and Johnson, W. C. 1993a. 'A spatial simulation model of hydrology and vegetation dynamics in semi-permanent prairie wetlands', *Ecol. Appl.* **3**, 279–293.

Poiani, K. A. and Johnson, W. C. 1993b. 'Potential effects of climate change on a semi-permanent prairie wetland', *Climatic Change*, **24**, 213–232.

Poiani, K. A., Johnson, W. C., and Kittel, T. G. F. 1995. 'Sensitivity of a prairie wetland to increased temperature and seasonal precipitation changes', *Wat. Resour. Bull.* **31**, 283–294.

Poiani, K. A., Johnson, W. C., Swanson, G. A., and Winter, T. C. 1996. 'Climate change and northern prairie wetlands: simulations of long-term dynamics', *Limnol. Oceanogr.*, **41**, 871–881.

Postel, S. L., Daily, G. C., and Erhlich, P. R. 1996. 'Human appropriation of renewable fresh water', *Science*, **271**, 785–788.

Powell, J. W. 1878. 'Lands of the arid regions of the United States', *Congress of the United States (3rd session) in the House of Representatives.* Government Printing Office, Washington, D.C.

Rabeni, C. F. 1996. 'Prairie legacies — fish and aquatic resources', in Samson, F. B. and Knopf, F. L. (Eds), *Prairie Conservation: Preserving North America's most Endangered Ecosystem.* Island Press, Covello., California. pp. 111–124.

Radle, N. J., Keister, C. M., and Battarbee, R. W. 1989. 'Diatom, pollen, and geochemical evidence for the paleosalinity of Medicine Lake, S. Dakota, during the Late Wisconsin and early Holocene', *J. Paleolimnol.* **2**, 159–172.

Reeves, C. C., Jr. 1966. 'Pluvial lake basins of West Texas', *J. Geol.*, **74**, 269–291.

Richman, M. B. and Lamb, P. J. 1985. 'Climatic pattern analysis of three- and seven-day summer rainfall in the central United States: some methodological consideration and a regionalization', *J. Climate Appl. Meteorol.*, **24**, 1325–1343.

Richman, M. B. and Lamb, P. J. 1987. 'Pattern analysis of growing season precipitation in southern Canada', *Atmos.–Ocean*, **25**, 137–158.

Richman, M. B., Lamb, P. J., and Angel, J. R. 1991a. 'Multivariate statistical analysis of precipitation associated with North American storm tracks', *Fifth Conference on Climate Variations.* American Meteorological Society, Boston. pp. 47–53.

Richman, M. B., Lamb, P. J., and Angel. J. R. 1991b. 'Relationships between monthly precipitation over central and eastern North American and the Southern Oscillation', *Fifth Conference on Climate Variations.* American Meteorological Society, Boston. pp. 151–158.

Riebsame, W. E., Changnon, S. A., Jr., and Karl, T. R. 1990. *Drought and Natural Resources Management in the United States: Impacts and Implications of the 1987–1989 Drought*, Westview Press, Boulder.

Ritchie, J. C. 1976. 'The late-Quaternary vegetational history of the Western Interior of Canada', *Can. J. Bot.*, **54**, 1793–1818.

Ropelewski, C. F. and Halpert, M. S. 1986. 'North American precipitation and temperature patterns associated with the El Nino/Southern Oscillation (ENSO)', *Mon. Weath. Rev.*, **114**, 2352–2362.

Ropelewski, C. F. and Halpert, M. S. 1987. 'Global and regional scale precipitation patterns associated with the El Nino/Southern Oscillation', *Mon. Weath. Rev.*, **115**, 1606–1626.

Ropelewski, C. F. and Halpert, M. S. 1989. 'Precipitation patterns associated with the high index phase of the Southern Oscillation', *J. Climate*, **2**, 268–284.

Ropelewski, C. F. and Halpert, M. S. 1996. 'Quantifying Southern Oscillation–precipitation relationships', *J. Climate*, **9**, 1043–1059.

Rosen, B. H., Adams, P., and Lal, H. 1995. 'A conceptual model for the assessment of depressional wetlands in the prairie pothole region', *Wetlands Ecol. Manage.*, **3**, 195–208.

Rosenzweig, C. 1990. 'Crop response to climate change in the southern Great Plains: a simulation study', *Prof. Geogr.*, **42**, 20–39.

Rosenzweig, C. and Hillel, D. 1993. 'The dust bowl of the 1930s: analog of greenhouse effect in the Great Plains?', *J. Environ. Qual.*, **22**, 9–22.

Rosenzweig, C. and Parry, M. L. 1993. 'Potential impacts of climate change on world food supply: a summary of a recent international study', in Kaiser, H. M. and Drennen, T. E. (Eds), *Agricultural Dimensions of Global Change*, St. Lucie Press, Delray Beach. pp. 87–116.

Sauchyn, D. J. 1990. 'A reconstruction of Holocene geomorphology and climate, western Cypress Hills, Alberta and Saskatchewan', *Can. J. Earth Sci.*, **27**, 1504–1510.

Schmandt, J. and Clarkson, J. 1992. 'Introduction: global warming as a regional issue', in Schmandt, J. and Clarkson, J. (Eds), *The Regions and Global Warming.* Oxford University Press, Oxford. pp. 3–11.

Schwartz, F. W. and Gallup, D. N. 1978. 'Some factors controlling the major ion chemistry of small lakes: examples from the prairie parkland of Canada', *Hydrobiologia*, **58**, 65–81.

Shjeflo, J. B. 1968. 'Evapotranspiration and the water budget of prairie potholes in North Dakota', *US Geol. Surv. Prof. Paper 585-B.* Reston, Virginia, 49 pp.

Skaggs, R. H. 1975. 'Drought in the United States, 1932–1940', *Ann. Assoc. Am. Geogr.*, **65**, 391–402.

Smale, M. A. and Rabeni, C. F. 1995. 'Influences of hypoxia and hyperthermia on fish species composition in headwater streams', *Trans. Am. Fish. Soc.*, **124**, 711–725.

Steiert, J. 1995. *Playas. Jewels of the Plains.* Texas Tech. University Press, Lubbock.

Stewart, R. E. and Kantrud, H. A. 1971. 'Classification of natural ponds and lakes in the glaciated prairie region', *Bur. Sport Fish. Wildl. Resource. Pub. No. 92.* US Department of the Interior, Washington, D.C.

Stockton, C. W. and Meko, D. M. 1983. 'Drought recurrence in the great Plains as reconstructed from long-term tree-ring records', *J. Climate Appl. Meteorol.*, **22**, 17–29.

Sublette, J. E. and Sublette, M. S. 1967. 'The limnology of playa lakes of the Llano Estacado, New Mexico and Texas', *Southwest. Nat.*, **12**, 369–406.

Sulzman, E. W., Poiani, K. A., and Kittel, T. G. F. 1995. 'Modeling human-induced climatic change: a summary for environmental managers', *Environ. Managem.* **19**, 197–224.

Swanson, G. A. 1988. 'Aquatic habitats of breeding waterfowl', in Hook, D. D. (Ed.), *Ecology and Management of Wetlands*, Vol. 1. *Ecology of Wetlands.* Timber Press, Portland. pp. 195–202.

Swanson, G. A. and Duebbert, H. F. 1989. 'Wetland habitats of waterfowl in the Prairie Pothole Region', in van der Valk, A. G (Ed.), *Northern Prairie Wetlands.* Iowa State University Press, Ames. pp. 228–267.

Swanson, G. A., Winter, T. C., Adomaitis, V. A., and LaBaugh, J. W. 1988. 'Chemical characteristics of prairie lakes in south-central North Dakota: their potential for influencing use by fish and wildlife', *Fish Wildl. Tech. Rep. 18.* US Fish Wildlife Service, Washington, D.C. 44 pp.

Teller, J. T. and Last, W. M. 1981. 'Late Quaternary history of Lake Manitoba, Canada', *Quat. Res.*, **16**, 97–116.

Timms, B. V. and Hammer, U. T. 1988. 'Water beetles of some saline lakes in Saskatchewan', *Can. Field Nat.* **102**, 246–250.

Trenberth, K. E., Branstator, G. W., and Arkin. P. A. 1988. 'Origins of the 1988 North American drought', *Science*, **242**, 1640–1645.

Trew, D. 1983. 'An approach for assessing the impact of lake level fluctuations on lake trophic status', in Cold Lake–Beaver River Water Management Study. Alberta Environment, Edmonton. Appendix N.

Turner, R. E. and Rabalais, N. N. 1991. 'Changes in Mississippi River water quality in this century', *BioScience*, **41**, 140–147.

Vance, R. E. and Mathewes, R. W. 1994. 'Deposition of modern pollen and plant macro-remains in a hypersaline prairie lake basin', *Can. J. Bot.*, **72**, 539–548.

Vance, R. E., Mathewes, R. W., and Clague, J. J. 1992. 'A 7000-year record of lake-level change on the northern Great Plains: a high resolution proxy of past climate', *Geology*, **20**, 879–882.
van der Valk, A. G. (Ed.), 1989. *Northern Prairie Wetlands*. Iowa State University Press, Ames.
Van Stempvoort, D. R., Edwards, T. W. D., Evans, M. S., and Last, W. M. 1993. 'Paleohydrology and paleoclimate records in a saline prairie lake core: mineral, isotope and organic indicator', *J. Paleolimnol.*, **8**, 135–147.
Varza, D. and Covich, A. P. 1995. 'Population fluctuations within a spring community', *J. Kansas Entomol. Soc.* **68**(2), 42–49.
Waggoner, P. E. and Schefter, J. 1990. 'Future water use in the present climate', in Waggoner, P. E. (Ed.), *Climate Change and U.S. Water Resources*. Wiley Interscience, New York. pp. 19–39.
Waring, G. A. 1965. 'Thermal springs of the United States and other countries of the world: a summary', (Revised by R. R. Blankenship and R. Bentall). *US Geol. Surv. Prof. Pap.* 492., Reston, Virginia.
Watts, W. A. and Bright, R. C. 1968. 'Pollen, seed, and mollusk analysis of a sediment core from Pickerel Lake, northeastern South Dakota', *Geol. Soc. Am. Bull.*, **79**, 855–876.
Watts, W. A. and Wright, H. E. 1966. 'Late-Wisconsin pollen and seed analysis from the Nebraska Sandhills', *Ecology*, **47**, 202–210.
Wells, P. V. 1970. 'Postglacial vegetational history of the Great Plains', *Science*, **167**, 1574–1582.
Wiche, G. J., Hoetzer, S. M., and Rankl, J. G. 1986. 'Hydrology of the Devils Lake Basin, Northeastern North Dakota', Water Resources Investigation 3. North Dakota State Water Commission. Bismark. 86 pp.
Winter, T. C. 1977. 'Classification of the hydrologic settings of lakes in the North Central United States,' *Wat. Resour. Res.*, **13**, 753–767.
Winter, T. C. 1989. 'Hydrologic studies of wetlands in the northern prairie', in van der Valk, A. G. (Ed.), *Northern Prairie Wetlands*. Iowa State University Press, Ames. pp. 16–54.
Winter, T. C. 1995. 'A landscape approach to identifying environments where ground water and surface water are closely interrelated', in Charbeneau, R. J. (Ed), Groundwater Management. Proceedings Internat. Groundwater Management Symp., San Antonio, Texas, Amer. Soc. Civil Engineers, New York, p. 139–144.
Winter, T. C. and Carr, M. R. 1980. 'Hydrologic setting of wetlands in the Cottonwood Lake Area, Stutsman Co., North Dakota', *US Geol. Surv. Wat. Resource. Invest. 80-99.* Reston, Virginia.
Winter, T. C. and Rosenberry, D. O. 1995. 'The interaction of ground water with prairie pothole wetlands in the Cottonwood Lake Area, east-central North Dakota, 1979–1990. *Wetlands*, **15**, 193–211.
Winter, T. C. and Woo, M,-K. 1990. 'Hydrology of lakes and wetlands', in Wolman, M. G. and Riggs, H. C. (Eds), Surface Water Hydrology, *Geol. Soc. Amer.*, Boulder, Colorado, v.0–1., pp. 159–187.
Woo, M.-K., and Rowsell, R. D. 1993. 'Hydrology of a prairie slough', *J. Hydrol.*, **146**, 175–207.
Woo, K. and Winter, T. C. 1993. 'The role of permafrost and seasonal frost in the hydrology of northern wetlands in North America', *J. Hydrol.*, **141**, 5–31.
Wright, H. E. 1985. 'Pollen diagram from the Nebraska Sandhills and the age of the dunes', *Quat. Res.*, **24**, 115–120.
Zale, A. V., Leslie, D. M., Jr., Fisher, W. L. and Merrifield, S. G. 1989. 'The physicochemistry, flora, and fauna of intermittent prairie streams: a review of the literature,' US Fish and Wildlife Service, *Biological Report*, **89**(5). pp. 1–44.

9

SENSITIVITY OF AQUATIC ECOSYSTEMS TO CLIMATIC AND ANTHROPOGENIC CHANGES: THE BASIN AND RANGE, AMERICAN SOUTHWEST AND MEXICO

NANCY B. GRIMM,[1] ARTURO CHACÓN,[2] CLIFFORD N. DAHM,[3] STEVEN W. HOSTETLER,[4] OWEN T. LIND,[5] PETER L. STARKWEATHER[6] AND WAYNE W. WURTSBAUGH[7]

[1] *Department of Zoology, Arizona State University, Box 871501, Tempe, AZ 85287-1501, USA*
[2] *University of Michoacan, Privada de Huan Escutia No. 100, Colonia Chapultec Norte, Morelia, Michoacan, C.P. 58260, Mexico*
[3] *Department of Biology, University of New Mexico, Albuquerque, NM 87131, USA*
[4] *US Geological Survey, ERL-C, 200 SW 35th St., Corvallis, OR 97333, USA*
[5] *Department of Biology, Baylor University, Waco, TX 76798, USA*
[6] *Department of Biological Sciences, University of Nevada-Las Vegas, 4505 Maryland Parkway, Las Vegas, NV 89154-4014, USA*
[7] *Department of Fisheries and Wildlife, Utah State University, Logan, UT 84322-5210, USA*

ABSTRACT

Variability and unpredictability are characteristics of the aquatic ecosystems, hydrological patterns and climate of the largely dryland region that encompasses the Basin and Range, American Southwest and western Mexico. Neither hydrological nor climatological models for the region are sufficiently developed to describe the magnitude or direction of change in response to increased carbon dioxide; thus, an attempt to predict specific responses of aquatic ecosystems is premature. Instead, we focus on the sensitivity of rivers, streams, springs, wetlands, reservoirs, and lakes of the region to potential changes in climate, especially those inducing a change in hydrological patterns such as amount, timing and predictability of stream flow.

The major sensitivities of aquatic ecosystems are their permanence and even existence in the face of potential reduced net basin supply of water, stability of geomorphological structure and riparian ecotones with alterations in disturbance regimes, and water quality changes resulting from a modified water balance. In all of these respects, aquatic ecosystems of the region are also sensitive to the extensive modifications imposed by human use of water resources, which underscores the difficulty of separating this type of anthropogenic change from climate change. We advocate a focus in future research on reconstruction and analysis of past climates and associated ecosystem characteristics, long-term studies to discriminate directional change vs. year to year variability (including evidence of aquatic ecosystem responses or sensitivity to extremes), and studies of ecosystems affected by human activity.

INTRODUCTION

The Basin and Range, Southwest and Mexico Region (hereafter, BRSWM) encompasses a wide band of mountain and lowland in western North America from the state of Oregon (48° N) in the US south to Oaxaca (18° N) in Mexico (Fig. 1 in Leavesley *et al.*, 1997). The region includes all of the major North American deserts, and is in general bordered by and dotted with mountainous terrain. This wide range of topographic relief (Plate 1 in Leavesley *et al.*, 1997) strongly influences both the horizontal and vertical spatial distribution of precipitation. Precipitation in subalpine watersheds of mountain ranges (many over

3000 m) is often 4–10 times higher than in the adjoining basin lowlands. Except for the often ephemeral outflows of lowland springs, perennial streams in this area arise at high elevation.

The value to humans of aquatic resources in arid and semi-arid regions such as the BRSWM is obvious. Less often considered are the values of aquatic ecosystems as foci of biodiversity. Springs and stream–riparian ecosystems in the BRSWM support a high diversity of plant and animal life that are dependent on these oases of water and food resources in the arid and semi-arid surroundings. Biodiversity in Mexico ranks fifth globally; this is probably owing to its varied topography and climate, but also, in part, to the high degree of endemism associated with isolated aquatic ecosystems. The importance of BRSWM aquatic ecosystems, both as water resources and as foci of biodiversity, argues for the need to understand fully their dynamics in a changing climate, as well as with changing pressure from expanding human populations.

Variability and unpredictability are characteristics of the region's hydrology and climatology. Neither hydrological nor climatological models for the region are sufficiently developed to describe the magnitude or direction of change in response to increased carbon dioxide; thus, an attempt to predict specific responses of aquatic ecosystems is premature. The objectives of this paper are to highlight the key features of BRSWM aquatic ecosystems that are likely to be most sensitive to climate change, to discuss the interaction of climate change with anthropogenic change that is already occurring, to review briefly the current status of climate and hydrology models as applied to the region, and to provide a few recommendations for future research that will be crucial for understanding these interactions in this particular region.

HYDROCLIMATOLOGY OF THE REGION

Temperature

The amplitude of the annual temperature cycle of the BRSWM increases from south (range =< 5°C) to north (range = 25°C) as a result of changes in source regions for the air masses that affect the BRSWM. Mean January temperatures range latitudinally from over 25°C in the extreme south to −5°C in the north; July temperatures vary in response to elevation to a greater extent than to latitude and range from 15 to 20°C in high plateaus and mountains to 35–40°C in hot desert lowlands (Plates 3 and 4 in Leavesley *et al.*, 1997).

Precipitation

The climate of the BRSWM ranges from semi-arid to arid, with locally mesic climates occurring at high elevations. Annual precipitation varies considerably both in space and time throughout the region. In the northern area, winter and early spring precipitation delivered by cyclonic storms originating in the Pacific dominates. Annual precipitation ranges from 250 to 750 mm. Precipitation increases sharply at the north-eastern boundary of the region with up to 1500 mm annually over the higher elevations of the western humid and subhumid slopes of the Rocky Mountains (Price, 1979; Plate 2 in Leavesley *et al.*, 1997). The southern area is predominantly arid and precipitation is dominated by monsoonal flows that carry moisture from the Gulf of Mexico. Average annual precipitation for the area ranges from below 100 mm in the Mohave and Sonoran deserts to as much as 800 mm in the western Sierra Madre in Mexico. Annual rainfall seldom exceeds 250 mm in the valleys, where long dry periods extending to 150–210 days have been recorded in Hermosillo, Mexico (Ibarra *et al.*, 1995). The southernmost part of the region is mainly humid to tropical with average annual precipitation reaching 2000 mm.

Runoff

Because of the wide range of land use, temperature, vegetation cover and sources of runoff (i.e., snowmelt, direct rainfall), runoff variability is very pronounced on both regional and subregional scales. According to Bedinger and Sargent (1989), 'most of the region's runoff is generated by precipitation occurring on the intermediate elevations of the higher basin areas and the lower parts of the mountains'. The scarcity of protective vegetation cover, high temperatures and low antecedent soil moisture at low altitudes combine to produce maximum rates of evaporation in lowlands. Consequently, runoff is lower in the region's valleys

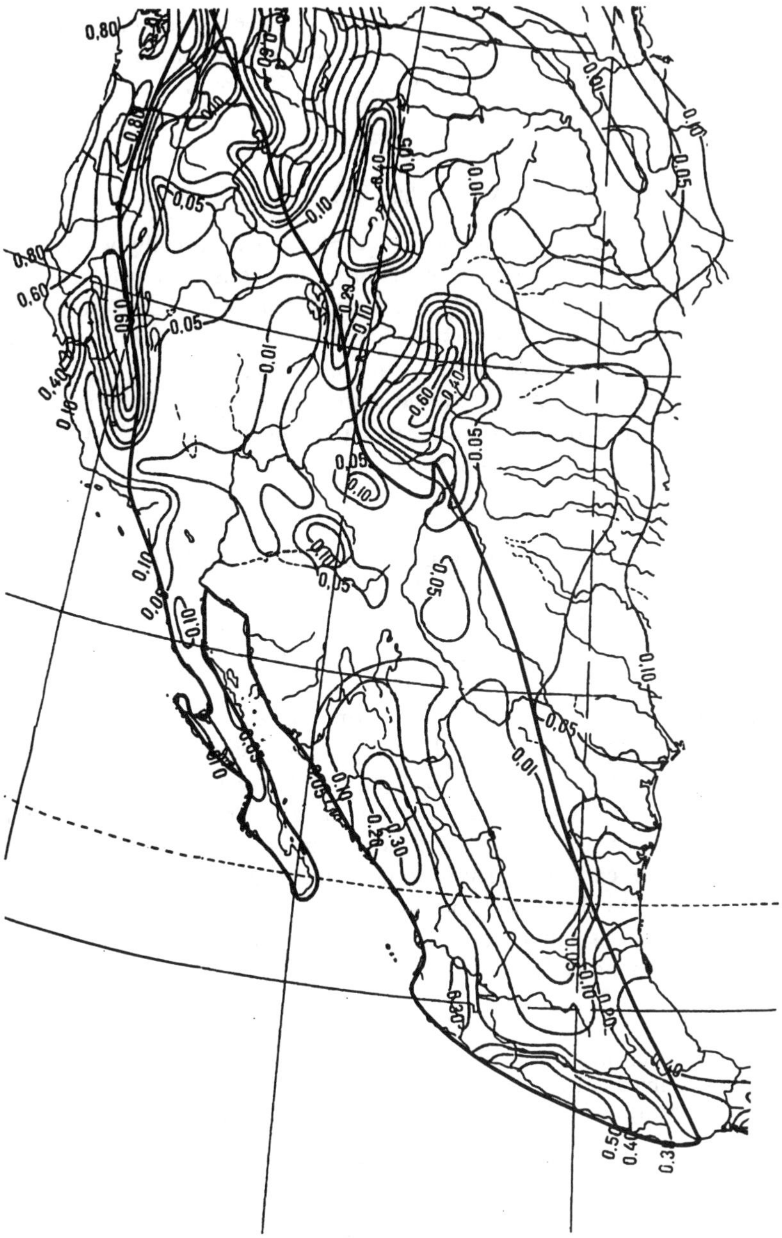

Figure 1. Map of western North America showing runoff ratios (runoff/precipitation) in the Basin and Range, Southwest and Mexico region (enclosed by thick line)

compared with its mountains, and is but a small fraction of precipitation. Runoff ratios (RO/P) are as low as 0·01 in some endorheic basins of northern Mexico and generally do not exceed 0·2 throughout the lowlands in much of the region (Figure 1). Furthermore, because most of the watersheds and subwatersheds in the region are characterized by high length to width ratios, mountain fronts play an important role in recharging the underlying aquifers, especially from spring snowmelt. This recharge represents an additional contribution of mountain precipitation to lowland stream flow within the BRSWM (Price, 1979).

Because of the combined effects of the above characteristics, average annual runoff in the southern part of the BRSWM varies between 2·5 mm in lowland desert areas to 50 mm at higher elevations. In contrast, average annual runoff in the northern BRSWM varies from 20 mm in the valleys to 125 mm at the highest mountain elevations.

OVERVIEW OF AQUATIC ECOSYSTEMS

A diversity of aquatic ecosystems can be found in the BRSWM, especially when elevational transects are considered. Nevertheless, certain region-wide generalities are consistent with its water-poor nature. Few natural lakes occur in the BRSWM, yet the region contains some of the largest reservoirs in North America. All large rivers in the region are impounded. Other natural aquatic ecosystems of the BRSWM are isolated, temporary waters are common (such as ephemeral and intermittent streams, tinajas and playas), and nearly all sizable aquatic ecosystems are stressed by the increasing demands for urban and agricultural water supplies. Ecotones such as riparian zones are particularly evident in the BRSWM because they represent zones where starkly contrasting ecosystems meet (Fisher, 1995).

Springs, Streams and Rivers

Streams and rivers of the BRSWM are spring fed or are derived from runoff (snowmelt in the north, rainfall in the south) from montane regions bordering these arid and semi-arid environments (Price, 1979; Minshall *et al.*, 1989). Perennial, large rivers of the region support much lower mean annual flows than those in more mesic regions with comparably sized catchments (Table I). The tendency for low flows is exacerbated by consumptive use throughout the basins. Prior to extensive impoundment, most rivers exhibited extreme variation in discharge over annual, decadal and longer time-scales. Mid-sized streams of the region that have not been impounded retain their 'flashy' hydrological characteristics. Many are spatially intermittent, flowing in summer only in bedrock-dominated segments where shallow alluvium permits surface flow. In general, constrained channel reaches with minimal sediment deposition are commonly gaining segments (i.e. discharge increases with downstream flow) of the streams where groundwater emerges into the channel. Conversely, unconstrained reaches with depositional features are sections of stream where groundwater recharge occurs and stream flow decreases. Springs sustain their permanent flow through aquifer discharge in high elevation montane regions and, in some smaller catchments of desert lowlands, by way of connection to artesian or other underground waters.

Table I. Comparison of largest rivers of the BRSWM with the Illinois River and the Columbia River, in terms of basin area and annual discharge. Colorado River discharge is at Hoover Dam (almost no runoff from the Colorado River reaches its mouth)

River	Basin area (km^2)	Annual discharge ($\times 10^6$ m^3)
Río Santiago	33 800	2100
Río Fuerte	36 600	5900
Río Yaqui	70 000	2800
Río Nazas	85 500	1700
Río de las Balsas	106 000	13 900
Snake River	267 000	40 000
Río Grande/Bravo	457 000	18 000
Colorado River	629 000	10 000
Comparison:		
Illinois River	65 527	18 300
Columbia River	668 000	230 000

High *in situ* primary production, the paucity of perennial flow, and the importance of disturbance are general attributes of streams in this dryland region. Lotic ecosystems of the northern BRSWM exhibit high rates of primary production and many may be autotrophic (i.e.$P_G > R$; Minshall, 1978; Cushing, 1996). High levels of solar radiation reaching the stream surface, through discontinuous riparian canopy cover, promotes generally high rates of aquatic primary production (Cushing and Wolf, 1984; Cushing, 1996). This feature is even more strongly expressed in the hot desert portions of the region, where high annual temperatures, long growing seasons and wide channels (with nearly year-long light saturation) result in high rates of both primary and secondary production (Busch and Fisher, 1981; Fisher *et al.*, 1982; Fisher and Gray, 1983; Jackson and Fisher, 1984). The food base in streams of both subregions is algae and aquatic macrophytes, with collector–gatherers feeding primarily on autochthonous detritus (Fisher and Gray, 1983; Gaines *et al.*, 1989).

Streams and rivers of the BRSWM are subjected to intense natural disturbance. Hydrological extremes, of both flood and drought, are the principal disturbance agents. Flash floods in the hot deserts of south-western USA (and presumably, in the more southerly desert regions) shape channel morphology, scour and deposit sediment and remove both instream and riparian biota (Fisher *et al.*, 1982; Grimm and Fisher, 1989; Stromberg *et al.*, 1993). Hot desert streams are also subject to extensive drying. Drying results in stranding and concentration of biota and loss of surface water. In some cases, hyporheic flow persists beneath the dry sediment surface (Stanley and Valett, 1992; Stanley and Boulton, 1995). In the northern subregion of the BRSWM, rain on snow events or thunderstorms can generate catastrophic floods. The few long-term databases on discharge for these ecosystems suggest an average of 3–4 years between such severe spates (Cushing and Gaines, 1989), compared with a more frequent occurrence in the south (for example, 2–9 flash floods per year occur in dry and wet years, respectively, in Sycamore Creek, Arizona; Grimm, 1993).

Lakes, Reservoirs and other Lentic Ecosystems

Standing water of the intermountain West are varied in geographical setting, basin origin and present day limnological conditions. The relatively few natural lakes or other open waters are of four types: large saline lakes that are remnants of far more extensive late-Pleistocene limnetic systems, temporary waters of desert basin lowlands (playas), montane/subalpine lakes and terminal wetlands with annual inundation but strong seasonality.

Many lakes of the BRSWM are closed, and lake levels therefore reflect the imbalance between annual precipitation and evaporation. They are thus sensitive to climate variability that would be of minor influence in systems with outflows (Cole, 1974; Hostetler and Giorgi, 1995). The size, depth and salinity of large saline lakes exhibit substantial temporal fluctuations caused by climate. For example, the area of Utah's Great Salt Lake, the largest saline lake in the region, more than doubled over 26 years between a drought and flood period (Stephens, 1990). Thermal stratification is limited in Great Salt Lake and many other lowland lakes because they are so shallow.

Hundreds of lowland valleys throughout the Great Basin, Mohave, Chihuahuan and Sonoran deserts can hold shallow, ephemeral lakes or *playas* (Snyder, 1962; Mifflin and Wheat, 1979). Some playas lie in the relict basins of late-Pleistocene pluvial lakes. Others are hydrologically isolated and have small watersheds, thus they have no geological record of limnetic permanence. The existence of lakes in these valleys depends on local (watershed) precipitation, playa surface evaporation (Hartland-Rowe, 1972) and, except in endorheic basins (especially prevalent in the Great Basin and north-central Mexico), alluvial hydrological drainage (Snyder, 1962).

Permanent montane lakes in the Great Basin/Mojave (northern) portion of the BRSWM lie at elevations of 2800 m or more. These lakes, which are essentially snowmelt- or spring-fed ponds, are found in isolated central mountain ranges, are generally no more than a few hectares in surface area, and have small watersheds such as those of cirque basins. These characteristics lead to high hydrological variability. Inflows peak during spring runoff and, in some, outflow may occur for only a few weeks of the year. The ponds display transient stratification as local convective storms produce regular episodes of severe wind shear and thorough mixing.

Montane lakes are more abundant in the southern BRSWM (Mexico). Most of these lakes are warm polymictic and exhibit minimal seasonal variation in temperature. Seasonal variation in other limnological properties is coupled to wet–dry seasons and dilution and evaporative concentration of solutes. Like lakes in the northern subregion, these lakes are very sensitive to changes in water input. For example, endorheic Lake Patzcuaro has exhibited extensive variation in level on both 10^3-year and 10^1-10^2-year time-scales (Chacón, 1993; O'Hara, 1993).

The few wetlands in the region include highly seasonal freshwater wetlands and saline marshes (Hofstetter, 1983). Wetlands associated with the littoral zones of lakes are poorly developed, owing to large excursions in lake level that result from variation in inflows (as in the northern terminal lakes). Riverine marshes, or ciénegas, were common in the south-western USA and northern Mexico until the mid-1800s, but these have all but vanished (Hendrickson and Minckley, 1984).

In addition to the natural lentic systems, many reservoirs, ranging in size from small stock ponds to some of the largest in North America, such as those of the lower Colorado River (Lakes Powell, Mead, Mohave and Havasu), are found in the BRSWM. The reservoirs of the Colorado River have conventional warm monomictic pelagia with largely familiar, cosmopolitan plankton biotas, yet are set within an arid climate and desert landscape. Many reservoirs are highly productive, with consequent hypolimnetic oxygen depletion during stratification. Intra- and inter-annual drawdowns of the reservoirs for irrigation and flood control can greatly reduce, or even dry up, shallow reservoirs. Additionally, the frequent water level fluctuations associated with the storage and use of water restrict the development of littoral zone plants and their associated animal communities.

PAST CLIMATE CHANGE AND ANTHROPOGENIC CHANGE

Climate

The geological record of climate since the last glacial maximum (21 000 years ago) indicates that the central and northern portions of the BRSWM were affected by relatively large climatic changes, i.e. shifts in the average climate (e.g. Thompson *et al.*, 1994). These late-Pleistocene climate changes have been associated with changes in synoptic-scale atmospheric circulation, characterized primarily by a southward displacement of the polar jet stream (westerly storm track) over western North America that was forced by the height and thermal effects of the Laurentide Ice Sheet (e.g. Kutzbach and Guetter, 1986; COHMAP Members, 1988). The climatological effect of the displaced storm track was a general enhancement of moisture throughout the Great Basin (Thompson *et al.*, 1994). Geological evidence from the southern portions of the Basin and Range (e.g. Van Devender *et al.*, 1987) suggests that the region was affected less by these late-Pleistocene climate changes than it was by Holocene climatic variability resulting, for example, from ENSO (El Niño–Southern Oscillation) events or from atmospheric circulation patterns associated with anomalous pressure patterns in the eastern Pacific (Enzel *et al.*, 1989, 1992; Ely *et al.*, 1993). More recent climate variations and flood events in the Southwest (e.g. the past 500 years) have also been attributed to anomalous circulation patterns (Redmond and Koch, 1991; Ely *et al.*, 1994). Interannual variation in stream flow over the historical record (past 50 years) has also been linked to both El Niño and La Niña events in the American Southwest (Molles and Dahm, 1990; Kahya and Dracup, 1994).

Anthropogenic Changes in Aquatic Ecosystems

Although the region includes some of the large rivers in the US and Mexico, these rivers are not large enough to meet the ever increasing, and rather fluctuating, demand for water. Rapid growth of urban areas, agriculture and industry, especially in the second half of this century, has placed significant stress on the region's water resources. Although the USGS water use estimates for 1985 (Solley *et al.*, 1988) indicate a 10% drop in freshwater consumption from 1980 levels, they remain at double the 1950 estimates.

One of the most obvious anthropogenic changes is the extensive impoundment of rivers to capture mountain runoff for municipal use, irrigation and hydroelectric power generation throughout the BRSWM.

Impoundment and diel, seasonal and interannual variation in water releases from these reservoirs have greatly altered the dynamics of the rivers below them (Ward and Stanford, 1979, 1983). Impoundment creates lentic ecosystems where none previously existed, and these ecosystems are now a common, if incongruous, feature of the desert landscape. Artificial control of surface water, whether for instream or offstream (diversion) uses, affects both the quantity and the quality of freshwater. The thermal properties of reservoirs can be influenced by the changing area–depth characteristics resulting from such controls. Additionally, return flow from irrigation diversion introduces pesticides and nutrients into the aquatic system.

The near-full utilization of surface water resources in the region has fostered an increasing reliance on groundwater resources. Until recently, groundwater and surface water were treated as two distinct resources. Consequently, continued groundwater pumping caused a significant drop in the water levels of several floodplain aquifers, leading to noticeable changes in the flow patterns of some perennial streams. In some cases, such as the Santa Cruz River of southern Arizona, perennially flowing stream reaches became intermittent or ephemeral (Arizona Department of Water Resources, 1995).

The Santa Cruz River is a striking example of anthropogenically induced changes in freshwater ecosystems, and a detailed examination of its history provides a good illustration of the effects of human manipulation on rivers in arid regions such as the BRSWM. Average annual precipitation in the 22 200 km^2 basin is 450 mm. The river flows southwards towards Mexico from its headwaters at the San Rafael Valley (south-eastern mountains of Arizona, altitude 2800 m), then flows westwards in Mexico for 56 km, and re-enters the US 10 km east of Nogales. In Mexico, most of the perennial flow of the river is captured by wells from the aquifer directly connected to the stream. Some of the captured water is returned to the stream through effluent discharge from Nogales.

Recently (May 1995), and under the title 'The Way We Were', the *Tucson Citizen* newspaper published a photograph dating to the late 1800s of a recreational lake called Silver Lake near Tucson, Arizona. The lake was an artificial lake generated by diverting water from the nearby perennial Santa Cruz River. Today the lake and, of course, the river reach are both dry because of the effect of groundwater pumping from the underlying floodplain aquifer. After 1940, the high rate of urban and agricultural development in the Tucson basin and subsequent lowering of the groundwater table (100 m below the river bed) caused further changes in stream characteristics. The elimination of marshes and wet meadows, which occupied short perennial reaches, resulted in the longitudinal expansion of arroyos, thus connecting the stream reaches into a continuous channel along the 80 km reach in the Tucson basin. Channel degradation continues as the result of floodplain development. In some areas the channel bed has been lowered four metres below its 1940 level.

According to the Arizona Department of Water Resources (1995), a wetter weather cycle starting in 1960, declining groundwater pumping in the past decade and, most importantly, the establishment of the Nogales treatment plant, are all contributing factors to the observed recovery of groundwater levels in the upper Santa Cruz Basin. This recovery has resulted in the establishment of a cottonwood–willow (*Populus fremonti* and *Salix goodingii*) riparian community along the river north of the Mexican border. Unlike the historical Santa Cruz, however, the flow in the river is derived primarily from treated effluent.

Where they occur, natural lakes are also used for water supply. In the southernmost portion of the BRSWM in Mexico, most lakes and reservoirs have experienced greatly decreased storage volumes over the past 20 years. Most of this is a result of increased demand by both a rapidly expanding population and, perhaps more importantly, a great increase in economic growth. Water demand in Mexico is closely tied to the economy. Fluctuations in the volume of Lake Chapala, Mexico over a 60-year period provide an excellent example of the difficulty in separating the magnitude of change wrought by human consumptive water use from climatic effects on hydrology.

Lake Chapala is possibly the oldest (middle Pleistocene; Clements, 1963) North American large lake. It lies in an east–west valley at 1524 m elevation. Its watershed is large (basin : lake area = 47), from which it derives nearly two-thirds of its water input, the remainder coming from direct rainfall (Limón and Lind, 1990). Changes in Lake Chapala brought about increased human demand are superimposed on climate-driven variation. In 1990, the volume of Lake Chapala was approximately one-fourth the mean long-term volume. Between 1974 and 1982, direct rainfall inputs to the lake increased slightly, but riverine inputs decreased by 45%. Lake volume declined by 40%. Limón *et al.* (1989) proposed that increased water use in the river basin, and not climate change, was responsible.

The decline in Lake Chapala's storage volume between 1974 and 1983 was coincident with the economic expansion produced by the demand for Mexican oil. With the crash of the oil market in 1982 the decline in lake storage volume stopped for five years, but the lake did not recover. The decline resumed in 1987 and continued to the minimum of record in 1990. As needs for water are no longer met by the natural lakes of the region, a large expenditure on reservoir construction and groundwater extraction is underway. Anthropogenic change, rather than climate variation, is especially implicated in lake level decline, since the years considered were characterized by a large number of warm episodes (El Niño-like conditions), which, at least in the American Southwest, have been linked to increased rainfall and runoff (e.g. Molles and Dahm, 1990).

Water quality deteriorated in Lake Chapala during the period of storage volume decline. As a result of the combined effects of reduced riverine inputs, control of releases through the outflow and relative increase in output to evaporation, the concentration of total dissolved solids increased by over 100%, and phosphorus increased by almost 400% (Limón *et al.*, 1989). However, this increase in salts did not produce increased algal growth and eutrophication. As the lake level declined, resuspension of sediments increased and light limitation prevented algal blooms. Because of concern over the diminished water supply experienced in the 1970s and 1980s, extra effort was taken to protect the quality of existing supplies; a series of 16 secondary waste water treatment plants were constructed around Lake Chapala between 1985 and 1992.

The examples from Lake Chapala and the Santa Cruz River clearly illustrate the tremendous changes in aquatic ecosystems that have occurred as a result of population pressure. Distinguishing between changes in aquatic ecosystems of the BRSWM caused by climate and those attributable to other anthropogenic effects will be difficult, especially against a background of high natural variability. Even if montane lakes of Mexico were sensitive to climate change, it is likely that the pressures from human populations have already induced changes in their structure and functioning that preclude detection of past responses to climate change. This may be true of permanent lentic ecosystems throughout the BRSWM. On the other hand, many temporary or smaller permanent systems are, by definition, on the brink of existence; their sensitivity is extreme and the record of a changed climate will be clearly indicated by their persistence. Finally, because of the pervasive effect of humans on naturally variable rivers like the Santa Cruz, it is unlikely that BRSWM rivers and streams will readily serve as sensitive bellwethers of climate change in the future. Rather, climate change will most likely be superimposed upon the substantial modifications of lotic ecosystems that have already occurred (Carpenter *et al.*, 1992).

FUTURE CLIMATE

Climate Models

Atmospheric general circulation models (GCMs) have been widely used to generate climate expectations for both past and future climates (e.g. Saltzman, 1990; Schneider *et al.*, 1990; Washington, 1992, and references therein). The results from GCM simulations of past climate (the past 10^5 years) suggest, on global, hemispherical, and continental scales, the range and type of changes that have occurred as a result of, for example, changes in earth–solar geometry and atmospheric composition. Predictions of future climate (the next 10^2 years) are based on the projected response to increased trace gas concentrations in the atmosphere. At the scale of the BRSWM, the models lack the ability to simulate details of regional climate (Grotch, 1988; Mearns *et al.*, 1990; Grotch and McCracken, 1991). Present computer power is inadequate to allow GCMs to be run at resolutions fine enough to provide the accuracy required for climatic and hydrological information related to aquatic ecosystems (Hostetler, 1994).

A major problem with GCMs is their inability to simulate convective precipitation (e.g. Washington, 1992; Hostetler, 1994), which is a large component of the annual precipitation in many parts of the BRSWM. Although advances have been made in *diagnostic* modelling of atmosphere–ocean phenomena such as El Niño, GCMs cannot model El Niños outside the period of data record (e.g. Cane, 1992 and references therein). Because of these problems, climate variables representing atmospheric conditions that lead to high magnitude precipitation and flood events or droughts in the BRSWM cannot be obtained from GCMs.

Hydrological Models

The BRSWM encompasses climates ranging from continental in the north, to arid desert in the southwestern USA, to tropical in southern Mexico, and is perhaps the most hydrologically diverse of all regions. This diversity is characterized by: (1) a large range of precipitation and temperature gradients; (2) the dependency of much of the region on high elevation precipitation for water supply; (3) surface–groundwater interactions associated with hydrologically closed basins; and (4) a very high degree of alteration of the natural surface water system through flow regulation and water use patterns.

Although current state of the art in both GCMs and hydrological models is quite advanced, progress is needed to fill the gap in scale between the two types of models (Hostetler, 1994). Physically based hydrological models, i.e. those in which the physics of flow, water storage and energy are accounted for mathematically, can be used to model small catchment areas that are homogeneous relative to the region as a whole. These models, which use time-series of meteorological parameters as input, can make use of output from climate models. However, the heterogeneity of precipitation, geology and land surface, compounded with the general problem that climate–hydrological interactions and processes are poorly understood in the region, provide major obstacles to aggregating small area models up to the regional scale.

Statistical models, such as time-series models, provide an alternative to physically based hydrological models. Time-series models are calibrated by using observed records and can be used to associate atmospheric circulation patterns with runoff. The calibrated models can be used to explore periods of flood or drought in the historical records (e.g. Ely *et al.*, 1993). Such investigations provide valuable insights into past climatic and hydrological variations, but time-series models cannot be used to make forecasts or extrapolations without violating the assumptions (e.g. stationarity) on which the calibrations are made.

Thus, the task of modelling hydrological responses to climate change remains a challenge. To gain an ability to project the effects of climate change on the region, we need to assess the influences of warming on precipitation (including its variability) through atmospheric modelling, and to investigate the role of vegetation feedbacks (such as evapotranspiration), soil characteristics and geology. This will require further research on development and the application of new and existing models and technology.

POTENTIAL RESPONSES OF AQUATIC ECOSYSTEMS

Approach

The problems discussed here preclude the matching of predicted climate with anticipated ecosystem changes in the BRSWM. Before we can make reasonable ecological predictions, we must know which aquatic ecosystems, or components of those ecosystems, are sensitive to climate. We present an overview of the sensitivities of BRSWM aquatic ecosystems to broad climate changes of four types (Table II): (1) overall changes in hydrological parameters, arising from enhanced or reduced precipitation with or without altered evapotranspiration, that lead to a reduction or increase in net basin supply (P–E); (2) changes in seasonality of precipitation and runoff; (3) changes in variability of precipitation and runoff; and (4) changes in temperature. Changes in runoff variability are viewed by some as potentially more disruptive of ecosystems than changes in mean conditions (Walker, 1991; Poff, 1992). Table II lists the properties that we identified as most sensitive to change for each ecosystem type, and the elements of climate, hydrology or anthropogenic stress to which they are most likely to be sensitive.

Table II. Properties of ten types of aquatic ecosystem in the BRSWM that are expected to be most sensitive to change. Third column lists the major climate, hydrology or anthropogenic variables (*) to which these properties are likely to respond. NBS = net basin supply (P-E); P/RO = precipitation and runoff. Numbers in parentheses refer to the temporal scale (yr) of P/RO variability to which the ecosystem would be most sensitive, and may apply to either the recurrence interval of extreme events or the duration of time periods over which frequency of extreme events is altered

Ecosystem type	Property	Sensitive to
Spring-fed streams	Permanence	NBS
		Groundwater recharge
		Groundwater withdrawal*
		P/RO seasonality
	Thermal and hydrological stability	NBS
		Temperature
		P/RO variability ($\times 10^2$ yr)
Lowland/desert streams	Flow permanence	P/RO seasonality
		NBS
		Groundwater recharge
		Groundwater withdrawal*
	Geomorphological structure	P/RO variability (10^2–10^3 yr)
	Riparian zone: recruitment and stability	Flow permanence
		Groundwater withdrawal*
		P/RO variability (10^1–10^2 yr)
Large rivers	Discharge	NBS (upper basin)
		Consumptive use*
		Regulation*
	Water quality	Diversion*
		Impoundment*
		Spatial variability in RO
Riparian zones or riverine wetlands	Permanence, biotic structure	Regulation*
		Groundwater withdrawal*
		P/RO variability
Lowland lakes, wetlands	Existence, permanence	NBS
		Groundwater withdrawal*
		P/RO variability (10^1–10^2 yr)
	Salinity/water quality	NBS
		P/RO variability (10^0–10^1 yr)
Lowland springs	Existence, permanence	NBS
		Groundwater recharge
		Groundwater withdrawal*
	Thermal stability	Temperature
Playas	Frequency of inundation, existence	NBS
		P/RO variability (10^1–10^2 yr)
Subalpine lakes (e.g. Great Basin)	Water balance	NBS
		P/RO seasonality
	Thermal stratification	Temperature
		P/RO seasonality
Montane lakes (e.g. Mexico)	Water balance	NBS
		Diversion*
		Consumptive use*
	Water quality	NBS
		Diversion*
		Consumptive use*
Reservoirs	Nutrient availability	P/RO seasonality
		P/RO variability (10^{-1}–10^0 yr)
	Habitat availability	NBS
		Consumptive use*
		P/RO seasonality

Streams, Rivers and Riparian Zones

Hydrology. The precipitation regime, particularly how precipitation translates to runoff, is paramount in its effect on stream and river ecosystems of the BRSWM. A small change in mean annual precipitation can yield large changes in surface flow from an arid or semi-arid catchment. For example, a decrease in mean annual precipitation of 10% can reduce surface runoff by 30–50% (Nemec, 1986; Dahm and Molles, 1992). For streams fed by springs, the impact of a changing precipitation regime will be strongly linked to groundwater recharge in the source areas. Lags of months to years, and longer, can occur between a change in aquifer recharge and the response of down-slope springs. Cessation of flow and conversion of perennial stream segments to intermittent reaches would be a likely by-product of diminished recharge of groundwater from an overall decrease in annual precipitation. This would be exacerbated by decreases in cool season precipitation in the northern BRSWM, as most aquifer recharge there derives from snowmelt during seasons when evapotranspiration rates are lower.

The timing and form of precipitation and runoff are influenced by temperature changes. If warming occurs, the likelihood of intense flooding from rain on snow events in the late winter or spring may increase. Such events are the cause of many of the catastrophic floods in the northern, cool deserts of the BRSWM that have strong effects on both riparian vegetation and stream biota (Baker, 1990; Cushing and Gaines, 1989). Pacific Ocean frontal storms of late winter and early spring occasionally provide heavy inputs of precipitation to the valleys and mountains of the northern BRSWM. A warmer climate could bring much of this moisture in the form of rain rather than snow, increasing the likelihood of catastrophic flooding and causing earlier removal of the snowpack, which in turn may increase the susceptibility of these ecosystems to extreme drying during the warm seasons of the year.

Timing is also critical in hot desert regions and in the seasonal rivers of Mexico. More streams of a given drainage area are dry in summer in the western deserts (Mohave, western portions of the Sonoran Desert) because they experience few summer monsoon storms. In contrast, streams that typically flow only in response to summer precipitation (Chihuahuan Desert) are less likely to have perennial segments because evapotranspiration is so high during summer. Although the southern portion of the BRSWM has more distinct wet and dry seasons, the combined effects of brief periods of intense rainfall, increased evaporation resulting from deforestation and reduced infiltration owing to heavy silt loads have led to the increased incidence of severe floods (Chacón and Múzquiz, in press). This change has occurred during this century; shifts in seasonality in the future could either alleviate or exacerbate the effect. In Sycamore Creek, an intensively studied, mid-sized stream of the northern Sonoran Desert, annual flow is dictated more by winter precipitation amounts than by summer rains, although rain falls in both seasons (Grimm and Fisher, 1992). During the five driest years of this stream's 30-year record, winter flows were slight, and accelerated drying conditions (indicated by $\geqslant$200-day periods without floods) were prevalent an average of 167 days, or 46% of the year (Grimm, 1993). Extended drying can lead to loss of habitat for aquatic organisms, slower recovery following rewetting and loss of aquatic ecosystem 'functions' such as nutrient retention and instream and riparian production (e.g. Stanley and Fisher, 1992; Stanley *et al.*, 1994).

Ecosystem response. Changes in hydrological factors, such as increased or decreased spate frequency and magnitude, changes in timing of high and low discharge events, changes in frequency of drying events or prolongation of drying, can affect the ecology of streams and rivers in numerous ways. Flash floods in hot deserts can obliterate much of the stream biota, but recovery is usually rapid (from a few weeks to 1–2 months). This high resilience is attributable to high temperatures and light (supporting high primary production) and a fauna characterized by extraordinarily rapid development (Gray and Fisher, 1981). Similarly, biotic communities of the northern BRSWM are disturbance prone but capable of rapid colonization and growth following disturbances (Cushing and Wolf, 1984; Cushing and Gaines, 1989). The stability of hyporheic (subsurface) habitats in these disturbance prone ecosystems may, in some cases, provide a refuge for many species, and in others has led to the development of a distinct and unique hyporheic faunal component (e.g. Boulton *et al.*, 1992).

Because lotic ecosystems of the BRSWM are disturbance prone, their resilient biotas may be less sensitive to increased variability than those of more constant ecosystems. But larger scale changes in fluvial systems can result from extended periods of altered precipitation and runoff variability. An arroyo-cutting and ciénega-draining episode of the late 1800s provides an example (see Hastings and Turner, 1965). When explorers first visited the Southwest, they encountered lush aquatic and riparian vegetation in extensive marshes (ciénegas) along streams and rivers. A period of erosion, beginning *ca.* 1880, drained most of these ciénegas, leaving steep-walled, channelized, and incised arroyos (Hastings and Turner, 1965). There is general agreement that climatic variation (a period of high intensity rainfall) was at least partially responsible for the arroyo-cutting era (Cooke and Reeves, 1976; Graf, 1979). Thus, changes in precipitation variability, in this case the frequency of large storms, can alter the geomorphological structure of streams. This, in turn, has profound consequences for ecosystem structure and functioning (Hendrickson and Minckley, 1984).

Riparian zones are also sensitive to precipitation and runoff variability, perhaps on a somewhat shorter time-scale. Although extreme events can prove damaging to extant forests, a period of wet weather and frequent floods also provides 'windows of recruitment' for some tree species (Baker, 1990; Stromberg *et al.*, 1991).

Consumption of water in the lower Colorado Basin is higher than that now supplied by reservoirs or rivers. This basin is thus sensitive to modifications in water gain or loss (Waggoner and Schefter, 1990). However, changes in lowland systems resulting from climate change will (except for evaporative losses) result from effects seen upstream and at higher elevation (i.e. in the upper Colorado Basin). Snowmelt represents almost 90% of annual runoff in the lower Colorado River Basin, thus 10–15% of the land area (above 2800 m) supplies three-quarters of the total hydrological flow (Schaake, 1990). At an intermediate scale, the region is subject to apparent non-equilibrium processes such as ENSO circulations (Molles *et al.*, 1992). The upper basin therefore has high stream flow variability (and is sensitive to climatic change) but the lower basin progressively integrates that variation through storage and thus greatly reduces climatic signals (Rhodes *et al.*, 1984; Gleick, 1990). Thus, upstream flow regulation and hydrological conditions, and climatic responses of upstream terrestrial ecosystems have primary control of lower Colorado basin hydro/limnological systems. In contrast, climate change occurring within the lower basin of the Colorado River should elicit very little detectable response against the background of anthropogenic manipulation.

Lowland Lakes, Playas and Wetlands

Cole (1974) suggests several aspects of desert limnology that are applicable to this region in terms of sensitivity to climate change. Since many of the systems are either permanently or seasonally closed basins, inflow and evaporation rates (usually negatively correlated) affect the water balance of lakes, which influences electrolyte concentrations and chemistry. Evaporation rates from desert lakes are high. For example, direct atmospheric losses from Pyramid and its companion pluvial relict, Walker Lake, are approximately 125 cm/year; Lake Mead (surrounded by the Mojave Desert) evaporates up to 200 cm/yr (Mifflin and Wheat, 1979; Galat *et al.*, 1981) and central Arizona reservoirs nearly 300 cm/yr (Sellers and Hill, 1974). Local precipitation inputs to desert lakes are insignificant relative to remote up-slope watershed rain and snow (Cole, 1974). Thus, the ontogeny of desert lakes, and in some cases the very existence of desert lakes, will be determined by patterns of regional and watershed climate modification.

Lakes in the Great Basin may be among the most vulnerable to climate change in the US, because precipitation and evaporation are nearly balanced (Gleick, 1990). The case of the Great Salt Lake illustrates this susceptibility very clearly at a variety of time-scales. In the wet Pleistocene, the Great Salt Lake covered 49 000 km^2, extended into three states and had an abundant freshwater fauna (Benson *et al.*, 1990). A 153-year historical record of lake level (Figure 2) demonstrates how sensitive the closed basin lakes are to shorter term climatic changes. In 1963 the lake had shrunk to only 2470 km^2, salts were concentrated to 280 g/l and the only higher organisms present were herbivorous brine shrimp and brine flies (Stephens, 1990). From 1963 to 1986 the lake level rose 6 m, and salinities dropped to 50 g/l. This allowed several predacious invertebrates to invade (see Figure 3). The predators eradicated the filter-feeding brine shrimp

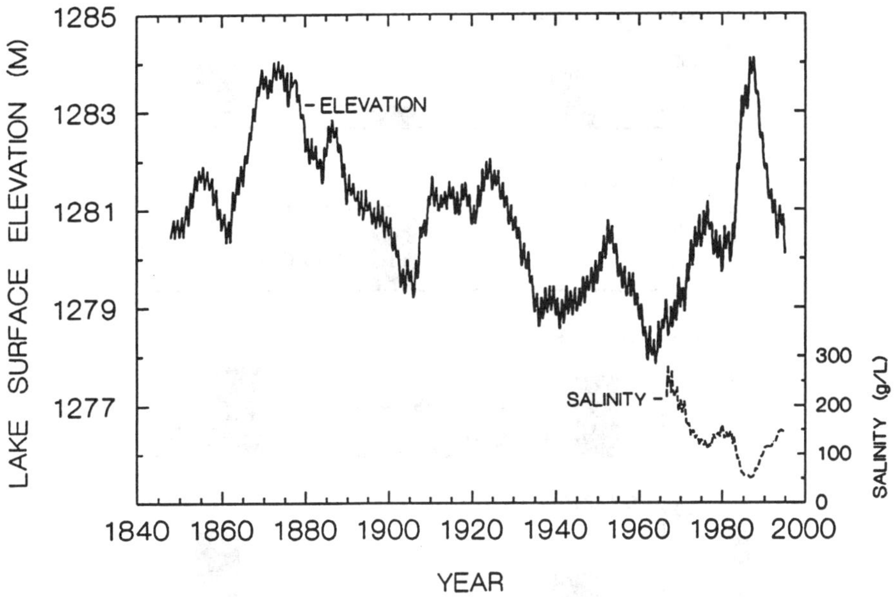

Figure 2. Changes in surface elevation (1847-1994) and salinity (1966-1994) of the Great Salt Lake, Utah, resulting from climatic changes. Note the inverse relationship between lake elevation and salinity in this terminal lake. Data provided by the Utah Geological Survey

(*Artemia franciscana*) and decreased overall herbivore biomass by more than an order of magnitude. This allowed algal levels to increase 20-fold, resulting in decreases in soluble nutrients and water clarity. Decreased light penetration reduced heating in the monimolimnion of the lake. Birds dependent on the brine shrimp in the lake became rare. Thus, climatically induced changes in runoff over a two-year period profoundly affected the food-web, chemistry and physics of the lake (Wurtsbaugh and Berry, 1990; Wurtsbaugh, 1992).

One recent modelling projection of the water and energy balances of saline Pyramid Lake under a warmer climate indicated potential changes in the timing of stratification (Hostetler and Giorgi, 1995). In one sensitivity test, the lake ceased annual turnover altogether. Like most lakes in the BRSWM, Pyramid Lake's water balance, especially inflow volume and seasonality, are largely controlled by anthropogenic watershed diversions (Galat *et al.*, 1981; Hostetler and Giorgi, 1995), confounding accurate interpretation of climatic effects.

Smaller closed basin lakes and spring pools could also be affected by future climatic change because the size, salinity and even the existence of each system is tightly coupled to runoff (Neal, 1975). A regional climatic change that decreased precipitation might eliminate many of the smaller water bodies and the aquatic communities dependent on them. Many isolated spring systems that harbour endemic fishes would be at risk (Minckley and Deacon, 1991). Conversely, a sustained increase in precipitation might increase the number and size of standing lakes. This can be beneficial for aquatic organisms, but may destroy human developments built in lake bottoms during drier times. Salinity changes would probably lead to altered community structures (Stephens. 1990), as described for the larger Great Salt Lake.

The sensitivity of playas to climate is implicit. Currently ephemeral lakes could become permanent with an increase in precipitation (Mifflin and Wheat, 1979), whereas drier and warmer conditions would lengthen the intervals between inundations and decrease the episodic durations of open water. Many species of playa invertebrates may be limited by long periods of desiccation in certain playa basins (Belk and Cole, 1975) and others by temperature extremes (Horne, 1972). These physiological constraints, coupled with changes in life history timing, may shift the geographical distribution of some species or community types in the face of

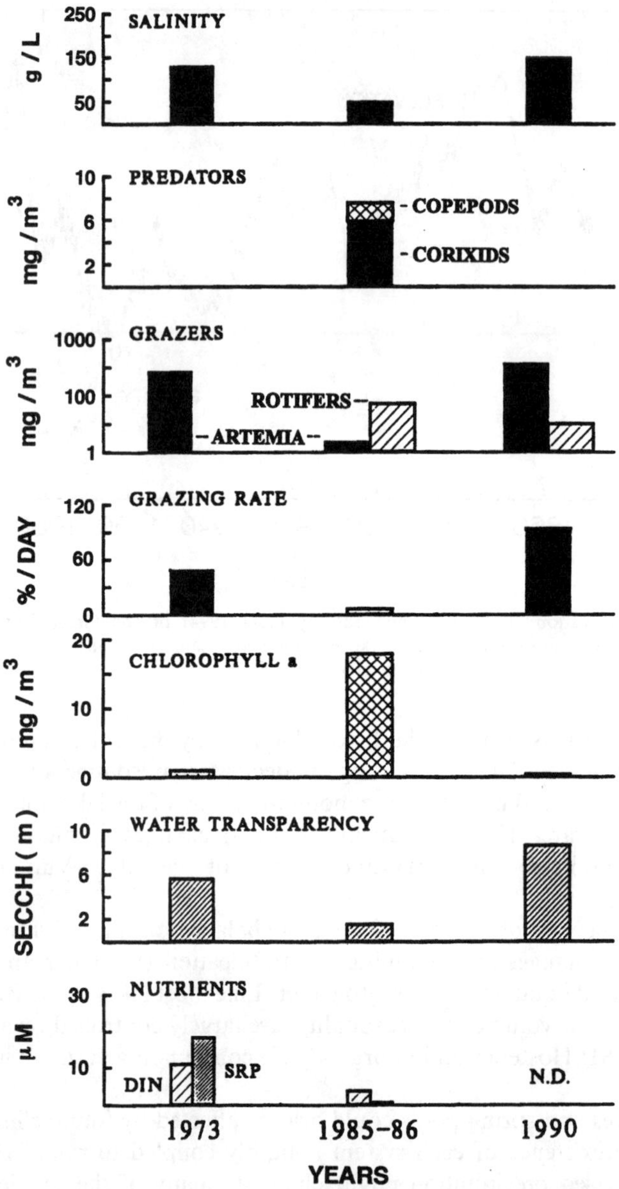

Figure 3. Changes in the Great Salt Lake ecosystem resulting from climatic changes that modified the runoff and consequently the salinity of this terminal lake. Decreasing salinity between 1973 and 1985 allowed invertebrate predators (*Trichocorixa verticalis*, *Diaptomus connexus*, *Cletocampus albuquerquensis*) to invade, which in turn initiated a trophic cascade within the lake. During the dry period that followed, salinity increased and the ecosystem reverted to its former state. Grazing rate = % of water column filtered per day by the zooplankton community; SRP = soluble reactive phosphorus; DIN = nitrate- + nitrite- + ammonium-nitrogen; N.D. = no data. Modified from Wurtsbaugh (1992)

climate change. On the other hand, the evolutionary history of the dominant playa faunal groups (eubranchiopods and ostracods) extends back more than 100 Ma (Tasch, 1980); contemporary climate change seems unlikely to threaten the taxa.

Wetlands, as complex free-standing aquatic systems or as wet/dry ecotones (*sensu* Risser, 1990), may (Franzen, 1994; Neilson, 1993) or may not (Noble, 1993) respond strongly to climate change. Natural wetlands in the desert, however, will probably be influenced, because their existence is continuously threatened by both the inherent variability of the regional climate and anthropogenic pressures. Wetlands in the intermountain region are usually subject to human management (diversion and regulation) which is driven by economic and political forces and, to a lesser degree, by climatic or hydrological patterns. This is clearly demonstrated in the history of the Stillwater Marsh–Lahontan Valley wetland complex, which has experienced high magnitude interannual variations in inflow and wetland water levels that are only partially coincident with changes in regional hydrological patterns (Yardas, 1994). Primary productivity, biodiversity, organic deposition and flux and nutrient cycling are dependent on hydrological patterns (Gosselink and Turner, 1978; Hogenbirk and Wein, 1991) and could be altered by shifts in seasonality or changes in net basin supply.

The southern portion of the BRSWM has a number of shallow endorheic basins, which are the remnants of an extensive late-Pliestocene lake system (Clements, 1963). These lakes are only moderately saline and have extensive wetlands along their perimeters. They are important wildlife habitats with large numbers of native and seasonally migrant birds. Increased water to these lakes might expand littoral wetlands, whereas a decrease in water supply could eliminate them as significant wildlife habitat.

High elevation systems, such as the numerous subalpine lakes in isolated mountain ranges of the BRSWM, have limnological characteristics that make them susceptible to climate change. Their small watersheds are not well buffered hydrologically and their inflows are limited to nearby snowmelt or small springs. Interannual variability can be very severe, and the open water season short. Any sustained and significant change in temperature and/or precipitation will modify the water balance, lake persistence and seasonal characteristics.

A regional climate change that affected net basin supply would have larger effects on small reservoir systems in the BRSWM than on large ones. For example, increased reservoir drawdown would decrease the mean annual size of each system, reducing available habitat for fishes, and perhaps decreasing oxygen levels in the more restricted hypolimnetic strata (Ploskey, 1986). Nutrient and organic matter inputs to reservoirs, like natural lakes, occur in pulses associated with extreme events or seasonal runoff; we therefore expect these systems to be sensitive to shifts in precipitation and runoff variability and seasonality. Larger reservoirs, like those of the Colorado River discussed earlier, are less likely to be sensitive to local climate change, because the characteristics of these systems represent integration of upstream and extraregional hydrological processes.

Temperature effects

Although we have proposed that hydrological factors are the most likely to effect change in aquatic ecosystems of the BRSWM (as influenced by changes in precipitation and runoff), the southernmost tropical subregion may be sensitive to changes in the thermal regime. This tropical zone normally has a very small temperature variation, both on daily and seasonal bases. For example, annual mean water column temperature in Lake Chapala is 21·7°C, but maximum seasonal temperature range during the 60-year period of record for this lake was 6°C. Organisms that have evolved under such constant conditions would be expected to be stenotherms. A strong biotic response to warming is likely in relatively constant thermal environments such as these, where large changes in temperature or temperature variability have not been experienced in ecological or recent evolutionary time. The surface temperature of Lake Pátzcuaro, Mexico has increased in comparison with past records (Chacón, 1993), and there is some evidence that spawning by native silversides (Pisces: Atherinidae), once a year-round event, is now restricted to the winter months (T. A. Chacón, personal observation).

In the mountainous and semi-arid region of the southern BRSWM, aquatic ecosystems are extremely isolated and endemism is common. There are also unique species flocks of native fishes in lakes located at different altitudes (Echelle and Echelle, 1984). Temperature changes might affect such communities both by disturbing the interspecies relationships within the flocks, and, possibly, by migration of species up or down the altitude gradient with rising or falling temperatures. Such migration, however, is less likely in this region, where few water holes are separated by vast expanses of desert.

In the northern subregion, the biotic effects of temperature change would most likely result from runs of abnormally warm weather that drive the stream temperature above thermal tolerances for those aquatic organisms adapted to cold desert and semi-arid environments. Cushing (1996) and Minshall (1978) show a relatively small variation in temperature at the source areas of spring-fed perennial reaches of a stream, but wider temperature ranges downstream. The downstream reaches of these ecosystems would be more likely to reflect warmer air temperatures and stenothermal organisms may be replaced by eurythermal species.

Changes in insolation and water temperature would also have an effect on lentic systems in the northern part of the region. Recent sensitivity experiments suggest that the duration of ice cover on dimictic lakes might be reduced and that summer temperatures would be warmer (Hostetler and Giorgi, 1995). As a result, increased insolation reaching the lake surfaces would most likely increase the annual productivity. Cold water species, particularly recreationally important salmonids, might disappear from lakes and reservoirs that warm, because many are already at their upper temperature limit during summer, and are restricted from deeper layers by anoxia (Utah Department of Health, 1982).

RECOMMENDATIONS

Aquatic ecosystems may be most effectively managed in the context of global climate change if both the more pressing anthropogenic threats and the occurrence of extreme events are considered and incorporated into management plans. Research and monitoring activities should focus on: (1) reconstruction and analysis of past climates and associated ecosystem characteristics; (2) continuation and/or initiation of long-term studies to discriminate directional change vs year to year variability; and (3) studies of ecosystems affected by human activity. When using a time-series approach, we should seek evidence of aquatic ecosystem responses or sensitivity to extremes (for example, wet or dry periods, years of especially high or low frequency of disturbance events). Such an effort would benefit from a coordinated study by hydrologists, climatologists, geomorphologists and ecologists using complementary techniques, and should focus on the identification of sensitive ecosystem components.

In this extreme and variable environment, a climate change signal may not be detectable against the background of extensive human manipulation; yet, we need a better understanding of how such manipulation has and will continue to alter the structure and function of aquatic ecosystems. To incorporate human-influenced ecosystems, we advocate a comparative ecosystem approach wherein relatively unaltered ecosystems are compared with highly manipulated ones.

ACKNOWLEDGEMENTS

We thank the US Environmental Protection Agency, the US Geological Survey, the American Society of Limnology and Oceanography and the North American Benthological Society for sponsoring and organizing the symposium that provided the inspiration for this paper. Thanks to Soroosh Sorooshian for providing the material on changes in the Santa Cruz River. Support for N.B.G. during manuscript preparation was possible through grants from the National Science Foundation.

REFERENCES

Arizona Department of Water Resources, 1995. *Riparian Protection Program*', Legislative Report. ADWR, Phoenix, AZ.

Baker, W. L. 1990. 'Climatic and hydrologic effects on the regeneration of *Populus angustifolia* James along the Animas River, Colorado', *J. Biogeogr.*, **17**, 59–73.

Bedinger M. S. and Sargent, K. A., 1989. 'Introduction, studies of geology and hydrology in the basin and range province, southwestern United States, for isolation of high level radioactive wastes', *USGS Open File Report 84–738, Professional paper; 1370-(A)*. U.S. Dept. of the Interior, Geological Survey, Denver, CO.

Belk, D. and Cole, G. A. 1975. 'Adaptational biology of desert temporary-pond inhabitants', in Hadley, N. F. (Ed.), *Environmental Physiology of Desert Organisms*. Dowden, Hutchinson & Ross, Stroudsburg. pp. 207–226.

Benson, L. V., Currey, D. R., Dorn, R. I., Lajoie, K. R., Oviatt, C. G., Robinson, S. W., Smith, G. I., and Stine, S. 1990. 'Chronology of expansion and contraction of four Great Basin lake systems during the past 35,000 years', *Palaeogeogr. Palaeoclimatol. Palaeoecol.*, **78**, 241–286.

Boulton, A. J., Valett, H. M., and Fisher, S. G. 1992. 'Spatial distribution and taxonomic composition of the hyporheos of several Sonoran Desert streams', *Arch. Hydrobiol.*, **125**, 37–61.

Busch, D. E. and Fisher, S. G. 1981. 'Metabolism of a desert stream,' *Freshwat. Biol.*, **11**, 301–307.

Cane, M. A. 1992. 'Tropical Pacific ENSO models: ENSO as a mode of the coupled system', in Trenberth, K. E. (Ed.), *Climate System Modeling*. Cambridge University Press, Cambridge. pp. 583–616.

Carpenter, S. R., Fisher, S. G., Grimm, N. B., and Kitchell, J. F. 1992. 'Global change and freshwater ecosystems', *Annu. Rev. Ecol. Syst.*, **23**, 119–139.

Chacón, T. A. 1993. 'Lake Pátzcuaro, México: watershed and water quality deterioration in a tropical high-altitude Latin American lake', *Lake Resevoir Manag.*, **8**, 37–47.

Chacón, T. A. and Múzquiz, I. L. E. 1997. 'Climatic trends, water balance and lake levels in a tropical high altitude lake', *Quat. Res.*, in press.

Clements, T. 1963. 'Pleistocene history of Lake Chapala, Jalisco, México', in Clements, T. (Ed.), *Essays in Marine Geology in Honor of K. O. Emery*. University Southern California Press, Los Angeles. pp. 35–49.

COHMAP Members, 1988. 'Climatic changes of the last 18,000 years: observations and model simulations', *Science*, **241**, 1043–1052.

Cole, G. A. 1974. 'Desert limnology', in Brown, G. W. (Ed.), *Desert Biology*, Vol. 2. Academic Press, New York. pp. 423–486.

Cooke, R. U. and Reeves, R. W. 1976. *Arroyos and Environmental Change in the American South-West*. Clarendon Press, Oxford. 213 pp.

Cushing, C. E. 1996. 'The ecology of cold desert spring-streams', *Arch. Hydrobiol.*, **135**, 499–522.

Cushing, C. E. and Gaines, W. L. 1989. 'Thoughts on recolonization of endorheic cold desert spring-streams', *J. North Am. Benthol. Soc.*, **8**, 277–287.

Cushing, C. E. and Wolf, E. G. 1984. 'Primary production in Rattlesnake Springs, a cold desert spring-stream', *Hydrobiologia*, **114**, 229–236.

Dahm, C. N. and Molles, M. C., Jr. 1992. 'Streams in semi-arid regions as sensitive indicators of global climate change', in Firth, P. and Fisher, S. (Eds), *Global Climate Change and Freshwater Ecosystems*. Springer-Verlag, New York. pp. 250–260.

Echelle, A. A. and Echelle, A. F. 1984. 'Evolutionary genetics of a "species flock": atherinid fishes on the Mesa Central of Mexico', in Echelle, A. and Kornfield, I. (Eds), *Evolution of Fish Species Flocks*. University of of Maine at Orono Press, Orono. pp. 93–110.

Ely, L. L., Enzel, Y., Baker, V. R., and Cayan, D. R. 1993. 'A 5000-year record of extreme floods and climate change in the southwestern United States', *Science*, **262**, 410–412.

Ely, L. L., Enzel, Y., and Cayan, D. R. 1994. 'Anomalous North Pacific circulation and large winter floods in the southwestern United States', *J. Climate*, **7**, 977–987.

Enzel, Y., Cayan, D. R., Anderson, R. Y., and Wells, S. G. 1989. 'Atmospheric circulation during Holocene lake stands in the Mojave Desert: evidence of regional climate change', *Nature*, **341**, 44–48.

Enzel, Y., Brown, W. J., Anderson, R. Y., McFadden, L. D., and Wells, S. G. 1992. 'Short-duration Holocene lakes in the Mojave River Drainage Basin, Southern California', *Quat. Res.*, **38**, 60–73.

Fisher, S. G. 1995. 'Western coterminous United States', in Cushing, C. E., Minshall, G. W., and Cummins, K. W. (Eds), *Ecosystems of the World (River and Stream Ecosystem)*. Elsevier, Amsterdam. pp. 57–78.

Fisher, S. G. and Gray, L. J. 1983. 'Secondary production and organic matter processing by collector macroinvertebrates in a desert stream', *Ecology*, **64**, 1217–1224.

Fisher, S. G., Gray, L. J., Grimm, N. B., and Busch, D. E. 1982. 'Temporal succession in a desert stream ecosystem following flash flooding', *Ecol. Monogr.*, **52**, 93–110.

Franzen, L. G. 1994. 'Are wetlands the key to the ice-age cycle enigma?', *Ambio*, **23**, 300–308.

Gaines, W. L., Cushing, C. E., and Smith, S. D. 1989. 'Trophic relations and functional group composition of benthic insects in three cold desert streams', *Southwest. Nat.*, **34**, 478–482.

Galat, D. L., Lider, E. L., Vigg, S., and Robertson, S. R. 1981. 'Limnology of a large, deep North American terminal lake, Pyramid Lake, Nevada, U.S.A'., *Hydrobiologia*, **82**, 281–317.

Gleick, P. H. 1990. 'Vulnerability of water systems', in Waggoner, P. E. (Ed.), *Climate Change and U.S. Water Resources*. Wiley-Intescience, New York. pp. 223–240.

Gosselink, J. G. and Turner, E. 1978. 'The role of hydrology in freshwater wetland ecosytems', in Good, R. E., Whigham, D. E., and Simpson, R. L. (Eds), *Freshwater Wetlands*. Academic Press, New York. pp. 63–86.

Graf, W. L. 1979. 'The development of montane arroyos and gullies', *Earth Surf. Proc. Landf.*, **4**, 1–14.

Gray, L. J. and Fisher, S. G. 1981. 'Postflood recolonization pathways of macroinvertebrates in lowland Sonoran Desert stream', *Am. Midl. Nat.*, **106**, 249–257.

Grimm, N. B. 1993. 'Implications of climate change for stream communities', in Kareiva, P., Kingsolver, J., and Huey, R. (Eds), *Biotic Interactions and Global Change*. Sinauer Associates, Inc., Sunderland. pp. 293–314.

Grimm, N. B. and Fisher, S. G. 1989. 'Stability of periphyton and macroinvertebrates to disturbance by flash floods in a desert stream', *J. North Am. Benthol. Soc.*, **8**, 293–307.

Grimm, N. B. and Fisher, S. G. 1992. 'Responses of arid land streams to changing climate'. in Firth, P. and Fisher, S. (Eds), *Global Climate Change and Freshwater Ecosystem*. Springer-Verlag, New York. pp. 211–233.

Grotch, S. L. 1988. 'Regional intercomparisons of general circulation model predictions and historical climate data', Report TR041. US Department of Energy, Washington, D.C.

Grotch, S. L. and MacCracken, M. C. 1991. 'The use of general circulation models to predict regional climate change', *J. Climate*, **4**, 286–303.

Harland-Rowe, R. 1972. 'The limnology of temporary waters and the ecology of euphyllopoda'. in Clark, R. B. and Whooton, R. J. (Eds), *Essays in Hydrobiology.* University of Exeter Press, Exeter. pp. 15–31.
Hastings, J. R. and Turner, R. M. 1965. *The Changing Mile.* University of Arizona Press, Tucson. 317 pp.
Hendrickson, D. A. and Minckley, W. L. 1984. 'Cienegas — vanishing climax communities of the American Southwest', *Desert Plants*, **6**, 131–175.
Hofstetter, R. H. 1983. 'Wetlands in the United States', in Gore, A. J. P. (Ed.), *Mires: Swamp, Bog, Fen and Moor. Ecosystems of the World 4B.* Elsevier, Amsterdam. pp. 201–244.
Hogenbirk, J. C. and Wien, R. W. 1991. 'Fire and drought experiments in northern wetlands: a climate change analogue', *Can. J. Bot.*, **69**, 1991–1997.
Horne, F. R. 1972. 'Some effects of temperature and oxygen concentration on phyllopod ecology', *Ecology*, **52**, 343–347.
Hostetler, S. W. 1994. 'Hydrologic and atmospheric models — the problem of discordant scales', Guest editorial, *Climatic Change*, **27**, 345–350.
Hostetler, S. W. and Giorgi, F. 1995. 'Effects of a 2 × CO_2 climate change on two large lake systems: Pyramid Lake, Nevada, and Yellowstone Lake, Wyoming', *Global and Planetary Change*, **10**, 43–54.
Ibarra, F. A., Cox, J. R., Martin, M. H., Crowl, T. A., Post, D. F., Miller, R. W., and Rasmussen, G. A. 1995. 'Predicting buffelgrass survival, organic carbon, and soil color in Mexico', *Soil Sci. Soc. Am. J.*, **59**, 1120.
Jackson, J. K. and Fisher, S. G. 1986. 'Secondary production, emergence, and export of aquatic insects of a Sonoran Desert stream', *Ecology*, **67**, 629–638.
Kahya, E. and Dracup, J. A. 1994. 'The influences of Type 1 El Niño and La Niña events on streamflows in the Pacific Southwest of the United States', *J. Climate*, **7**, 965–976.
Kutzbach, J. E. and Guetter, P. J. 1986. 'The influences of changing orbital parameters and surface boundary conditions on climate simulations for the past 18 000 years', *J. Atmos. Sci.*, **43**, 1726–1759.
Leavesley, G. H., Turner, K., D'Agnese, F. A., McKnight, D. 1997. 'Regional delineation of North America for the assessment of freshwater ecosystems and climate change', *Hydrol. Process.*, **11**, 819–824.
Limón, J. and Lind, O. 1990. 'The management of Lake Chapala, México', *Lake Resevoir Manage.*, **6**, 61–70.
Limón, J., Lind, O., Vodopich, D., Doyle, R., and Trotter, B. 1989. 'Long- and short-term variation in the physical and chemical limnology of large, shallow, turbid tropical lake (Lake Chapala, México)', *Arch. Hydrobiol./Suppl. 83*, **1**, 57–83.
Mearns, L. O., Gleick, P. H., and Schneider, S. H. 1990. 'Climate forecasting', in Waggoner, P. E. (Ed.), *Climate Change and U.S. Water Resources.* Wiley-Interscience, New York. pp. 87–138.
Mifflin, M. D. and Wheat, M. M. 1979. *Pluvial Lakes and Estimated Pluvial Climates of Nevada*, Nevada Bureau of Mines and Geology Bulletin 94. 57 pp. University of Nevada, Reno, NV.
Minckley, W. L. and Deacon, J. E. 1991. *Battle Against Extinction: Native Fish Management in the American West.* University of Arizona Press, Tucson. 517 pp.
Minshall, G. W. 1978. 'Autotrophy in stream ecosystems', *BioScience*, **28**, 767–771.
Minshall, G. W., Jensen, S. E., and Platts, W. S. 1989. 'The ecology of stream and riparian habitats of the Great Basin region: a community profile', *Biological Report 85 (7.24).* US Fish and Wildlife Service. Washington, D.C. 142 pp.
Molles, M. C. and Dahm, C. N. 1990. 'A perspective on El Nino and La Nina — global implications for stream ecology', *J. North Amer. Benthol. Soc.*, **9**, 68–76.
Molles, M. C., Jr., Dahm, C. N., and Crocker, M. T. 1992. 'Climatic variability and streams and rivers in semi-arid regions', in Robarts, R. D and Bothwell, M. L. (Eds), *Aquatic Ecosytems in Semiarid Regions: Implications for Resource Management*, NHRI Symposium Series 7. Environment Canada, Saskatoon. pp. 197–202.
Neal, J. T. 1975. *Playas and Dried Lakes: Occurrence and Development.* Dowden, Hutchinson and Ross, Inc., Stroudsburg. 411 pp.
Nemec, J. 1986. *Hydrologic Forecasting.* D. Reidel Publishing Co., Boston. 239 pp.
Noble, I. R. 1993. 'A model of responses of ecotones to climate change', *Ecol. Appl.*, **3**, 396–403.
O'Hara, S. L. 1993. 'Historical evidence of fluctuations in the level of Lake Pátzcuaro, Michoacán, México over the last 600 years', *Geograph. J.*, **159**, 51–62.
Ploskey, G. R. 1986. 'Effects of water-level changes on reservoir ecosystems, with implications for fisheries management', in Hall, G. E. and Van Den Avyle, M. J. (Eds), *Resevoir Fisheries Management: Strategies for the 80's.* American Fisheries Society, Bethesda. pp. 86–97.
Poff, N. L. 1992. 'Regional hydrologic response to climate change: an ecological perspective', in Firth, P. and Fisher, S. (Eds), *Global Climate Change and Freshwater Ecosystems.* Springer-Verlag, New York. pp. 88–115
Price D, 1979. 'Summer appraisal of the water resources of the Great Basin', *Proceedings, RMAG-UGA-Basin and Range Symposium.*
Redmond, K. T. and Koch, R. W. 1991. 'Surface climate and streamflow variability in the western United States and their relationship to large-scale circulation indices', *Wat. Resour. Res.*, **27**, 2381–2399.
Rhodes, S. L., Ely, D., and Dracup, J. A. 1984. 'Climate and the Colorado River: the limits to management', *Bull. Am. Meteorol. Soc.*, **65**, 682–691.
Risser, P. G. 1990. 'The ecological importance of land-water ecotones', in Naiman, R. J. and Decampo, H. (Eds), *The Ecology and Management of Aquatic — Terrestrial Ecotones*, Man and the Biosphere Series, Vol. 4. UNESCO and Parthenon Publishing Group, Paris, France. pp. 7–21.
Saltzman, B. 1990. 'Three basic problems of paleoclimatic modeling: a personal perspective and review', *Climate Dynam.*, **5**, 67–78.
Schaake, J. C. 1990. 'From climate to flow', in Waggoner, P. E. (Ed.), *Climate Change and U.S. Water Resources.* Wiley-Interscience, New York. pp. 177–206.
Schneider, S. H., Gleick, P. H., and Mearns, L. O. 1990. 'Prospects for climate change', in Waggoner, P. E. (Ed.), *Climate Change and U.S. Water Resources.* Wiley-Interscience, New York. pp. 41–73.
Sellers, W. D. and Hill, R. H. 1974. *Arizona Climate.* University of Arizona Press, Tucson.

Snyder, C. T. 1962. 'A hydrologic classification of valleys in the Great Basin, western United States', *Bull. Int. Assoc. Sci. Hydrol.*, **7**, 53–59.

Solley W. B., Pierce, R. R., and Merk, C. F. 1988. 'Trends in water use in the United States, 1950 to 1985', *Water use data for water resources management, Proceedings, AWRA Symposium*, Bethesda, Maryland.

Stanley, E. H. and Boulton, A. J. 1995. 'Hyporheic processes during flooding and drying in a Sonoran Desert stream: hydrologic and chemical dynamics', *Arch. Hydrobiol.*, **134**, 1–26.

Stanley, E. H. and Fisher, S. G. 1992. 'Intermittency, disturbance, and stability in stream ecosystems', in Robarts, R. D. and Bothwell, M. L. (Eds), *Aquatic Ecosystem in Semiarid Region: Implications for Resource Management*. NHRI Symposium Series 7. Environment Canada, Saskatoon. pp. 271–280.

Stanley, E. H. and Valett, H. M. 1992. 'Interactions between drying and the hypoheic zone in a desert stream', in Firth, P. and Fisher, S. (Eds), *Global Climate Change and Freshwater Ecosystems*. Springer-Verlag, New York. pp. 234–249.

Stanley, E. H., Buschman, D. L., Boulton, A. J., Grimm, N. B., and Fisher, S. G. 1994. 'Invertebrate resistance and resilience to intermittency in a desert stream', *Am. Midl. Nat.*, **131**, 288–300.

Stephens, D. W. 1990. 'Changes in the lake levels, salinity and the biological community of Great Salt Lake, Utah, USA (1847–1987)', *Hydrobiologia*, **197**, 139–146.

Stromberg, J. C., Patten, D. T., and Richter, B. D. 1991. 'Flood flows and dynamics of Sonoran riparian forests', *Rivers*, **2**, 221–235.

Stromberg, J. C., Richter, B. D., Patten, D. T., and Wolden, L. G. 1993. 'Response of a Sonoran riparian forest to a 10-year return flood', *Gt Basin Nat.*, **53**, 118–130.

Tasch, P. 1980. *Paleobiology of the Inverebrates*. John Wiley & Sons, New York.

Thompson, R. S., Whitlock, C., Bartlein, P. J., Harrison, S. P., and Spaulding, W. G. 1994. 'Climatic changes in the western United States since 18,000 yr. BP', in Wright, H. E., Kutzbach, J. E., Webb, W. F., Ruddiman, F. A., Street-Perrott, F. A., and Bartlein, P. J. (Eds), *Global Climates Since the Last Glacial Maximum*. University of Minnesota Press, Minneapolis. pp. 468–513.

Utah Department of Health, 1982. *State of Utah Clean Lakes Inventory and Classification*. Department of Health, Salt Lake City. 1031 pp.

Van Devender, T. R., Thompson, R. S., and Betancourt, J. L. 1987. 'Vegetation history of the deserts of southwestern North America: the nature and timing of the Late Wisconsin-Holocene transition', in Ruddiman, W. F. and Wright, H. E., Jr. (Eds), *North America and Adjacent Oceans During the Last Deglaciation*, Geological Society of America. pp. 323–352. Boulder, CO.

Wagonner, P. E. and Schefter, J. 1990. 'Future water use in the present climate', in Wagonner, P. E. (Ed.), *Climate Change and U.S. Water Resources*. Wiley-Interscience, New York. pp. 19–39.

Walker, B. H. 1991. 'Ecological consequences of atmospheric and climate change', *Climate Change*, **18**, 301–316.

Ward, J. V. and Stanford, J. A. 1979. *The Ecology of Regulated Streams*, Plenum Press, New York. 398 pp.

Ward, J. V. and Stanford, J. A. 1983. 'The serial discontinuity concept of lotic ecosystems', in Fontaine, T. D. and Bartell, S. M. (Eds), *Dynamics of Lotic Ecosystem*. Ann Arbor Science. Ann Arbor. pp. 29–42.

Washington, W. M. 1992. 'Climate-model responses to increased CO_2 and other greenhouse gases', in Trenberth, K. E. (Ed.), *Climate System Modeling*. Cambridge University Press, Cambridge. pp. 643–668.

Wurtsbaugh, W. A. 1992. 'Food-web modification by an invertebrate predator in the Great Salt lake (USA)', *Oecologia*, **89**, 168–175.

Wurtsbaugh, W. A. and Berry, T. S. 1990. 'Cascading effects of decreased salinity on the plankton, chemistry, and physics of the Great Salt Lake (Utah)', *Can. J. Fish. Aquat. Sci.*, **47**, 100–109.

Yardas, D. 1994. 'Acquiring water rights to protect aquatic resources: the evolving experience at Stillwater', in Wilcove, D. S. and Bean, M. J. (Eds), *The Big Kill: Declining Biodiversity in America's Lakes and Rivers*. Environmental Defense Fund. pp. 157–181.

10

WIDESPREAD EFFECTS OF CLIMATIC WARMING ON FRESHWATER ECOSYSTEMS IN NORTH AMERICA

DAVID W. SCHINDLER

Department of Biological Sciences, University of Alberta, Edmonton, Alberta T6G 2E9, Canada

ABSTRACT

Freshwaters in different regions show many similarities and differences in their responses to climatic warming. Bases for comparison include reports from regional committees, long-term records for several sites where climate has warmed in the past two decades and other human alterations to freshwaters that simulate some of the expected results of climatic warming, such as reservoir construction. Palaeoecological studies of freshwaters under climatic warming and differences in communities under different climatic regimes are also considered. Major changes in the physical, chemical and biological characteristics of lakes occur. Many of the changes to lakes and streams are the result of strong effects of climatic warming on terrestrial catchments. Inputs from catchments can be either dampened or amplified by in-lake processes, in some cases causing counter-intuitive responses, such as the acidification of streams but alkalinization of lakes in areas where supplies of base cations are limited. Consideration of land–water interactions and interactions between climatic warming and other human stresses are important for the accurate prediction of the effects of climatic change.

INTRODUCTION

The focus of other papers in this volume is to examine the effects that climatic warming has on freshwaters in different physiographical regions of North America (Figure 1 in Leavesley *et al.*, 1997). Because of the scarcity of real-time records, many of the regional summaries have relied heavily on modelled predictions based on global climate models (GCMs), which are themselves of questionable validity. In this chapter, I attempt a cross-regional summary, basing it on data from case histories rather than models, and rationalizing that many of the effects of changing climate on freshwater know no regional boundaries, i.e. they are either common to all lakes and streams, or to similar lakes and streams in different regions. The following summary is based on regional summaries, papers for this conference (Limnol. and Oceanogr., **41**(5), 1996), an earlier freshwater conference (Firth and Fisher, 1992) and from my own observations on lakes and streams in the Experimental Lakes Area (ELA) under climatic warming (Schindler *et al.*, 1990, 1992, 1996a,b, 1997). I have also considered other human perturbations that simulate some of the effects that climatic warming will have on freshwaters, long-term historical records and palaeoecological studies.

In particular, there are records from several sites in northern and western North America, during a period from approximately 1970 to 1990, when temperatures warmed by 1–2°C in approximately 20 years, following a period of relatively little change for the preceding 40–50 years. This warming has occurred at a rate similar to that predicted from many greenhouse models (GCMs), and has occurred over a time period that is a significant fraction of the period over which doubling of CO_2 is predicted to occur (40–70 years). Several types of human perturbations can be interpreted as climatic warming 'experiments', such as the effects of thermal effluents, and the dewatering of streams and changes in the levels and thermal structure of lakes caused by reservoir formation. Distributions of organisms and communities along temperature gradients caused by latitude and altitude also appear to form a reasonable basis for speculation about the long-term effects of climatic warming. I have de-emphasized the results of both GCMs and limnological models that are unconfirmed by data, for until such models are verified, they provide a questionable predictor of the consequences of climatic warming for freshwaters, particularly at cross-regional scales. In any case, models are discussed at some length in other papers from this symposium.

THE ELA DATA SET

Detailed, continuous records for weather, hydrological flows and many physical, chemical and biological characteristics have been kept for a number of catchments at the Experimental Lakes Area (ELA) in north-western Ontario since the late 1960s to early 1970s. The record is of particular interest for several reasons. First, the annual average air temperature at the site increased by 1.6°C between 1970 and 1990 (Schindler *et al.*, 1996a). This is among the highest rates of warming observed in North America in that period. Secondly, the lakes and their catchments have not been perturbed by direct human activity, other than whole-lake experiments conducted in a small number of lakes and parts of catchments. The only human habitation in the area is the ELA research site. While low concentrations of airborne pollutants clearly reach the lakes, and some are accumulated in sediments or biota, there is no evidence of acidification or other long-term human influence. The catchments of lakes and streams have never been logged or otherwise modified by humans. In brief, the ELA has escaped the problems faced in many areas, where it is difficult to disentangle the effects of even moderate climatic warming from other human perturbations. Thirdly, a parallel project, the Northern Ontario Lake Size Survey (NOLSS) has also been carried out in the area for several years, to determine whether there are size-related changes in lake processes that must be accounted for in extrapolating the results from ELA lakes to large lake systems (Fee and Hecky, 1992; Fee *et al.*, 1996). In many of the cases that follow, I shall use the ELA record as the basis for making comparisons with other areas. Finally, many of the long-term records for ELA have been compiled by single individuals who have devoted their entire careers to the ELA study. In many cases, measurements have been made in identical fashion throughout the study. Where new methods or instruments were introduced, periods of careful calibration allowed detailed comparisons of records.

EFFECTS OF CLIMATE ON HYDROLOGY

The hydrological cycle in North America is only a small part of the global cycle. It has been modified in many ways, by reservoir construction, channelization, irrigation and other land-use changes, in addition to climatic warming.

In most cases, climatic warming will cause less water to be found in the landscape (Hengeveld, 1990). Even where precipitation remains constant, higher temperatures and extended ice-free seasons usually lead to increases in the ratio of evaporation + evapotranspiration to precipitation. For example, at the Experimental Lakes Area (ELA) in north-western Ontario, average evaporation increased by an average of 35 mm per °C increase in annual air temperature, or 68 mm per degree increase in summer air temperature (Figure 1). For the 20-year period of record considered here, this is an average increase in evaporation + evapotranspiration of 9 mm/yr. In brief, it would require a considerable increase in precipitation to keep pace with the increase in evaporation, particularly in summer when increased temperature and evaporation rapidly increase the threat of forest fires in boreal landscapes. In turn, forest fires cause changes to hydrological regimes, with flows generally being higher than before fire, but less evenly distributed throughout the season for several years following fire (Schindler *et al.*, 1980).

As a result of the great increase in evaporation with increasing air temperature, unless climatic warming is accompanied by substantial increases in precipitation or extraordinary decreases in plant stomatal resistance from increasing atmospheric CO_2, we can anticipate that lake levels, water renewal rates, stream flows, the extent of and water levels in wetlands, soil moisture and groundwater levels will all decline. For western North America, Revelle and Waggoner (1983) conclude that declining precipitation would amplify the effect of increasing evapotranspiration. For annual precipitation less than 500 mm, which would include ELA in

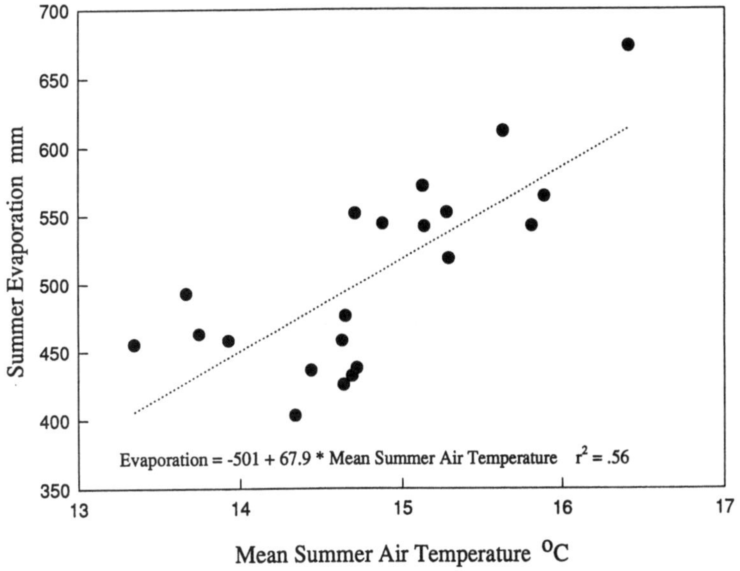

Figure 1. The relationship between mean summer air temperature and evaporation at the Experimental Lakes Area (ELA), 1970–1990. Note that a change in annual average air temperature of 1°C caused a decrease of 68 mm in annual evaporation

the late 1980s, they conclude that a warming of 2°C would cause a greater decline in stream flow than a 10% decrease in precipitation. Grimm *et al.* (1997) reach similar conclusions. Obviously, there are important implications for the physical and chemical environments of aquatic organisms, as well as for human uses of water. The problems would be most acute in areas that are already extremely water limited (Grimm *et al.*, 1997). Altogether, there is probably no other effect of climatic warming that will approach crisis proportions as quickly as the diminution of freshwater resources.

Less winter precipitation, earlier break-up and earlier spring snowmelt appear to cause spring stream flows to be lower and earlier at ELA and in many other areas of north-eastern North America and Alaska (Magnuson *et al.*, 1997, P. J. Dillon and M. McDonald, personal communication). Native elders in north-western Canada have noted similar changes in the northern Athabasca–Peace–Slave river systems, a region where climate has warmed by 1.7°C in the past 95 years (Table 1 in Hengeveld, 1995). These changes have caused major disruptions to traditional life-styles, with their heavy reliance on freshwaters for both travel and food.

In some years, the spring pulse may be all but eliminated and the onset of lake outflow delayed (Figure 2; P. J. Dillon, personal communication). In addition, at several sites the number of days that first-order streams are dry in summer has increased manyfold (Dillon *et al.*, 1994; Grimm *et al.*, 1997; Rahel *et al.*, 1996; Schindler *et al.* 1996a). Indeed, at ELA, first-order streams that flowed continuously during the ice-free season in the early 1970s were dry for up to 150 days during the summers of the late 1980s (Schindler *et al.*, 1996a). Under similar climatic conditions, groundwater inputs to lakes and streams are severely reduced (Grimm *et al.*, 1997; Webster *et al.*, 1996).

The lack of high spring flows also threatens the integrity of major deltaic ecosystems. In particular, the frequency and intensity of spring floods that rejuvenate wetland areas and perched lakes may be reduced (Prowse *et al.*, 1996). In northern areas, such floods are the result of a combination of high spring flows and ice jams during break-up, which can cause rises in river level of several metres. Climatic warming appears to cause both a reduction in spring flows and thinner, weaker ice which is less likely to cause damming. Some of the potentially affected areas are huge, such as the Peace–Athabasca Delta in Alberta, the Mackenzie River Delta (Rouse *et al.*, 1997). While ice jams are not a factor, similar problems relating to flow regulation are affecting the Atchafalaya Delta area of the Mississippi. As discussed later, the reduced flows resulting from

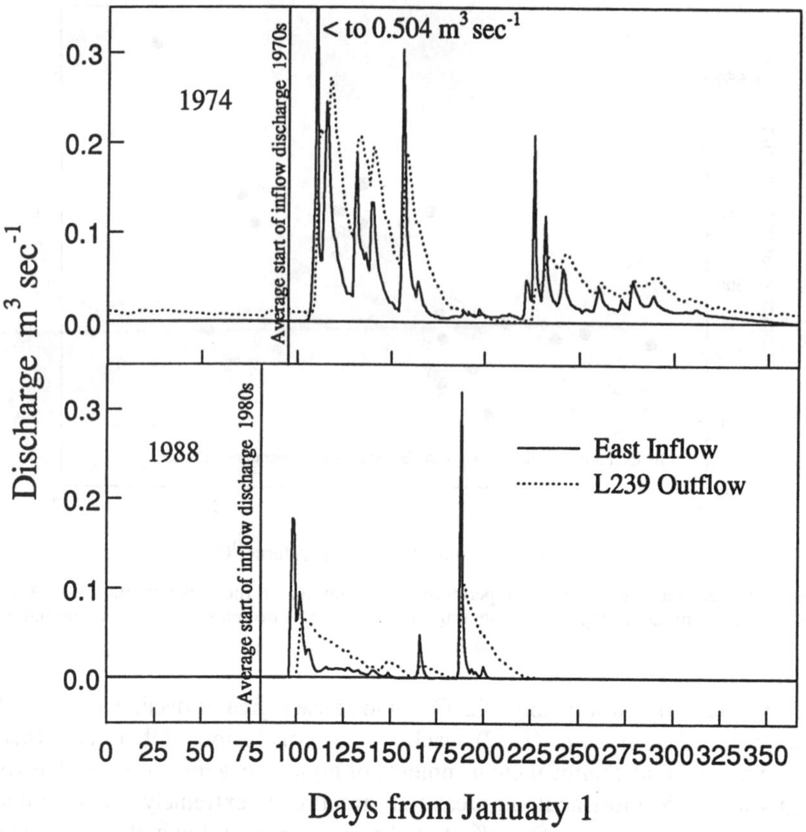

Figure 2. The spring snowmelt pulse in a first-order tributary to Lake 239 at ELA and in the lake outflow following a year with normal winter precipitation typical of the early 1970s, and a year of low snowfall typical of the late 1980s. Note the delay of several days between the onset of tributary flow and the onset of lake outflow in the later year, when the outlet was dry in spring and the late winter lake level had to rise several centimetres before outflow could begin

warmer, drier climates would be analogous in many ways to the results from building dams that damp out spring flows in order to store water in reservoirs for use at other times of the year.

Decreased river flows in winter in northern regions must also be of concern, for at many locations industrial effluents can form several percent of river flows under winter ice, increasing concentrations of contaminants and depleting oxygen. For example, the Athabasca River, which runs 1400 km from the Rocky Mountains to Lake Athabasca, often has low oxygen concentrations under winter low flow conditions, when long reaches of the river are covered with ice and snow. Oxygen can approach 5 ppm, the Alberta provincial guideline, below which several species of salmonid fish are threatened. The low oxygen values are caused by a combination of low oxygen tributaries draining petroliferous catchments and biological oxygen demand-rich effluents from pulp mills and small communities (Chambers *et al.*, 1996). If climatic change causes further declines in winter flows, it may be necessary to apply more stringent standards to effluents reaching northern rivers. Other aspects of northern hydrology are reviewed by Prowse and Ommanney (1990).

Dewatered streams resulting from damming of rivers also provide an analogue to channels receiving less water because of climatic warming. In general, river channels are complex manifestations of the amount and distribution of water energy over long time periods, and any factors that substantially alter flow patterns can be expected to produce major changes in channel morphology, regardless of region (Williams, 1978; Kellerhals and Church, 1989; Williams and Wolman, 1984; Poff, 1992; Grimm and Fisher, 1992), and in the habitats for freshwater species (Ligon *et al.*, 1995). In the case of the Southern Indian Lake reservoir in semi-permanent permafrost of northern Manitoba, dewatering of the lower Churchill River caused considerable alteration in channel morphometry, ice dynamics and bank stability (Newbury *et al.*, 1984). Reservoir projects are widespread in North America, and it might prove fruitful to study the effects of their altered flows as surrogates for the effects of climate change in several regions. Obviously, climatic warming would cause reductions in the amount of hydroelectric power that could be generated, as well as exacerbating the effects of reservoirs on flow regimes, as described above.

During the past 30 years, glaciers in much of western Canada and the US have lost 10–50% of their mass (Melack *et al.*, 1997). Data on terminus locations suggest that losses also occurred during the previous 80 years (Brugman, 1992). Glacial meltwater is often the main supply for streams and rivers in mountain areas, causing highest water flows to occur in early or midsummer, depending on latitude (Hauer *et al.*, 1997; Melack *et al.*, 1997). If these glaciers disappear completely, many mountain streams may be completely dry in midsummer. Where glaciers remain, spring melt pulses may occur earlier and midsummer flows remain higher as climate warms, changing aquatic habitats (Melack *et al.*, 1997). If climatic warming brings great increases in cloud cover and precipitation at high elevation, as described by Williams *et al.* (1996) for the Rocky Mountains of Colorado, these effects could be further amplified. At present, long-term records for high altitudes are very scarce, so that it is impossible to tell how widespread such effects may be.

THE EFFECTS OF EL NIÑO

Several investigators have reported correlations of drought, early ice-out, changed circulation patterns and/or warmer temperatures with El Niño events (Strub *et al.*, 1985; Robertson, 1989; Reycraft and Skinner, 1993; Anderson *et al.*, 1996; Dillon *et al.*, 1997; Covich *et al.*, 1997; Melack *et al.*, 1997). Typically, the strongest effects are seen in the mature phase of El Niño, when winter temperatures in waters of the Pacific Ocean are warmest. The strongest effects are seen in mid-March. El Niño affects freshwaters in several ways, including the seasonality of water flows, and mixing depths and nutrient concentrations, which in turn control primary production. The effects are discussed extensively in Covich *et al.*, 1997; Melack *et al.*, 1997; Magnuson *et al.*, 1997 and other papers in this issue.

Curiously, the El Niño effect is not discernible at ELA, while it appears to affect sites to the east and south (Anderson *et al.*, 1996; Dillon *et al.*, 1996). While the reason for this is not clear, it may be that stronger overall warming trend at ELA has simply overwhelmed the El Niño signal.

Obviously, regional assessment of how widespread El Niño effects may be on freshwaters requires long-term data sets. Records for river flows in many areas are probably sufficient to detect such effects, but only a few data sets have been assessed so far. Moreover, in many regions river flows are disrupted by a number of other human activities, making it difficult to isolate the El Niño signal.

EFFECTS OF CLIMATE ON PHYSICAL FEATURES OF LAKES AND STREAMS

Changes in lake water temperature and duration of the ice-free season

As climate warms, there is a general tendency for average lake temperatures to become warmer (Schindler *et al.*, 1990, 1996a; DeStasio *et al.*, 1996). Peculiarly, mean annual air temperature provides almost as good a predictor of lake temperature as air temperature during the ice-free season (Figure 3). It appears to be worth examining such relationships for other areas with good long-term data sets, for they may provide a reasonable basis for predicting the effects of further increases in air temperature on survival of key aquatic species. For example, DeStasio *et al.* (1996) predict changes in the distribution of many species of freshwater fish and other organisms as the result of warming water temperatures and changes in thermal stratification. Similarly, Eaton *et al.* (1996) used Stefan and Preud'homme's (1993) empirical evidence that stream water temperature would increase by 0.9°C for each degree of warming in air temperature. Our results suggest that extrapolations may be more variable for lakes, with greater increases in integrated temperature occurring in

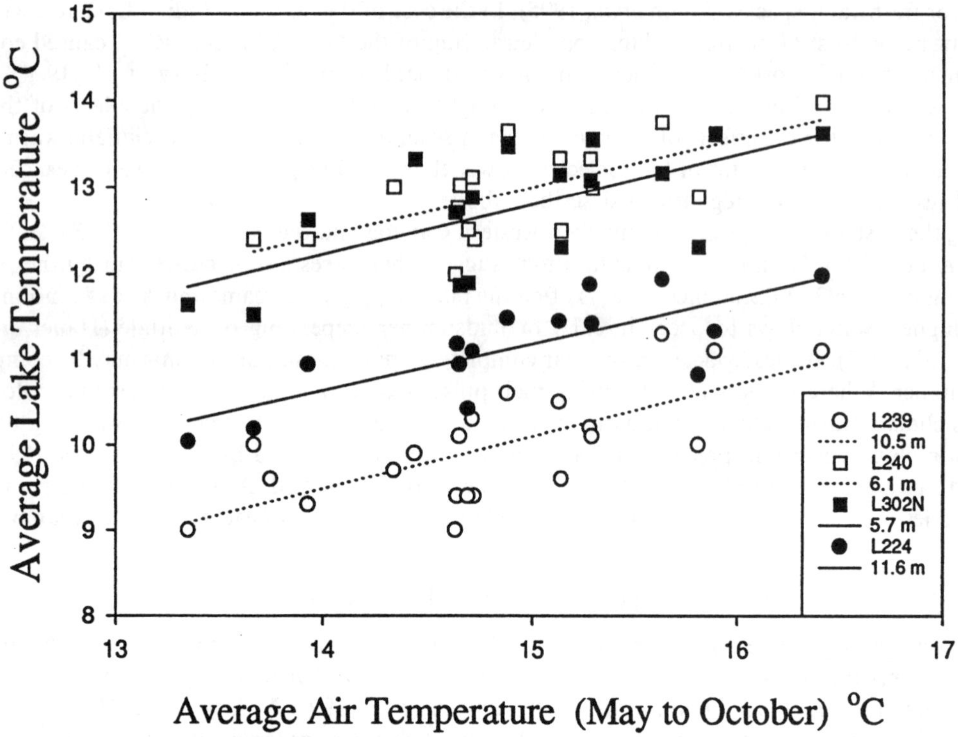

Figure 3. The relationships between average air temperature in the ice-free season and volume-weighted average lake temperature in the ice-free season for several lakes at ELA. Lake temperatures are calculated from several temperature profiles in the ice-free season. The slopes are not significantly different ($P = 0.05$) for ELA lakes, but intercepts decrease significantly with increasing mean depth

deeper lakes. For example, mean summer water temperatures in Lake 239 increased about 10% more per degree rise in air temperature than in the shallower Lake 240 (Figure 3). Similar relationships are evident for other ELA lakes, with the relationship between summer air temperature and water temperature represented by a series of lines with similar slopes, but different intercepts. In general, the volume-weighted mean seasonal water temperature is warmer for lakes with shallower mean depths, but the rate of warming for all lakes in the same area is nearly identical. During the increasingly warm 1970s and 1980s in northern North America, increases in the length of the ice-free season have been observed at several sites in the northern US and Canada (Comb, 1990; Schindler *et al.*, 1990, 1996a; Etkin, 1991; Hanson *et al.*, 1992; Robertson *et al.*, 1992; Dillon *et al.*, 1994; Anderson *et al.*, 1996; Assel and Robertson, 1996; McDonald *et al.*, 1996; Schindler *et al.*, 1996a). At all of the above sites, the increase in ice-free season occurs almost entirely in the spring. This fits with climatic patterns, for the most pronounced period of warming has occurred in late winter and early spring (Hengeveld, 1991). As mentioned above, at many sites, the most extreme warming appears to be during mature phases of El Niño (Anderson *et al.*, 1996). It seems likely that earlier seasonal changes in albedo resulting from lower snowfalls and earlier spring melts contribute to this pattern.

In mid-temperate latitudes, the frequency of winters without ice cover will undoubtedly increase, perhaps eliminating dimixis altogether in some lakes. For example, total elimination of ice cover on Lake Mendota, Wisconsin, is predicted to occur as the result of doubling of atmospheric CO_2. Such predictions are supported by past observations of the duration of ice cover on the lake (DeStasio *et al.*, 1996).

Changes in thermal stratification

As mentioned above, at mid-temperate latitudes, where the duration of ice cover is usually short, dimictic lakes may become warm monomictic, i.e. in circulation throughout the winter. In more northerly latitudes, cold monomictic water bodies may become stratified in summer. Because of the greater density difference between surface and bottom waters, stratification in large dimictic lakes might be expected to become stronger and shallower (McCormick, 1990; DeStasio *et al.*, 1996).

In contrast to the above predictions, in the small lakes at ELA thermoclines deepened and weakened during the warming period of the 1970s and 1980s, despite increasing surface temperatures. This discrepancy occurred largely because the declines in dissolved organic carbon (DOC) accompanying climatic warming allowed solar energy to penetrate deeper, causing greater buoyancy in subthermocline strata (Schindler *et al.*, 1996a). Similar results were reported by Mazumder and Taylor (1994). An analogous deepening of thermoclines has been observed in acidified lakes by declines in DOC (Malley *et al.*, 1982; Dillon *et al.*, 1984; Shearer *et al.*, 1987; Schindler *et al.*, 1996b).

A detailed examination of the relative roles of lakes size and transparency in thermocline formation by Fee *et al.* (1996) indicates that the effects of increasing transparency on thermocline depth would be confined to lakes less than about 500 ha in area, since, in such lakes, the size-determined thermocline is typically only a few metres in depth (Gorham and Boyce, 1989). In larger lakes, size alone dictates a thermocline depth that is too deep to allow significant solar energy to penetrate the thermocline, even in the clearest lakes of north-western Ontario. As a result, in large lakes light penetration plays little role in thermocline depth. This example illustrates the critical need for size-dependent scaling when extrapolating results between lakes of different sizes.

The combined increases in ice-free season, thermocline depth and epilimnion temperatures act in concert on the thermal regimes of lakes to increase greatly degree–days, which affects the rates of hatching and development, shortens the life-cycle and alters the zoogeographical distribution of many species of aquatic organisms (Ward and Stanford, 1982; Oswood *et al.*, 1992; Schindler *et al.*, 1996a).

Changes in optical environments

If small lakes become clearer, there are a number of implications in addition to the above-mentioned thermocline deepening. First, euphotic zones will certainly becomes deeper, owing to the deeper penetration of photosynthetically active wavelengths. While this will probably not affect phytoplankton productivity in severely nutrient-limited lakes (Schindler *et al.*, 1996a), it would be expected to affect lakes that are largely light limited (for example Fee *et al.*, 1996).

There is also some concern for increased exposure of aquatic organisms to UV-B. In small lakes, declines in DOC would exacerbate the effects of increased UV-B expected from depletion of stratospheric ozone. Climatic warming at ELA caused DOC to decline by 15–25% (Schindler *et al.*, 1996b, 1997). In lakes that initially had average (300–600 μM) DOC concentrations for boreal regions, the changes in UV penetration were only about 30%. However, owing to the negative exponential relationship between DOC and UV penetration (Scully and Lean, 1994), equivalent declines in DOC in lakes that originally had 200–300 μM concentrations would cause much greater increases in UV penetration. Significant UV radiation can penetrate up to several metres in the clearest boreal lakes, as well as a high percentage of alpine and arctic lakes (Scully and Lean, 1994; Schindler *et al.*, 1996b; Williamson *et al.*, 1996). The effects of climatic warming, via changes in DOC, may cause greater increases in the UV exposure of aquatic organisms than changes in the stratospheric ozone layer (Schindler *et al.*, 1996b). In shallow ponds and streams, and in alpine, subalpine and arctic lakes, climatic warming may contribute to an increasingly hostile UV environment via its effects on DOC.

If thermoclines become shallower, as predicted for clear larger lakes (Magnuson *et al.*, 1997), exposure to UV or to the toxic chemical by-products of UV–DOC interactions could also increase, since planktonic organisms would be confined to a narrower zone near the lake surface.

Changes in stream temperature

In addition to the hydrological changes already mentioned, it appears that stream temperatures will closely track air temperatures, making it rather easy to predict changes in biota that have known temperature tolerances (Eaton *et al.*, 1996; Magnuson *et al.*, 1997; Rahel *et al.*, 1996). Stefan and Preud'homme (1993) found that stream temperatures increased by an average of approximately 0.9°C for every degree increase in air temperature. An exception appears to be where groundwater is the major water source for small streams; it tracks annual average air temperature rather closely, keeping streams cooler in summer than they would otherwise be (Meisner, 1990a). However, decreases in groundwater flow are also expected from climatic change, and may allow such streams to become warmer as well. In boreal regions and deciduous forests of the eastern US, increased forest fires are expected to follow climatic warming (Price and Rind, 1994; Kurz *et al.*, 1995). Forest fires may accelerate the warming of streams, since, in hot fires, even the riparian vegetation that shields streams from direct sunlight can be burned. In such circumstances, increases in stream temperatures may be several degrees, as documented in studies of streams exposed by clear-cut logging.

As discussed in detail by Grimm *et al.* (1997) and Rouse *et al.* (1997), the character of many rivers is also a function of maximum flows, seasonality of flows and ice cover characteristics, so that physical characteristics are strongly shaped by climate.

CHANGES IN CHEMICAL CHARACTERISTICS

Changes in the chemistry of streams

In general, lower stream flows appear to cause lower exports of chemicals from catchments (Dillon *et al.*, 1994; Schindler *et al.*, 1996a). Concentrations of many chemical constituents are higher during drought periods, probably as a result of increased contact time with soils (Schindler *et al.*, 1992; Hinton *et al.*, 1997; Webster *et al.*, 1996), and decreased exports are largely a result of decreased flows. In particular, lower exports of phosphorus at ELA affect the nutrient status of receiving waters (Schindler *et al.*, 1996a), and declining export of DOC was the primary cause of the declining DOC in ELA lakes as climate warmed. Similarly, Meyer and Pulliam (1992) showed empirically that a decline in phosphorus export occurred in Bear Brook, New Hampshire, and that DOC exports from Ogeechee River, Georgia, both declined with declining annual flow.

Less information is available for the effects of climatic warming on nitrogen. At ELA, export from catchments of all nitrogen fractions increased for a few years following forest fires (Schindler *et al.*, 1980; Bayley *et al.*, 1992b), but after several years during which climatic warming and drying continued, nitrogen exports declined (Schindler *et al.*, 1996a). Lower exports of silica, presumably the result of lower weathering rates, may have implications for the proportion of diatoms in the algal flora, and changes in the ratio of silica to other ions may affect species composition (Kilham *et al.*, 1996). At ELA, decreases in exports of base cations were greater than those in strong acid anions, resulting in decreasing alkalinity (acidification) of streams (Schindler *et al.*, 1996a). At both ELA and Dorset, Ontario, sites, particularly acidic stream flow pulses follow long periods of drought. These appear to be caused by the reoxidation of sulfur, accumulated in the catchments by reduction of sulfate from atmospheric precipitation during preceding years. In particular, *Sphagnum* and associated vegetation appear to play a major role in this cycle (Bayley *et al.*, 1992a; Lazerte, 1993). In some cases, the effect has been strong enough to delay the recovery of acidified lakes following reductions in acidic precipitation (Keller *et al.*, 1992; Dillon *et al.*, 1996). Bayley *et al.* (1992a, b), Schindler *et al.* (1992, 1996a), Dillon *et al.* (1994) and Clair *et al.* (1997) provide further examples of the effects of climatic warming and drying on stream chemistry. Clow and Ehrman (1996) and Williams *et al.* (1996) show that wetter climate in the Rocky Mountains causes increased weathering, in agreement with the above interpretations.

In boreal regions, forest fires may alter long-term patterns of chemical exports. At ELA, fire increased the export of most chemical constituents over those from unburned basins (Schindler *et al.*, 1987; Bayley *et al.*, 1992a, b). Comparing burned with unburned catchments revealed that the effects of fire on stream flow and nutrients lasted for only a few years, but in the case of base cations and strong acid anions they lasted for several years to over 15 years. However, the effects of even extreme fires were rather weak compared with the general decline in exports of chemicals from all basins caused by reduced flows during the prolonged drought (Schindler *et al.*, 1996a).

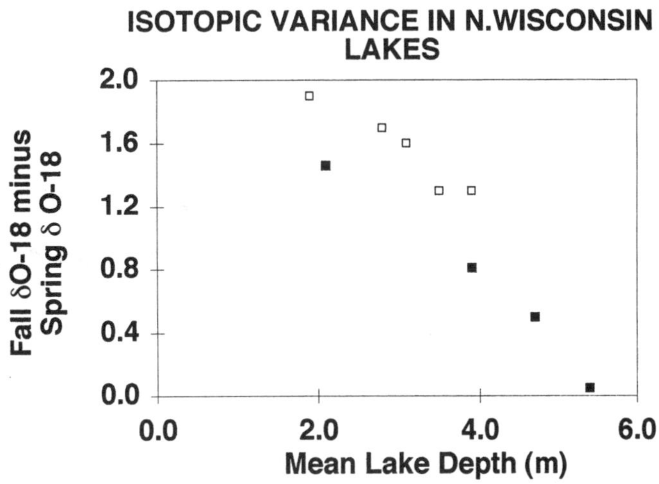

Figure 4. The differences between spring and autumn values of ^{18}O for several seepage lakes in northern Wisconsin, illustrating the greater variability in shallower lakes. Data from Bowser and Ackerman (1993) (■) and Krabbenhoft *et al.* (1994, (□). The effects of increasing mean depth under constant annual water loading ($m^{-2}\ yr^{-1}$) would be analogous to the effects of reduced water loading for lakes of constant mean depth. Both would be expected to cause reduced seasonal and interannual variability. Figure supplied by Dr Carl Bowser, University of Wisconsin, Madison

Changes in the chemistry of lakes

There are important differences in the responses of streams and lakes to changing climate. Despite reduced chemical yields from catchments, concentrations of conservative chemicals like sodium and chloride in lakes increase as climate warms. This is because decreased stream flows and increased evaporation at lake surfaces cause increased water retention times, which in turn increase the retention of chemical constituents. In general, doubling the water renewal time (or halving the 'water loading') is analogous to doubling the input of a conservative chemical in its effect on the concentration at steady state. Water renewal times have been shown to have critical effects on eutrophication (Dillon, 1975; Vollenweider, 1975; Schindler *et al.*, 1978), on retention of sulfate and nitrate (Baker *et al.*, 1986; Kelly *et al.*, 1987) and on conservative substances (Schindler *et al.*, 1987). For example, water renewal times of ELA lakes have increased nearly fourfold as a result of the relatively small warming and drying (Schindler *et al.*, 1996a).

Longer water retention times would also decrease lakes' abilities to rid themselves of persistent dissolved chemicals, including toxins, and contribute to lower seasonal and interannual variability (see, for example, Kratz *et al.*, 1991). That such declines in variability will occur can be deduced from seasonal or interannual data collected from adjacent lakes with different mean depths but similar water loading, where variability is much less in lakes with greater mean depths, hence longer flushing times (for example, Figure 4).

Biological or chemical reactivity may cause exceptions to the expected increases, since longer water residence may allow in-lake removal processes more time to work. For example, phosphorus, DOC and silica concentrations decline in lakes as well as in streams at ELA, presumably because longer water residence times increase their exposure to biological removal processes (Schindler *et al.*, 1996a). Sulfate, of intermediate reactivity owing to microbial sulfate reduction and precipitation as iron or organic sulfides at the mud–water interface, reaches steady state well before it is expected to from changes in water renewal time (Kelly *et al.*, 1987). In contrast, calcium, which diffuses from sediments where mineralization is occurring, increases in concentration with increasing water residence (Figure 5).

The net result of the differing reactivity of calcium and sulfate with changing water renewal is that the alkalinity of lakes increases under climatic warming, which contrasts with the loss of alkalinity observed in streams. Similar sediment sources of base cations and sinks for strong acid anions have been observed in the Catskills and in eastern Ontario (Schiff and Anderson, 1986), and it appears that such effects would be

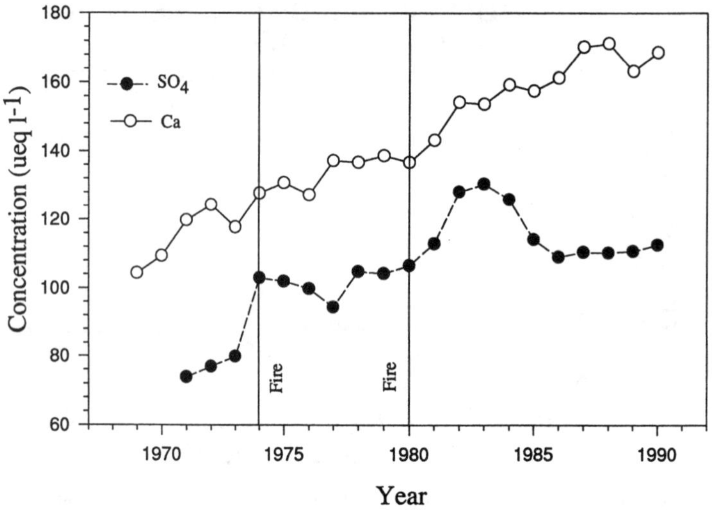

Figure 5. Sulfate and calcium concentrations with time in Lake 239 at ELA. Note that concentrations of sulfate appear to reach an asymptote after the effects of fire disappear, although water renewal times continue to increase throughout the 1980s. The asymptote is the result of increasing in-lake sulfate reduction, which increases with both increased sulfate concentration and increased water renewal time. In contrast, calcium, which is supplied from lake sediments as well as inflows, continues to increase, and would probably do so for the near future as climate warms, for concentrations are well below the solubility products with sulfate or carbonate

widespread. Lakes more alkaline than their surface inputs have been widely observed (Schindler, 1986), including those in well-buffered areas (Kilham, 1982).

There are, however, some reported exceptions to these patterns. P. Mulholland (personal communication) reports that phosphorus concentrations increase in south-eastern lakes as the result of declining water renewal. This must be the result of increased release from anoxic sediments, and decreased efficiency of phosphorus sedimentation, probably because phytoplankton are phosphorus saturated. Dillon *et al.* (1996), Yan *et al.* (1996) and Keller *et al.* (1992) reported that recovery was delayed in several lakes that had been acidified by atmospheric deposition, owing to increasing reoxidation of sulfur to sulfate. The implications of these changes for biological processes are discussed below.

As mentioned above, declines in DOC are expected to cause changes in a number of chemical processes. Increased photochemical production of peroxides, superoxides and hydroxyl radicals may be of considerable importance to both chemical and biological processes, owing to their high reactivity and toxicity (Cooper *et al.*, 1994). The ability of DOC to participate in flocculation and removal of metals and nutrients may also decline (Weilenmann *et al.*, 1989).

The longer periods of summer stratification are also predicted to cause increased hypolimnetic anoxia in the lower Great Lakes (Magnuson *et al.*, 1997). Observations made during 1983, an exceptionally warm year, generally support this prediction (Schertzer and Sawchuk, 1990). At ELA, prolonged climatic warming has caused slight increases in hypolimnetic oxygen deficits, but the trend is not statistically significant (Schindler *et al.*, 1996a). The increase appears to occur in a few years just after the 1980 fire, when the higher oxygen deficit may be caused by increased airborne inputs, bedloads (Beaty, 1995) and direct inputs of twigs, leaves and branches in the few years following fire.

Seepage lakes: special considerations

Seepage lakes show quite different response patterns to drainage lakes. For lakes high in aquifers, reduced groundwater inflows may cause chemical concentrations in lakes to decline, since precipitation on lake surfaces becomes of greater relative importance. In contrast, increased contact times between groundwater and soils may cause increased concentrations of chemicals in groundwater entering lakes lower in aquifers

(Webster *et al.*, 1996). An example of how the complexity of groundwater flow can affect lake chemistry is the case of alkalinity. In northern Wisconsin, decreased flows of groundwater rich in acid-neutralizing capacity (ANC) were caused by drought. The result was a considerable net acidification of a seepage lake, accompanied by declines in calcium and magnesium (Webster *et al.*, 1990). In lakes where groundwater is the predominant source of chemical inputs, even small changes in precipitation and temperature can cause large changes in the chemistry of lakes (Crowe, 1993; Grimm *et al.*, 1997).

Closed basin lakes

Lakes in endorheic basins are common in several regions of the western US and Canada, as well as on all other continents except Europe. Several articles discuss their chemical and biological character, which are particularly affected by water balance (Covich *et al.*, 1997; Grimm *et al.*, 1997; Melack *et al.*, 1997; Rouse *et al.*, 1997). Lakes with closed basins are particularly vulnerable to climatic warming. Because they have no outflow and are in generally arid regions where groundwater inputs and outputs are small, water levels and chemical concentrations respond quickly once evaporation exceeds water inputs. Even in droughts of the magnitude of the 1970s and 1980s, many smaller water bodies of western North America dried up or decreased greatly in size, greatly reducing habitats for aquatic creatures, including waterfowl.

Dramatic chemical changes occur when evaporation of lake water concentrates chemicals enough to exceed the solubility products of different ions. Typically, calcium carbonate is precipitated first. Although the chemical composition of an individual lake depends on interactions of lake elevation, groundwater composition and climate, typically, lake waters are dominated by sodium sulfate following precipitation of calcium carbonate. As solubility for these salts is exceeded, sodium chloride predominates in the final stages of evaporation. Evans and Prepas (1996) attributed changes in the chemistry of closed basin lakes of Alberta during a prolonged drought to changes in groundwater chemistry. However, the decrease in calcium and increase in sodium at high chemical concentrations suggests that, instead, the solubility products for calcium and carbonate are simply exceeded, a possibility that was not addressed. Last (1992) gives a detailed review of the factors controlling chemical concentrations in 500 lakes of western Canada, and papers in *Hydrobiologia*, **158**, (1988) describe features of saline lakes throughout the world. As lakes become more saline, DOC typically increases very rapidly. However, instead of the coloured DOC originating from the humic catchments of eastern lakes, it is typically uncoloured, perhaps as the result of decades or even centuries of photobleaching, for the water renewal time in closed basin lakes is often a century or more (Curtis and Adams, 1995). While productivity of saline and hypersaline lakes is often high, biodiversity generally declines as lakes increase in salinity, and only a handful of species are able to survive in the most saline waters (Melack *et al.*, 1997).

There is a considerable literature on the effects of past climate change on the hydrology, chemistry and biology of lakes in endorheic basins, and a complete review cannot be done here. Some examples of particular interest include evidence that Lake Manitoba, one of the largest lakes on the continent outside the Great Lakes, was dry on several occasions between 9000 and 4500 BP, under climatic conditions that are estimated to have been only about 1°C warmer than in the twentieth century (Last and Teller, 1983). Mineralogical, chemical, stable isotope and subfossil studies have shed further light on the characteristics of lakes under past climatic conditions (see, for example, Last and Slezak, 1988; Fritz, 1990; Fritz *et al.*, 1991; Walker *et al.*, 1991).

Alpine and arctic lakes

In arctic and alpine lakes, winter cryoconcentration can be a major mechanism increasing chemical concentrations in lake waters (Schindler *et al.*, 1974). As a result, at high latitudes, chemical concentrations in lakes can be nearly double those in inflows. If climatic warming causes ice cover to become thinner, this mechanism will become of lesser importance and lakes may actually become more dilute, depending on interactions between the degree of decreased cryoconcentration and increased water retention times. Similar considerations probably apply to alpine lakes, although, because of the greater depths of insulating snow cover in most alpine regions, a lesser degree of cryoconcentration is expected. For example, in alpine lakes of the Canadian Rockies, ice depths seldom exceed one metre, whereas in arctic lakes, ice thicknesses of 2–3 metres are not uncommon (Schindler *et al.*, 1974 and unpublished data).

CLIMATE CHANGE AND WETLANDS

The catchments described above that export greater amounts of sulfur under drought are generally covered with *Sphagnum*, whether they are upland or wetland areas. Declines in the water table of wetlands may also cause changes in the fluxes of greenhouse gases. For example, as water tables decline, fluxes of CO_2 to the atmosphere increase and fluxes of methane decrease (Gorham, 1991; Roulet *et al.*, 1994). No estimate of the net effect of such changes on global warming is available at present.

Wetlands in arid regions of western Canada were largely absent during the mid-Holocene, under climatic conditions less extreme than those predicted to occur from increased greenhouse warming (Zoltai and Vitt, 1990). Declining recharge of groundwaters and first-order streams, and declining waterfowl production are among the effects expected from reduction in wetland areas. Moore *et al.* (1997) discuss in detail the adverse effects of wetland disappearance expected for the north-eastern region.

In addition, wetlands have been shown to be a significant global carbon sink (419 Pg; Gorham, 1991). Lake sediments also contain a large amount of carbon (120 Pg), a fact that has only recently been realized (Molot and Dillon, 1996). The importance of both has been ignored in global carbon budget calculations.

CHANGING CLIMATE AND AQUATIC ORGANISMS

Climatic change may cause changes in the production, biomass, life-cycles and community composition of aquatic ecosystems. For example, as mentioned above, the climatic warming at ELA caused a decline in the phosphorus concentrations in lakes. A parallel decline in chlorophyll concentrations was observed (Schindler *et al.*, 1996a), although both phytoplankton biomass and diversity increased slightly (Schindler *et al.*, 1990; D. L. Findlay, personal communication). While the depth of the euphotic zone, and the extent of the zone inhabited by deep algal blooms increased, there was no corresponding climate-related change in either total area or subthermocline rates of production (Schindler *et al.*, 1996a). In contrast, Byron and Goldman (1990) found that climatic warming caused increases in both phytoplankton production and standing crop in Castle Lake.

Field observations show that high summer water temperatures and longer periods of stable stratification can promote dominance by filamentous cyanobacteria (George and Harris, 1985; George *et al.*, 1990; Sommer *et al.*, 1986). Laboratory studies indicate similar effects of warmer temperatures on the outcome of competition for nutrients. At high temperatures (17 and 24°C), cyanobacteria and chlorophytes were favoured, whereas at low temperatures (<14°C) diatoms were favoured over a broader range of nutrient conditions (Tilman *et al.*, 1986).

In periphyton communities at ELA, no climate-related changes in productivity or standing crop were observed, although warmer temperatures appeared to cause a slight increase in the proportion of filamentous green algae such as *Mougeotia* and *Zygnema* spp. (M. A. Turner, personal communication).

Studies of thermally polluted lakes may also provide insights into the effects of climatic warming on organisms. For example, Patalas (1970) found that Lake Lichenskie, Poland, which had its temperature increased by 6°C by effluent from a thermal power plant, had double the phytoplankton production of adjacent Lake Mikorzynskie. Secchi disc visibility declined. The biomass of herbivorous zooplankton changed very little, but production and turnover rates (production/biomass) of zooplankton were substantially greater in the warmer lake. Carnivorous zooplankton were lower in the heated lake, although their production and P:B increased slightly.

The spawning, hatching and survival of juveniles of many organisms may be affected by changing temperatures or decreased water flows. Recruitment of freshwater species is known to be affected by water temperature (Trippel *et al.*, 1991; Chen and Folt, 1996). Those species that can produce resting eggs or diapausal stages may also be better able to survive periods of drought or extreme temperature (see, for example, Hairston, 1996). These and other changes may affect the composition of aquatic communities. Moore and co-workers (Moore and Folt, 1993; Moore *et al.*, 1996) found that water temperatures >25°C caused reductions in body size of zooplankton populations, and that zooplankton assemblages were increasingly dominated by smaller forms at higher temperatures. The shift in size structure was mediated both directly and indirectly by other changes in the community, such as increased abundance of cyanobacteria and increased planktivory at higher temperature. More than 25% of the lakes in the north-eastern US had epilimnion temperatures above 25° in 1991, and the effect may be even stronger in more southerly lakes.

Some changes are also suggested by the distribution and biomass of organisms with respect to temperature. Patalas (1975) observed that the warmer North American Great Lakes generally had more diverse zooplankton assemblages than colder ones. Warmer lakes also had higher zooplankton biomass. Patalas and Salki (1984) showed that zooplankton species assemblages changed markedly when changes in flow patterns caused by impoundment resulted in cooling of waters in the northern part of Southern Indian Lake. Average abundance and biomass declined by 60 and 40%, respectively. Presumably, warming of the system would cause the opposite to occur.

In a study of the zoogeographical distribution of zooplankton between latitude 45° and 55° N, Patalas (1990) found that maximum species richness occurred where mean ice-free season temperatures were about 15°C and that species richness declined at warmer temperatures. The region of maximum richness was where a northern cold water faunal assemblage and a southern warm water fauna overlapped. Climatic warming would presumably shift the region of maximum richness northwards.

Lamontagne and Paul (unpublished data) found that higher elevation lakes in the Canadian Rockies had less diverse zooplankton communities than those at lower elevations, and that many species only tolerated temperature ranges of 5–10°C. Similar observations for invertebrate communities in streams in mountainous areas have been observed at several sites [reviewed by Schindler (1990); see also Hauer and Stanford (1994)].

Climatic warming may also accelerate the completion of life-cycles. In arctic and alpine waters, many species require two or more years to complete their life-cycles, whereas in warmer waters, one or more generations per year are observed for the same species. One such example is the amphipod *Gammarus lacustris*, which has a three-year cycle in alpine lakes of Alberta, but requires only one year at lower altitudes (F. Wilhelm, personal communication). Several authors have already used observed distributions and thermal habitats to predict climate-related habitat changes for aquatic species. For example Meisner (1990a,b) predicted a decline in both low latitude and low altitude habitats for brook trout in eastern North American streams under a warming climate. Flebbe (1994), Eaton *et al.* (1996) and Rahel *et al.* (1996) advanced similar argument in predicting the decline in fish habitats of streams. Rahel *et al.* (1996) also predicted increasing fragmentation and isolation of cold water fish populations at high elevations, if high temperatures prevent them from inhabiting the interconnecting streams at lower altitudes.

Magnuson *et al.* (1990), DeStasio *et al.* (1996) predict a general increase in the thermal niches of cold water species of north temperate lakes. The tendency is expected to be less in smaller lakes. Based on known thermal tolerances, Meisner *et al.* (1987) predicted that no species would be eliminated from the Great Lakes, although in contrast to the predictions of Magnuson *et al.*, they expected salmonid and coregonid populations to diminish as cold water habitats declined, compensated by increased cyprinids, esocids, centrarchids and ictalurids. Schindler *et al.* (1990, 1996a) found a decrease in cold water thermal habitats for animals in small boreal lakes, because of the general thermocline deepening caused by decreasing DOC concentrations, as described earlier. They also observed a slight increase in the hypoxic zone at lake bottoms, further decreasing cold water habitat. The differences between these studies are related to the size of lakes and the relative importance of DOC-controlled light penetration through the thermocline, as discussed above and in Fee *et al.* (1996).

Special concern appears to be in order for populations of cold stenotherms in cold monomictic lakes, where there are no deep summer refugia. Many of the large, productive lakes of north-central Canada are isothermal in summer. They contain important stocks of lake trout, lake whitefish, cisco and other cold stenothermic fishes, often supported by *Mysis*, *Pontoporeia* and a host of other cold stenothermic invertebrates. In midsummer, the temperatures in many of these lakes already exceed the preferred temperatures for these species and occasionally approach lethal values. Unless thermal stratification is initiated as a result of climatic warming, it is possible that even a few degrees increase in temperature could cause declines or even extirpation of certain populations. Suitable replacements, even if desirable, would probably have to be introduced by humans, since natural invasion rates are slow.

However, the behavioural and physiological adaptations of organisms may confound these predictions. Most thermal tolerance data for fish have been generated under laboratory conditions, where relatively few individuals were tested. Field populations appear to have less rigid temperature tolerances. For example, Snucins and Gunn (1995) found that lake trout were able to inhabit a small lake that had midsummer temperatures about lethal values by clustering in cold seeps from the catchment, punctuated by occasional feeding forays into warm regions. In brief, habitat heterogeneity and behaviour allowed the fish to survive. It should be noted that under climatic warming, the temperature of groundwater would also be expected to increase, so that in some cases refugia would also become less common. Sellers (1995) found a wide range in the temperature tolerances of both *Mysis* and lake trout among small lakes at ELA. In one lake, lake trout chose to remain in the epilimnion at temperatures of over 20°C throughout the summer, even though cold, well-oxygenated hypolimnion waters were easily accessible! Populations in other lakes had temperature preferences of 15°C and higher. Clearly, the temperature preferences and tolerances of aquatic organisms requires further investigation.

Of course, in some areas habitats would either disappear altogether as climate warms, or cease to support some species owing to changes in salinity or ionic composition (Covich *et al.*, 1997; Grimm *et al.*, 1997; Melack *et al.*, 1997; Rouse *et al.*, 1997).

EFFECTS ON NON-NATIVE SPECIES

The spread of some introduced species appears to be temperature dependent. Johnson and Evans (1990) considered the occurrence of several warm winters in the late 1940s to be responsible for the invasion of the Great Lakes by white perch (*Morone americana*), indigenous to the Atlantic coast. In the western mountains, the native bull trout (*Salvelinus confluentus*) is threatened by the introduction of the eastern brook trout (*Salvelinus fontinalis*), which has already taken over much of the bull trout's habitat in montane areas. Brook trout appear to be prevented from invading high altitude bull trout habitats by cold water temperatures. Presumably, this limitation would be reduced by climatic warming, causing further extirpation of bull trout, which are already considered to be threatened on the eastern slopes of the Rockies.

Many of the non-native animal and plant invaders of the southern US are subtropical taxa such as *Melaleuca*, *Tilapia* and *Eichhornia*. The northern limits of their ranges are generally determined by minimum winter temperatures, and they will probably expand their distributions northwards if climate warms. This topic is discussed in more detail by Mulholland *et al.* (1997) and Lodge (1993).

Martin and Lefebvre (1995) predict that climatic warming could cause the spread of malaria to higher latitudes. Despite the effects of warming on evaporation, the seasonal habitat for *Anopheles* mosquitoes is predicted to increase. The effect of climate change on other diseases that are water-borne or have water-dependent vectors has not been considered.

INTERACTIONS BETWEEN WARMING CLIMATES AND OTHER HUMAN ACTIVITIES

Acidification

Several interactions have already been inferred above. For example, climatic warming may enhance the acidification of streams and delay recovery of acidified lakes (Dillon *et al.*, 1997; Yan *et al.*, 1996). At first glance, this appears to contradict the observation at ELA that the alkalinity of lakes increased with climatic warming, but it must be remembered that the sulfate deposition at Dorset has been several times that at ELA, so that much higher inputs of strong acid can follow droughts at Dorset. This is illustrated by the much larger acid pulses from catchments during summer storms following periods of drought at Dorset than at ELA, which is obvious in comparing the results of Bayley *et al.* (1992a,b) and Lazerte (1993) for the two sites. The studies of Murdoch and Stoddard (1992) suggest that, in regions of high nitrate deposition, such summer acid pulses may also be caused by releases of nitric acid from catchments.

Key interactions affecting lakes can also occur in the atmosphere. Recent models suggest that the full effect of climatic warming caused by greenhouse gases may be being attenuated by the sulfate aerosol content of the troposphere (Charlson and Wigley, 1994). This implies that we must choose between controlling the acidification of freshwaters and poorly buffered catchments and more rapid climatic warming.

The declining DOC concentrations caused by climate warming as described above may be exacerbated by acidification. Experimental acidification of Lake 302S to pH 4.5 caused loss of 80% of DOC (Schindler *et al.*, 1996b). Effler *et al.* (1985) and Driscoll and van Dreason (1993) report similar losses of DOC for acidified lakes in the Adirondacks, and Yan *et al.* (1996) give several references for similar observations in eastern Ontario. Acidified lakes in eastern Ontario have DOC concentrations as low as 40 μM (P. J. Dillon, personal communication). The implications are discussed below.

Eutrophication

Where nutrient loadings are from point sources, the effect of lower water flows will exacerbate eutrophication problems, even in the absence of increased nutrient inputs. As mentioned above, water renewal is as important as nutrient loading (Dillon, 1975; Vollenweider, 1975, 1976; Schindler *et al.*, 1978, 1987). In addition, oxygen saturation occurs at lower concentration in warmer waters, exacerbating the effects of oxygen consuming effluents and lower flows on the oxic habitats of lakes and streams. In addition, the warmer hypolimnions and longer periods of summer stratification may exacerbate hypolimnetic oxygen problems (Stefan *et al.*, 1993). However, where nutrient inputs are from diffuse sources the relationships become more complicated, since the inputs may decline with decreasing water flows. This may be particularly important in estuaries, where stratification decreases with declining freshwater inflows, leading to less anoxia as well as less eutrophication (Moore *et al.*, 1997). For lakes the effects may be quite complicated, since declining water inflows may cause both declining nutrient inputs and increasing water residence, which would have counteracting effects, as discussed by Schindler *et al.* (1996a).

Climate change, acidification and stratospheric ozone depletion

Recently, there has been considerable concern for the increased exposure of freshwater organisms to UV radiation, resulting from stratospheric ozone depletion (Williamson and Zagarese, 1994). In freshwater lakes, protection from UV radiation is provided by coloured DOC. As described above, both climatic warming and acidification cause declines in DOC. They thus act in concert with stratospheric ozone depletion to increase the UV exposure of freshwater communities. For example, a decline in DOC of 15–25% caused by climatic warming and drought caused a near doubling in the volume of lakes 239 and 240 that were exposed to >1% UV. Acidification of Lake 302S caused an 80% decline in DOC, to values <100 μM. As a result, the volume of Lake 302S above the isopleth for 1% of incident UV increased from <10% to about 50% (Schindler *et al.*, 1996b). Williamson *et al.* (1996) and Yan *et al.* (1996) reach similar conclusions. Owing to the negative exponential relationship between DOC and UV penetration (Scully and Lean, 1994; Williamson *et al.*, 1996), lakes with under 300 μM DOC are particularly vulnerable to decreases in DOC. In the boreal areas of Ontario, this would include about 20% of all lakes (Neary *et al.*, 1990). Low DOC lakes are even more common in arctic and alpine areas, where allochthonous sources of DOC are low. In extreme cases, DOC may be less than 10 μM, and absorption of UV wavelengths approaches that of distilled water (Figure 6).

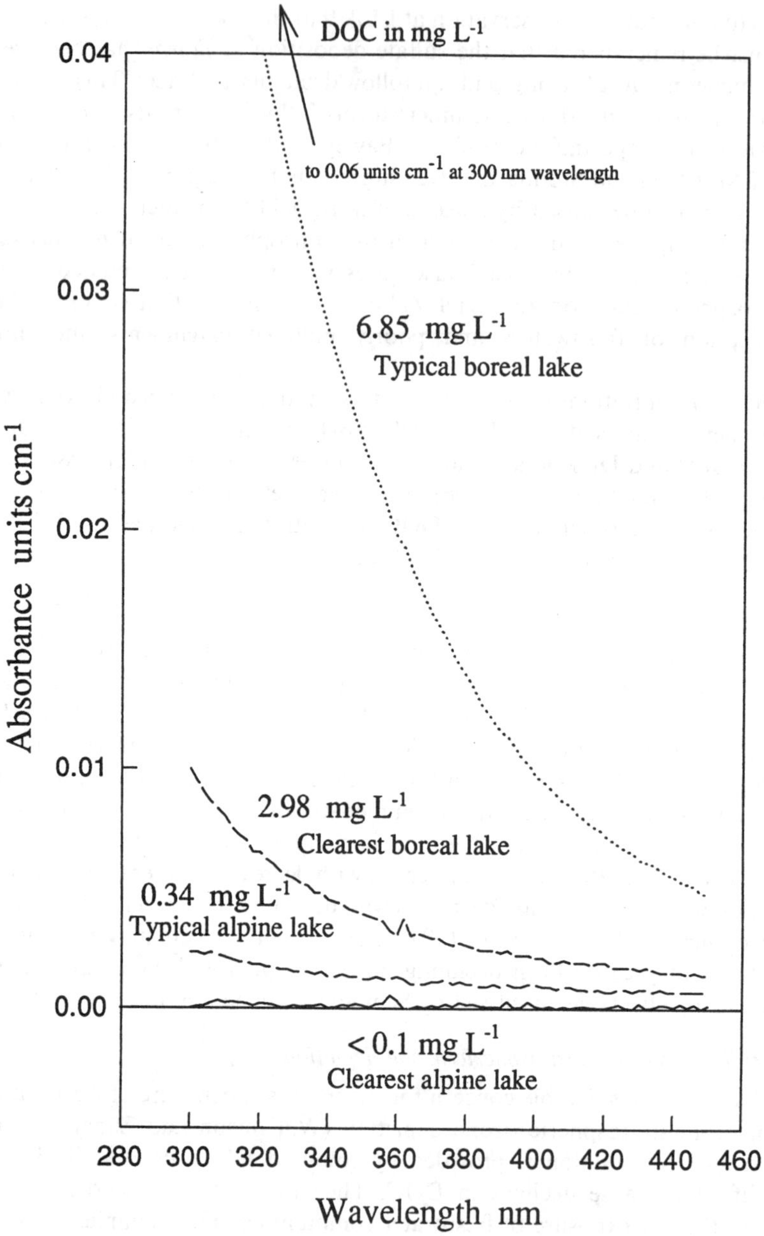

Figure 6. Absorbance of solar radiation, UV and short visible wavelengths, in several oligotrophic lakes, illustrating the key role of DOC in absorbing UV radiation. DOC is indicated in mg/L^{-1}. The alpine lakes are at approximately 2400 and 2900 m, well above tree-line in the Rocky Mountains of Banff National Park, Alberta. The boreal lakes are at the Experimental Lakes Area in north-western Ontario. Note that the absorption spectrum for the clearest alpine lake is barely distinguishable from distilled water

Conversely, increasing UV exposure may exacerbate DOC declines, owing to photolytic decomposition (Molot and Dillon, personal communication). In the process, a number of potentially toxic chemical compounds are formed, including hydroxyl radicals, superoxides and hydrogen peroxide (Cooper *et al.*, 1994). Whether such chemicals have toxic effects on exposed aquatic biota is not presently known. Bothwell *et al.* (1994) observed an increase in periphyton in artificial flumes under increasing UV-B exposure. The effect was caused by the avoidance of high UV environments by grazing chironomids. Such indirect effects of climatic warming have been little explored, in either aquatic or terrestrial studies. Williamson *et al.* (1996) predicted that increasingly inhospitable UV environments for zooplankton in shallow, clear lakes may result from declines in DOC. Little and Fabacher (1994) showed that UV-B exposure of the magnitude expected in shallow alpine streams left some trout species susceptible to fungal infections, which decreased their survival.

Climate change and reservoirs

The lower flows caused by a warmer, drier climate would be expected to exacerbate the rate of impoundment caused by industrial development alone. Several effects would be expected, in addition to those mentioned above, on the timing and magnitude of river flows. In some regions, reservoirs are the most prominent bodies of standing water, and all large rivers have been impounded (Grimm *et al.*, 1997). As a result, the natural, extremely seasonal, flow rates of these systems have been greatly altered. Increased evaporation with climatic warming would aggravate a regional water shortage that is occurring already.

In the case of the Peace–Athabasca Delta, mentioned briefly above, the absence of spring floods appears to result from a combination of the building of Bennett Dam in British Columbia and the effects of climatic warming. The resulting lack of ice jams and spring floods has caused low-lying lakes of the Athabasca Delta to be nearly flood free for over 20 years. As a result, willows and other woody vegetation have invaded areas that were previously riparian wetlands. The dramatic changes in vegetation have in turn caused declines in muskrat, fish and large mammal populations. In brief, the entire riparian community has been severely disrupted (Carbyn *et al.*, 1993). The result has been the near total collapse of the livelihood that once sustained large populations of aboriginal people.

Ice dam formation is a complex result of the interactions of channel morphology, spring melt patterns and ice thickness. So far, no one has predicted the effects of climatic warming on ice dynamics, to say nothing of secondary effects such as those mentioned above. Instead, the human response has been predictable in the case of the Peace–Athabasca system: large, costly engineering studies are now underway to attempt to devise structures that will renew the flooding to the delta, by freezing the river to the bottom to create an ice dam!

The importance of spring floods is not confined to northern regions. In the case of the Atchafalaya Delta in Louisiana, the elimination of spring floods from the Mississippi River by a variety of control structures has prevented the addition of the thin layer of sediments to the delta that, in the past, compensated for subsidence of the delta into the Gulf of Mexico. As a result, the delta has suffered rather large losses in land area (Baumann *et al.*, 1984). In brief, reduced spring flows would probably both aggravate the problems caused by damming of reservoirs and produce effects analogous to damming on undammed rivers.

The formation of reservoirs may also exacerbate the greenhouse problem, by increasing fluxes of methane and CO_2 to the atmosphere (Rudd *et al.*, 1993; Kelly *et al.*, 1994). The effect is much greater than the simple replacement of a terrestrial or wetland environment, which is a net sink for greenhouse gases, with an aquatic surface, and emissions are a result of the decomposition of inundated vegetation.

Mercury released from flooded terrestrial vegetation and soils, coupled with increased methylation, also causes increases in the concentration of mercury in fish. In many areas, concentrations exceed marketing or health limits after impoundment (Bodaly *et al.*, 1993; Rosenberg *et al.*, 1995). The area of additional water surface caused by impoundment in North America is already roughly equal to forming a second Lake Ontario (Rudd *et al.*, 1993). Globally, such effects may be even more important, since large-scale water diversion and impoundment have occurred in many areas (see, for example, Rozengurt, 1992).

Climate change and toxins

The interactions between climate and toxins will be manyfold and complex. For example, decreased water inputs will undoubtedly both reduce the delivery of pesticides from polluted lands to water, and increase their retention in lakes, owing to increased water retention. Increased water temperatures may cause both increased revolatilization and decomposition of pesticides, PCBs (polychlorinated biphenyls) and other semi-volatile organics, both reducing concentrations in southern lakes and facilitating their movement to cooler northern waters (Mackay and Wania, 1995).

Warmer temperatures may both increase the ingestion and uptake of pollutants (Moore *et al.*, 1997), and dilute the concentration in fishes owing to accelerated growth (reviewed by Schindler *et al.*, 1995). In brief, because of the complex interactions of climate, pollutants of various characteristics and different aquatic communities, there will probably be a wide variety of responses from aquatic ecosystems. Moore *et al.* (1997) discuss other possible interactions.

Climatic warming may also act with other human perturbations to change the mercury cycle in freshwaters. Methylation increases and demethylation decreases as water temperature increases (Ramlal *et al.*, 1993). Increased light penetration causes increased losses of methyl mercury by photolytic conversion to Hg^0 and subsequent loss to the atmosphere (Sellers *et al.*, 1996). Drying of catchments may greatly decrease exports of mercury to freshwaters, since it is complexed with DOC (Miskimmin *et al.*, 1992; St Louis *et al.*, 1994). In-lake methylation would also be expected to increase as DOC declines (Miskimmin *et al.*, 1992). However, raising of water levels causes increased net mercury methylation (Rudd, 1995), resulting in increased contamination of fishes, many of which are the same cold stenothermic species that may be negatively affected by warming of lakes. The high mercury values are in turn passed on to fish-eating birds and mammals. Food chain transfers appear to increase with higher temperatures (Moore *et al.*, 1997). In addition to the above changes, atmospheric concentration and deposition of Hg are also increasing (Slemr and Langer, 1992; Lockhart, 1995). Together, the results of these perturbations are difficult to predict, and warrant further study. It should be noted that mercury burdens by aboriginal humans relying heavily on aquatic organisms for food are already approaching unacceptable levels (Lockhart, 1995; Wheatley and Paradis, 1995).

Population growth and climate change

As climate changes, increasing human use of water for drinking, agriculture and industry will aggravate water problems, both by removal and befoulment. Water removal will be particularly detrimental to water quality by reducing dilution of effluents, and thermal discharges into smaller water masses will reduce thermal habitats for many species. The effects are predicted to be particularly great in the south-western and south-eastern US (Grimm *et al.*, 1997; Mulholland *et al.*, 1997). In the US Southwest, increasing human populations are already exceeding the capacity of surface waters and groundwater aquifers to sustain many human activities (Grimm *et al.*, 1997), and it is clear that unless both water use and population are strongly regulated, a crisis situation is only a few years away.

TRANS-REGIONAL PROBLEMS

In some cases, climatic warming in one region may affect water resources in another. For example, the water resources of much of the Great Basin and southern Alberta are largely rivers that originate on the eastern slopes of the Rocky Mountains. Changes in the amount and timing of precipitation, snowmelt and glacial runoff can therefore have enormous consequences for these otherwise arid areas. For example, Clow *et al.* (1996) and Williams *et al.* (1996) observed an increase in precipitation and a decrease in temperature in the Rocky Mountains during the same period when widespread drought and temperature increases were observed at lowland sites to the north and east. To date, only one long-term record for the Rocky Mountains shows a statistically significant increase in temperature in the past century (Luckman, 1990; see Hauer *et al.*, 1997). There are no long-term records or models that allow us to predict how changes in the climate of alpine areas might be coupled to changes at lower altitudes, and it is obvious that such couplings are of critical importance in any assessment of the effects of climatic warming on water resources.

Other examples of the trans-regional effects on water resources are large rivers. The Atchafalaya and other wetlands of the lower Mississippi depend on delivery of sediments, as described earlier. Flows in the Ohio and Missouri river basins are important sources of water and sediments, having greater effects on the wetlands of the Mississippi Delta than climatic shifts within the region. Similarly, reduced flows in the headwaters of the Canadian Rockies have exacerbated the above-mentioned lack of flooding in the Peace–Athabasca Delta. Reductions in spring flooding are also predicted to reduce the number, level and frequency of flooding in lakes of the Mackenzie Delta (Marsh and Lesack, 1996). The enormous amount of heat carried northwards by the Mackenzie River allows relatively moderate climates to persist to the arctic coast (Gill, 1974). Trees grow to the mouth of the Mackenzie, whereas in areas away from the river, the tree-line is hundreds of kilometres to the south.

DISCUSSION

Our most complete information about the effects of climatic warming on lakes is for region 1, where strong limnological research and long-term monitoring has occurred for several decades. This summary makes it obvious that strong, long-term limnological programmes in other regions are essential to obtaining a more complete regional picture of the effects of climate change on freshwater.

Many of the effects of climatic warming and drought are counter-intuitive, and would be unknown without direct research and monitoring of freshwaters. For example, without long-term observations, it would have been impossible to predict the thermocline migration caused by declining DOC, the acidification of streams while the lakes in the same region become more alkaline and the differences in response of primary production between lakes where climatic warming perturbs chemical and physical regimes in different ways. The acidification of headwater seepage lakes caused by climatic warming while lakes further down the aquifer become more alkaline could not have been predicted from first principles. Despite many years and international expenditure of over a billion dollars on acid precipitation research, the discoveries that climate warming would delay or even reverse the recovery of lakes from acid precipitation, and the combined effects of acidification, climatic warming and stratospheric ozone depletion on freshwaters were made from long-term monitoring data, not from sophisticated high-tech simulations or laboratory experiments. For the foreseeable future, the collection and analysis of long-term data sets, and the opportunistic use of other perturbations as surrogates for climate change, appears to be the most reliable way to discover the consequences of climatic change for freshwaters.

Unfortunately, for large regions of the continent there are no long-term studies of freshwaters. It appears to be particularly important to begin such studies in arid areas, where water supplies are already limiting to human activities (Grimm *et al.*, 1997), or have been limiting under only slightly drier conditions in the recent past (Covich *et al.*, 1997). Also, the responses of waters in mountainous areas of the west to human-induced global warming appear to be particularly hard to predict. These areas are extremely important in supplying water to more arid adjacent lands (Grimm *et al.*, 1997; Hauer *et al.*, 1997). Tragically, at the very time when such data sets are critically needed, many US and Candian agencies are phasing out or cutting back their freshwater programmes.

Some final thoughts

In my view, there has been entirely too much focus on whether or not the recent trends for warming climate are caused by humans or part of a natural cycle. In either case, it is essential to anticipate the consequences for aquatic ecosystems and the human endeavours that depend on them. Palaeolimnological studies indicate that, in the past, even natural warming and drying of climates has had dire consequences for lakes and wetlands. Long-term records show that the recent changes in duration of ice cover are not unique; they represent a continuation, perhaps slightly accelerated, of climatic warming that extends back at least 200 years (Rannie, 1983).

Much of our focus to date has been on the development of global-scale GCM models to predict the effects of increasing greenhouse gases and other anthropogenic changes. But such models will be of very limited use on regional and local scales for many years to come. There are other approaches that may prove more fruitful for determining the consequences for regional water supplies.

There appear to be many opportunities for combined studies of climate records and water flows, for such records exist in most regions, and in most cases periods of prolonged warming and drought would provide some insights into at least the trends expected from increased greenhouse warming. Moore *et al.* (1997) suggest the study of lakes in urban heat islands, which are already several degrees warmer than surrounding areas, to deduce changes expected from climatic warming. Decamps (1993) argues that a historical perspective is necessary to predict the interactions of climate change and human activity on river landscapes. Of course, palaeoecological methods allow us to study the consequences for freshwater of changes in both climate and human patterns of settlement.

In water-rich North America, there is a tendency to underrate the importance of secure supplies of freshwater. Freshwater is the natural resource that is currently in shortest supply for human needs in many regions of the world. At the Stockholm Water Conference in 1991, I found that representatives of Third World countries regarded water quality problems to be of secondary interest, for in areas where water quantity is of immediate concern for survival, concerns about pollutants and pathogens become secondary. The result is inevitably an increase in waterborne diseases and other water-related health problems in such areas. But even in relatively water-rich North America, water quantity is a problem in arid and overpopulated regions of the Southwest, and water quality has become a major issue in populous parts of the East, despite relatively abundant water. Research on water quality in the basin of the Athabasca, Peace and Slave rivers, considered remote and underdeveloped by most standards, showed that most communities had drinking water contaminated to unacceptable levels with pathogens, trihalomethanes, suspended particulates and other substances (Wrona *et al.*, 1996). It is clear that we face a crisis in freshwater: it is here now in some areas, while in others it is only a few years away. It is critical that we have the scientific basis for dealing with future water shortages, contamination and biodiversity issues, regardless of whether the problem is natural or man-made.

ACKNOWLEDGEMENTS

Production of this manuscript was supported by NSERC Operating Grant no. 89673. Brian Parker produced many of the figures. John Eaton, Diane McKnight, John Magnuson, Marianne Moore and Patrick Mulholland provided insightful reviews. Diane also allowed early access to abstracts for the Leesburg Conference and pre-publication copies of manuscripts from the conference. Peter Dillon supplied early drafts of his manuscripts on delays in recovery of lakes from acidification and photolytic degradation of DOC. Carl Bowser kindly produced Figure 4 from data for lakes in northern Wisconsin. Margaret Foxcroft prepared the manuscript and references.

REFERENCES

Anderson, W., Robertson, D. M., and Magnuson, J. J. 1996. 'Evidence of recent warming and El Niño related variations in ice breakup of Wisconsin lakes', *Limnol. Oceanogr.*, **41**(5), 815–821.

Assel, R. A., and Robertson, D. M. 1996. 'Changes in winter air temperatures near Lake Michigan during 1851–1993, as determined from regional lake-ice records', Poster, presented at ASLO/NABS Symposium, Regional Assessment of Freshwater Ecosystems and Climate Change in North America, October 24–26, 1994, Xerox Training Center, Leesburg, VA.

Baker, L. A., Brezonik, P. L., and Pollman, C. D. 1986. 'Model of internal alkalinity generation: sulfate retention component', *Wat. Air Soil Pollut.*, **31**, 89–94.

Baumann, R. H., Day, J. W., Jr., and Miller, C. A. 1984. 'Mississippi deltaic wetland survival: sedimentation versus coastal submergence', *Science*, **224**, 1093–1095.

Bayley, S. E., Schindler, D. W., Parker, B. R., Stainton, M. P., and Beaty, K. G. 1992a. 'Effects of forest fire and drought on acidity of a base-poor boreal forest stream: similarities between climatic warming and acidic precipitation', *Biogeochemistry*, **17**, 191–204.

Bayley, S. E., Schindler, D. W., Beaty, K. G., Parker, B. R., and Stainton, M. P. 1992b. 'Effects of multiple fires on nutrient yields from streams draining boreal forest and fen watersheds: nitrogen and phosphorus', *Can. J. Fish. Aquat. Sci.*, **49**, 584–596.

Beaty, K. G. 1995. 'Sediment transport in a small stream following two successive forest fires', *Can. J. Fish. Aquat. Sci.*, **51**, 2723–2733.

Bodaly, R. A., Rudd, J. W. M., Fudge, R. J. P., and Kelly, C. A. 1993. 'Mercury concentrations in fish related to size of remote Canadian Shield lakes', *Can. J. Fish. Aquat. Sci.*, **50**, 980–987.

Bothwell, M. L., Sherbot, D. M. J., and Pollock, C. M. 1994. 'Ecosystems response to solar ultraviolet-B radiation: influence of trophic-level interactions', *Science*, **265**, 97–100.

Bowser, C. J., and Ackerman, J. 1993. 'Isotopic variation in lake-groundwater systems: extending the isotope based ($d^{18}O$, dD) mass budget technique for lakes, and comparisons with solute based budgets', *Abstr. Am. Geophys. Union*, Dec. 1993.

Brugman, M. M. 1992. 'Search for trends in glacier mass balance from western Canada', in *Using Hydrometric Data to Detect and Monitor Climatic Change*, NHRI Symposium No. 8. National Hydrology Research Institute, Saskatoon. pp. 233–244.

Bukaveckas, P. A., and Driscoll, C. T. 1991. 'Effects of whole-lake base addition on the optical properties of three acidic clearwater lakes', *Can. J. Fish. Aqaut. Sci.*, **48**, 1030–1040.

Byron, E. R., and Goldman, C. R. 1990. 'The potential effects of global warming on the primary productivity of a subalpine lake', *Wat. Resour. Bull.*, **26**, 983–989.

Carbyn, L. N., Oostenbrug, S. M., and Anions, D. W. 1993. *Wolves, bison and the dynamics related to the Peace–Athabasca Delta in Canada's Wood Buffalo National Park*. Canadian Circumpolar Institute, University of Alberta, Edmonton.

Chambers, P. A., Pietroniro, A., Scrimgeour, G. J., and Ferguson, M. 1996. 'Assessment and validation of modelling under-ice dissolved oxygen using DOSTOC, Athbasca River, 1988 to 1994', *Report 95*. Northern River Basins Study, Edmonton, Alberta.

Charlson, R. J. and Wigley, T. M. L. 1994. 'Sulfate aerosol and climatic change', *Scientific Am.*, February, 48–57.

Chen, C. Y., and Folt, C. L. 1996. 'Consequences if fall warming for zooplankton overwintering success', *Limnol. Oceanogr.*, **41**(5), 1077–1086.

Clair, T. A., and Ehrman, J. M. 1996. 'Variation in discharge and dissolved organic carbon and nitrogen export from terrestrial basins with changes in climate: a neural network approach', *Limnol. Oceanogr.*, **41**(5), 921–927.

Clow, D. W., Williams, M. W., Baron, J. S., Mast, M. A., and Campbell, D. H. 1996. 'Response of concentrations of weathering products in surface waters to variations in annual precipitation at selected high-elevation sites in the western US'. Presented at the ASLO/NABS Symposium, *Regional Assessment of Freshwater Ecosystems and Climate Change in North America, October 24–26, 1994*, Xerox Training Center, Leesburg, VA.

Comb, D. G. 1990. 'Ice out in Maine', *Nature*, **347**, 510.

Cooper, W. J., Shao, C., Lean, D. R. S., Gordon, A. S., and Scully, F. E., Jr. 1994. 'Factors affecting the distribution of H_2O_2 in surface waters', in Baker, L. A. [Ed.], *Environmental Chemistry of Lakes and Reservoirs*, Advances in Chemistry Series 237. American Chemical Society. Washington, D.C. pp. 391–422.

Covich, A. P., Fritz, S. C., Lamb, P. J., Marzolt, R. D., Matheus, W. J., Porani, K. A., Drepas, E. E., Richman, M. B., and Winter, T. C. 1997. 'Potential effects of climate change on aquatic ecosystems of the Great Plains of North America', *Hydrol. Process.*, **11**, 993–1021.

Crowe, A. S. 1993. 'The application of a coupled water-balance-salinity model to evaluate the sensitivity of a lake dominated by groundwater to climatic variability', *J. Hydrol.*, **141**, 33–73.

Curtis, P. J. and Adams, H. E. 1995. 'Dissolved organic matter quantity and quality from freshwater and saline lakes in east-central Alberta', *Biogeochemistry*, **30**, 59–76.

Decamps, H. 1993. 'River margins and environmental change', *Ecol. Appl.*, **3**, 441–445.

DeStasio, B. T., Hill, J. M., Kleinhans, N. P., Nibbelink, N. P., and Magnuson, J. J. 1996. 'Potential effects of global climate change on small north temperate lakes: physics, fish and plankton', *Limnol. Oceanogr.*, **41**(5), 1136–1149.

Dillon, P. J. 1975. 'The phosphorus budget of Cameron Lake, Ontario: the importance of flushing rate to the degree of eutrophy of lakes', *Limnol. Oceanogr.*, **20**, 28–39.

Dillon, P. J., Yan, N. D., and Harvey, H. H. 1984. 'Acidic deposition: effects on aquatic ecosystems', in Straub, C. P. [Ed.], *Critical Reviews in Environmental Control*. CRC Press, Inc., Boca Raton, USA. pp. 167–194.

Dillon, P. J., Hutchinson, B. A., and Futter, M. 1994. 'Long-term changes in meteorologic, hydrologic and physical factors in lakes and catchments, central Ontario: comparison with the Experimental Lakes Area (ELA)', presented at the ASLO/NABS Symposium, *Regional Assessment of Freshwater Ecosystems and Climate Change in North America, October 24–26, 1994*, Xerox Training Center, Leesburg, Va.

Dillon, P. J., Molot, L. A., and Futter, M. 1997. 'The effect of El Nino-related drought on the recovery of acidified lakes', *Environ. Monit. Assess.*, in press.

Driscoll, C. T., and Van Dreason, R. 1993. 'Seasonal and long-term temporal patterns in the chemistry of Adirondack lakes', *Water Air Soil Pollut.*, **67**, 319–344.

Eaton, J. G. and Scheller, R. M. 1996. 'Effects of climate warming on fish thermal habitat in streams of the United States', *Limnol. Oceanogr.*, **41**(5), 1109–1115.

Effler, S. W., Schafran, G. C., and Driscoll, C. T. 1985. 'Parititioning light attenuation in an acidic lake', *Can. J. Fish. Aquat. Sci.*, **42**, 1707–1711.

Etkin, D. A. 1991. 'Break-up in Hudson Bay: its sensitivity to air temperatures and implications for climate warming', *Climatol. Bull.*, **24**, 21–34.

Evans, J. C. and Prepas, E. E. 1996. 'Potential effects of climate change on ion chemistry and phytoplankton communities in prairie saline lakes', *Limnol. Oceanogr.*, **41**(5), 1063–1076.

Fee. E. J. and Hecky, R. E. 1992. 'Introduction to the Northwest Ontario Lake Size Series (NOLSS)', *Can. J. Fish. Aqat. Sci.*, **49**, 2445–2459.

Fee, E. J., Hecky, R. E., Kasian, S. E., and Cruikshank, D. 1996. 'Effects of lake size, water clarity and climatic variability on mixing depths in Canadian Shield Lakes, *Limnol. Oceanogr.*, **41**(5), 912–920.

Firth, P. and Fisher, S. G. (Eds), 1992. *Global Climate Change and Freshwater Ecosystems*. Springer-Verlag, New York.

Flebbe, P. A. 1994. 'Fragmentation of trout habitat in the southern Appalachian mountains and the future under global warming', Presented at the ASLO/NABS Symposium, *Regional Assessment of Freshwater Ecosystems and Climate Change in North America*, October 24–26, 1994, Xerox Training Center, Leesburg, VA.

Fritz, S. C. 1990. 'Twentieth-century salinity and water-level fluctuations in Devils Lake, North Dakota: test of a diatom-based transfer function', *Limnol. Oceanogr.*, **35**, 1771–1781.

Fritz, S. C., Juggins, S., Battarbee, R. W., and Engstrom, D. R. 1991. 'Reconstruction of past changes in salinity and climate using a diatom-based transfer function', *Nature*, **352**, 706–708.

George, D. G. and Harris, G. P. 1985. 'The effect of climate on long-term changes in the crustacean plankton of Lake Windermere, U.K.', *Nature*, **316**, 536–539.

George, D. G., Hewitt, D. P., Lund, J. W. G., and Smyly, W. J. P. 1990. 'The relative effects of enrichment and climate change on the long-term dynamics of *Daphnia* in Esthwaite Water, Cumbria, *Freshwat. Biol.*, **23**, 55–70.

Gill, D. 1974. 'Significance of spring breakup to the bioclimate of the Mackenzie River delta', in Reed, J. C., and Sater, J. E. (Eds), *The Coast and Shelf of the Beaufort Sea*. Arctic Inst. of N. America, Arlington. pp. 543–548.

Gorham, E. 1991. 'Northern peatlands: role in the carbon cycle and probable responses to climatic warming', *Ecol. Appl.*, **1**, 182–195.

Gorham, E. and Boyce, F. M. 1989. 'Influence of lake surface area and depth upon thermal stratification and the depth of the summer thermocline', *J. Great Lakes Res.*, **15**, 233–235.

Grimm, N. B. and Fisher, S. G. 1992. 'Responses of arid-land streams to changing climate', in Firth, P. and Fisher, S. G. (Eds), *Global Climate Change and Freshwater Ecosystems*. Springer-Verlag, New York. Chapter 10.

Grimm, N. B., Chacón, A., Dahm, C. N., Lund, O. T., Starkweather, P. L., and Wurtsbaugh, W. W. 1997. 'Sensitivity of aquatic ecosystems to anthropogenic and climate changes: the Basin and Range, American Southwest and Mexico', *Hydrol. Process.*, **11**, 1023–1041.

Hairston, N. 1996. 'Zooplankton egg banks as biotic reservoirs in changing environments', *Limnol. Oceanogr.*, **41**(5), 1087–1095.

Hanson, H. P., Hanson, C. S., and Yoo, B. H. 1992. 'Recent Great Lakes ice trends', *Bull. Am. Meteorol. Soc.*, **73**, 577–584.

Hauer, F. R. and Stanford, J. A. 1994. 'Global climate change and its effect on complex distribution and abundance patterns of macroinvertebrates in northern Rocky Mountain streams', presented at the ASLO/NABS Symposium, *Regional Assessment of Freshwater Ecosystems and Climate Change in North America, October 24–26, 1994*, Xerox Training Center, Leesburg, VA.

Hauer, F. R., Baron, J. S., Campbell, D. H., Fausch, K. D., Hostetler, S. W., Leavesley, G. H., Leavitt, R. R. McKnight, D. M. and Stanford, J. A. 1997. Assessment of climate change and freshwater ecosystems of the Rocky Mountains, US and Canada. *Hydrol. Process.* **11,** 903–924.

Hengeveld, H. G. 1990. 'Global climate change: implications for air temperature and water supply in Canada', *Trans. Am. Fish. Soc.*, **119**, 176–182.

Hengeveld, H. 1991. *Understanding atmospheric change*, Environment Canada SOE Rept. 91-2. Ottawa.

Hengeveld, H. 1995. Understanding atmospheric change, 2nd Ed., Environment Canada SOE Report, 95–2, Ottawa, 68 pp.

Hinton, M. J., Schiff, S. L., and English, M. C. 1997. 'The significance of storms for the concentration and export of dissolved organic carbon from two Precambrian Shield catchments', *Biogeochemistry*, **36**, 67–88.

Johnson, T. B. and Evans, D. O. 1990. 'Size-dependent winter mortality of young-of-the-year white perch: climate warming and invasion of the Laurentian Great Lakes', *Trans. Am. Fish. Soc.*, **119**, 301–313.

Keller, W., Pitblado, J. R., and Carbone, J. 1992. 'Chemical responses of acidic lakes in the Sudbury, Ontario, area to reduced smelter emissions', *Can. J. Fish. Aquat. Sci.*, **49**(Suppl. 1), 25–32.

Kellerhals, R. and Church, M. 1989. 'The morphology of large rivers: characterization and management', *Can. J. Fish. Aquat. Sci.*, **106**, 31–38.

Kelly, C. A., Rudd, J. W. M., Hesslein, R. H., Schindler, D. W., Dillon, P. J., Driscoll, C. T., Gherini, S. A., and Hecky, R. E. 1987. 'Prediction of biological acid neutralization in acid-sensitive lakes', *Biogeochemistry*, **3**, 129–140.

Kelly, C. A., Rudd, J. M. W., St Louis, V. L., and Moore, T. 1994. 'Turning attention to reservoir surfaces: a neglected area in greenhouse studies', *EOS*, **75**, 332–333.

Kilham, P. 1982. 'Acid precipitation: its role in the alkalization of a lake in Michigan', *Limnol. Oceanogr.*, **27**, 856–867.

Kilham, S. S., Theriot, E. C., and Fritz, S. C., 1996. 'Linking planktonic diatoms and climate change in the large lakes of the Yellowstone ecosystem using resource theory , *Limnol. Oceanogr.*, **41**(5), 1052–1062.

Krabbenhoft, D. P., Bowser, C. J., Kendall, C., and Gat, J. R. 1994. 'Use of oxygen-18 and deuterium to assess the hydrology of groundwater-lake sytems', in Baker, L. A. (Ed.), *Environmental Chemistry of Lakes and Reservoirs*, Advances in Chemistry Series 237. American Chemical Society. Washington, D.C. pp. 67–90.

Kratz, T. K., Benson, B. J., Blood, E. R., Cunningham, G. L., and Dahlgren, R. A. 1991. 'The influence of landscape position on temporal variability in four North American ecosystems', *Am. Nat.*, **138**, 355–378.

Kurz, W. A., Apps, M. J., Beukemia, S. J., and Lekstrum, T. 1995. '20th Century carbon budget of Canadian forests', *Tellus*, **47B**, 170–177.

Last, W. M. 1992. 'Chemical composition of saline and subsaline lakes of the northern Great Plains, western Canada, *Int. J. Salt Lake Res.*, **1**, 47–76.

Last, W. and Slezak, L. A. 1988. 'The salt lakes of western Canada: a paleolimnological overview', *Hydrobiologia*, **158**, 301–316.

Last, W. M. and Teller, J. T. 1983. 'Holocene climate and hydrology of the Lake Manitoba basin', in Teller, J. T. and Clayton, L. (Eds), *Glacial Lake Agassiz*. Geol. Assn. Canada Spec. Paper 26. pp. 333–353.

Lazerte, B. D. 1993. 'The impact of drought and acidification on the chemical exports from a minerotrophic confier swamp', *Biogeochemistry*, **18**, 153–175.

Leavesley, G. H., Turner, K., D'Agnese, F. A., and McKnight, D. 1997. 'Regional delineation of North America for the assessment of freshwater ecosystems and climate change', *Hydrol. Process.*, **11**, 819–824.

Ligon, F. K., Dietrich, W. E., and Trush, W. J. 1995. 'Downstream ecological effects of dams', Bioscience, **45**, 183–192.

Little, E. E. and Fabacher, D. L. 1994. 'Comparative sensitivity of rainbow trout and two threatened salmonids, Apache trout and Lahontan cutthroat trout to ultraviolet-B radiation', *Arch. Hydrobiol. Limnol.*, **43**, 217–226.

Lockhart, W. L. 1995. 'Implications of chemical contaminants for aquatic animals in the Canadian Arctic: some review comments', *Sci. Tot. Environ.*, **160/161**, 631–641.

Lodge, D. M., 1993. 'Species invasions and deletions: Community effects and responses to climate and habitat change', in Kareiva, P. M., Kingsolver, J. G., and Huey, R. B. (Eds), *Biotic Interactions and Global Change*. Sinauer Associates, Sunderland. pp. 367–387.

Luckman, B. H. 1990. 'Mountain areas and global change: a view from the Canadian Rockies', *Mountain Res. Devel.*, **10**, 183–195.

Mackay, D. and Wania, F. 1995. 'Transport of contaminants to the Arctic: partitioning, processes and models', *Sci. Tot. Environ.*, **160/161**, 25–38.

Magnuson, J. J., Meisner, J. D., and Hill, D. K. 1990. 'Potential changes in the thermal habitat of Great Lakes fish after global climate warming', *Trans. Am. Fish. Soc.*, **119**, 254–264.

Magnuson, J. J., Assel, R. A., Bouser, C. J., Dillon, P. J., Eaton, J. G., Evans, H. E., Fee, E. J., Hall, R. I., Mortsch, L. R., Schindler, D. W., Quinn, F. H., and Webster, K. H. 1997. 'Potential effects of climate changes on aquatic systems: Laurentian Great Lakes and Precambrian Shield Region', *Hydrol. Process.*, **11**, 825–872.

Malley, D. F., Findlay, D. L., and Chang, P. S. S. 1982. 'Ecological effects of acid precipitation on zooplankton', in D'Itri, F. M. [Ed.], *Acid Precipitation: Effects on Ecological Systems*. Ann Arbor Science Publishers, Ann Arbor. pp 297–327.

Marsh, P. and Lesack, L. F. W. 1996. 'The hydrologic regime of perched lakes in the Mackenzie Delta: potential responses to climate change', *Limnol. Oceanogr.*, **41**(5), 849–856.

Martin, P. H. and Lefebvre, M. G. 1995. 'Malaria and climate: sensitivity of malaria potential transmission to climate', *Ambio*, **24**, 200–207.

Mazumder, A. and Taylor, W. D. 1994. 'Thermal structure of lakes varying in size and water clarity', *Limnol. Oceanogr.*, **39**, 968–976.

McCormick, M. J. 1990. 'Potential changes in thermal structure and cycle of Lake Michigan due to global warming', *Trans. Am. Fish. Soc.*, **119**, 183–194.

McDonald, M. E., Hershey, A. E., and Miller, M. C. 1996. 'Global warming impacts on lake trout in arctic lakes', *Limnol. Oceanogr.*, **41**(5), 1102–1108.

Meisner, J. D. 1990a. 'Effect of climatic warming on the southern margins of the native range of brook trout, *Salvelinus fontinalis*, *Can. J. Fish. Aquat. Sci.*, **47**, 1065–1070.

Meisner, J. D. 1990b. 'Potential loss of thermal habitat for Brook Trout, due to climatic warming, in two southern Ontario streams', *Trans. Am. Fish. Soc.*, **119**, 282–291.

Meisner, J. D., Goodier, J. L., and Regier, H. A. 1987. 'An assessment of the effects of climate warming on Great Lakes basin fishes', *J. Great Lakes Res.*, **13**, 340–352.

Melack, J. M., Dozier, J., Goldman, C. R., Greenland, D., Mitner, A. M., and Naimah, R. J. 1997. 'Effects of climate change on inland waters of the Pacific Coastal Mountains and Western Great Basin of North America', *Hydrol. Process.*, **11**, 971–992.

Meyer, J. L. and Pulliam, W. M. 1992. 'Modification of terrestrial-aquatic interactions by a changing climate', In Firth, P. and Fisher, S. G. (Eds), *Global Climate Change and Freshwater Ecosystems*. Springer-Verlag, New York. Chapter 8.

Miskimmin, B. M., Rudd, J. W. M., and Kelly, C. A. 1992. 'Influence of dissolved organic carbon, pH and microbial respiration rates on mercury methylation and demethylation in lake water, *Can. J. Fish. Aquat. Sci.*, **49**, 17–22.

Molot, L. and Dillon, P. J. 1996b. 'Storage of terrestrial carbon in boreal lake sediments and evasion to the atmosphere', *Glob. Biogeochem. Cycles*, **10**, 483–492.

Moore, M. V. and Folt, C. L. 1993. 'Zooplankton body size and community structure: effects of thermal and toxicant stress', *Trends Ecol. Evol.*, **8**, 178–183.

Moore, M. V., Folt, C. L., and Stemberger, R. S. 1996. 'Consequences of elevated temperatures for zooplankton assemblages in temperate lakes', *Arch. Hydrobiol.*, **135**, 289–319.

Moore, M. V., Pace, M. L., Mather, J. R., Murdoch, P. S., Howarth, R. W., Folt, C. L., Chen, C. Y., Hemond, H. F., Flebbe, P. A., and Driscoll, C. T. 1997. 'Potential effects of climate change on freshwater ecosystems of the New England/Mid-Atlantic region', *Hydrol. Process.*, **11**, 925–948.

Mulholland, P. J., Best, G. R., Coutant, C. C., Hornberger, G. M., Meyer, J. L., Robinson, P. J., Stenberg, J. R., Turner, R. E., Vera-Herrera, F., and Wetzel, R. G. 1997. 'Effects of climatic change on freshwaters ecosystems of the south-eastern United States and Gulf Coast of Mexico', *Hydrol. Process.*, **11**, 949–970.

Murdoch, P. S. and Stoddard, J. L. 1992. 'The role of nitrate in the acidification of streams in the Catskill Mountains of New York', *Wat. Resour. Res.*, **28**, 2707–2720.

Neary, B. P., Dillon, P. J., Munro, J. R., and Clark, B. J. 1990. 'The acidification of Ontario lakes: an assessment of their sensitivity and current status with respect to biological damage', Limnology Section, Dorset Research Centre, Ontario Environment.

Newbury, R. W., McCullough, G. K., and Hecky, R. E. 1984. 'The Southern Indian Lake impondment and Churchill River diversion', *Can J. Fish. Aquat. Sci.*, **41**, 548–557.

Oswood, M. W., Milner, A. M., and Irons, J. G. III. 1992. 'Climate change and Alaskan rivers and streams', in Firth, P. and Fisher, S. G. (Eds), *Global Climate Change and Freshwater Ecosystems*. Springer-Verlag, New York. Chapter 9.

Patalas, K. 1970. 'Primary and secondary production in a lake heated by thermal power plant', in Gallagher, G. A. (Ed.), *The Environmental Challenge of the 70s*. Inst. of Environ. Sciences. Ht. Prospect, IL. pp. 267–271.

Patalas, K. 1975. 'The crustacean plankton communities of fourteeen North American great lakes', *Verh. Int. Verein. Limnol.*, **19**, 504–511.

Patalas, K. 1990. 'Diversity if the zooplankton communities in Canadian lakes as a function of climate', *Verh. Int. Verein. Limnol.*, **24**, 360–368.

Patalas, K. and Salki, A. 1984. 'Effects of impoundment and diversion on the crustacean plankton of Southern Indian Lake', *Can. J. Fish. Aquat. Sci.*, **41**, 613–637.

Poff, N. L. 1992. 'Regional hydrologic response to climate change: an ecological perspective', in Firth, P. and Fisher, S. G. (Eds), *Global Climate Change and Freshwater Ecosystems*. Springer-Verlag, New York. Chapter 5.

Price, C. and Rind. D. 1994. 'The impact of a 2 × CO_2 climate on lightning-caused fires', *J. Climate* **7**, 1484–1494.

Prowse, T. D. and Ommanney, C. S. L. (Eds), 1990. *Northern Hydrology: Canadian Perspectives*, NHRI Science Report no. 1. Minister of Supply and Services, Ottawa.

Prowse, T. D. *et al.*, 1996. Synthesis report no. 1. Northern River Basins Study, Edmonton, Alberta.

Rahel, F. J., Keleher, C. J., and Anderson, J. L. 1996. 'Potential habitat loss and population fragmentation for coldwater fish in the North Platte River drainage of the Rocky Mountains: Response to climate warming', *Limnol. Oceanogr.*, **41**(5), 1116–1123.

Ramlal, P. S., Kelly, C. A., Rudd, J. W. M., and Furutani, A. 1993. 'Sites of methyl mercury production in remote Canadian Shield lakes', *Can J. Fish. Aquat. Sci.*, **50**, 972–979.

Rannie, W. F. 1983. 'Breakup and freezeup of the Red River at Winnipeg, Manitoba Canada in the 19th century and some climatic implications', *Climatic Change*, **5**, 283–296.

Revelle, R. R. and Waggoner, P. E. 1983. 'Effects of carbon dioxide-induced climatic change on water supplies in the western United States', in *Changing Climate: Report of the Carbon Dioxide Assessment Committee*. National Academy Press. pp. 419–432.

Reycraft, J. and Skinner, W. 1993. 'Canadian ice conditions: an indicator of climate variability', *Climate Perspect.*, **15**, 9–15.

Robertson, D. M. 1989. 'The use of water temperature and ice cover as climatic indicators', *PhD Thesis*, University of Wisconsin, Madison.

Robertson, D. M., Ragotzkie, R. A., and Magnuson, J. J. 1992. 'Lake ice records used to detect historical and future climate changes', *Climatic Change*, **21**, 407–427.

Rosenberg, D. M., Bodaly, R. A., and Usher, P. J. 1995. 'Environmental and social impacts of large hydro-electric development: who is listening?' Global Environmental change, **5**, 127–148.

Roulet, N., Moore, T., Bubier, J., Comer, N., and Waddington, J. M. 1994. 'Climate change and northern peatland ecosystems', Presented at the ASLO/NABS Symposium, *Regional Assessment of Freshwater Ecosystems and Climate Change in North America', October 24–26, 1994*, Xerox Training Center, Leesburg, VA.

Rouse, W. R., Douglas, M. S. V., Hecky, R. E., Hershery, A. E., Kling, G. W., Lesack, L., Marsh, P., McDonald, M., Nicholson, B. J., Roulet, N. T., and Smol, P. 1997. 'Effects of climate change on the freshwaters of arctic and subarctic North America', *Hydrol. Process.*, **11**, 873–901.

Rozengurt, M. A. 1992. 'Alteration of freshwater inflows', in Stroud, R. H. (Ed.), *Stemming the Tide of Coastal Fish Habitat Loss. Proc. Symp. on Conservation of Coastal Fish Habitat*, Baltimore, MD, March 7–9, 1991. Natural Coalition for Marine Conservation, Inc. Savannah, GA.

Rudd, J. W. M., 1995. 'Sources of methyl mercury to freshwater ecosystems: a review', *Wat. Air Soil Pollut.*, **80**, 697–713.

Rudd, J. W. R., Harris, R., Kelly, C. A., and Hecky, R. E. 1993. 'Are hydroelectric reservoirs significant sources of greenhouse gases?', *Ambio*, **22**, 246–248.

Schertzer, W. M. and Sawchuk, A. M. 1990. 'Thermal structure of the Lower Great Lakes in a warm year: implications for the occurrence of hypolimnion anoxia', *Trans. Am. Fish. Soc.*, **119**, 195–209.

Schiff, S. L. and Anderson, R. F. 1986. 'Alkalinity production in epilimnetic sediments: acidic and non-acidic lakes', *Wat. Air Soil Pollut.*, **31**, 941–948.

Schindler, D. W. 1986. 'The significance of in-lake production of alkalinity', *Wat. Air Soil Pollut.*, **30**, 931–944.

Schindler, D. W. 1990. 'Natural and anthropogenically imposed limitations to biotic richness in freshwaters', in Woodwell, G. (Ed.), *The Earth in Transition: Patterns and Processes of Biotic Impoverishment*. Cambridge University Press, Cambridge. pp. 425–462.

Schindler, D. W., Welch, H. E., Kalff, J., Brunskill, G. J., and Kritsch, N. 1974. 'Physical and chemical limnology of Char Lake, Cornwallis Island (75° N. Lat.)', *J. Fish. Res. Bd. Can.*, **32**, 1673–1674.

Schindler, D. W., Fee, E. J., and Ruszczynski, T. 1978. 'Phosphorus input and its consequences for phytoplankton standing crop and production in the Experimental Lakes Area and in similar lakes', *J. Fish. Res. Bd Can.*, **35**, 190–196.

Schindler, D. W., Newbury, R. W., Beaty, K. G., Prokopowich, J., Ruszczynski, T., and Dalton, J. A. 1990. 'Effects of a windstorm and forest fire on chemical losses from forested watersheds and on the quality of receiving streams', *Can. J. Fish. Aquat. Sci.*, **37**, 328–334.

Schindler, D. W., Hesslein, R. H., and Turner, M. A. 1987. 'Exchange of nutrients between sediments and water after 15 years of experimental eutrophication', *Can. J. Fish. Aquat. Sci.*, **44** (Suppl. 1), 26–33.

Schindler, D. W., Beaty, K. G., Fee, E. J., Cruikshank, D. R., DeBruyn, E. D., Findlay, D. L., Linsey, G. A., Shearer, J. A., Stainton, M. P., and Turner, M. A. 1990. 'Effects of climatic warming on lakes of the central boreal forest', *Science*, **250**, 967–970.

Schindler, D. W., Bayley, S. E., Curtis, P. J., Parker, B. R., Stainton, M. P. and Kelly, C. A. 1992. 'Natural and man-caused factors affecting the abundance and cycling of dissolved organic substances in precambrian shield lakes', *Hydrobiologia*, **229**, 1–21.

Schindler, D. W., Kidd, K. A., Muir, D., and Lockhart, L. 1995. 'The effects of ecosystem characteristics on contaminant distribution in northern freshwater lakes', *Sci. Tot. Environ.*, **160/161**, 1–17.

Schindler, D. W., Bayley, S. E., Parker, K. G., Beaty, K. G., Cruikshank, D. R., Fee, E. J., Schindler, E. U., and Stainton, M. P. 1996a. 'The effects of climatic warming on the properties of boreal lakes and streams at the Experimental Lakes Areas, northwestern Ontario', *Limnol. Oceanogr.*, **41**(5), 1004–1017.

Schindler, D. W., Curtis, P. J., Parker, B. R., and Stainton, M. P. 1996b. 'Consequences of climate warming and lake acidification for UV-B penetration in North American boreal lakes', *Nature*, **379**, 705–708.

Schindler, D. W., Curtis, P. J., Bayley, S. E., Parker, B. R., Beaty, K. G., and Stainton, M. P. 1997. 'Climate-induced changes in the dissolved organic carbon budgets of boreal lakes', *Biogeochemistry*, **36**, 9–28.

Scully, N. M. and Lean, D. R. S. 1994. 'The attenuation of ultraviolet radiation in temperate lakes', *Arch. Hydrobiol. Limnol.*, **43**, 135–144.

Sellers, T. J. 1995. 'The distribution of lake trout, *Salvelinus namaycush*, and opossum shrimp *Mysis relicta*, in small boreal lakes with respect to temperature, dissolved oxygen, and light', *MSc Thesis*, University of Alberta, Edmonton, Alberta.

Sellers, P., Kelly, C. A., Rudd, J. W. M., and MacHutchon, A. R., 1996. 'Photodegradation of methylmercury in lakes', *Nature*, **380**, 694–697.

Shearer, J. A., Fee, E. J., DeBruyn, E. R., and DeClercq, D. R. 1987. 'Phytoplankton primary production and light attenuation responses to the experimental acidification of a small Canadian Shield lake', *Can. J. Fish. Aquat. Sci.*, **44**, 83–90.

Slemr, F. and Langer, E. 1992. 'Increase in global atmospheric concentrations of mercury inferred from measurements over the Atlantic Ocean', *Nature*, **355**, 434–437.

Snucins, E. J. and Gunn, J. M. 1995. 'Coping with a warm environment: behavioral thermoregulation by lake trout', *Trans. Am. Fish. Soc.*, **124**, 118–123.

Sommer, U., Gliwicz, Z. M., Lampert, W., and Duncan, A. 1986. 'The PEG-model of seasonal succession of planktonic events in fresh waters', *Arch. Hydrobiol.*, **106**, 433–485.

Stefan, H. G. and Praud'homme, E. B. 1993. 'Stream temperature estimation from air temperature', *Wat. Resour. Bull.*, **29**, 27–45.

Stefan, H. G., Hondzo, M., and Fang, X. 1993. 'Lake water quality modelling for projected future climate scenarios', *J. Environ. Qual.*, **22**, 417–431.

St Louis, V. L., Rudd, J. W. M., Kelly, C. A., Beaty, K. G., Bloom, N. S., and Flett, R. J. 1994. 'Importance of wetlands as sources of methyl mercury to boreal forest ecosystems', *Can. J. Fish. Aquat. Sci.*, **51**, 1065–1076.

Strub, P. T., Powell, T., and Goldman, C. R. 1985. 'Climate forcing: effects of El Niño on a small temperate lake', *Science*, **227**, 55–57.

Tilman, D., Kieesling, R., Sterner, R., Kilham, S. S., and Johnson, F. A. 1986. 'Green, bluegreen and diatom algae: taxonomic differences in competitive albility for phosphorus, silicon and nitrogen', *Arch. Hydrobiol.*, **106**, 437–485.

Trippel, E. A., Eckmann, R., and Hartmann, J. 1991. 'Potential effects of global warming on whitefish in Lake Constance, Germany', *Ambio*, **20**, 226–231.

Vollenweider, R. A. 1975. 'Input–output models, with special reference to the phosphorus loading concept in limnology', *Schweiz. Z. Hydrol.*, **37**, 53–84.

Vollenweider, R. A. 1976. 'Advances in defining critical loading levels for phosphorus in lake eutrophication', *Mem. Ist. Ital. Idrobiol.*, **33**, 53–83.

Walker, I. R., Smol, J. P., Engstrom, D. R., and Birks, H. J. B. 1991. 'An assessment of Chironomidae as quantitative indicators of past climatic change', *Can. J. Fish. Aquat. Sci.*, **48**, 975–987.

Ward, J. V. and Stanford, J. A. 1982. 'Thermal responses in the evolutionary ecology of aquatic insects', *Annu. Rev. Entomol.*, **27**, 97–117.

Webster, K. E., Newell, A. D., Baker, L. A., and Brezonik, P. L. 1990. 'Climatically induced rapid acidification of a softwater seepage lake', *Nature*, **347**, 374–376.

Webster, K. E., Kratz, T. K., Bowser, C. J., Magnuson, J. J., and Rose, W. J. 1996. 'The influence of landscape position on lake chemical responses to drought in northern Wisconsin, U.S.A.', *Limnol. Oceanogr.*, **41**, 977–984.

Weilenmann, U., O'Melia, C. R., Stumm, W. 1989. 'Particle transport in lakes: models and measurements', *Limnol. Oceanogr.*, **34**, 1–18.

Wheatley, B. and Paradis, S. 1995. 'Exposure of Canadian aboriginal peoples to methylmercury', *Wat. Air Soil Pollut.*, **80**, 3–11.

Williams, G. P. 1978. 'The case of the shrinking channels — The North Platte and Platte Rivers, Nebraska', *USGS Circular no. 781.*

Williams, G. P. and Wolman, M. G. 1984. 'Downstream effects of dams on alluvial rivers', *USGS Prof. Paper 1286*, USGS, Washington, D.C. 83 pp.

Williams, M. W., Losleben, M., Caine, N., and Greenland, D. 1996. 'Changes in climate and hydrochemical responses in a high elevation catchment in the Rocky Mountains, U.S.A.', *Limnol. Oceanogr.*, **41**, 939–946.

Williamson, C. E. and Zagarese, H. E. 1994. 'The impact of UV-B radiation on pelagic freshwater ecosystems', *Arch. Hydrobiol. Limnol.*, **43**, IX–XI.

Williamson, C. E., Stemberger, R. S., Morris, D. P., Frost, T. M., and Paulsen, S. 1996. 'Ultraviolet radiation in North America lakes: attenuation estimates from DOC measurements and implications for plankton communities', *Limnol. Oceanogr.*, **41**, 1024–1034.

Wrona, F. J., Gummer, W., Cash, K. J., and Crutchfield, K. 1996. 'Cumulative impacts within the Northern River Basins', *Northern River Basins Study, Synthesis report No. 11.* Edmonton.

Yan, N. D., Keller, W., Scully, N. M., Lean, D. R. S., and Dillon, P. J. 1996. 'Increased UV-B penetration in a lake owing to drought-induced acidification', *Nature*, **381**, 141–143.

Zoltai, S. C. and Vitt, D. H. 1990. 'Holocene climatic change and the distribution of peatlands in western interior Canada', *Quat. Res.*, **33**, 231–240.

INDEX

Note: Page numbers in *italics* refer to tables; those in **bold** type refer to diagrams.

Index compiled by Colin Will